THE EDIBLE CITY RESOURCE MANUAL

Richard Britz, et. al

William Kaufmann, Inc. • Los Altos, California

Library of Congress Cataloging in Publication Data

Main entry under title:

The Edible city resource manual.

   Bibliography: p.
    1. Agriculture.   2. Urban agriculture.   3. School
farms.   4. Agriculture—Oregon—Willamette Valley.
5. Urban agriculture—Oregon—Willamette Valley.
I. Britz, Richard, 1941–
S501.2.E33   1981   338.1'9'091732   81-8187

Printed on entirely recycled
paper. No trees were destroyed
in the making of this book.

ISBN 0-913232-97-1

10 9 8 7 6 5 4 3 2 1 *Experimental edition*

Printed in the United States of America

Publisher's Note

I first saw *The Edible City Resource Manual* in March 1980 as a set of some 400 pages of copy, draped in sequence around three walls of a small library room in a building on the Mills College campus in Oakland, California. Richard Britz and William Higginson, whom I had just met for the first time, explained—carefully but enthusiastically—the aims, concepts, organization and evolution of their Edible City/Edible Earth projects. Within two hours, we had reached an agreement to produce a relatively inexpensive "experimental edition" for worldwide distribution. Never mind its incompleteness, its mixtures of type, script and art styles, its still imperfect state; the inspirational and practical values embodied in *The Edible City* seemed to me so fundamentally important and so urgently in need of widespread availability that I recommended that we forgo much of the customary time-consuming prepublication reviewing, editing, design, production and other procedures and get this book into print. Thus, you are seeing *The Edible City* in book form almost as I first saw it—literally as an "off-the-wall" production.

Many readers will note that *The Edible City* is a most appropriate new title for us to add to our list as our company approaches the beginning of its second decade of publishing activity. It continues our tradition of innovative, experimental publishing, and it represents another unique attempt by a group of individuals to develop approaches and models for healthy living and working arrangements for all of us who live together in an increasingly complex world. Our first book, *The California Tomorrow Plan*, was published in the summer of 1972, and it received worldwide acclaim. In its Foreword, I expressed the belief that "it is an important book, one which has profound implications that will be apparent to anyone who reads it carefully. It addresses itself effectively and in an innovative way not only to our most pressing environmental problems, but also to related social and economic imbalances in our society. Most significantly, it emphasizes problem solving and offers specific, workable, constructive programs of action." Although the focus of *The Edible City* is also regional—in this instance on the Willamette Valley in Oregon—we are confident, just as we were in publishing *The California Tomorrow Plan*, that its applications are universal.

We look forward to receiving readers' comments. All of the reactions, experiences, suggestions, reviews, and other comments that we receive will be carefully considered by the authors for incorporation into future versions of *The Edible City*.

William Kaufmann

Preface

This book is a resource manual. We hope it is useful for you while we are changing our homes and neighborhoods into gardens, farms, orchards, and forests.

This book is an experiment. A prototype in its own sort of way, in that it's basically a progress report; it was compiled at the conclusion of a five-year period of experimentation and applied research involving lots of people. The work in this book has primarily been supported by three major sources of energy: the National Center for Appropriate Technology in Butte, Montana (a division of C.S.A.), the University of Oregon Department of Landscape Architecture and the University of Oregon Urban Farm, and the Edible City Resource Center, Inc., a non-profit educational and research division of Edible Earth, Inc. We worked on the grant from N.C.A.T for/with the Whiteaker Neighborhood Community Council, the Neighborhood Advisory Group, and lots of neighbors; we hope to continue working at these varied and diverse levels on a continuing basis . . . we have moved into the neighborhood.

The people who worked together to produce this book are numerous, and our networks number into the thousands. Community people, volunteers, paid and unpaid staff, students, faculty, and consultants are all deeply appreciated. The cooperative nature of our work leaves much of it fuzzy and open for interpretation—just the way we would have it. We are still learning. We have needed energy, organization, enthusiasm, and humor to continue our course. Any returns from this work will go directly into experimental housing-food cluster block farms and the Whiteaker economy.

Richard Britz

Contributors

ORGANIZATION & WRITING

Richard Britz
Barbara Britz
Linda Smiley
Jeff & Nanci Wilson
Kate Goodrich
Skeeter Duke
Barbara McPhail
Larry Parker
Amity Foundation
Jim Klein
Noel Prchal
Chris Gum
Tom Bettman
Dean Baker
Hugh Prichard
Tam Shawgo
Frank Theis
Laurel Lyon
Charlie Sundberg
Lloyd Lindley III

PHOTOGRAPHY

Richard Britz
Jeff Koonce
Barbara Britz
Mollie Favor-Miller
Bradley Miller

DRAWINGS

Richard Britz
Paul Wilbert
Vickie Saugen
Faye Cummins
Barbara Britz
Jim Klein
William Higginson
Alan Green
Dorene Steggel

Larry Parker
"Mert"
Jeff Koonce
Mary Truax
Noel Prchal
Sharon Ledbury
Cindy Girling
Liz Lardner
Fred Patch
Gary Hoyt
Frank Theis
Laurel Lyon
Charlie Sundberg
Bob Downing
Don High
Garth Ruffner

CONSULTANTS

Reggi Norton, Nutrition Counselor
Barbara Weinstein, Nutrition Counselor
Eco-Alliance, Corvallis, Oregon
William Higginson, Edible Earth, Inc.

GRAPHICS, PASTE-UP, and TYPING

Richard Britz
William Higginson
Barbara Britz
Sandra Langstor

Contents and Synopses

Introductory drawings and essays work together to describe the colonization of the farmland by the City and the simultaneous diminution of the City's ability to support life. In a call to action, the authors are united in the recommendation for policies and plans to reorganize priorities to maintain a renewable and continuous life support system shared by urban and rural alike.

In this section, general trends of local agriculture and agribusiness are discussed, and ways of utilizing and supporting locally self-reliant food supply techniques are described. Concepts developed include the neighborhood food distribution network, the neighborhood food center, urban gleaning potentials, issues of basic food and nutrition, and an extensive school farms program. These systemic and adaptable concepts are intended to unite urban and rural peoples and provide mechanisms to help maintain the continuity of local and regional agriculture through the process of education and outreach. Inherent in this section is the challenge to banks and lending institutions to reinvest in economically viable Willamette Valley agriculture.

In this section proposals and projects supportive of biologically healthy urban environments are put forth. A neighborhood-scale program for recycling, solar housing clusters, and integrated food production is developed which is intended to counteract the progressive depreciation of the urban biomass. Utilizing the existing block configuration of many American cities, transformation principles are described graphically and in written form; these maximize solar gain, integrate diverse land uses and densities, and develop an evolutionary urban structure.

This section is a case study application (neighborhood-wide and project-based) of the principles developed in Section I and II above. Here the neighborhood is divided into five semi-autonomous sectors by various highways and geographic barriers, and projects are designed and built with neighbors to reintegrate and renovate community productive systems. Six neighborhood farms are described in various stages of development, and block farm projects are developed to logistically and progressively support these larger scale economic endeavors. The concluding portion of the book describes the neighborhood orchards program, one which weaves a web of fruit and nut trees throughout the neighborhood and reclaims the orchard remnants by-passed by urbanization's first sweep. The riverfront necklace of publicly maintained crop trees supplements the six neighborhood farms and these together form a parent structure for both the perimeter farmland and the urban block farm residents.

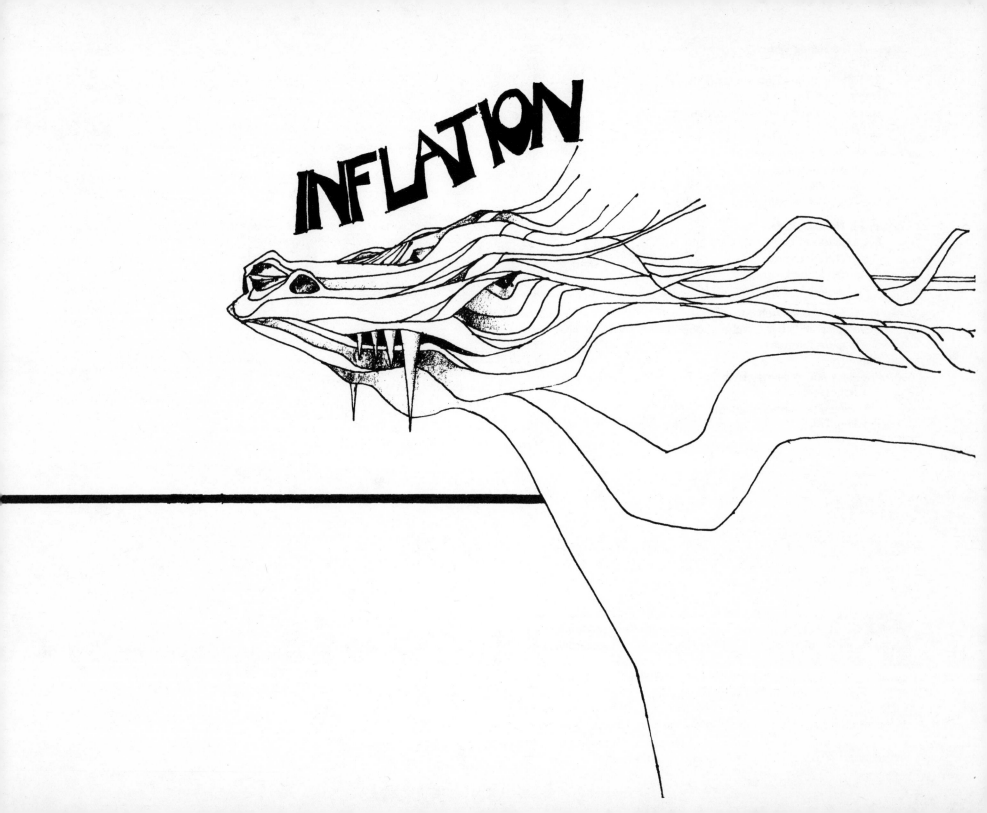

Introduction

Our goal in this book is to assist in developing maximum food production where we live—in our homes, in clusters of homes, in alleys behind our homes, converted streets in front of our homes, and larger public lands in close proximity to our homes. It is our premise that reconsideration and reorganization of our existing systems begins at home and in our places of learning; in discussion with close friends and our immediately-adjacent clusters of neighbors, and that the evolution of change stands its best chance by springing from our own personal environments. Due to the nature of the presently existing economic climate, we have rejected the "business approach" to simply "make jobs" or provide immediate income for few, and rather, have assessed the problem as systemic and not easily assessed or corrected by the few alone.

We hope this work will provide specific guidelines, suggestions, and models for home "turf" modification which address the need for personal space conservation and conversion without injuring the sacrosant identity of the single family home. In addition to preserving its best qualities, we recommend that cooperative and mutually supportive clusters of friends work together for the goal of neighborhood self-reliance, which we believe, through mutual and interdependent support, will improve our communities' health, well-being, and economic stability.

This work is comprised of several sections, each of which we hope represents a seed from which increased local self-reliance might grow.

The first section recommends that neighbors organize around and support existing Willamette Valley small farmers through cooperative neighborhood bulk-buying, gleaning, and/or bartering labor for food. It includes site specific locations of why, where, and how to get food locally, what stores within your neighborhood stock these foods now, how groups utilize efficiently-distributed food canning and preservation centers.

Additionally, section one identifies appropriate food values nutritionally, discusses existing food systems critically, and recommends a school "farms" curriculum to counter media propaganda regarding food and food-related issues.

The second section recommends the concurrent development of a strong urban agricultural network, consisting primarily of many evolving "block farms" (blocks being the essence of most middle-sized American cities) through the conversion of private property "rights" into semi-private, semi-public arrangements for mutual gain. This step then can evolve into the organizational structure capable of sustaining continuously productive "neighborhood farms" which primarily abut the river and utilize its immediately-adjacent class one agricultural soils. Lacing throughout both scales of organization and places are renovated and new fruit and nut orchards.

We aim to achieve a measure of urban self-support and deepen the benefits of the understanding of such self-support. The concepts of imperialism, with its attendant forms of colonization and extraction, begin with the city feeding off of the country and this drive, this archetypal motivation toward conquest and submission, must be checked in ourselves, now and here where we live.

Urban agriculture gives us new selves, new friends at all levels of the foodweb, and encourages us not to waste the cores of our own apples but to recycle them into edible cities where we will be proud to raise our next crops and our next generations.

So we advance our cause by planting: urban farms, neighborhood gardens, edible street trees, urban woodlots, local orchards. We consider ourselves permanent residents, working to heal the tear in our psychic tissue and promoting life over death, sweat as a virtue, manure as a resource, farming as a religion of soil, water, sun, and plant worship, and organic recycling as the essence of life. In the perimeter areas we glean our fields after the mechanical harvest, and amid dying plant bodies find abundant and continuing resources that we can both harvest and replant, (without poisons this time) so that agriculture in its fullest and healthiest sense will be the heritage of our successors. By pairing respect for the land and economic advantage, we look forward to the future with enthusiasm and believe it is bright.

Prelude to an Urban Farm

The gentle people who have retreated here in Eugene, Oregon, and the hardy people who settled are in the process of forming bonds and alliances to protect the soils and the plants that enable them to grow. These alliances are forming between the disadvantaged urban neighborhood low-income groups (especially the young and old on no or fixed incomes) and the farmers on the perimeter of the expanding metropolitan organism.

People to Preserve Agricultural Land, Organically Grown, Inc., the *Small Farmer's Journal, Growers Market* and other food co-ops, *Amity Foundation, Whiteaker Neighborhood Project Self-Reliance, Lane County Housing and Community Development School Gardens Program,* the *City of Eugene Community Gardens Program,* the *Edible City Resource Center* and the *University of Oregon Urban Farm* are all groups spearheading the counterattack. And while some regional support is provided by our distant friends in *Tilth* and the *Ecotope* group, we do not fail to realize that it is up to us, and us alone, to achieve our goals.

Ravenous and greedy developers are everywhere, and we must place aside our planting tools for our weapons of intellect and skill in organizing to repel them with our bare hands and hearts. They have already breached the wall of our urban services boundary and are consuming our farmland inside and outside its borders. Their advance "advisors" have done their jobs well and have weakened our defenses by working "with" us. U.S.D.A. county extension "agents", "officers" of the U.S. Forest Service, and banks and lending "agents" of all descriptions, have often succeeded in "aiding" farmers and foresters to replace their measures of a good healthy life with indices of productivity by hooking them into an entropic spiral of impoverishing economics. They have encouraged the dissolution of balanced ecosystem management. The tragic result is a diminution of farmland quantity and quality as agricultural productivity is eroded by urbanization.

Elsewhere, we have watched as real and primary commodities such as food and fibre economic support systems have been isolated and choked off in favor of the *image* of less work and higher pay provided by jobs in industries such as metals and electronics, which are non-renewable resources. When real estate speculation and construction (with what lumber is left) threaten to replace farming and forestry in our productive paradise, we call for help. It has become painfully evident that our national emphasis on consumption has dominated our emphasis on production.

The core of our dilemma is in our cities themselves, and it is there that we must begin. We in the neighborhoods send messages to the perimeter and they in turn give us their support. As we value and use it for our own food production, we warn them that their farms and forests are in imminent danger from explosive growth without a sustainable economic support base. We work to reject this threat. We write letters to our perimeter neighbors to mentally "pull back" into the city, into powerful urbanrural coalitions, and establish direct marketing systems, local economics, and jobs for people and animals to perform with dignity and pride.

Urban agriculture is at its core a conception of rebirth. It is the seed of the already consumed apple dropping into the cracks of our concrete and its remaining consciousness to sprout life anew among burned-out, economically pillaged cities which have become so ugly, so depleted, so un-life-like that they demean even the rats and dogs that live there. It is the meaning and hope that sprouts where it is really needed to provide images of pride and beauty in the midst of clutter and disorder. It is the renewable and regenerative spirit at work and it spirals and revolves around and among the institutions which somehow have overlooked us. We urban farmers have persevered by acquiring root-holds to grow a healthier society free of aggressive and short-sighted competition and destruction. By intensifying our diversity we are seeking out the abundance of the energy source most appealing to us—the sun.

Urban agriculture fosters life amid wasted atmospheres of lead, asbestos, and particulate matter swirling out from our national fetish, the automobile. We advocate greater local understanding at smaller scales emphasizing bicycle and pedestrian traffic and we even spend time stooping on our hands and knees to attend closely our soils, our waters and our plants. We become attached to the freshness, the tenderness, and the continuities inherent in paying close attention to living things. But to do these things we must exemplify a more simple and direct life and become more locally self-reliant, by stabilizing our spheres of impact, by working as cells alone and in small groups and consolidating our energies by planting our neighborhoods. And as we work we watch the "regular" food run out. So we must also work to preserve what agricultural and open land that remains on the perimeter and recycle wasted mentalities that choose to raise houses rather than food—for greed is a function of scarcity and we seek a world of abundance. Joint actions in the urban core and at the urban boundary skin are necessary and provide for us the potential consolidation of both the poor and the disenfranchised small farmer, in a coalition that has the potential to once again re-dignify physical labor and maximize the probability of a renewable and continuing life support system.

We indicate caution, for as the urban becomes more like the rural with lands serving the people directly, our counterparts with different values will push even more strongly to pave and build over the rural, and restraint methods such as urban services boundaries, municipal agricultural land purchase through city and county governments, land trusts, public and private, and organic farm-city belts become more and more necessary. We aim to achieve a measure of urban self-support and deepen the benefits of the understanding of such self-support. The concepts of imperialism, with its attendant forms of colonization and extraction, begin with the city feeding off of the country and this drive, this archetypal motivation toward conquest and submission, must be checked in ourselves, now and here where we live.

Urban agriculture gives us new selves, new friends at all levels of the foodweb, and encourages us not to waste the cores of our own apples but to recycle them into edible cities where we will be proud to raise our next crops and our next generations.

So we advance our cause by planting: urban farms, neighborhood gardens, edible street trees, urban woodlots, local orchards. We consider ourselves permanent residents, working to heal the tear in our psychic tissue and promoting life over death, sweat as a virtue, manure as a resource, farming as a religion of soil, water, sun, and plant worship, and organic recycling as the essence of life. In the perimeter areas we glean our fields after the mechanical harvest, and amid dying plant bodies find abundant and continuing resources that we can both harvest and replant, (without poisons this time) so that agriculture in its fullest and healthiest sense will be the heritage of our successors. By pairing respect for the land and economic advantage, we look forward to the future with enthusiasm and believe it is bright.

SOME BASIC FACTS ABOUT MODERN AGRICULTURE

AGRICULTURE IS THE NATION'S LARGEST INDUSTRY, WITH THREE-FIFTHS OF THE CAPITAL ASSETS OF ALL MANUFACTURING IN THE UNITED STATES. AGRICULTURE IS THE LARGEST EMPLOYER IN THE UNITED STATES, HIRING SOME 14-17 MILLION PEOPLE TO WORK IN JOBS FROM THE FIELDS TO THE MARKET PLACE. ONE FARMWORKER PRODUCES ENOUGH FOOD AND FIBER FOR 56 PEOPLE. THE UNITED STATES IS THE WORLD'S LARGEST EXPORTER OF FARM PRODUCTS WHICH INCLUDES A 10.5 BILLION DOLLAR NET GAIN IN THE AGRICULTURAL BALANCE OF PAYMENTS (1977). IT IS INTERESTING TO NOTE THAT IN 1977 THERE WAS A 40.2 BILLION DOLLAR DEFICIT IN NON-AGRICULTURAL TRADE.

AGRICULTURE IN OREGON

OREGON'S FARMS RANK HIGHLY AMONG THE NATION'S FARMS IN MANY CROPS. FOR INSTANCE, OREGON RANKS:

1ST IN PEPPERMINT AND RYEGRASS SEED;
2ND IN SNAP BEANS AND STRAWBERRIES;
3RD IN SWEET CHERRIES, ONIONS, PEARS, AND SWEET CORN.

RESEARCH SHOWS THAT FARM SALES OF ONE BILLION DOLLARS WILL GENERATE THREE BILLION DOLLARS WITHIN THE OREGON ECONOMY BY THE MULTIPLIER EFFECT. IN 1977 OREGONIANS SPENT 14.6% OF THEIR INCOME ON FOOD. THIS AVERAGE IS COMPARED TO 17.7% FOR THE NATION AS A WHOLE; TO 24% IN FRANCE; 27% IN GERMANY; 31% IN THE UNITED KINGDOM; 33% IN JAPAN; 40% IN RUSSIA; AND 45% IN POLAND. OBVIOUSLY, OUR AGRICULTURAL LANDS REPRESENT A SUBSTANTIAL AND VALUABLE INVESTMENT FOR ALL AMERICANS. WE MUST REALLY CLARIFY HOW VALUABLE OUR AGRICULTURAL LANDS ARE.

HOW MUCH LAND ARE WE LOSING?

IN 1977 THERE WERE 345,417,000 ACRES OF AGRICULTURAL LANDS IN THE UNITED STATES (196,342,000 OF THAT WAS PRIME CROPLAND). ACCORDING TO THE NATIONAL AGRICULTURAL LANDS STUDY WHICH IS BEING CONDUCTED BY THE COUNCIL ON ENVIRONMENTAL QUALITY, WE LOSE THREE MILLION ACRES OF FARMLAND EVERY YEAR. THIS IS EQUIVALENT TO TWELVE SQUARE MILES OF AGRICULTURAL LAND CONVERTED TO NON-AGRICULTURAL USES EVERY DAY!

BUT WHAT ABOUT AGRICULTURAL LAND IN LANE COUNTY?

COMPARING CROP ACREAGES AND VARIETIES OF VEGETABLES GROWN IN LANE COUNTY IN 1935 WITH THOSE IN 1979, WE SEE AN OVERALL INCREASE IN ACREAGES UNDER PRODUCTION BUT A REDUCTION IN KINDS OF CROPS PRODUCED.

IN 1935, CROPS RAISED FOR CANNERY PROCESSING INCLUDED 15 VEGETABLE VARIETIES PLANTED ON 1,027 ACRES; IN 1979, CANNERY CROPS WERE REPRESENTED BY 7 VEGETABLE VARIETIES, BUT ON 8,478 ACRES. (BEANS AND SWEET CORN ALONE ACCOUNTED FOR 7,220 ACRES OUT OF THE 8,478.)

VEGETABLE ACREAGE IN PRODUCTION FOR FRESH MARKET SALE DROPPED DRASTICALLY. IN 1939 THERE WERE 25 VEGETABLE VARIETIES GROWN ON 736 ACRES. IN 1979, THIS HAD DROPPED TO 10 VARIETIES GROWN ON 245 ACRES. WE TOTALLY LOST COMMERCIAL PRODUCTION OF 15 DIFFERENT KINDS OF VEGETABLES, INCLUDING KIDNEY BEANS, BROCCOLI, CAULIFLOWER, SPINACH, TURNIPS, CELERY, CANTELOUPE, LETTUCE, PARSNIPS, RADISHES, RUTABAGAS, WATERMELON, BEETS, CUCUMBERS AND RHUBARB. COMMERCIAL

FRUIT PRODUCTION WAS ALSO LOST FOR RED RASPBERRIES, LOGANBERRIES, YOUNGBERRIES AND GOOSEBERRIES.

OF THE 50,000+ INCREASED ACREAGE PRODUCTION IN 1979, OVER 41,000 ACRES WAS FOR NON-FOOD PASTURELAND AND GRASS SEED PRODUCTION. OF THE REMAINING 9,000+ ACRES, OVER 7,600 WAS FOR GROWING FOUR CANNERY CROPS (BEANS, SWEET CORN, BEETS AND CARROTS).

THE FOLLOWING CHART FROM THE LANE COUNTY EXTENSION SERVICE SHOWS, IN PART, THE SHIFT IN MARKET PRIORITIES, THE REDUCTION OF VARIETY AND DIVERSITY, AND THE EMPHASIS ON PROCESSED OVER FRESH VEGETABLE FOODS. SOME CROPS (I.E. STRAWBERRIES) ARE NOT ACCOUNTED FOR ON THE CHART.

LANE COUNTY ACREAGE IN PRODUCTION

	1935	1979
PRUNES	2,465	106
SWEET CHERRIES	1,300	476
SOUR CHERRIES	267 (CANNERY)	239
PEARS	455 (CANNERY)	31
PEACHES	223	120
APPLES	310	70
BLACKCAPS	3	102
EVERGREENS	3	
FILBERTS	1,110	2,400
WALNUTS	2,140	221

	CANNERY	FRESH	CANNERY	FRESH
ASPARAGUS	30	20	--	--
SNAP BEANS	145	20	3,270	30
KIDNEY BEANS	25		--	--
BEETS	270	40	780	--
BROCCOLI	3	2	--	--
CAULIFLOWER	4	5	--	--
CARROTS	230	60	420	15
CABBAGE	8	60	10	5
CUCUMBERS	1.5	12	8	--
SWEET CORN	140	100	3,950	100
SQUASH	40	30	--	25
RHUBARB	4	5	40	--
SPINACH	10	50	--	--
TOMATOES	115	75	--	25
TURNIPS	2	40	--	--
CELERY		5	--	--
CANTALOUPE		65	--	--
LETTUCE		25	--	--
ONIONS		5	--	5
GREEN ONIONS		5	--	1
PEAS		50	--	35
PARSNIPS		15	--	--
PEPPERS		15	--	4
RADISHES		20	--	--
RUTABAGAS		2	--	--
WATERMELON		10	--	--

	1935	1979
WHEAT	--	18,000
OATS	--	2,500
BARLEY	4,000	500
FIELD CORN	5,000	1,700
ALFALFA	9,000	2,500
PASTURES	--	25,000
GRASS SEED	--	16,530

HOW DOES URBAN DEVELOPMENT OCCUR IN THE WILLAMETTE VALLEY?

THE METROPOLITAN AREA PLANNING ADVISORY COMMITTEE (MAPAC) HAS SUG-
GESTED THREE "ALTERNATIVES" FOR GROWTH IN THE EUGENE/SPRINGFIELD
METROPOLITAN AREA. ALL OF THESE ALTERNATIVES SUGGEST THAT THE
URBAN GROWTH BOUNDARIES INCREASE INCREMENTALLY AT THE EDGES OF THE
CITY.

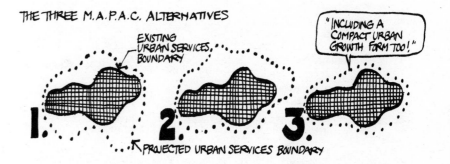

Preserving farms

Local control of political issues seems to be a tool that some people only want used if it serves their purposes. For example, look at land-use meetings where people call for the freedom to develop and sell their land as they wish, in any size parcel, for any purpose.

I agree that most zoning is a limitation and should be accompanied by tax or building options for landowners who have lost sale value. However, local control means local responsibility to everyone in Oregon, now and for future generations, and thus the public good demands that "fast buck" housing development of farmland be halted.

Too many people and too few resources are the main reasons we have land-use planning and history records the failures of societies which let these warning signs go unheeded.

We need all of our farmland, but merely zoning it as "exclusive farm use" or in 20-acre parcels is not enough. Making it possible to farm means lowering tax rates, developing local markets and promoting cooperative ownership of machinery to allow small farms to afford the cost. It also means producing low-cost fertilizer from agricultural waste and sewage sludge because petrochemicals are constantly rising in cost and decreasing in supply.

Economics are deceptive, but the small farm is failing today because of pressures brought to bear by those people most likely to profit from such failures. The meager attempt of land-use planning is to prevent such failures and keep our farmland available for a time when reason prevails, and not just profit.

EDD WEMPLE
319 Jefferson St.
Cottage Grove

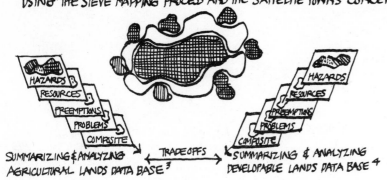

IDENTIFYING SUPPORTIVE ENVIRONMENTS FOR URBAN RURAL LANDSCAPES USING THE SIEVE MAPPING PROCESS AND THE SATELLITE TOWNS CONCEPT...

SUMMARIZING & ANALYZING AGRICULTURAL LANDS DATA BASE [3] ← TRADE OFFS → SUMMARIZING & ANALYZING DEVELOPABLE LANDS DATA BASE [4]

REALIZING THAT IT IS BOTH ECONOMICALLY AND ECOLOGICALLY HEALTHY TO MAINTAIN EXISTING AGRICULTURE CLOSE TO THE CITY AND TO SUPPORT EXISTING URBAN ENVIRONMENTS WITH MORE PEOPLE, IT IS CLEAR THAT THE SIEVE MAPPING CAN SHOW WHERE IN THE REGION THESE LANDSCAPES EXIST. MORE IMPORTANTLY, THE SIEVE MAPPING PROCEDURE CAN CLARIFY WHAT THE TRADE-OFFS ARE RELATIVE TO MAKING WISE LAND USE DECISIONS.

PROTECTING RESOURCES:

ONCE WE HAVE DETERMINED WHERE OUR URBAN ENVIRONMENTS ARE GOING TO BE, WE MUST LOOK CAREFULLY AT HOW THEY COEXIST WITH RESOURCE LANDS. WE MUST FIND WAYS TO PROTECT THESE VALUABLE LANDS. AFTER ALL, URBAN DEVELOPMENT CONTRIBUTES 100 TONS/ACRE OF SOIL LOSS EACH YEAR CAUSING A 10-15% DROP IN SOIL PRODUCTIVITY. THIS TRANSLATES INTO INCREASED FERTILIZER COSTS FOR THE FARMER, AND HIGHER PRICES AT THE TABLE. SOMETIMES IT SPELLS DISASTER FOR THE STRUGGLING YOUNG FARMER. IN THE SAME LIGHT, INCREASING DENSITIES IN URBAN CENTERS DOES NOT DECREASE THE COSTS OF HOUSING IN URBAN AREAS. IT IS THE ECONOMIC GAIN OF A FEW INVESTORS THAT USUALLY GET THE BENEFITS OF HIGH DENSITY, NOT THE HOMEOWNER AND TAXPAYER. AS DENSITY INCREASES, THE KIND OF SERVICES MUST CHANGE. THE INVESTORS DO NOT PAY FOR THAT. THEY DO NOT PAY FOR THE COST OF THE LOST SUNLIGHT AND CLEAR, CLEAN AIR EITHER. IT IS PLAIN THAT URBAN DEVELOPMENT MUST BE DONE CAREFULLY AND CONSCIENTIOUSLY.

STRUCTURING THE LANDSCAPE:

HOW THEN DOES THIS NEW REGIONAL SCHEME RELEASE THIS CONGESTIVE PRESSURE? INSTEAD OF RELIEVING THE SYMPTOMS OF OVERCROWDING, THE REGIONAL APPROACH LOOKS TO A FRAMEWORK WHICH STRUCTURES URBAN ENVIRONMENTS INTO A MORE NATURAL ORGANIZATION. IT IS BASED ON THE NATURAL CYCLES OF THE SUN, SOIL AND WATER WHICH SUPPORT LIFE.

SUBURBAN LANDS

IN SUBURBIA, THE MAJOR CONCERN IS THE CONVERSION OF PRODUCTIVE LAND, WHICH CONVERTS SOLAR ENERGY INTO STORED FOOD ENERGY, INTO URBANIZED LAND WHICH DOES NOT DO THIS. AS WAS DISCUSSED EARLIER, URBAN DEVELOPMENT CAUSES A 10-15% DROP IN PRODUCTIVITY OF OUR SOIL. PART OF THE PROBLEM IS DUE TO THE INCREASED RUNOFF SPEED OF LAND AREAS WHICH HAVE LARGE EXPANSES OF PAVEMENT. ONE OF THE MAJOR DETERMINANTS OF THE AMOUNT OF PAVED SURFACES IS THE AUTOMOBILE. THE SUBURBAN COMMUNITIES EVOLVED FROM ROADS WHICH CONNECTED THE SUBURB TO THE CITY. OFTENTIMES, ZONING EVOLVED FROM THE SELF-INTEREST OF

METROPOLITANISM

SINCE THE WILLAMETTE VALLEY WAS SETTLED AS AN AGRICULTURAL VALLEY, THE URBAN SETTLEMENTS TENDED TO SPRING UP IN CLOSE PROXIMITY TO THE MOST PRODUCTIVE AGRICULTURAL LANDS. THIS PRESENTS A UNIQUE SITUATION RELATIVE TO TODAY'S GROWING METROPOLITAN AREAS. THE INCREMENTAL GROWTH OF URBAN AREAS IS GOBBLING UP AGRICULTURAL LANDS (CLASS I-IV SOILS ACCORDING TO THE SOIL CONSERVATION SERVICE). THIS INCREMENTAL URBAN GROWTH AT THE CITY'S EDGE, TERMED "METRO-POLITANISM" ASSUMES THAT SPRAWL AND CONGESTION ARE INEVITABLE, AND IT TRIES TO PROVIDE FOR ITS OCCURANCE. THESE PROVISIONS OCCUR IN THE FORM OF TALLER BUILDINGS, WIDER STREETS AND HIGHWAYS, BIGGER MUNICIPAL WASTE PLANTS, AND MORE POWERFUL TRANSMISSION LINES, USUALLY AT THE EXPENSE OF RELATIVELY INEXPENSIVE URBAN OPEN SPACE. THE METROPOLITAN APPROACH TREATS THE SYMPTOMS INSTEAD OF THE PROBLEM ITSELF. THIS DOES NOT MAKE THE PROBLEM GO AWAY.

REGIONALISM

THERE IS ONE CLEAR ALTERNATIVE WHICH THE MAPAC PLANNERS ARE NEGLECTING. THAT IS THE CHOICE OF A REGIONAL DISPERSAL OF URBAN GROWTH INTO SUPPORTIVE URBAN ENVIRONMENTS EMBRACED AND SEPARATED BY AGRICULTURAL GREENBELTS. THE REGIONAL APPROACH TO URBAN GROWTH DOES NOT ACCEPT SPRAWL AND CONGESTION AS FACTS. IT SEEKS MOST OF ALL TO PROVIDE HEALTHY URBAN ENVIRONMENTS WHILE PRESERVING THE PRODUCTIVE CAPABILITY OF THE SOIL.

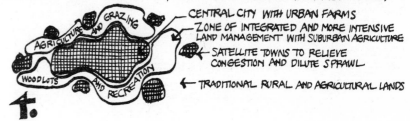

SATELLITE TOWNS CONCEPT APPLIED TO THE SOUTHERN WILLAMETTE VALLEY

- CENTRAL CITY WITH URBAN FARMS
- ZONE OF INTEGRATED AND MORE INTENSIVE LAND MANAGEMENT WITH SUBURBAN AGRICULTURE
- SATELLITE TOWNS TO RELIEVE CONGESTION AND DILUTE SPRAWL
- TRADITIONAL RURAL AND AGRICULTURAL LANDS

UNDERSTANDING THE COMPLEXITY OF THE ENVIRONMENT

REGIONALISM TRIES TO ACCEPT THE FACT THAT THE REGION IS THE BASIS FOR UNDERSTANDING THE COMPLEXITY OF THE ENVIRONMENT. THE REGIONAL-IST TRIES TO FIND WAYS TO RELEASE THE PRESSURES OF GROWTH AND SPRAWL WHICH ARE BOTH ECONOMICALLY AND ECOLOGICALLY UNHEALTHY. AN ENVIRONMENTAL ANALYTICAL PROCESS CAN BE USED TO EVALUATE AND SUMMARIZE LARGE AMOUNTS OF DATA AT ONE TIME. THE SIEVE PROCESS ALLOWS US TO WEIGH THE PROS AND CONS OF ANY CHOICE RELATING TO "DEVELOPMENT" AND "NON-DEVELOPMENT" OF A GIVEN PART OF THE LANDSCAPE.

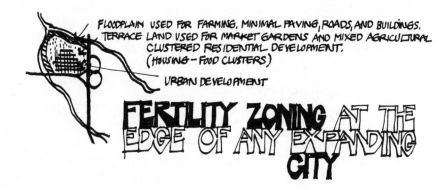

FLOODPLAIN USED FOR FARMING, MINIMAL PAVING, ROADS, AND BUILDINGS.
TERRACE LAND USED FOR MARKET GARDENS AND MIXED AGRICULTURAL
CLUSTERED RESIDENTIAL DEVELOPMENT.
(HOUSING - FOOD CLUSTERS)

URBAN DEVELOPMENT

FERTILITY ZONING AT THE EDGE OF ANY EXPANDING CITY

LANDHOLDERS WHO OWNED PROPERTY NEAR WHERE MAJOR HIGHWAY INTERCHANGES
WERE PLANNED. IN FACT, IT WAS OFTEN THE POLICY DECISION-MAKERS WHO
WOULD WIDEN ROAD DEVELOPMENT NEAR THEIR OWN PROPERTY.

A MORE REALISTIC WAY TO PLAN OUR CITIES, SUBURBS AND RURAL LANDS IS
TO USE NATURAL PRINCIPLES AND GEOMORPHIC CONDITIONS TO GUIDE DE-
CISION-MAKING INSTEAD OF ECONOMICS AND GREED. AS THE COMPLEXITY
OF THE ENVIRONMENT IS UNRAVELED, WE BEGIN TO SEE PATTERNS RELATED
TO HYDROLOGY, SOIL, AND VEGETATION. THESE ARE THE INHERENT QUALITIES
OF THE LANDSCAPE AND PRESENT AN UNALTERABLE BASIS FOR DECISION
MAKING.

IF THESE LANDS WERE TO BE INVENTORIED, THE MORE PRODUCTIVE AGRICUL-
TURAL LANDS CAN BE PROTECTED SO ALL MAY ENJOY THE BENEFITS OF THEIR
PRODUCTIVE CAPABILITY. THIS BENEFIT IS REALIZED NOT ONLY AS MONEY
IN THE BANK FROM CHEAPER-PRICED FOOD, BUT ALSO AS A SOURCE OF FRESH
AIR, CLEAN WATER, AND OPEN COUNTRYSIDE, IN PERPETUITY.

WITHIN THE REMAINING LANDS, THE ZONING AND DENSITY IS DETERMINED
BY THE SURFACES WHICH ARE IMPACTED. IT IS A FUNCTION OF SOIL,
WATER, SLOPE, VEGETATIVE CANOPY AND EXISTING LAND USES. RATHER THAN
ALLOWING A CERTAIN NUMBER OF HOUSES PER ACRE, WE NEED TO FIND A
METHOD OF ZONING A CERTAIN AMOUNT OF IMPERMEABLE SURFACES PER ACRE.
THIS WOULD THEN CLUSTER THE URBAN DEVELOPMENT IN RURAL AREAS, MINI-
MIZING THE IMPACTS ON THE SOIL RESOURCE.

THE DEVELOPMENT OF THE "CONURBATION" ALONG HIGHWAYS (AFTER GEDDES)

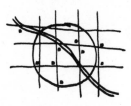 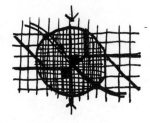 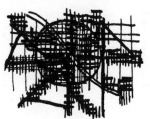

THE ORIGINAL FARMING
COMMUNITY....

INMIGRATION TO THE CITY...
OVERCROWDING DETERIORATION....

CAUSING URBAN
OUTMIGRATION TO THE
SUBURBS AND SPRAWL

THE ADVANTAGES ARE OBVIOUS. CLUSTERING OF HUMAN LIVING ENVIRONMENTS
ENHANCES THE URBAN EXPERIENCE WHILE PROTECTING THE PRODUCTIVE CAPA-
BILITY OF THE SOIL. ECONOMICALLY, THE INFRASTRUCTURE REQUIRED TO
SERVICE CLUSTERED DEVELOPMENTS WOULD ALSO BE LESS. TO MONITOR THIS

NEW ZONING, A SITE REVIEW COULD BE INITIATED FOR ALL NEW SUBURBAN
DEVELOPMENTS. THIS WOULD MAKE SURE THAT THE RESOURCES WOULD BE
PROTECTED.

WHAT NEEDS TO BE DISCUSSED FURTHER IS HOW THE NEW CLUSTERED LAND-
SCAPE PATTERNS TAKE INTO ACCOUNT ALL OF THE OTHER RELATED LAND USES
WHICH MIGHT REQUIRE IMPERMEABLE SURFACES. ROADS, SCHOOLS, SHOPPING
CENTERS AND INDUSTRIAL LAND ARE SOME EXAMPLES.

THE URBAN CONCEPT

SEE SECTION II:

DEVELOPING A STRONG URBAN AGRICULTURE NETWORK

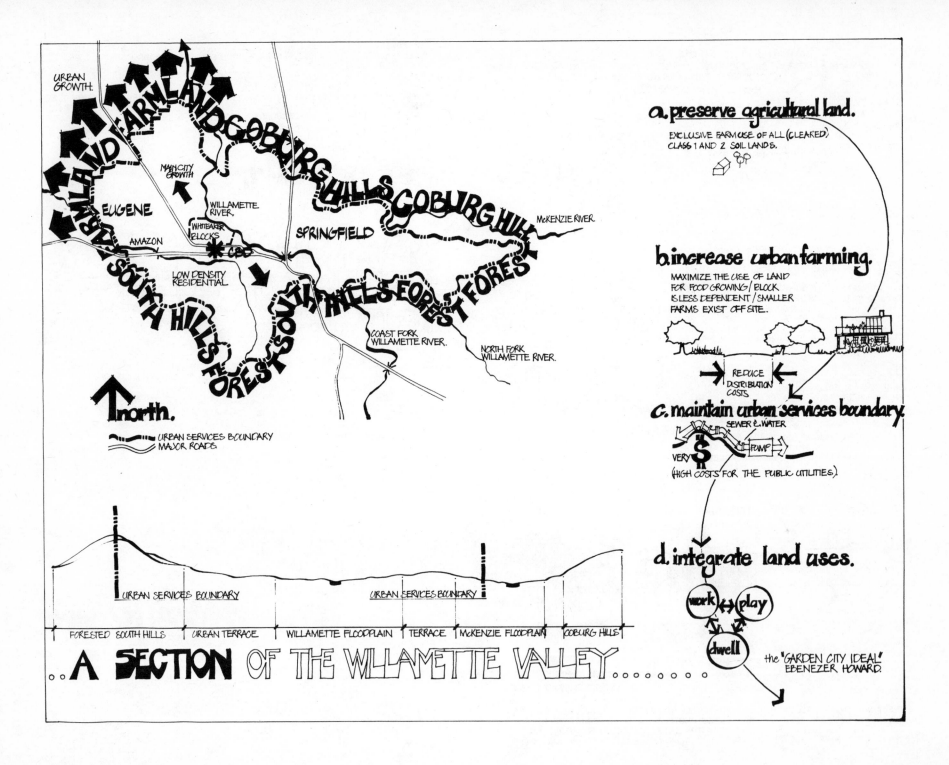

URBAN GROWTH.

FARMLAND FARMLAND GOBURG HILLS COBURG HILL

EUGENE

MAIN CITY GROWTH

WILLAMETTE RIVER

AMAZON

WHITEAKER BLOCKS

CBD

SPRINGFIELD

McKENZIE RIVER

LOW DENSITY RESIDENTIAL

SOUTH HILLS FOREST SOUTH HILLS FOREST FOREST

COAST FORK WILLAMETTE RIVER.

NORTH FORK WILLAMETTE RIVER.

north.

URBAN SERVICES BOUNDARY
MAJOR ROADS

URBAN SERVICES BOUNDARY

URBAN SERVICES BOUNDARY

FORESTED SOUTH HILLS | URBAN TERRACE | WILLAMETTE FLOODPLAIN | TERRACE | McKENZIE FLOODPLAIN | COBURG HILLS

...A SECTION OF THE WILLAMETTE VALLEY..........

a. preserve agricultural land.

EXCLUSIVE FARM USE OF ALL (CLEARED) CLASS 1 AND 2 SOIL LANDS.

b. increase urban farming.

MAXIMIZE THE USE OF LAND FOR FOOD GROWING / BLOCK IS LESS DEPENDENT / SMALLER FARMS EXIST OFF SITE.

REDUCE DISTRIBUTION COSTS

c. maintain urban services boundary.

SEWER & WATER

PUMP

VERY $

(HIGH COSTS FOR THE PUBLIC UTILITIES.)

d. integrate land uses.

work ↔ play
dwell

the "GARDEN CITY IDEAL" EBENEZER HOWARD.

AS THE REGIONAL EXPERIENCE UNFOLDS, WE BEGIN TO SEE THE IMPORTANCE OF STRUCTURING THE LANDSCAPE SO THAT THE LIVABILITY IS MAINTAINED. AN EFFICIENT FORM OF COMMUNITY STARTS TO EMERGE WHERE CLUSTERS OF LIVING ENVIRONMENTS ARE BUILT ON A FRAME OF OPEN SPACE WHICH CONSISTS OF ALLOTMENT GARDENS, SCHOOLS, PARKS, WOODLANDS, WATERWAYS AND FARMLANDS. THE FARMLANDS AND WOODLANDS EXIST AT DIFFERING SCALES FROM URBAN TO SUBURBAN TO RURAL. WHAT IS NEEDED IS A NATURAL ORGANIZATION WHICH IS FLEXIBLE TO THE CONSTANTLY CHANGING "ZEITGEIST" OF AN AREA. BY BUILDING CONGESTIVE SKYSCRAPERS AND HUGE FREEWAYS, WE ARE LIMITING THE MALLEABILITY OF THE REGION.

SINCE WE ARE ALREADY BLESSED WITH BOTH CONGESTION AND SPRAWL, WE MUST FIND WAYS TO BOTH RELIEVE THE CONGESTION AND DILUTE THE SPRAWL. IT IS OUR HYPOTHESIS THAT THE TWO ARE INTEGRALLY CONNECTED, AND NEED TO BE DEALT WITH AT THE SAME TIME. IT IS HERE, THEN, THAT WE DETAIL OUT THE PROCESS OF TRANSFORMATION.

PORTLAND OREGONIAN 21 FEBRUARY 1980

Sprawl continues to knock out crop lands

Buried in the report to the Congress from the Council on Environmental Quality, featuring the nation's troubled waters and air problems, was the discouraging finding that Americans still are busily trading prime farm lands for homes, shopping centers, highways and other non-agricultural projects.

A survey, called the National Agriculture Lands Study, found that in 1977 the United States had 345,417,000 acres of agricultural land. But when grazing, pastures and forested lands were subtracted, the nation's bank of crop lands contained only 196,397,000 acres. Last year, the CEQ reported, the nation lost 3 million acres of farmland, including 1 million acres of prime crop lands, to urban sprawlers and other farm predators.

These withdrawals, the CEQ warned, can only mean higher prices at the grocery counters. As agriculture is pushed away from urban centers by sprawl, the marketing of crops becomes more expensive. Also, the increased values of farms caused by urban land demands help boost food costs.

A tragic example of the misuse of land can be seen in Alaska's Matanuska Valley, site of a large-scale U.S. farming experiment in 1935 that brought the successful settlement of victims of the Great Depression north to work the rich, virgin lands. The valley is being gobbled up by sprawl spreading out from Anchorage like a giant oil slick. The farms, developed with so much pain and sacrifice, are losing their profitability, becoming only subdivided tracts and lots for those who no longer husband the land.

Protected by high mountains and the extremes of the climate, Matanuska Valley is virtually the only land in Alaska where it is possible to do much farming during the state's short growing season. Food that could have been raised in this fertile river valley now must be imported at great cost and expenditure of energy from the lower 48 states.

Such shortsighted decisions, while not always as dramatic as in the case of Alaska, are occurring in all the states, including Oregon where the fertile lands of the Willamette Valley are disappearing under the impact of planning delays and the pressure to build outward from urban areas rather than inward.

It is not enough just to have concerns for pure water and air. The human condition includes eating those things that can only be grown on crop lands.

*OTHER RESOURCES:

SIEVE MAPPING TECHNIQUES - JOE MEYERS, R.P.G. UNIV. OF OREGON

ECOTONAL AGRICULTURE - DICK GLANVILLE, THESIS UNIV. OF OREGON

NEW DESIGNS FOR OLD SUBURBAN NEIGHBORHOODS - JIM KLIEN
THESIS FORTHCOMING

WE BELIEVE WE MUST CONSERVE AS MUCH OF THE FARMLAND <u>WITHIN</u> THE CITY BOUNDARIES AS POSSIBLE BECAUSE THAT MAY BE ALL THAT'S LEFT IN THE WILLAMETTE VALLEY. THE RATE OF URBANIZATION AROUND THE SERVICE PERIMETERS OF THE WILLAMETTE VALLEY CITIES OF PORTLAND, SALEM, ALBANY, AND SMALL TOWNS AND WAYSIDES CLOSER TO EUGENE, SEEMS TO INDICATE THAT THE MAJORITY OF OUR PRECIOUS AGRICULTURAL SOILS, THE LAND ITSELF, AND OUR RENEWABLE ECONOMIC RESOURCE, IS BECOMING RAPIDLY EXTINCT. UNDER PORCHES, SIDEWALKS, CUL-DE-SACS, DRIVEWAYS, STREETS, AVENUES, FREEWAYS AND YES, PARKING LOTS, IS SLIDING OUR FUTURE OF WELL-BEING, LIFE AND NOURISHMENT. GONE INTO OUR EGOS, OUR COFFERS, AND OUR SHODDY SUBDIVISIONS IS A RESOURCE SO PRECIOUS AND DIFFICULT TO RECLAIM FROM URBANIZATION THAT WE WEEP TO THINK ABOUT IT, BUT DEMAND SOME WAY TO DO IT.

WE MUST RECLAIM IT! WE MUST DIG IT OUT FREE FROM THE POISONS, THE CRAP, THE CHEMICALS WE POUR ON IT; WE'VE GOT TO LIBERATE THE EARTH WE HAVE IN OUR OWN YARDS--RENTER OR OWNER ALIKE--AND LET IT BREATHE AND LIVE. RECLAIM OUR TURF AND STABILIZE! HOLD THE GROUND AND MAKE IT GROW -- STRETCH YOUR HOURS TO PLANT THE FUTURE, AND POSSIBLY WE CAN WALK PROUDLY INTO THE FUTURE WITH OUR HEADS UP AND PASS ON TO OUR CHILDREN AN OPPORTUNITY TO TOUCH THE EARTH.

WHILE WE LABOR ON THE EMERALD CANAL TO IRRIGATE OUR CROPS, AND CONCEIVE WAYS TO ALTER OUR ARCHAIC ZONING AND PLANNING REGULATIONS, AND TURN AROUND THE TAX STRUCTURE TO BENEFIT OUTSELVES AND OUR LAND, AND LOOK TOWARD OUR RIVERFRONT FARMS AND FISH PONDS IN THE OLD GRAVEL PITS, WE MIGHT IMAGINE OUR ALTERNATIVE, THE STENCH AND ROT OF ANOTHER SANTA CLARA VALLEY.

THE MOST IMPORTANT THING IN OUR PART OF THE VALLEY IS AGRICULTURE.

NOT TEXACO, NOT FORD MOTOR COMPANY, NOT GENERAL MOTORS OR WESTINGHOUSE, OR TECHTRONICS OR IBM, BUT SIMPLY OUR MOST BASIC AND MOST ENERGY STABLE AND LONG-RANGE FORECASTABLE RENEWABLE RESOURCE INCOME -- THE PROCESS OF PHOTOSYNTHESIS.

FOOD PRODUCTION HAS AN ENERGY CONVERSION EFFICIENCY ON THE ORDER OF 90% OR BETTER....; NO MINING NECESSARY; LITTLE TRANSPORT; LITTLE CONVERSION; LITTLE PACKAGING OR PROCESSING, JUST A VERY EFFICIENT AND VERY PRODUCTIVE PROCESS.

LET'S PROMOTE GROWTH BUT DO IT TOWARD RE-ESTABLISHING THE TRADITION OF THE WATER, THE LAND, THE PLANTS, AND THE ANIMALS WE DEPEND ON. AND LET'S PUT THEM IN VERY CLOSE PROXIMITY TO OURSELVES -- AND LEARN TO LIVE CLOSE AMONG OUR OWN NATURES, IN A WEB OF MUTUAL SUPPORT AND INTERDEPENDENCE.

SO HERE'S TO PHOTOSYNTHESIS, OUR FRIEND, OUR COMRADE IN LIFE, OUR SUPPORTER, OUR POLITICAL ALLY, AND LAST BUT NOT LEAST, OUR MOTHER, BROTHER, SON, DAUGHTER, AND FATHER -- OUR COMPANION IN LIFE....

"FREE" PHOTOSYNTHESIS!!

GARDEN CITIES

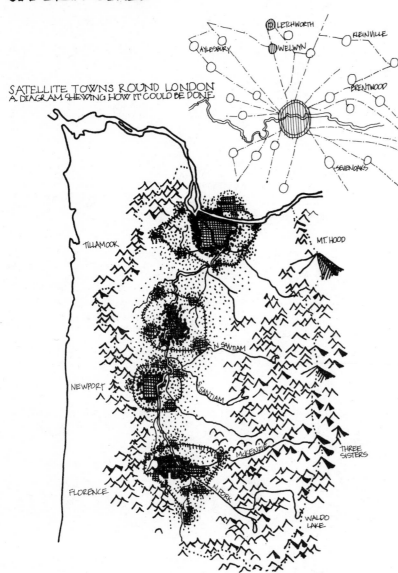

SATELLITE TOWNS ROUND LONDON
A DIAGRAM SHEWING HOW IT COULD BE DONE

LETCHWORTH
WELWYN
AYLESBURY
KLEINVILLE
BRENTWOOD
SEVENOAKS

TILLAMOOK
MT. HOOD
N. SANTIAM
SANTIAM
NEWPORT
McKENZIE
THREE SISTERS
FLORENCE
N. FORK
WALDO LAKE

IN RESPONSE TO GROWING HARSHNESS AND URBANITY ACCRUED DURING THE INDUSTRIAL REVOLUTION IN THE UNITED KINGDOM, A PROGRESSIVE MOVEMENT STIRRED AMONG SOME OF THE BEST THINKER-PLANNERS OF THE TIME, SUCH AS BUCKINGHAM, HOWARD, PARKER AND UNWIN, DEVELOPED THE IDEA OF THE **GARDEN CITY**. TODAY IN THE WILLAMETTE VALLEY, WE FACE A SIMILAR SITUATION. AS PRIME AGRICULTURAL LAND GETS SWALLOWED AROUND OUR CITIES BY INDUSTRIAL AND HOUSING DEVELOPMENTS, AN EVER INCREASING POPULATION STIMULATES MORE ECONOMICAL, CULTURAL AND PHYSICAL GROWTH.
WHERE SHALL THIS GROWTH, INEVITABLE AS IT MAY SEEM, BE DIRECTED?

IF WE LOOK TO OUR THREE MOST ESSENTIAL PREREQUISITES FOR LIFE; AIR, WATER AND FOOD — IT CAN BE SEEN THAT FOOD IS THE MOST SERIOUSLY JEOPARDIZED. OVER 8,000,000 ACRES, OR 34%, OF THE PRIME FARMLAND WAS CONVERTED TO URBAN, OR WATER USES BETWEEN 1967 AND 1975.* OUR SOIL FERTILITY, THE REAL CAPITAL OF THE U.S., IS BEING JEOPARDIZED BY URBAN TRANSFORMATION, HENCE EROSION, POLLUTION OF CHEMICAL FERTILIZERS AND "MALPRACTICED AGRIBUSINESS".

ON THE URBAN SCENE, OUR CITIES ARE BEING OVERBURDENED BY SPRAWL ECONOMIES, URBAN WASTELANDS, AND OVERTAXED INFRASTRUCTURAL SYSTEMS. A CITY CAN REACH A POINT WHERE IT HAS OVERGROWN ITS SUPPORT SYSTEMS AND DETERIORATES.

THESE COMMON ILLS THAT PLAGUE THE CITY AND THE COUNTRY SUGGEST A THEORY THAT RESTORES THE PEOPLE BACK TO THE LAND <u>AND</u> MORE OF AN **INTEGRATED** FOUNDATION FOR URBAN LIFE.

*USDA ECON RESEARCH SERVICE

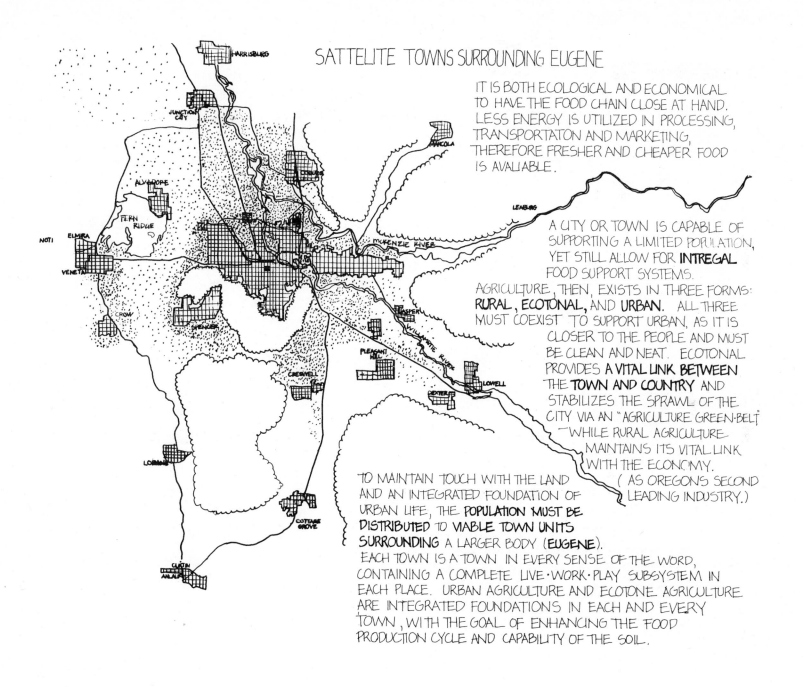

SATTELITE TOWNS SURROUNDING EUGENE

IT IS BOTH ECOLOGICAL AND ECONOMICAL TO HAVE THE FOOD CHAIN CLOSE AT HAND. LESS ENERGY IS UTILIZED IN PROCESSING, TRANSPORTATON AND MARKETING, THEREFORE FRESHER AND CHEAPER FOOD IS AVALIABLE.

A CITY OR TOWN IS CAPABLE OF SUPPORTING A LIMITED POPULATION, YET STILL ALLOW FOR **INTREGAL** FOOD SUPPORT SYSTEMS. AGRICULTURE, THEN, EXISTS IN THREE FORMS: **RURAL, ECOTONAL, AND URBAN.** ALL THREE MUST COEXIST TO SUPPORT URBAN, AS IT IS CLOSER TO THE PEOPLE AND MUST BE CLEAN AND NEAT. ECOTONAL PROVIDES **A VITAL LINK BETWEEN** THE **TOWN AND COUNTRY** AND STABILIZES THE SPRAWL OF THE CITY VIA AN "AGRICULTURE GREEN-BELT" — WHILE RURAL AGRICULTURE MAINTAINS ITS VITAL LINK WITH THE ECONOMY. (AS OREGONS SECOND LEADING INDUSTRY.)

TO MAINTAIN TOUCH WITH THE LAND AND AN INTEGRATED FOUNDATION OF URBAN LIFE, THE **POPULATION MUST BE DISTRIBUTED** TO **VIABLE TOWN UNITS SURROUNDING** A LARGER BODY (**EUGENE**). EACH TOWN IS A TOWN IN EVERY SENSE OF THE WORD, CONTAINING A COMPLETE LIVE·WORK·PLAY SUBSYSTEM IN EACH PLACE. URBAN AGRICULTURE AND ECOTONE AGRICULTURE ARE INTEGRATED FOUNDATIONS IN EACH AND EVERY TOWN, WITH THE GOAL OF ENHANCING THE FOOD PRODUCTION CYCLE AND CAPABILITY OF THE SOIL.

BUT WHAT CAN WE DO WHERE WE LIVE ?...

WELL, THE WILLAMETTE VALLEY IS A VERY FERTILE VALLEY, BLESSED WITH
A FINE, MILD CLIMATE AND SOME OF THE WORLD'S BEST SOILS. FLUSHED
NORTH BY THE WILLAMETTE RIVER, ALONG ITS BANKS, OVER LOTS OF YEARS
OF TIME, HAS BEEN THE TOPSOIL WHICH HAS FLOWED DOWN, WITH THE RAINS,
INTO AND ALONG THE BANKS OF THE MUDDY WILLAMETTE. HERE AND THERE
OLD BASALT INTRUSIONS WITH STICKY CLAYS, STILL CLINGING TO THEIR
BASES, SPROUT UP THROUGH THE RIVER'S BED, ERODING HITHER AND YON
AS THE RIVER WISHES. AND SOME OF THESE LUMPS FACE NORTH AND THEIR
OTHER SIDES SOUTH. AND AROUND THE SOUTH SIDE OF ONE, EUGENE
SKINNER, IN HIS WISDOM, PLANTED THE SEEDS FOR EUGENE'S FUTURE. ON
THAT RIVERBED, AT THE BASE OF THAT BUTTE IS THE MOUTH OF A GIANT
STREAM FLOW OF SOIL, THICK AND LOAMY, WELL-DRAINED AND RICH,
SUITABLE FOR GROWING DAMN NEAR ANYTHING -- AND NOW THAT AREA IS
NAMED WHITEAKER NEIGHBORHOOD. THERE ARE ONLY TWO MAJOR KINDS OF
SOILS IN WHITEAKER, THE STICKY BROWN GOO WHICH STICKS CLOSE TO
THE ROCKS AND TREES, AND THE DARK RICH BROWN FLUFFY SOIL BY THE
RIVER, BROUGHT THERE BY THE RIVER OVER QUITE AWHILE. EITHER ONE
WILL GROW LOTS OF FOOD, BUT ONE TAKES MORE WORK. LATER ON, WE
WILL DESCRIBE IN DETAIL HOW TO MAKE EACH SOIL MORE FERTILE.

THE NEW YORK TIMES, SATURDAY, JANUARY 5, 1980

A Farm At Every Garage

By Dean Baker

EUGENE, Ore. — It'll be 64 years in April since grandpa and my dad who was 12 years old then drove a buckboard across the frozen roadless prairie of eastern Montana and busted open the first farm ground ever on the Fort Peck Indian Reservation.

Farm kids like me have been through a revolution since then. Half my life ago I left the land for college, the city, the army. Now the cycle has turned and here I am in Eugene, Ore., a 36-year-old city dweller, a writer and a born-again urban farmer.

Ah, yes, the romance of the earth. I can hardly contain the old Montana of my dreams. Stage center: my sodbusting gramp, stalwart behind the plow among the sneering mustangers and the blanketed Assiniboins proudly piloting their new Model-T Fords. 1916. Days of heaven. Blizzard. Sagebrush. Silence. Big Sky.

Illusions.

"Oh, yeah," my stoic gramp would allow like Gary Cooper, "that eastern Montana, it's a next-year country." Next year the crop would thrive. Next year after rust, after grasshoppers, after hail, after the dust bowl — if it rained.

"Most of them that broke that land lost their shirts," dad says. No romance for him. He had a family-farm dream but it faded a decade ago when I — his only son — gave him the usual blank look when he asked me again whether he should sell our last piece of ground, 160 acres lying further west in Montana, on the east slope of the Rockies near Great Falls where he had moved when I was 3.

I shrugged and brushed my ducktail. Dad sold and was elected county assessor. I found the city, became a newspaper reporter. What happened is we, like 20 million other Americans between 1945 and 1970, left the farm for the city.

Only recently I found, though, that part of me never left the land. I brought it with me.

This discovery came to me slowly. After 12 years of frantic newspapering my mind-trips into gardens grew longer. Sometimes I forget typewriters and telephones. Finally I realized it was the dirt itself I missed. My wife and kids and I started growing our own food in our backyard and we found something besides dirt: The sun. Rain. Wind. Roots. Leaves.

Ah, yes, romance? No, and not an illusion. Here I think are seeds for a new farming revolution.

I've come to believe that farms belong inside cities. Urban farms belong in yards, on apartment-house roofs, in place of some streets, along rivers and drains. Fruit trees lining avenues. Tomatoes in window boxes. Carrots and lettuce in vacant lots.

This makes both social and economic sense — not only to me but to city planners like University of Oregon architecture professors Richard Britz and Jerry Diethelm, to farmers like Tom Bowerman and to political leaders like former Congressman Charles Porter. All these Eugene men are working to bring farms into our city, population 106,000.

City sod-busting. Imagine the fuel savings when New York neighborhoods grow and distribute their own vegetables instead of importing from Arizona. Farming vacant lots, as some neighbors already are doing in Eugene and other cities, can lead to an old-fashioned barn-raising ethic, healthy jobs for the unemployed, for jail prisoners and the elderly. Urban farming can stop waste;. kitchen scraps make compost. Cooperation can lead to neighbors joining together to raise rabbits, chickens and turkeys, then to form corporations to produce clean energy — small-scale solar power, gasohol, methane enough for a few blocks.

In Eugene, work has begun on conversion of a litter-strewn, five-mile-long concrete drainage ditch into a greenbelt through the city lined with high-density housing and tiny farms, greenhouses, parks, bike paths, water-power generators, even canoe and gondola ports like Venice. At the edges of the city, farmers like Mr. Bowerman advocate creation of self-reliant "agri-villages" to replace the no-man's-land between suburb and farm where tractors now ruin patio parties with dust and sheep are killed by stray city dogs.

So, like gramp and dad, suddenly I have an urge to open new land — though, unexpectedly, this new farmground is within the city, and farm leaders haven't yet recognized the urban farm movement.

Instead, Agriculture Secretary Bob Bergland worries that traditional farms are in trouble, are being gobbled up by corporations which fail to support small towns and cities, which in turn are financially troubled.

Part of the solution is to break new ground in the cities. Mr. Bergland hasn't yet seen that. Why? Perhaps because he, like many of us new urbanites, is blind to his own backyard.

Dean Baker is setting up an "integral urban house" where he will grow his own food and produce his own energy.

Section I

Local Food and Farming
(Hopefully Organic)

Chapter 1

Basic Food, Nutrition, and and Local Resources

BASIC FOOD AND NUTRITION

OUR PREMISES FOR THIS PORTION OF OUR WORK ARE 1) THAT COMMUNITIES WHICH INTEGRATE A DIVERSITY OF SOUND LOCAL FOOD PRODUCTION ARE SOCIALLY AND ECONOMICALLY MORE HEALTHY; 2) THAT FOODS THAT ARE GROWN LOCALLY, SEASONALLY, AND WITHOUT CHEMICAL FERTILIZERS AND PESTICIDES, ETC. ARE MORE HEALTHFUL FOODS THAN THOSE GROWN WITH THE USE OF SUCH CHEMICAL PRACTICES; AND 3) THAT A MORE DEVELOPED USE OF LOCALLY PRODUCED FOODS (WHETHER URBAN OR RURAL) WILL CREATE LESS RELIANCE ON CORPORATE AGRIBUSINESS FOOD SYSTEMS AND ECONOMIC CONTROL, AND WILL PROMOTE GREATER COMMUNITY RELIANCE ON THIS AREA'S OWN RESOURCES OF CLIMATE, LAND, AND PEOPLE.

LOCAL AGRICULTURE

A CALIFORNIA STUDY DONE IN 1977, ENTITLED THE FAMILY FARM IN CALIFORNIA: REPORT OF THE SMALL FARM VIABILITY PROJECT, SHOWED THE RELATIVE HEALTH OF AREAS HAVING A WIDE DIVERSITY OF SMALL FARMS TO BE GREATER THAN SIMILAR AREAS HAVING FEWER AND LARGER FARM HOLDINGS.

> "THE TASK FORCE FOUND THAT COMMUNITIES WITH ECONOMIES BASED ON FAMILY FARM AGRICULTURE ARE MORE PROSPEROUS AND DESIRABLE TO LIVE IN THAN ARE COMMUNITIES BASED ON LARGE-FARM AGRICULTURE. IT INTERPRETED THE EVIDENCE TO INDICATE THAT SMALL FARMS GENERATE MORE LOCAL BUSINESS, HIGHER LEVELS AND DIVERSITY OF EMPLOYMENT, AND SUPERIOR GOVERNMENTAL SERVICES....THIS MEANS THAT A PROPERLY IMPLEMENTED POLICY OF ENCOURAGING THE CREATION OF SOUNDLY STRUCTURED AND WELL MANAGED FAMILY FARMS CAN BE EXPECTED TO BRING TANGIBLE ECONOMIC AND SOCIAL BENEFITS TO THE RESIDENTS OF RURAL COMMUNITIES AND TO THE STATE AS A WHOLE."

OTHER BENEFITS, SUCH AS THE AVAILABILITY OF FARM-FRESH FOODS, THE HOLDING OF THE WILLAMETTE VALLEY CLASS I SOILS FOR FARMLAND USE, THE INTEGRATION OF DIVERSE PLANTS, ANIMALS, AND PEOPLE OF ALL AGES, BUILD OUR CONTINUITY WITH OUR OWN ROOTS AND BRING US CLOSER IN FEELING AND KNOWLEDGE WITH ONE OF OUR MOST BASIC NECESSITIES -- PROVIDING OURSELVES AND OUR COMMUNITIES WITH FOOD.

AGRIBUSINESS

AGRIBUSINESS DEFINES THE CORPORATE CONTROL OF FOOD PRODUCTION, A TOTAL AND DELIBERATE SHIFTING OF CONCEPTS, FROM AGRICULTURE TO AGRIBUSINESS. CONTROL BY LARGE CORPORATIONS OF FOOD PRODUCTION AND MARKETING WILL CONTINUE TO INCREASE THE COSTS OF FOOD TO THE CONSUMER WHILE DECREASING THE NUTRITIONAL QUALITY OF THE FOOD. IT IS OUR GOAL TO REVERSE THIS TREND.

TODAY, CORPORATE CONTROL OF FOOD PRODUCTION IS EXPANDING AT AN ALARMING RATE, TO WHERE A HANDFUL OF CORPORATIONS CONTROL A VERY LARGE PERCENTAGE OF THE AMERICAN FOOD SUPPLY. TODAY IT IS ESTIMATED THAT A MERE 50 CORPORATIONS NOW REAP 90 PERCENT OF THE PROFITS OF THE ENTIRE FOOD INDUSTRY. SOME COMMODITIES, SUCH AS BROILERS AND SUGAR BEETS, ARE ALMOST 100% PRODUCED UNDER CORPORATE CONTRACT TODAY, AND CITRUS FRUITS ALMOST 50%. LAMB IS IMPORTED INTO EUGENE SUPERMARKETS FROM ARKANSAS, CHICKENS FROM MISSISSIPPI, LETTUCE, CARROTS AND OTHER PRODUCE FROM CALIFORNIA AND OTHER STATES, ALTHOUGH ALL OF THESE FOODS GROW VERY WELL HERE -- AND ARE GROWN HERE ALREADY.

THESE SAME CORPORATIONS ARE INCREASING THEIR SPHERE OF CONTROL ALSO, TO EXTEND BEYOND JUST THE PRODUCTION STAGE OF FOOD, AND INTO OTHER AREAS SUCH AS SEED PRODUCTION; THE MANUFACTURE OF FARMING EQUIPMENT; FERTILIZER AND PESTICIDE PRODUCTION; HARVESTING; PROCESSING; PACKAGING, DISTRIBUTION AND MARKETING, AND ARE INCREASINGLY TRYING TO CONTROL DATA ON ALTERNATIVE APPROACHES, THUS FURTHER REMOVING CONTROL AND INPUT FROM THE SMALLER-SCALE AND MORE LOCAL SECTOR.

THE MOST IMMEDIATE IMPACT OF CORPORATE CONTROL OF FOOD PRODUCTION IS SIX-FOLD:

1. CORPORATE CONTROL REMOVES CONTROL OF FOOD QUALITY FROM THE CONSUMER. THE CONSUMER HAS NO CHOICE AS TO WHAT GOES INTO THE FOOD, OR WHAT IS REMOVED FROM IT, DURING PRODUCTION, PROCESSING, DISTRIBUTION OR MARKETING.

2. IT CONTROLS A MARKET THAT THE SMALL FARMER CANNOT COMPETE WITH, THUS FORCING HIM TO SELL OUT, OFTEN AT URBAN DEVELOP-MENT PRICES WHICH ARE FAR ABOVE THE MARKET VALUE OF AGRI-CULTURAL LAND. BETWEEN 1960 AND 1976, FOURTEEN HUNDRED FARMS FOLDED EACH WEEK IN THE UNITED STATES.

3. IT DEPLETES THE SOIL STRUCTURE THROUGH ITS EMPHASIS ON MONO-CULTURE (RAISING A SINGLE CROP ON LARGE TRACTS OF LAND), AND CREATES A SOIL AND PLANT DEPENDENCY UPON NON-RENEWABLE FOSSIL FUEL ENERGY TO SUPPLY THE FOOD FOR GROWTH, RESISTANCE TO DISEASE AND INSECT ATTACK. AN ARTICLE ENTITLED, "INDUS-TRIALIZATION OF FARMS THREATENS U.S. TOPSOIL" STATES THAT ..."THE GENERAL ACCOUNTING OFFICE WARNED THAT, BECAUSE OF EXCESSIVE EROSION, FARMS IN THE GREAT PLAINS, CORN BELT AND PACIFIC NORTHWEST ARE LOSING TOPSOIL AT A RATE WHICH THREATENS PRODUCTIVITY." SOILS LOST THROUGH THE PRACTICE OF MONOCULTURE ARE AS LOST AS THOSE SOILS BEING BUILT UPON BY EVER-EXPANDING URBAN BOUNDARIES.

4. IT IS RESPONSIBLE FOR UNNECESSARILY HIGH FOOD COSTS TO THE CONSUMER. THE HIDDEN COSTS INVOLVED IN THE OPERATION OF MECHANIZED AGRICULTURE AND INCREASING COSTS (AND SCARCITY OF) PETROLEUM FUELS FOR PRODUCTION, HARVESTING, PROCESSING, PACKAGING AND TRANSPORTATION TO MARKETS WILL CONTINUE TO SHOW UP IN THE COST THE CONSUMER PAYS FOR HIS FOOD. OF THE INCREASES IN FOOD COSTS IN THE SUPERMARKET FROM 1954 TO 1975, ONLY 6% OF THAT INCREASE WENT BACK TO THE FARMER, WHILE THE OTHER 94% FILTERED INTO THE CORPORATE STRUCTURE TO PAY FOR PROCESSING, MARKETING, ADVERTISING, ETC. IT IS FURTHER ESTIMATED THAT THE CORPORATE MONOPOLY OF FOOD MANU-FACTURE COSTS CONSUMERS AN ADDITIONAL $15 BILLION EACH YEAR IN ADDED "HIDDEN" FOOD COSTS.

5. IT IS A WASTEFUL SYSTEM. THIS ACCOUNTS FOR A PORTION OF THE "HIDDEN" COSTS MENTIONED ABOVE. ACCORDING TO A RECENT GENERAL ACCOUNTING OFFICE STUDY, SOME 137 MILLION TONS OF FOOD VALUED AT $31 BILLION WASTED IN 1974 ALONE. OF THIS AMOUNT, 47% DISAPPEARED BEFORE IT REACHED THE CONSUMER, "IN ROUNDED FIGURES, 16 PERCENT WAS LOST AT HARVEST, 20 PER-CENT AT THE FOOD DISTRIBUTOR LEVEL (WHOLESALE AND RETAIL), SEVEN PERCENT DURING STORAGE, TWO PERCENT DURING PROCESSING, AND ONE PERCENT DURING TRANSPORTING. THE COMBINED LOSS OF U.S. PRODUCED FOOD GRAINS, MEAT, SUGARS, OIL-SEEDS, VEGE-TABLES, FRUITS AND NUTS IN 1974 COULD HAVE FED AN ESTIMATED 49 MILLION PEOPLE. IN ADDITION TO THIS, APPROXIMATELY 66 MILLION ACRES OF LAND, NINE MILLION TONS OF FERTILIZER AND THE EQUIVALENT OF 461 MILLION BARRELS OF OIL WERE USED TO PRODUCE FOOD THAT WAS ULTIMATELY LOST."

6. IT ADVERSELY AFFECTS THE NUTRITIONAL QUALITY OF THE FOOD. FOODS ARE GENETICALLY MANIPULATED FOR HIGH PRODUCTION YIELDS. FOODS ARE ADAPTED AND CHANGED TO SUIT PACKAGING, DISTRIBUTION AND STORAGE REQUIREMENTS, NOT NUTRITIONAL REQUIREMENTS. TO MAKE FOOD AESTHETICALLY PLEASING (AND TO WITHSTAND THE RIGORS OF TRANSPORTATION AND STORAGE) WAXES, SPRAYS, COLORS AND OTHER CHEMICALS ARE USED, VERY FEW OF WHICH HAVE BEEN ADEQUATELY TESTED FOR HUMAN SAFETY. THE PROCESSING IN-VOLVED TO CREATE THE CONVENIENCE FOODS WE SEE ON THE SUPER-MARKET SHELF TODAY FURTHER REMOVES THE ORIGINAL FOODS FROM THEIR NUTRITIONAL HERITAGE; LEAVING US WITH NUTRITIONALLY-DEPLETED PRODUCTS OR ARTIFICALLY FORTIFIED ONES WHICH DO NOT RESEMBLE NATURE'S ORIGINAL INTENT.

NUTRITIONAL TRENDS

NUTRITION AND THE MARKET: IN THE EXISTING MONOCULTURE SYSTEM OF AGRICULTURE AND MARKETING, COMMUNITIES PRODUCE A LIMITED NUMBER OF CROPS, SHIPPING THESE TO DISTANT MARKETS IN OTHER PARTS OF THE COUNTRY OR WORLD, AND "IMPORTING" TO THEIR AREA THOSE PRODUCTS NOT GROWN LOCALLY. BECAUSE OF THIS, FRUITS AND VEGETABLES MUST BE PICKED BEFORE THEIR NATURAL MATURING TIME, BEFORE THE NUTRIENTS HAVE HAD A CHANCE FOR MAXIMUM NUTRITIONAL AND FLAVOR DEVELOPMENT. ANOTHER DETRIMENTAL EFFECT OF SHIPPING IS THAT PRODUCE BEGINS LOSING NUTRIENTS WHEN PICKED. FOR EXAMPLE, CORN WILL LOSE 50% OF ITS SUGAR WITHIN TWENTY-FOUR HOURS AT 75° F. THIS LOSS BEGINS ABOUT TWELVE MINUTES AFTER PICKING. MANY NUTRIENTS ARE WATER SOLUABLE AND ARE LOST IN THE WATER USED IN WASHING AFTER PICKING. TO WITHSTAND FURTHER NUTRITIONAL LOSSES, FRUITS AND VEGETABLES RECEIVE TREATMENTS OF WAXES, PRESERVERS, AND COLOR RETAINERS. SO, BY THE TIME THE FRUIT OR VEGETABLE REACHES ITS FINAL DESTIN-ATION IN THE SUPERMARKET, ITS NUTRITIONAL VALUE HAS BEEN DECREASED.

GRAINS ARE PROCESSED, MILLED AND BLEACHED. THIS GIVES THEM THE BENEFIT OF NOT SPOILING SO QUICKLY WHICH IS AN IMPORTANT CONSID-ERATION FOR PEOPLE OBSESSED WITH THE RICHES OF LONG SHELF LIFE. ALSO, THE BLEACHING "AGES" THE FLOUR QUICKLY, THUS REDUCING OR ELIMINATING THE NEED TO STORE GREAT AMOUNTS OF FLOUR THAT MUST AGE OVER LONG PERIODS OF TIME. HOWEVER, THE MILLING PROCESS REMOVES MANY NUTRITIONAL QUALITITES OF THE GRAIN (A FEW OF WHICH MAY BE SUBSTITUTED ARTIFICIALLY AT A LATER DATE). ACCORDING TO A STUDY MADE BY THE UNIVERSITY OF CALIFORNIA, COLLEGE OF AGRI-CULTURE, THE MILLING OF WHEAT REDUCES MINERALS AND VITAMINS AS FOLLOWS:

> MANGANESE, IRON AND THIAMIN: 80-98%; MAGNESIUM, PHOSPHORUS AND NIACIN 70-75%; COPPER AND RIBO-FLAVIN 65%; CALCIUM AND POTASSIUM 50%; PLUS DESTROYING MUCH OF THE VALUABLE VITAMIN B COMPLEX.

EVERYTHING POSSIBLE IS BEING DONE TO FOODS TO PROMOTE THE ALL-IMPORTANT "LONGER SHELF LIFE" OF THE GROCERY STORE FARE. STABILIZERS (CHEMICAL) ARE USED TO IMPROVE TEXTURE AND KEEPING QUALITY. OTHER CHEMICALS ARE ADDED IN THE FORM OF PRESERVATIVES, SUCH AS BHA AND BHT, SODIUM BENZOATE, SODIUM NITRITE. SODIUM NITRITE IS THE CHEMICAL ADDITIVE THAT PRESERVES PROCESSED MEATS (BOLOGNA, LUNCH MEATS, SAUSAGE, HOT DOGS, ETC.) AND GIVE THEM THEIR FRESH RED COLOR. BUT IT IS ALSO KNOWN THAT NITRITES CAN COMBINE WITH OTHER SUBSTANCES (AMINES) IN THE FOOD AND BODY AND PRODUCE NITROSAMINES WHICH ARE HIGHLY CARCINOGENIC OR TUMOR CAUSING. IT IS ESTIMATED THAT 90% OF THE CANCERS IN PEOPLE ARE CAUSED BY CHEMICALS, MANY OF WHICH COULD BE ELIMINATED BY REMOVING CHEMICAL CARCINOGENS FROM OUR FOOD, OR CHOOSING FOODS THAT DO NOT CONTAIN THEM.

NUTRITIONAL CONSUMPTION TRENDS: THAT AMERICANS ARE MOVING AWAY FROM RESPONSIBLE CHOICES IN THE NUTRITION OF THE FOODS THEY EAT HAS BEEN SHOWN BY MONITORING THE CHANGE IN THE CONSUMPTION PATTERN OF FOOD IN THE UNITED STATES. IN EARLY 1978, THE SELECT COMMITTEE ON NUTRITION AND HUMAN NEEDS PUBLISHED EATING IN AMERICA: DIETARY GOALS FOR THE UNITED STATES, IN WHICH THEY DOCUMENTED TRENDS IN THE EATING HABITS OF AMERICANS FROM 1910 TO 1976. FIGURE 1 INDI-CATES THE TREND IN THE CONSUMPTION OF FRESH AND PROCESSED FRUITS AND VEGETABLES, ALONG WITH THE STATEMENT THAT:

> "IT APPEARS THAT INCREASED CONSUMPTION OF FRESH FRUITS AND VEGETABLES, PARTICULARLY THE HIGH-NUTRIENT FORMS, WOULD BE BENEFICIAL FOR MANY PERSONS IN NEED OF DIETARY

IMPROVEMENT. EDUCATING CONSUMERS, PARTICULARLY THOSE OF LOW INCOMES, TO THE GREATER ADVANTAGE OF THE MOST ECONOMICAL AND MOST NUTRITIOUS FRUITS AND VEGETABLES, WOULD OFFER A GREAT POTENTIAL FOR DIETARY IMPROVEMENT."

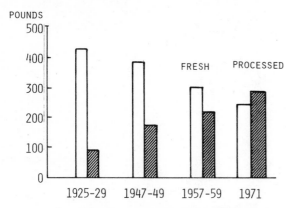

FIGURE 1: TRENDS IN CONSUMPTION OF FRESH VS. PROCESSED FRUITS AND VEGETABLES (PER CAPITA, PER YEAR)

THE REPORT ALSO STATES THAT FRESH PRODUCE IN THE GROCERY STORES HAS ABOUT THE SAME NUTRITIONAL EQUIVALENT AS FROZEN VARIETIES DUE TO "NUTRIENT-DEPLETION IN SHIPPING AND STORAGE...", AND INDICATES THE NUTRITIONAL SUPERIORITY OF PRODUCE CONSUMED DIRECTLY FROM THE GARDEN.

FIGURE 2 SHOWS THAT WHEREAS FOOD STARCHES USED TO PROVIDE THE GREATER MAJORITY OF THE BODY'S CARBOHYDRATE ENERGY, THIS ENERGY IS NOW SUPPLIED BY SUGAR. THESE AND OTHER TRENDS IN CONSUMPTION

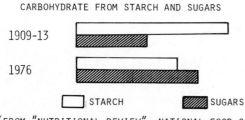

(FROM "NUTRITIONAL REVIEW", NATIONAL FOOD SIUTATIONS, CFE (ADM.) 299-9, JAN. 1975. PRELIMINARY DATA FOR 1976 UNPUBLISHED. AGRICULTURAL RESEARCH SERVICE, U.S.D.A.)

FIGURE 2

PATTERNS (I.E., FATS, SALT) LEAD DIRECTLY TO DIET-RELATED PROBLEMS AND DISEASES SUCH AS STROKES, OBESITY, CANCER, DIABETES. THE RE-SULTS OF THE TRENDS HAVE LED THE RESEARCHERS OF EATING IN AMERICA: DIETARY GOALS FOR THE UNITED STATES TO RECOMMEND SPECIFIC STEPS TO CHANGE THE TRENDS THEMSELVES. THEY STATE THAT THESE HEALTH PROBLEMS

ARE NOT SOLVED SIMPLY BY PROVIDING MORE MEDICAL CARE, BUT BY ADDRESSING THE FUNDAMENTAL REASONS UNDERLYING INADEQUATE DIETS. THEY ARE BEGINNING TO LOOK AT THE SOCIAL IMPLICATIONS OF FOOD, ITS PRESENTATION AND PREPARATION, AND THE PSYCHOLOGICAL EFFECTS OF DELIVERY AND SERVING METHODS.

ADVERTISING PLAYS AN OVERWHELMING ROLE IN PEOPLE'S FOOD CHOICES, AND IS DIRECTLY TIED IN WITH TRENDS IN FOOD CONSUMPTION AND THE PROLIFERATION OF FAST FOOD SERVICE ESTABLISHMENTS AND THE AT-HOME PREPARATION OF "CONVENIENCE" MEALS. ADVERTISING ALSO ACCOUNTS FOR A PORTION OF EACH FOOD DOLLAR THE CONSUMER PAYS AT THE GROCERY STORE. THOSE SAME 50 CORPORATIONS MENTIONED EARLIER (AS CON-TROLLING NINETY PERCENT OF THE FOOD INDUSTRY PROFITS) ALSO CONDUCT 75% OF ALL MEDIA ADVERTISING. THIS ADVERTISING IS DONE FOR SPECIFIC CORPORATE PRODUCTS RATHER THAN FOR PROMOTING NUTRITIONAL HEALTH, AS SHOWN BY THE FOLLOWING QUOTE FROM A STUDY PREPARED BY LYNNE MASOVER AND DR. JEREMIAH STAMLER OF NORTHWESTERN UNIVERSITY MEDICAL SCHOOL.

"WHEN THIS OUTLAY OF FOOD ADVERTISING IS JUXTAPOSED WITH WHAT IS KNOWN ABOUT THE PREVALENCE IN THE UNITED STATES OF MALNUTRITION...IT IS REASONABLE TO CONCLUDE THAT ON WEEKDAYS OVER 70 PERCENT AND ON WEEKENDS OVER 85 PERCENT IS NEGATIVELY RELATED TO THE NATION'S HEALTH NEEDS."

A GREAT DEAL OF EMPHASIS IS PLACED ON "CONVENIENCE" (QUICK-COOKING OR EASY-TO-PREPARE FOODS SUCH AS T.V. DINNERS AND PRE-PACKAGED STOVE-TOP MEALS, ETC.), PACKAGING THAT CALLS ATTENTION TO SELLING GIMMICKS RATHER THAN NUTRITION (CEREALS THAT ARE SUGAR COATED OR FILLED WITH MARSHMALLOWS, AND COME WITH TOYS AS ADDED INDUCERS). THESE ITEMS ARE AIMED AT CHILDREN; AND T.V. AND SUPERMARKETS BECOME PUSHERS FOR FOODS THAT HAVE VERY LITTLE FOOD VALUE.

THE SCHOOL FARMS AND CURRICULA DEVELOPING NOW IN WHITEAKER NEIGH-BORHOOD WILL SERVE, IN PART, AS A COUNTERFORCE TO MEDIA CONTROL OF NUTRITIONAL TRENDS, AND IS DIRECTED TOWARD AN UNDERSTANDING OF FOOD, ITS GROWTH AND CYCLES, AND THE RELATIONSHIP OF FOOD IN THE CHILD'S NURTURING ENVIRONMENT.

USING LOCAL RESOURCES

ALL OF THE REASONS POINTED OUT EARLIER IN THIS WORK IDENTIFY THE BENEFITS OF ALTERING OUR FOOD BUYING HABITS TOWARDS MORE ECONOMIC- ALLY AND NUTRITIONALLY SOUND FOODS, AND FOR DEVELOPING AND USING OUR OWN ABUNDANT LOCAL RESOURCES FOR FOOD PRODUCTION.

THE WHITEAKER AREA HAS A WIDE RANGE OF FOOD STORES--RETAIL, WHOLE- SALE, AND PROCESSING--AS SHOWN ON THE FOLLOWING MAP. A FEW OF THESE STORES PROMOTE SALE OF FOODS WHICH HAVE BEEN GROWN BY LOCAL FARMERS, AND SUPPORT PROCESSING OF FOODS FREE FROM HARMFUL ADDI- TIVES. A FEW OTHERS OFFER ECONOMIC AIDS IN PROVIDING STAPLE ITEMS--FLOUR, GRAINS, DRIED FRUIT, ETC.--IN UNPACKAGED BULK FORM. THIS ELIMINATES THE COST OF PRODUCING AND USING PACKAGING MATER- IALS AT ONE END, AND THE WASTE OF THROWING THEM AWAY AT THE OTHER. (SEE ALSO SECTION ON BULK-BUYING).

CHANGING BUYING HABITS

ALTERING OUR FOOD BUYING HABITS AND IMPROVING OUR NUTRITIONAL CONSUMPTION CAN BEGIN WITHIN THE FRAMEWORK OF THE PRESENT MARKET SYSTEM IN AND AROUND THE WHITEAKER NEIGHBORHOOD, BY BUYING FOODS THAT HAVE BEEN GROWN LOCALLY, THAT ARE FREE OF CHEMICAL ADDITIVES AND PESTICIDES, AND THAT HAVE NOT BEEN OVERPROCESSED. ONE WAY IS TO SUBSTITUTE MORE WHOLESOME FOODS FOR THOSE THAT HAVE LESS NU- TRITIONAL VALUE. FOR EXAMPLE:

GRAINS: USE WHOLE GRAINS FOR FLOURS, CEREALS, AND BREADS. WHOLE GRAINS HAVE BEEN MILLED WITH THE GERM AND THE OUTSIDE BRAN INTACT (THEREBY PRESERVING THE VITAMINS AND MINERALS WHICH ARE OTHERWISE LOST IN PROCESSING). GENERALLY, THIS MEANS BROWN BREADS, BROWN FLOURS AND BROWN RICE.

LOCAL SOURCES FOR WHOLE GRAIN FOODS INCLUDE SOLSTICE BAKERY, 350 E. 3RD, FOR DIFFERENT KINDS OF BREAD ITEMS. (SOLSTICE BREADS ARE ALSO SOLD IN OTHER LOCAL OUTLETS)

UNTREATED FLOUR, CEREALS, AND OTHER WHOLE GRAINS CAN BE FOUND AT MANY FOOD CO-OPS AND MARKETS, SUCH AS THE COMMUNITY NATURAL FOODS AND GENERAL STORE, AND GROWER'S MARKET IN WHITEAKER, AS WELL AS OTHER STORES IN EUGENE (WILLAMETTE PEOPLE'S FOOD CO-OP, KIVA, NEW FRONTIER, STARFLOWER, ETC.) SEE SECTION ON BULK BUYING.

SWEETENERS: THOSE WHICH HAVE HAD MINIMUM REFINEMENT WILL HAVE A HIGHER NUTRITIONAL VALUE, AS SHOWN IN THE CHART BELOW, FROM DIET FOR A SMALL PLANET BY FRANCES MOORE LAPPE.

LOOK FOR HONEY OR MOLASSES OR BROWN SUGAR AS A REPLACEMENT FOR WHITE GRANULATED SUGAR. HONEY IS RAISED LOCALLY AROUND EUGENE BUT BEEKEEPERS DO NOT SELL IT IN THE MARKETS BECAUSE OF THE HIGH GOVERNMENTAL REQUIREMENTS AND RESTRICTIONS ON PROCESSING FOR MARKET. HOWEVER, SOME HONEY RAISERS (FROM THE QUEENRIGHT BEEKEEPERS COOPERATIVE, AND OTHER INDIVIDUAL RAISERS) DO SELL THEIR HONEY, USUALLY WHEN IT IS HARVESTED IN THE FALL AND EARLY WINTER. LOOK FOR LOCAL HONEY AT THE SATURDAY MARKET AND AT THE FARMERS MARKET, BOTH LOCATED IN THE BLOCK BORDER-

ING 7TH AND OAK STREETS. ALSO CHECK THE "MARKET BASKET" SECTION OF THE EUGENE REGISTER-GUARD FOR OTHER AVAILABLE LOCAL HONEY.

SUGARS, HONEY AND MOLASSES COMPARED

	WHITE SUGAR (GRANULATED)	BROWN SUGAR (BEET OR CANE)	MOLASSES THIRD EXTRACTION OR BLACKSTRAP	HONEY (STRAINED OR EXTRACTED)	MAPLE SUGAR
MINERALS	mg	mg	mg	mg	mg
CALCIUM	0	85	684	5	143
PHOSPHORUS	0	19	84	6	11
IRON	0.1	3.4	16.1	0.5	1.4
SODIUM	1.0	30	96	5	14
POTASSIUM	3.0	344	2927	51	242
VITAMINS					
THIAMINE	0	0.01	0.11	trace	--
RIBOFLAVIN	0	0.03	0.19	0.04	--
NIACIN	0	0.2	2.0	0.3	--

(SOURCE: "COMPOSITION OF FOODS," AGRICULTURE HANDBOOK, NO. 8, U.S.D.A.)

JUICES: LOOK FOR UNFILTERED UNPRESERVED WHOLE FRUIT JUICES RATHER THAN FRUIT "DRINKS". IN WHITEAKER, THE GENESIS JUICE COOPERATIVE PROCESSES JUICES USING ORGANIC LOCAL PRODUCE WHENEVER IT IS AVAILABLE, TO BE SOLD IN LOCAL MARKETS. THIS MIGHT BE A GOOD OUTLET FOR SOME OF THE EXTRA, UNUSED FRUIT FROM THE ABUNDANT NEIGHBORHOOD TREES, OR EXTRA VEGETABLES FROM THE GARDENS.

SOYFOODS: SOYBEAN IS A LOW-COST SOURCE OF VERY HIGH QUALITY PROTEIN, AND CAN BE EATEN IN SO MANY FORMS--FROM SPROUTS TO ROASTED SEEDS TO TOFU--THAT EVERYONE CAN FIND A FAVORITE WAY TO EAT IT. ITS COST AND PROTEIN CONTENT MAKE IT A VERY DESIRABLE SUBSTITUTE FOR MEAT. TOFU IS MADE IN THE NEIGH- BORHOOD BY SURATA SOYFOODS, 518 OLIVE, AND MARKETED LOCALLY AT THE COMMUNITY STORE, AND GROWER'S MARKET IN WHITEAKER, AND AT OTHER STORES IN EUGENE. SURATA ALSO PROVIDED EDU- CATIONAL INFORMATION ON THE NUTRITIONAL QUALITIES OF SOY, DIFFERENT SOY PRODUCTS, AND HOW TO PREPARE THEM.

FOR AVOIDING ADDITIVES IN PACKAGED FOODS, READING THE LABEL IS A GOOD START. BEGIN BUYING THOSE FOODS THAT HAVE THE FEWEST, OR BETTER YET, NO CHEMICAL ADDITIVES. AVOID THE PROCESSED MEATS THAT CONTAIN SODIUM NITRITE AND SODIUM NITRATE. THE CUSTOM MEAT MARKET IN WHITEAKER PROVIDES MEATS WHICH ARE PROCESSED AT THEIR SITE USING SALT AND SMOKE CURES. THEY BEGAN PROVIDING THIS SERVICE SEVERAL YEARS AGO AS A RESPONSE TO CUSTOMER REQUESTS FOR NITRITE/NITRATE- FREE MEATS. THEY NOW ESTIMATE THAT SALE OF THESE MEATS COMPRISE ABOUT 75% OF THEIR BUSINESS. IN ADDITION, THEIR MEATS--PORK, BEEF AND CHICKEN--COME FROM LOCAL PRODUCERS IN THE EUGENE AND CRESWELL AREAS. ALL PROCESSING IS DONE AT THEIR MARKET LOCATION, 577 PEARL.

BULK BUYING

BULK BUYING DOES NOT MEAN YOU HAVE TO PURCHASE AND STORE FOOD IN 50-100 POUND LOTS (ALTHOUGH YOU CAN DO THIS IF YOU WANT TO). RATHER, IT MEANS THAT THE FOOD SOLD IS NOT PRE-PACKAGED AND YOU BUY THE AMOUNT YOU NEED FROM LARGE LOTS. IT OFFERS A WAY TO

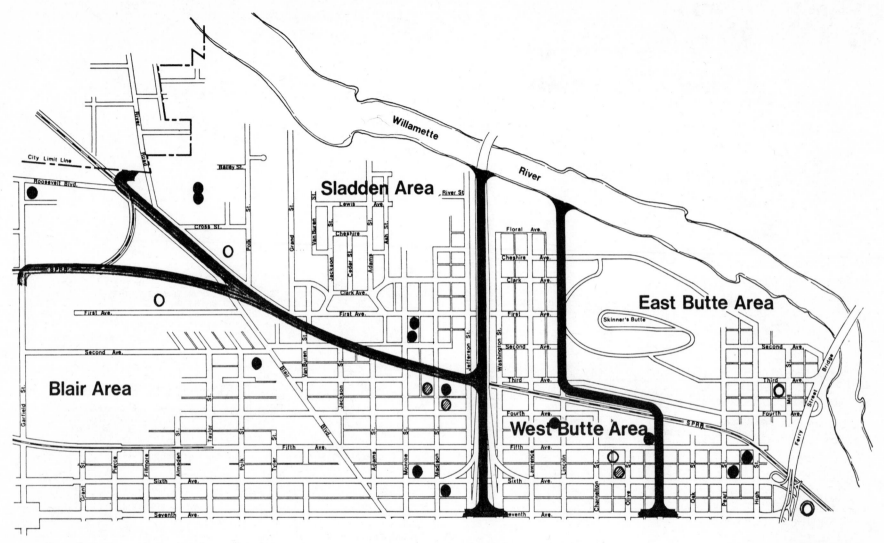

Sladden Area

East Butte Area

Blair Area

West Butte Area

Willamette River

<u>FOOD DISTRIBUTORS IN WHITEAKER</u>

● RETAILERS

◍ WHOLESALERS

○ PROCESSORS

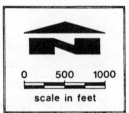

0   500   1000

scale in feet

PURCHASE IN QUANTITIES CONVENIENT TO YOUR HOUSEHOLD'S USE--A DINNER'S WORTH OF MACARONI, OR A 6-MONTH SUPPLY. YOUR OWN BAGS, BOXES OR JARS ARE REUSABLE FOR STORAGE OR FOR USE WHEN YOU PURCHASE MORE FOOD. THE ADVANTAGES OF BULK BUYING ARE ITS CONVENIENCE, ITS ELIMINATION OF WASTEFUL PACKAGING, AND ITS LOWER COST. ANOTHER ADVANTAGE IS THAT BULK BUYING THROUGH A BUYING CLUB ORGANIZED OF PEOPLE WITHIN NEIGHBORING BLOCKS WILL FURTHER LOWER COST OF THE FOOD PURCHASED. SEE ORGANIZING A BUYING CLUB.

THE FOLLOWING STUDY, DONE FOR THE EDIBLE CITY RESOURCE CENTER BY ECO-ALLIANCE IN CORVALLIS, SHOWS COMPARISON PRICES FOR BUYING SELECTED FOOD ITEMS IN THEIR PACKAGED AND THEIR NON-PACKAGED FORMS:

SURVEY OF FOOD/PACKAGING COSTS -- COMPARISON OF BULK-SOLD AND PREPACKAGED COMMODITIES

IN A SURVEY OF LOCAL FOOD-PURCHASING OUTLETS IN THE WHITEAKER NEIGHBORHOOD, A NOT-SO-SURPRISING FACT WAS UNCOVERED: FOODS PURCHASED FROM BULK, UNPACKAGED, ARE CONSISTENTLY LESS EXPENSIVE THAN PREPACKAGED FOODS. SOMETIMES THE DIFFERENCE IS DRAMATIC, WITH BULK FOOD PRICES RANGING IN AVERAGES FROM $.02 TO $.90 PER POUND LESS EXPENSIVE.

THE SURVEY CONSISTED OF VISITS TO SIX LOCAL FOOD RETAIL OUTLETS, THREE SELLING FOOD FROM BULK, THREE SELLING PACKAGED FOODS. THE OUTLETS INCLUDED: (BULK) THE COMMUNITY STORE, GROWER'S MARKET, AND CONSUMER WAREHOUSE; (PACKAGED) SAFEWAY, ONE O'CLOCK MARKET, AND BILLY'S MARKET. ALL STORES WERE SURVEYED BY A RESEARCHER TAKING PRICES PER POUND FOR SELECTED FOOD ITEMS. RESULTS ARE LISTED IN TABLE A. PRICES FOR PACKAGED FOODS WERE TAKEN FOR THE SMALLEST AND LARGEST QUANTITY AVAILABLE COMMON TO AS MANY STORES AS POSSIBLE. BULK FOOD PRICES WERE TAKEN DOWN AS READ ON CONTAINERS.

PRICE AVERAGES ARE GIVEN FOR EACH COMMODITY (TABLE B) AS INDICATIONS OF DIFFERENCE, NOT AS ACTUAL PRICES. AN EXAMPLE OF DIFFERENCES INCLUDES A $.70/LB DIFFERENCE BETWEEN THE SMALLEST AND LARGEST AVAILABLE PACKAGES OF MEDIUM CHEDDAR CHEESE. AND A $.37/LB. DIFFERENCE BETWEEN THE LARGEST AND UNPACKAGED CHEDDAR. THE ACCUMULATIVE DIFFERENCE FROM SMALLEST TO UNPACKAGED: $1.07/LB.

SEVERAL FACTORS CONTRIBUTE TO THE ADDED EXPENSES OF BUYING PREPACKAGED FOOD. PACKAGES REQUIRE MATERIALS, LABOR, AND ENERGY TO MANUFACTURE AND TRANSPORT, ADDING TO THE SELLING COST. PACKAGES ARE FREQUENTLY USED AS FORMS OF ADVERTISING, MAKING THEM EVEN MORE COSTLY. SMALLER PACKAGES FOR FOOD ITEMS ALMOST ALWAYS MEAN A MORE CONCENTRATED USE OF MANUFACTURING MATERIALS, LABOR, ENERGY, AND MONEY PER POUND OF ACTUAL FOOD DELIVERED, MAKING THEM THE MOST EXPENSIVE.

BULK SELLING OF FOODS IS CONSISTENTLY LESS EXPENSIVE FOR SEVERAL REASONS. THE PACKAGING COST PER NET WEIGHT POUND OF FOOD IS MUCH LOWER; FEWER MATERIALS, LESS LABOR AND ENERGY, AND LESS PACKAGING REDUCE SHIPPING WEIGHT, AND STORES USE FEWER EMPLOYEE HOURS FOR STOCKING SHELVES AND PRICING. BUYING FOOD IN THIS MANNER ALMOST ALWAYS PERMITS THE CONSUMER TO PURCHASE EXACTLY THE AMOUNT DESIRED.

BULK FOODS PURCHASED THROUGH COOPERATIVES (OR FOOD BUYING CLUBS) OFTEN RESULT IN SAVINGS. MEMBERS CONTRIBUTE SMALL AMOUNTS OF TIME, LABOR, OR MONEY IN RETURN FOR BUYING THEIR FOOD AT COSTS BELOW THE AVERAGE RETAIL CHARGE.

WHEN PURCHASING FOOD ON A BUDGET, BULK FOODS PROVIDE AN EXCELLENT SOURCE OF SAVINGS. FOOD STORED IN PROPER CONTAINERS AFTER PURCHASE WILL ALSO STAND A BETTER CHANCE OF REMAINING FRESH AND EDIBLE THAN FOODS STORED IN CARDBOARD CARTONS OR OPEN SACKS.

BULK BOUGHT FOODS ARE STORED AT YOUR HOME IN THE SAME MANNER AS PACKAGED FOODS. YOUR EXTRA JARS OR PLASTIC CONTAINERS ARE IDEAL BECAUSE THEY DO NOT ALLOW INSECTS TO GET INTO THE FOODS. THESE CONTAINERS CAN ALSO BE CONTINUALLY REUSED. THE GLASS HOUSE AT 24TH AND HILYARD HAS A CONTINUAL STOCK OF RECYCLED BOTTLES AND JARS -- ALSO CHECK GARBAGIO'S AND BRING RECYCLING. WHEN KEPT IN CLEAR GLASS OR PLASTIC, FOODS SHOULD BE STORED IN A COOL CUPBOARD OUT OF THE LIGHT. AS WITH ALL FOODS, LIGHT AND HEAT HASTEN DETERIORATION AND CAUSE UNDUE LOSS OF SOME OF THE VITAMINS. THE REFRIGERATOR OR EVEN FREEZER, IS EXCELLENT FOR STORAGE OF FLOURS, NUTS, GRAINS AND NOODLES, ALTHOUGH THESE CAN BE KEPT VERY SATISFACTORILY ON A COOL DARK SHELF ALONG WITH THE DRIED FRUITS AND HERBS. REFRIGERATOR STORAGE IS PREFERABLE FOR THE OILS AND PEANUT BUTTER.

BULK ITEMS CAN BE BOUGHT AT RETAIL STORES LIKE THE COMMUNITY NATURAL FOODS AND GENERAL STORE (THE COMMUNITY STORE) AT 444 LINCOLN. THEY HAVE A GOOD SELECTION OF SEEDS, NUTS, GRAINS, FLOURS, DRIED FRUITS, OILS, SYRUPS AND OTHER SWEETENERS, PEANUT BUTTER, TOFU, NOODLES AND OTHER PASTAS. IN ADDITION, THEY ALSO CARRY OTHER ITEMS SUCH AS MILK, BREAD, JUICES AND PRODUCE. AN ADDED BENEFIT IS THAT ALL OF THE FOODS THEY SELL ARE CERTIFIABLY ORGANIC.

ANOTHER WHITEAKER MARKET SELLING BULK GRAINS, FLOURS, SEEDS, AND LEGUMES (SOME OF WHICH ARE ORGANIC) IS THE LANE COUNTY FEED AND SEED CO. AT 532 OLIVE. THEY HAVE A WIDE VARIETY OF ITEMS, AND MOST CAN BE BOUGHT BY THE POUND, OR IN 25, 50 OR 100-POUND LOTS. OTHER FOOD STORES IN EUGENE WHICH SELL FOOD IN BULK ARE THE WILLAMETTE PEOPLE'S FOOD COOP, THE KIVA, NEW FRONTIER, THE UNIVERSITY FOOD-OP, SUNDANCE, AND STARFLOWER.

FOOD BUYING CLUBS

ANOTHER WAY TO BUY FOODS IN BULK IS THROUGH A BUYING CLUB SUCH AS GROWER'S MARKET, A MEMBER COOPERATIVE. GROWER'S MARKET IS OPEN ONE DAY A WEEK, THURSDAY, AND DOES ALL OF ITS FOOD HANDLING ON THAT DAY. EACH HOUSEHOLD PURCHASING FOOD THROUGH THE MARKET HAS THE OBLIGATION TO WORK ONE HOUR FOR EACH TIME THEY ORDER. BECAUSE THEY ALL WORK AS VOLUNTEER MEMBERS OF THE COOPERATIVE, THEY ARE ABLE TO KEEP THE COST OF THE FOOD TO 15% OVER WHOLESALE COST. FOOD QUALITY IS HIGH -- WHENEVER POSSIBLE THEY BUY LOCALLY-GROWN OR UNSPRAYED PRODUCE. ORDERS ARE TURNED IN ON WEDNESDAY, THEN THE AMOUNT OF FOOD NEEDED TO FILL THE ORDERS IS DETERMINED. FOOD IS PURCHASED FROM LOCAL WHOLESALERS AND FARMERS, AND DELIVERED TO THE GROWER'S MARKET SITE ON THURSDAY MORNING. VOLUNTEER MEMBERS WEIGH-IN THE FOOD AND DETERMINE THE UNIT PRICE. THEN THE FOOD IS BOXED, BY ORDER, THURSDAY AFTERNOON, AND IS READY TO BE PICKED UP LATE THURSDAY AFTERNOON.

ORDER BLANKS CAN BE PICKED UP FROM THE GROWER'S MARKET OFFICE, 454 WILLAMETTE, OR CALL THEM FOR MORE INFORMATION, 687-1145.

STARFLOWER CO. 885 McKINLEY, A WAREHOUSE COOPERATIVE, IS AN EXCELLENT OUTLET FOR NEIGHBORHOOD GROUPS TO BUY IN BULK FORM. STARFLOWER IS A MAJOR SUPPLIER TO THE LOCAL FOOD COOPS AND NATURAL FOOD STORES IN EUGENE AS WELL AS OTHER RETAIL OUTLETS IN THE NORTHWEST. THEIR INVENTORY CONSISTS OF LEGUMES, DAIRY PRODUCTS, FLOURS, GRAINS, JUICES, OILS, PASTA, SWEETENERS, TOFU, BOTANICALS, NUTS AND SEEDS, PLUS VARIOUS OTHER FOOD ITEMS.

FOR SOME PEOPLE, IT WOULD BE EXTREMELY HELPFUL TO BE ABLE TO BUY AT WHOLESALE PRICES, AND STARFLOWER PROVIDES A MEANS TO DO THIS. HOUSEHOLDS OR BLOCK GROUPS FORM A BUYING CLUB AND TOGETHER PLACE ONE LARGE ORDER, WHEN NEEDED (USUALLY ABOUT ONCE A MONTH). TO BEGIN A BUYING CLUB, INTERESTED PEOPLE SHOULD MEET TO DECIDE THE GROUP'S COMMITMENT TO THIS TYPE OF BUYING, AND TO DETERMINE BUYING

BULK BUYING VS. PACKAGED FOODS

TABLE A
FOOD COMMODITIES, SOURCES, AND PRICES

BULK COMMODITY	BULK PRICE PER POUND			PACKAGED PRICE PER POUND SMALLEST (LARGEST)** QUANTITY AVAILABLE		
	1	2	3	4	5	6
UNBLEACHED WHITE FLOUR	-	.22	-	.29(.22)**	-	.34(.28)
WHOLE WHEAT FLOUR	-	.19	.26*	.34	-	-
LONG GRAIN WHITE RICE	-	-	-	.63(.44)	.54	.58
BROWN RICE	.59	.33	.44	.49	-	.72
PINTO BEANS	-	.50	.49*	.68(.56)	-	.69
RED BEANS	.29	-	.40*	.69(.62)	.77	-
GREEN SPLIT PEAS	.59	-	.45*	.39	-	-
LENTILS	.58	.68	.69*	.85	-	-
POPCORN	.35	-	.39*	1.23(.31)	1.27(.45)	1.29(.42)
BROWN SUGAR	-	-	-	.51(.52)	.55	.51
HONEY	-	-	.79	1.52(1.17)	1.58 (1.56)	1.86(1.34)
REGULAR ROLLED OATS	.35	.26	.31	.68(.53)	.93	.75
REGULAR GRANOLA	1.10	1.21	1.23	1.34(1.22)	1.25	1.33
CHEDDAR CHEESE (MED.)	-	1.95	1.98	3.00(2.10)	2.30(2.44)	3.33(2.44)
SWISS CHEESE	-	2.82	2.74	4.77(2.69)	2.95	3.80(2.81)

*ORGANIC
**DENOTES PRICE/LB IN LARGEST AVAILABLE QUANTITY

TABLE B
FOOD PRICE AVERAGES

COMMODITY	PRICE OF BULK AVERAGE PER POUND	COST OF SMALLEST (LARGEST) QUANTITY AVAILABLE PER POUND
UNBLEACHED WHITE FLOUR	.22	.32 (.25)
WHOLE WHEAT FLOUR	.22	.34
LONG GRAIN WHITE RICE	NA	.58 (.44)
BROWN RICE	.45	.60
PINTO BEANS	.50	.69 (.56)
RED BEANS	.35	.73 (.62)
GREEN SPLIT PEAS	.52	.39
LENTILS	.65	.85
POPCORN	.37	1.26 (.39)
BROWN SUGAR	NA	.52 (.51)
HONEY	.79	1.65(1.35)
REGULAR ROLLED OATS	.31	.79 (.53)
REGULAR GRANOLA	1.18	1.31(1.22)
CHEDDAR CHEESE (MED.)	1.97	3.04(2.33)
SWISS CHEESE	2.75	4.29(3.84)

BULK STORES
1. CONSUMER WAREHOUSE   225 RIVER ROAD   484-1144
2. GROWER'S MARKET   454 WILLAMETTE   687-1145
3. COMMUNITY STORE   444 LINCOLN   345-1856

PACKAGE STORES:
4. SAFEWAY   849 W. 6TH   345-0640
5. ONE O'CLOCK MARKET   689 W. 6TH   343-1723
6. BILLY'S MARKET   111 MONROE   686-1521

CAPACITY AS A GROUP. STARFLOWER OFFERS <u>WHOLESALE</u> PRICES TO SUCH BUYING CLUBS WHICH ARE ESTABLISHED WITH A NAME AND A CHECKING ACCOUNT IN THE NAME OF THE CLUB. THIS ALLOWS GROUPS TO BUY BULK STAPLES AT 15-20% LESS THAN MOST GROCERY STORE PRICES. THE MINIMUM ORDER FOR BUYING AT WHOLESALE PRICES IS $100, MAKING ONE OR TWO BLOCKS THE IDEAL SIZE FOR ORGANIZING INTO A BUYING CLUB. USING LOCAL BLOCKS ALSO FACILITATES EASY DISTRIBUTION OF THE FOOD TO PARTICIPATING HOUSEHOLDS.

A BUYING CLUB COULD BE ORGANIZED IN CONJUNCTION WITH THE BLOCK OR NEIGHBORHOOD FARMS TO PROVIDE MEMBERS WITH STAPLE FOOD ITEMS, COMPLEMENTING THE FRESH PRODUCE GROWN ON THE FARMS. MEMBERS OF THE EDIBLE CITY RESOURCE CENTER ARE INTERESTED IN PARTICIPATING AND HELPING TO FORM SUCH A GROUP, AND WILL CONTINUE DISCUSSION WITH NEIGHBORS PAST THE END OF THIS PLANNING STUDY.

HARVEST DATES FOR MOST EUGENE-AREA FOOD CROPS
(MAY VARY ACCORDING TO WEATHER)

	JAN.	FEB.	MAR.	APR.	MAY	JUNE	JULY	AUG.	SEPT.	OCT.	NOV.	DEC.
SPINACH												
ASPARAGUS												
ONION, GREEN												
CAULIFLOWER												
LETTUCE												
RHUBARB												
PEAS												
STRAWBERRIES												
ONIONS, DRY												
RASPBERRIES												
CHERRIES, SWEET												
ARTICHOKE, GLOBE												
MUSTARD GREENS												
BLUEBERRIES												
CHERRIES, SOUR												
PEACHES												
BEANS, GREEN												
BLACKBERRIES												
SQUASH, SUMMER												
POTATOES												
BROCCOLI												
CUCUMBERS												
CORN, SWEET												
PEPPERS												
SQUASH, WINTER												
PEARS												
APPLES												
CARROTS												
EGGPLANT												
TOMATOES												
LIMAS												
CABBAGE												
CHARD												
KALE												
ARTICHOKE, JERUSALEM												
GARLIC												
BEETS												
LEEKS												
GRAPES												
PRUNES												
FILBERTS												
WALNUTS												
TURNIPS												
CELERY												

USE AS MANY FOODS AS YOU CAN FIND THAT HAVE BEEN PRODUCED LOCALLY, THUS AVOIDING THE EARLY HARVESTING AND TREATMENTS INVOLVED IN SHIPPING.

<u>FARMERS MARKET</u>

REINSTITUTION OF THE LANE COUNTY FARMERS MARKET IN 1979 PROVIDES ONE OUTLET FOR PURCHASE OF LOCAL PRODUCE. MANY OF THE FARMERS PARTICIPATING OFFER UNSPRAYED OR UNTREATED PRODUCE (SUCH AS FUNKE,

FRYBERGER, STANLEY, AND AMACHER). OTHER ORGANIC FARMERS WILL BE ENCOURAGED TO PARTICIPATE IN FUTURE MARKETS. THE FARMERS MARKET WILL BE OPEN AGAIN THIS YEAR BEGINNING IN MAY AND WILL CONTINUE THROUGH LATE FALL (NOVEMBER). PRODUCE IS SOLD DIRECTLY FROM THE TRUCKS, EACH OF WHICH IS ALLOTED "STALL" SPACE. THE LOCATION OF THE MARKET IS THE ALLEY JUST WEST OF THE SATURDAY MARKET SITE, 7TH AND OAK, IN DOWNTOWN EUGENE.

<u>U-PICK</u>

SOME LOCAL FARMERS PARTICIPATE EACH YEAR IN U-PICK PROGRAMS, OFFERING THEIR FIELDS TO BE PICKED BY LOCAL RESIDENTS. THIS ELIMINATES THE HARVESTING AND TRANSPORTATION COSTS TO THE FARMER OF GETTING THE CROP TO THE MARKET. AT THE SAME TIME, IT ALLOWS THE CONSUMER TO BUY FIELD-FRESH PRODUCE AT LOWER-THAN-STORE PRICES, AS MUCH AS HE OR SHE CAN USE OR PRESERVE.

ORGANIZATION ON A BLOCK SCALE OF TRIPS TO U-PICK FIELDS, IN CONJUNCTION WITH GLEANING AND FOOD PRESERVATION (DISCUSSED BELOW) CAN MAKE BETTER USE OF THESE FOODS IN THE FIELDS. AS IT IS NOT ECONOMICAL FOR EACH INDIVIDUAL TO DRIVE ALONE TO FIELDS, SIX OR EIGHT HOUSEHOLDS CAN GO TOGETHER IN A VAN OR TRUCK. POOLING TRANSPORTATION, TOOLS (LADDERS, BUCKETS, ETC.) AND LABOR WILL FURTHER REDUCE COSTS.

THE FOLLOWING <u>FRESH PRODUCE BUYER'S GUIDE</u>, PUBLISHED BY THE <u>FRESH MARKET GROWERS OF LANE COUNTY</u>, SHOWS THE CROPS AVAILABLE, AS WELL AS WHEN AND WHERE THEY CAN BE PICKED. CHECK THE "MARKET BASKET" SECTION ALSO, IN THE REGISTER-GUARD CLASSIFIED ADS FOR FARMS THAT

DIRECTORY

	where to go	when	which crops		where to go	when	which crops
1	MEADOWS' VINEYARD (369-2333) 25700 Powerline Rd., 3 miles S of Halsey.	Daily. 10 A.M.-8 P.M. Orders accepted beginning May 1 for Sept.-Oct. harvest. Advance orders recommended.	European (wine) and American (juice) grapes.	27	BABBS (688-3274) 715 Hunsaker Lane, Santa Clara.	Mon.-Sat. 8 A.M.-6 P.M. Closed Sun. U-Pick.	Apples, peaches, sweet cherries.
2	B.J. BROWN (847-5545) 5015 Hulbert Lake Rd., 5 miles W of Junction City on Hwy. 99 W and ½ mile S on Hulbert Lake Rd.	Daily. 7 A.M.-7 P.M. Picked. U-Pick.	Peas, strawberries, sweet corn.	28	COBURG STRAWBERRY PATCH (344-3681) 91471 N Coburg Rd.	Daily. 7 A.M.-2:30 P.M. Picked. U-Picked.	Strawberries, raw honey.
3	BLISS U-PICK CHERRIES (998-2579) 2 miles N on Noraton Rd. from the Lancaster-Hwy 99 E intersection. The next driveway past Adams Lane.	Closed Fri. 4 P.M.-Sun. 8 A.M. Other days 7 A.M.-7 P.M.	Raspberries (limited), sour cherries, sweet cherries.	29	PERRY'S PRODUCE (342-8416) 3804 Coburg Rd., N Lane at intersection Coburg Rd. and Game Farm Rd.	Daily. Picked. U-Pick.	Beets, cabbage, peas, cukes, cauliflower, corn, onions, peppers, squash, tomatoes, parsnips, rutabaga.
4	BLISS BERRY FARM (998-6156) 96585 Adams Lane, Junction City.	8 A.M.-Dark. Closed Sat. U-Pick.	Blackberries (Marion), blackcaps, raspberries.	30	JOHNSON VEGETABLE & FRUIT STAND (343-9594) 89733 Armitage Rd., ½ mile before Armitage Park on Coburg Rd., turn right on N Game Farm Rd., follow signs.	Daily. May-Sept. 9 A.M.-6 P.M.; Sept.-Dec. 10 A.M.-5 P.M. Some orders accepted. Picked. U-Pick.	All fruits and vegetables.
5	MERLE HENTZE (998-6075) 30065 Hentze Lane, 2 miles E of Junction City.	7 A.M.-8 P.M. Picked on order. U-Pick.	Raspberries, strawberries.	31	COCKERLINE BERRY FARM (747-8822) 3651 Game Farm Rd., Springfield, turn left at E end of Beltline, last house on Game Farm Rd.	Daily. 8 A.M.-5 P.M. Picked. U-Pick.	Strawberries, corn, sugar peas, walnuts.
6	DETERING ORCHARDS (995-6341) 7 miles N Coburg on Harrisburg-Coburg Rd.	Daily. 8 A.M.-6 P.M. Aug.-Nov. Picked. U-Pick.	Apples, cider, honey, peaches, pears, plums, sour cherries, sweet cherries, peppers, corn, tomatoes, squash, etc.	32	TAYLORS', WILLIAM & VIRGINIA (746-0483) E end of Beltline at Game Farm Rd.	Daily during season. 8 A.M.-8 P.M. Advance orders filled. U-Pick.	Blackberries (Marion).
7	BEA'S BERRIES (998-2800) N of Applegate Market at Cheshire, next to Long Tom River.	Daily. 7 A.M.-7 P.M. Picked. U-Pick.	Blackberries (Marion) Raspberries.	33	PURCELL'S BLUEBERRIES (746-0040) 1537 Hayden Bridge Rd., Springfield.	Mon.-Sat. 8 A.M.-7 P.M. when berries available. Picked. U-Pick. Call first.	Blueberries.
8	WHITE OAKS ORCHARD (998-8299) 92842 River Rd. (mile post 4), Junction City.	Daily except Sunday. Picked.	Sour cherries, sweet cherries.	34	LITTLE PINE (726-8861) 3560 Hayden Bridge Rd., Springfield.	Daily. 8 A.M.-8 P.M. Picked. U-Pick. Call in advance. Ph. hours 6-8 a.m., 6-8 p.m.	Apples, blackberries, blueberries, grapes, pears, plums, quince, raspberry, sour & sweet cherries, walnuts.
9	ZUMWALT FARMS (998-2001) 7 miles N Santa Clara, right on Hayes Lane ½ mile left to Zumwalt Lane.	Weekdays 8 A.M.-6 P.M. Weekends 1 P.M.-6 P.M. Advance orders and U-Pick. Some organically grown.	Apples, blackberries, (Evergreen) pears, prunes, raspberries, most vegetables, honey, herbs.	35	LOG CABIN ORCHARD (747-4324) 3820 Hayden Bridge Rd., Springfield.	Daily. 8 A.M.-8 P.M. Picked. U-Pick.	Apples, peaches, pears, honey, walnuts, beans, cabbage, cauliflower, cukes, sugar pod peas, squash, corn, tomatoes.
10	HENDERSON FARMS (998-2257) 92451 River Rd. (Lone Pine Dr.)	Daily. 9 A.M.-Dusk Picked. U-Pick.	Apples, peaches, pears, sweet cherries, plums, beans, cukes, squash, corn, tomatoes, honey, strawberries, pumpkins.	36	SMITH'S BLUEBERRIES (747-7779) 37256 Camp Creek Rd., Springfield.	Daily during crop season. 9 A.M.-7 P.M. Picked. U-Pick.	Blueberries, filberts.
11	MERL WATTS BERRIES AND PRODUCE (998-8842) 92355 River Rd. (Maple Dr.), Junction City.	Daily except Sun. Stand-9 A.M.-dusk. U-Pick-hours vary with crop. Orders taken for Marion, Evergreen, cukes.	Blackberries (Marion & Evergreen), Raspberries, cukes, peppers, squash, tomatoes, corn, pumpkins.	37	GRIER FARMS (726-6131) 1342½ N 66th St., Springfield, first left immediately after second bridge on N 66th.	Tues.-Sat. 9 A.M.-6 P.M. Sun. 1-4 P.M. Call ahead advised. Special price break for advance orders. Picked. U-Pick.	Strawberries, beans, beets, broccoli, cukes, carrots, cauliflower, edible pod peas, corn, tomatoes, pumpkins, corn.
12	BUD ANDREWS (688-3455) 30636 Lone Pine Dr., 4½ miles N Santa Clara.	Please call. Picked. U-Pick.	Peaches, pears.	38	HALBERT'S BLUEBERRY FARM (896-3928) 44382 McKenzie Hwy., Leaburg, ¼ mile E of mile post No. 23.	Mon.-Sat. 10 A.M.-6 P.M Closed Sun. Picked. U-Pick.	Blueberries
13	AMACHER'S GEMS (688-8284) 91922 River Rd., Junction City.	Daily during crop season. 8 A.M.-8 P.M. or by order. Organically grown.	Apples, pears.	39	CUPP FARM (Winger) (747-9325) 3901 Harmon Lane, Springfield, E Main to 32nd, S to Jasper Rd., 39th off Jasper Rd.	Mon.-Sat. 9 A.M.-3 P.M. Other hours and Sun. by appointment. Picked. U-Pick.	Pole beans, corn, tomatoes, strawberries.
14	ALVADORE DRYER (689-4159) or 688-7577) Turn off Hwy. 99 left on Clear Lake Rd. (4½ miles), then right on Alvadore Rd. (1½ miles) to Alvadore.	Daily in season. 9 A.M.-6 P.M. Closed Sun. after Nov. 1. Picked. U-Pick.	Apples, peaches, plums, prunes (fresh & dried), sweet cherries, filberts, walnuts.	40	FLYNN'S ORCHARDS (746-2771) 34133 Seavey Loop Rd., ½ mile off Franklin Blvd.	Mon.-Sat. 8 A.M.-8 P.M. Closed Sun. U-Pick.	Apples, peaches, pears, sweet cherries, cabbage, cukes, peppers, corn, tomatoes, grapes.
15	L. L. OGLES ORCHARD (688-4106) 27210 Louden Lane, Alvadore area.	Daily during crop season. 8 A.M.-6 P.M. Picked.	Peaches, sweet cherries.	41	EVONUK FARMS (746-3982) 34377 Seavey Loop, 1 mile off Franklin Blvd.	Mon.-Sat. 8 A.M. until picked out. Closed Sun. Picked. U-Pick bring containers.	Strawberries
16	RATLIFF'S ORCHARD (689-1236) 27143 Orchard Rd., Alvadore area.	Daily. Check for sale hours and Picked or U-Pick.	Apples, peaches, plums, prunes, sour cherries, sweet cherries, strawberries, corn, tomatoes.	42	HORN OF PLENTY PRODUCE (746-2763) or (746-3982) 34596 Seavey Loop, 1½ miles off Franklin Blvd. Formerly Brougher's Produce.	Mon.-Sat. 8A.M.-8 P.M. Closed Sun. Picked. U-Pick. Orders for picked produce. Call before coming on other.	Apples, peaches, beans, beets, carrots, cukes, edible pod peas, corn, tomatoes, squash, etc.
17	ELLIOTT'S ROYAL ANN CHERRIES (688-1864) 89929 Shore Lane, S end of road. Fern Ridge area.	Daily. 8 A.M.-Dark. U-Pick.	Sweet cherries.	43	BUSS ORCHARD (747-5792) 33512 Cherry Hill Lane, Second driveway to left off Dillard Rd.	Daily. 8 A.M.-9 P.M. U-Pick.	Raspberry, sour and sweet cherries.
18	BUSH'S FERN VIEW FARMS (998-6805) or (935-7047) 90536 Territorial Rd.	Daily. 8 A.M.-Dark. Picked. U-Pick.	Peaches, tomatoes, filberts.	44	ROBBERSON ORCHARD (747-9747) 84907 Edenyale Rd., off Hwy. 58, Pleasant Hill.	Daily. 7 A.M.-7 P.M. Picked.	Apples, peaches, pears, sweet cherries, filberts.
19	ROYAL BLUEBERRY PATCH (689-1836) 28718 Royal Ave.	Weekdays 7 A.M.-7 P.M. Sat. & Sun. Afternoons. Picked. U-Pick.	Blueberries.	45	DECOU'S TINKER ROAD ORCHARD (747-1140) 36425 Tinker Rd., Pleasant Hill, ¼ mile off Hwy. 58 down Enterprise Rd.	Mon.-Sat. 10 A.M.-5 P.M. Closed Sun. Picked. U-Pick.	Apples, peaches, pears, prunes, sour and sweet cherries.
20	ALLEN'S (688-9518) 4600 River Rd., ¼ mile S Beacon Dr., left side of road.	Mon.-Sat. 7 A.M.-7 P.M. Closed Sun. U-Pick.	Blackberries (Marion).	46	SWANS (746-1018) 84020 Brown Rd., between Dexter and Pleasant Hill at end of Brown Rd.	Mon.-Sat. Closed Sun. Call before coming. U-Pick.	Blackberries (Marion), strawberries, corn, tomatoes.
21	TOM'S ORCHARD (688-8976) 909 Beacon Dr. E, 2½ miles N Santa Clara.	Daily. 8 A.M.-Dusk. Picked. Advance orders only on pitted pie cherries.	Apples, apple cider, peaches, pears, sour cherries, sweet cherries.	47	FALL CREEK FARM & NURSERY (937-2973) 39318 Jasper-Lowell Rd., mid-way between Fall Creek and Unity.	Daily except Wed. Picked. U-Pick.	Blueberries
22	RIVERBROOK FARMS (688-0586) 1225 E Beacon Dr., Eugene.	Daily. 9 A.M.-9 P.M. U-Pick.	Sweet cherries.	48	BROCK'S RASPBERRIES (895-2325) 92 N 4th, Creswell, white house, two storied.	Daily. 8 A.M.-6 P.M. Picked (on order). U-Pick.	Raspberries
23	CORWIN FARMS PRODUCE, 4190 River Rd.	Daily. 8 A.M.-Dark. Picked. U-Pick.	Apples, peaches, strawberries, beans, peas, broccoli, cabbage, corn, cauliflower, onions, potatoes, tomatoes.	49	HANSEN'S BERRIES & PRODUCE (895-4857) (343-4355, or 895-4287) ¼ mile S on Sears Rd. from Cloverdale Rd., Creswell.	Daily. 7 A.M.-6 P.M. Picked. U-Pick.	Beans, cukes, peas, pumpkins, corn, tomatoes, blueberries, strawberries.
24	SPRING CREEK BLUEBERRY FARM (688-2962) 101 Spring Creek Dr., 1½ miles N of Santa Clara.	Mon.-Sat. 7 A.M.-6 P.M. Picked. U-Pick.	Blueberries, sour cherries.	50	GLENN WICKS (895-2788) 82682 N River Dr., Creswell, E on Cloverdale Rd., over I-5 at Creswell exit, to Y, S on River Dr.	Daylight during season. U-Pick. Picked on advance order.	Sweet cherries
25	THOMPSON'S FRUIT & PRODUCE (688-0725) 3910 River Rd., 1 mile N Santa Clara.	Mon.-Sat. 8 A.M.-7 P.M. Sun. P.M. Picked. U-Pick.	Apples, peaches, pears, prunes, raspberries, sour cherries, sweet cherries, most vegetables, filberts.	51	SUN & MOON HERB (942-5192) 30973 Kenady Lane, Cottage Grove.	Call prior to visiting. All herbs grown organically including 200 var. live plants.	Herbs
26	FORTNER'S FRUIT AND CHRISTMAS TREE FARM (688-0866) 777 Irvington Dr.	Wed.-Sat. 8 A.M.-6 P.M. Closed Sun., Mon., Tues. Picked. Adv. orders for Aug. Gravensteins and winter apples.	Apples				

MAY NOT BE LISTED IN THE PRODUCE GUIDE. PICKING FROM FIELDS THAT
HAVE NOT BEEN SPRAYED, SUCH AS ZUMWALTS, AMACHERS, RIVERBROOK, THE

ROYAL BLUEBERRY PATCH, AND OTHERS, WILL ELIMINATE POTENTIALLY
HARMFUL CHEMICALS FROM YOUR FOOD WHILE SUPPORTING LOCAL ECOLOGI-
CALLY SOUND AGRICULTURAL PRACTICES. LET FARMERS KNOW THAT YOU
PREFER PICKING FROM ORGANIC FIELDS.

GLEANING

GLEANING MEANS GATHERING AND SHARING CROPS THAT REMAIN AFTER HAR-
VEST. ALL FIELDS, WHEN HARVESTED, LEAVE SOME OF THE CROP BEHIND
-- TOMATOES THAT WEREN'T QUITE RIPE ENOUGH, CORN THAT HAD NOT
FULLY DEVELOPED, LEFT IN THE FIELD BY THE HAND PICKERS; OR BEANS,
CARROTS OR CHERRIES OVERLOOKED BY MECHANICAL HARVESTING MACHINES.

ANY FIELD CAN BE GLEANED IF THE FARMER IS WILLING TO HAVE PEOPLE
COME ONTO HIS FARM. MOST FARMERS ARE HAPPY TO SEE THE PRODUCE
USED, AND ARE GLAD TO HAVE GROUPS OF PEOPLE COME. OTHER FARMERS
BENEFIT BY THE FREE LABOR PROVIDED BY GLEANERS WHEN RAIN RUINS A
CROP FOR MARKET AND IT MUST BE PICKED IMMEDIATELY TO MAKE ROOM
FOR A NEW CROP. A SMALL FINANCIAL BENEFIT TO THE FARMERS IS THE
10%-OF-COST DEDUCTION THEY ARE ALLOWED ON THEIR OREGON STATE
INCOME TAX RETURN. IN THESE WAYS, GLEANING BENEFITS THE FARMERS,
THE GLEANERS, AND THOSE OTHERS THAT ALSO SHARE THE HARVEST.

GLEANING REQUIRES TRANSPORTATION TO THE FIELDS, PHYSICALLY-ABLE
PEOPLE TO PICK CROPS (FROM TREES TO UNDER THE GROUND) AND
COOPERATIVE FARMERS. THE FARMER (AS WELL AS THE GLEANER) APPRE-
CIATES A DEGREE OF ORGANIZATION, AND KNOWING THE GROUP WHO WILL
BE DOING THE GLEANING AND WHEN THEY WILL BE COMING. ORGANIZATION
SHOULD INCLUDE A TELEPHONE NUMBER WHERE A TELEPHONE COORDINATOR
CAN BE REACHED WHEN THE FARMER CALLS TO SAY HIS FIELD IS READY
FOR HARVESTING. OFTEN, THE BEST TIME TO GLEAN IS TO FOLLOW
DIRECTLY BEHIND THE HARVESTING MACHINE, SO BEING ABLE TO GET THE
GROUP MOBILIZED QUICKLY IS AN ADVANTAGE, AS THE FARMER DOES NOT
ALWAYS KNOW IN ADVANCE THE EXACT TIME OF THIS HARVEST. IF THE
FIELD DOES NOT NEED IMMEDIATE ACTION, THE GROUP CAN FOLLOW A
REGULAR TIME SCHEDULE FOR MOST GLEANING. IN GENERAL, 2-3 HOURS
PICKING TIME IN THE FIELD IS ENOUGH AND MOST PEOPLE ARE READY TO
LEAVE. IN OUR CLIMATE, MUCH OF THE GLEANING IS DONE IN HOT SUMMER
WEATHER AND IT IS HELPFUL TO BE ABLE TO GO OUT EARLY IN THE
MORNINGS AND COME BACK ABOUT NOON BEFORE THE SUN GETS TOO HOT.

THIS ALSO ALLOWS TIME IN THE AFTERNOON FOR DISTRIBUTION OF THE
FOOD GLEANED TO OTHER NEIGHBORHOOD RESIDENTS. FOOD SHOULD BE
DISTRIBUTED IMMEDIATELY, ENSURING FRESHNESS AND ELIMINATING THE
COST AND MAINTENANCE OF WAREHOUSE OR FREEZER SPACE.

WHEN THE FARMER CALLS, THE TELEPHONE COORDINATOR FINDS OUT WHAT
CROP IS AVAILABLE, AND DIRECTIONS TO THE FARM. HE/SHE ALSO
ARRANGES WITH THE FARMER THE TIME THEY WILL COME, WHERE TO PARK
AND HOW MANY PEOPLE TO BRING (DOES HE HAVE THREE APPLE TREES OR
FIFTEEN ACRES OF CARROTS). SOMETIMES IT CAN TAKE SEVERAL TRIPS
TO PICK ALL THE PRODUCE IN LARGE FIELDS. THEN THIS INFORMATION
IS PASSED ON IN A TELEPHONE CHAIN SYSTEM TO THE OTHER GLEANERS
UNTIL A LARGE ENOUGH GROUP HAS BEEN REACHED. GLEANERS USUALLY
MEET IN A CENTRAL LOCATION AND GO TO THE FIELD AS A GROUP. THIS
FOSTERS IDENTITY OF THE GROUP AND AT THE SAME TIME ELIMINATES
TRANSPORTATION COSTS FOR THE INDIVIDUAL AND PARKING PROBLEMS AT
THE FIELD.

EUGENE AREA GLEANING PROGRAMS

TWO GLEANING PROGRAMS HAVE BEEN ORGANIZED IN LANE COUNTY DURING THE
PAST TWO YEARS. EACH WAS ORGANIZED AROUND THE NEEDS OF LOW-INCOME
PEOPLE, SENIOR CITIZENS AND HANDICAPPED PEOPLE. THOSE WHO ARE
PHYSICALLY ABLE TO PICK GO TO THE FIELDS. THEY KEEP WHAT THEY
CAN USE OF THE AMOUNT PICKED AND DONATE THE REST TO PEOPLE OR
FAMILIES WHO ARE UNABLE TO PICK.

THE FIRST GLEANING PROGRAM, THROUGH THE LANE COUNTY COMMUNITY FOOD
BANK, HAS BROUGHT IN NEARLY 100,000 POUNDS OF FOOD DURING THE
SUMMERS OF 1978 AND 1979. MOST OF THIS FOOD CAME FROM ABOUT A
DOZEN FARMS, PLUS A NUMBER OF CITY RESIDENTS WITH HOME GARDENS
AND FRUIT OR NUT TREES. THIS PROGRAM HAS BEEN STAFFED BY A PAID
COORDINATOR AND CETA POSITIONS, PLUS MANY VOLUNTEER GLEANERS.
PLANS INCLUDE LESS EMPHASIS ON GLEANING IN 1980 AND AFTER, MAKING
IT MORE DESIRABLE AND NECESSARY FOR THE NEIGHBORHOOD TO ORGANIZE
AROUND THIS VAST FOOD RESOURCE.

THE SECOND GLEANING PROGRAM, SAVE OUR SENIORS, OPERATES OUT OF
SPRINGFIELD ON VOLUNTEER EFFORT ALONE--NO PAID POSITIONS--AND
HAS REDISTRIBUTED ALMOST 140,000 POUNDS OF GLEANED PRODUCE DURING
1979 ALONE. THIS ALSO COMES FROM A RELATIVELY SMALL NUMBER OF
FARMERS, ALONG WITH HOME GARDEN DONATIONS.

THESE AMOUNTS ARE TREMENDOUS AND HAVE PROVIDED HUNDREDS OF LOW-
INCOME AND SENIOR CITIZEN FAMILIES WITH FRESH PRODUCE. THIS
REPRESENTS ONLY A FRACTION OF FOOD STILL LEFT IN THE FIELDS.
ADDED TO THIS IS THE FOOD WITHIN THE NEIGHBORHOOD ITSELF THAT IS
NOT BEING USED, THAT IS LEFT ON THE BUSHES, TREES OR GROUND --
THE UNPICKED FRUIT AND NUT TREES, THE BLACKBERRY BUSHES ALONG
THE RIVERWAY, AND THE EXCESS IN PEOPLE'S GARDENS.

URBAN GLEANING POTENTIAL

TO SHOW RESIDENTS ATTENDING AN APPROPRIATE TECHNOLOGY CONFERENCE
IN EUGENE IN 1979 HOW THEY COULD TAKE ADVANTAGE OF LOW-COST
SOURCES OF FOOD SUCH AS URBAN GLEANING, A MEMBER OF THE EDIBLE
CITY RESOURCE STAFF CONDUCTED A GLEANING PROJECT OF FOOD, MOSTLY
WITHIN THE WHITEAKER AREA. AS A RESULT, AT THE CONFERENCE, FOUR
WHEELBARROWS AND FOUR LARGE BASKETS OF FOOD, TOTALING 451.5
POUNDS, WERE WHEELED IN. THE RETAIL VALUE OF THIS "FREE" FOOD
WAS OVER $200. THE FOOD WAS GATHERED IN 11 HOURS DURING ONE
WEEK IN SEPTEMBER. THE MAIN GOAL WAS TO SEE HOW MUCH FOOD ONE
PERSON COULD FIND IN ONE WEEK'S TIME. THE OBJECTIVES WERE AS
FOLLOWS:

1. TO SPEND TWO HOURS DAILY, MAXIMUM, OF LEISURE TIME.

URBAN GLEANING

1 PERSON HARVEST TIME--2 HRS/DAY
FOR 1 WEEK IN LATE SEPTEMBER

FOOD	TIME	POUNDS	WITHIN WHITEAKER	OUTSIDE WHITEAKER	RESIDENT LAND	PUBLIC LAND	ABANDONED	APPROXIMATE COST
APPLES	5 HR	133-1/2	X		3		X	65.17
GRAPES	20 MIN	15		X	1			10.50
QUINCE	45 MIN	50	X			X		
WALNUTS	2 HR	14	X			X		19.46
FILBERTS	1-1/2 HR	9	X	X	1	X	X	12.51
CARROTS	10 MIN	3-1/2	X			X	X	1.20
SQUASH	10 MIN	73	X			X	X	50.37
TOMATOES	1-1/2 HR	61	X	X		X	X	35.79
CORN	10 MIN	18	X	X		X	X	7.20
CUCUMBERS	10 MIN	7	X	X		X	X	1.05
PUMPKIN	10 MIN	46	X			X	X	3.68
PEPPERS	10 MIN	6	X	X		X		4.14
DRY BEANS	10 MIN	1	X			X	X	
SUNFLOWERS	5 MIN	10	X			X	X	
CANTALOUPE	2 MIN	4-1/2	X			X	X	
TOTALS	11 HRS	451-1/2	(13)	(6)	5	(13)	(3)	$200.00+

2. TO USE FOOT OR BICYCLE TRANSPORTATION (NO AUTOMOBILE).

3. TO FIND OUT WHAT FOOD WAS AVAILABLE.

4. WHERE TO FIND IT.

5. HOW TO GET IT.

6. WHAT TO DO WITH IT.

WHAT FOODS ARE AVAILABLE: THE FOLLOWING CHART SHOWS THE FOOD GLEANED FROM WITHIN THE CITY IN LATE SEPTEMBER OF 1979. PEARS, BEANS, AND OTHER COMMON SUMMER GARDEN VEGETABLES ARE FOODS WHICH ARE HARVESTABLE BUT LESS ABUNDANT AT THIS TIME. IT IS IMPORTANT TO NOTE THAT MOST OF THESE FOODS WOULD NOT BE SOLD IN MOST SUPERMARKETS, BECAUSE OF THEIR IMPERFECTIONS, BLEMISHES AND IN-CONSISTENCIES OF SIZE, SHAPE, COLOR, ETC. HOWEVER, BECAUSE OF FRESHNESS ALONE, THIS FOOD CAN BE CONSIDERED NUTRITIONALLY SUPERIOR TO THE SAME ITEMS FOUND IN GROCERY STORES. THIS FOOD TASTES GOOD AND HAS NOT BEEN TREATED WITH "SHELF-LIFE PRESERVATIVES" OR HARMFUL CHEMICALS.

WHERE TO FIND IT AND HOW TO GET IT:

1. NEIGHBORS CAN BE CALLED OR VISITED TO SEE IF YOU CAN HELP THEM HARVEST IN EXCHANGE FOR A SHARE OF THE PRODUCE. THIS WORKS PARTICULARLY WELL WITH ELDERLY PEOPLE WHO MAY WELCOME A CHANCE TO GET THEIR FRUIT PICKED.

2. ENCOURAGE RESIDENTS TO LIST AVAILABLE EXCESS FOOD IN THEIR NEIGHBORHOOD NEWSLETTER (OR OTHER SYSTEM SUCH AS MENTIONED BELOW IN NO. 5), ALONG WITH RESIDENT'S CONDITIONS FOR THE FOOD BEING PICKED (I.E., TIME, QUANTITY, ETC.). THE GARDENERS IN THE NORTH POLK COMMUNITY GARDEN PUT EXTRA FOOD ON A TABLE OUTSIDE THE GARDEN. PEOPLE ON THE BIKE PATH COULD STOP AND TAKE WHAT THEY NEEDED.

3. FIND OUT WHO OWNS "ABANDONED" ORCHARDS WHERE IT IS OBVIOUS THAT THE TREES ARE NOT BEING CARED FOR, AND ASK IF YOU CAN HAVE THE FRUIT OR NUTS.

4. GLEAN FROM STREET TREES -- WHEN EXCESS IS OVER GROUND (IF ON PUBLIC LAND) WHERE TREE MAINTENANCE PROGRAMS ARE NOT ENFORCED.

5. ORGANIZE A NEIGHBORHOOD-WIDE OR SECTOR-WIDE GLEANING GROUP TO MONITOR THE MOST LIKELY GLEANING SPOTS IN THE IMMEDIATE AREA. THIS WILL HELP ELIMINATE RUSHES OF PEOPLE COMING TO ONE'S DOOR FOR FOOD, OR MANY PHONE CALLS. AN "ADOPT-A-TREE" PROGRAM HAS BEEN PROPOSED WITHIN THE NEIGHBORHOOD, TO MATCH PEOPLE WHO HAVE EXTRA FOOD WITH THOSE WHO WANT IT.

6. USEFUL EQUIPMENT: BIKE CART; BASKETS, BOXES, PLASTIC BAGS, OR BELT BUCKET (2-QT. OLIVE OIL CAN WITH HOOK TO ATTACH TO YOUR BELT OR PANTS); LADDER IF NEEDED; POLE WITH HOOK FOR REMOVING HARD-TO-GET FOOD.

   THE WHITEAKER PROJECT SELF-RELIANCE TOOL LIBRARY, 315 MADISON, HAS BIKE CART, LADDER, AND FOOD PRESERVATION EQUIPMENT FOR LOAN AT NO COST.

WHAT TO DO WITH IT:

1. IMMEDIATE TABLE USE.

2. STOCK UP BY PRESERVING FOODS (I.E., CANNING, DRYING, PRESERV-ING, FREEZING). NEIGHBORS ARE ARRANGING AT THIS TIME FOR DRYING AND CANNING EQUIPMENT AND USE OF FACILITIES AT VARIOUS NEIGHBORHOOD LOCATIONS.

3. LANE COUNTY FOOD BANK, 135 E. 6TH STREET, EUGENE, WILL DIS-TRIBUTE EXCESS FOOD TO LOCAL SENIORS, LOW-INCOME FAMILIES, AND HANDICAPPED PEOPLE. SHARE WITH OTHERS IN YOUR AREA.

4. BARTER OR TRADE WITH FRIENDS AND NEIGHBORS.

NEIGHBORHOOD ORGANIZATION AROUND THESE SOURCES OF UNUSED FOOD WOULD PROVIDE MUCH FOOD TO NEIGHBORHOOD RESIDENTS, AND TO THOSE WHO WOULD MOST BENEFIT FROM ITS USE.

ORGANIZATION FOR GLEANING

AN EXCELLENT WAY HAS PRESENTED ITSELF TO WHITEAKER NEIGHBORS IN-TERESTED IN GLEANING. SAVE OUR SENIORS HAS SUGGESTED WE USE THEIR ALREADY-DEVELOPED ORGANIZATION AND JOIN WITH THEIR PROGRAM. TO BEGIN WITH, NEIGHBORS CAN VOLUNTEER TO GO WITH OTHER GLEANERS TO THE FIELDS, BRING HOME WHAT THEY CAN USE AND DONATE THE REST TO OTHER NEIGHBORHOOD PEOPLE WHO ARE UNABLE TO GO TO THE FIELDS TO PICK. THIS IS A YEAR-ROUND, ON-GOING ACTIVITY, AND, AS SHOWN BY THE PRECEDING FIGURES, CAN RESULT IN VAST QUANTITIES OF FRESH PRODUCE BEING BROUGHT INTO THE NEIGHBORHOOD. VERY OFTEN THE FOOD IS FROM AN ENTIRE UNSPRAYED ORCHARD OR FIELD DONATED WHOLE AS BEING "UNMARKETABLE" BECAUSE OF IMPERFECTIONS IN SIZE OR BEAUTY, OR PERHAPS THE CROP WAS NOT PICKED AT ITS OPTIMUM TIME AND IS AGAIN CONSIDERED "UNMARKETABLE" AS BEING PAST ITS PRIME.

STAFF FROM THE EDIBLE CITY RESOURCE CENTER IS INTERESTED IN FOLLOW-ING THROUGH ON A GLEANING PROGRAM WITH WHITEAKER RESIDENTS AND WILL DONATE TIME TO ITS DEVELOPMENT. THIS PHASE IS BEING FOLLOWED UP IN CONJUNCTION WITH PROPOSED FOOD PRESERVATION FACILITIES FOR MAXIMUM USE OF FOODS BOTH GLEANED FROM FARMERS AND HOME-PRODUCED WITHIN THE NEIGHBORHOOD.

FOOD PRESERVATION AND EDUCATION CENTERS

NEIGHBORHOOD FOOD PRESERVATION CENTERS AT VARIOUS LOCATIONS THROUGH-OUT THE NEIGHBORHOOD CAN REVOLVE AROUND SEVERAL FUNCTIONS. WE SUGGEST THEY BE ORGANIZED IN CONJUNCTION WITH BLOCK FARM DEVELOP-MENT AND ATTENTION PAID TO SPECIFIC USE NEEDS OF EACH AREA.

1. EXISTING COMMUNITY FACILITIES WITH A STOVE, SINK AND WORK SPACE CAN BE USED FOR ACTUAL PRESERVATION OF FOODS, WHETHER BY CANNING OR DRYING, OR BY PROVIDING ROOT CELLAR SPACE.

2. THIS SPACE CAN ALSO BE USED TO PROVIDE EDUCATION ON DIFFERENT METHODS OF PUTTING FOODS BY -- TECHNICAL INFORMATION AND SUPERVISION ON USING A PRESSURE COOKER, WORKSHOPS ON HOW TO BUILD AND USE A SOLAR DRYER, HOW TO MAKE A ROOT CELLAR OR PIT FOR OVERWINTERING ROOT CROPS.

3. IT CAN BE A CENTRAL LOCATION FOR A "BULLETIN-BOARD" EXCHANGE, FOR POSTING INFORMATION ON WHERE EXCESS FOOD IS IN THE NEIGH-BORHOOD AT ANY GIVEN TIME, THUS MATCHING THE "HAVES" WITH THE "HAVE NOTS".

4. IT CAN BE A DROP-OFF POINT FOR THE USE AND REDISTRIBUTION OF GLEANED FOODS. AND,

5. IT CAN BE A PLACE TO MEET, TO EXCHANGE RECIPES, FOODS, LABOR AND IDEAS.

MIDWAY IN THIS STUDY A ROUGH PROPOSAL WAS DRAFTED BY EDIBLE CITY RESOURCE CENTER STAFF, BASED ON A MODEL DEVELOPED BY THE LINN-BENTON COMMUNITY FOOD AND NUTRITION PROGRAM IN CORVALLIS, OREGON. THIS PROPOSAL, ATTACHED, WAS SUBMITTED TO THE NATIONAL CENTER FOR APPROPRIATE TECHNOLOGY (NCAT), VIA THE WHITEAKER COMMUNITY COUNCIL, AND HAS BEEN TENTATIVELY ADOPTED FOR IMPLEMENTATION FOLLOWING THIS CURRENT GRANT PERIOD. THE PLANS INCLUDE A HALF-TIME FOOD COORDINATOR FOR THE NEIGHBORHOOD. THIS PERSON WOULD PROVIDE COORDINATION BETWEEN WHITEAKER'S BLOCK AND NEIGHBORHOOD FARM ACTIVITIES, INTEGRATE GEOGRAPHIC AREAS AND DIVERSE AGE AND ECONOMIC LEVELS THROUGHOUT THE NEIGHBORHOOD, AND PROVIDE COORDIN-ATION AMONG OTHER APPROPRIATE TECHNOLOGY GROUPS IN AND AROUND WHITEAKER FOR WORKSHOPS AND OTHER EDUCATIONAL ACTIVITIES RE-GARDING FOOD. IN ADDITION, SOME FUNDS HAVE BEEN ALLOTTED FOR FOOD PRESERVATION EQUIPMENT AND MATERIALS. THE REST OF THE SUPPORT FOR IMPLEMENTING THE CENTERS IS EXPECTED TO COME FROM NEIGHBORHOOD MOBILIZATION ON A VOLUNTEER BASIS, AS HAS ALREADY BEGUN.

THE LANE COUNTY AGRICULTURAL EXTENSION SERVICE, HOMEMAKER DIVISION, WAS CONTACTED FOR EDUCATIONAL SUPPORT, AND THEY HAVE AGREED TO HELP NEIGHBORS DEVELOP AN INFORMATION NOTEBOOK ON FOOD PRESERVA-TION TECHNIQUES, AND TO PROVIDE SOME ON-SITE SUPERVISION ON A WEEKLY BASIS. THE OREGON STATE EXTENSION SERVICE MAY REPEAT A TRAINING PROGRAM FOR FOOD PRESERVATION SUPERVISORS, AND IF SO, WE PROPOSE THAT SOME OF OUR POTENTIAL STAFF PARTICIPATE.

POTENTIAL COMMUNITY FACILITIES AND NEIGHBORHOOD ORGANIZATION BY SECTOR

WITHIN WHITEAKER'S GEOGRAPHIC SECTORS, SEVERAL EXISTING FACILITIES HAVE BEEN IDENTIFIED AS POTENTIAL FOOD PRESERVATION AND EDUCATION CENTERS, AND ARE BEING PURSUED IN VARIOUS DEGREES.

IN THE EAST BUTTE AREA, THE CELESTE CAMPBELL SENIOR CENTER, 3RD AND HIGH STREETS, HAS INDICATED ITS WILLINGNESS TO COOPERATE WITH THE NEIGHBORS IN THIS PROGRAM. INITIAL CONTACT HAS BEEN MADE WITH THE DIRECTOR OF THE CENTER BY STAFF OF THE EDIBLE CITY RESOURCE CENTER. THE CAMPBELL CENTER IS A LARGE ONE-STORY MEETING AND ACTIVITY BUILDING SPONSORED BY THE CITY'S PARKS AND RECREATION DEPARTMENT. THE BUILDING IS IN FAIRLY CONSTANT USE DURING THE WEEK WITH REGU-LAR SENIOR ACTIVITIES PLUS OTHER NEIGHBORHOOD ACTIVITIES SCHEDULED ON A REQUEST BASIS. DURING THE WEEKENDS, HOWEVER, THE BUILDING IS NOT USED, AND IT IS DURING THIS TIME WE PROPOSE TO USE THE KITCHEN AND WORK SPACE FACILITIES. THE KITCHEN IS LARGE ENOUGH TO ACCOM-MODATE APPROXIMATELY 4-6 PEOPLE, AND CONTAINS A 4-BURNER STOVE, A STERILIZER FOR JARS, SEVERAL LARGE KETTLES PLUS OTHER MISCEL-LANEOUS KITCHEN UTENSILS, AND A SINK. A LARGE MEETING ROOM ADJOINS THE KITCHEN WHICH CAN SERVE AS ADDITIONAL WORK SPACE AS WELL AS FOR EDUCATIONAL PURPOSES.

EDIBLE CITY RESOURCE CENTER STAFF HAVE USED THIS SPACE SEVERAL TIMES DURING THE PAST YEAR FOR WORKSHOPS ON WATERBATH CANNING, USING A PRESSURE COOKER, AND FREEZING AND DRYING. THE SPACE WOULD BE AVAILABLE ONLY ON WEEKENDS AT THIS TIME, BUT THIS LIMITATION IS MORE THAN OFFSET BY THE POTENTIAL OF SENIOR INVOLVEMENT THROUGH THEIR INTEREST AND KNOWLEDGE OF PRESERVING FOOD. FOOD PRESERVA-TION (ESPECIALLY AT A SENIOR ACTIVITY CENTER) IS ONE OF THE QUICKEST POTENTIAL WAYS TO INTEGRATE DIFFERENT AGE LEVELS WITHIN THE NEIGHBORHOOD AND TO BE ABLE TO DRAW ON THE VAST EXPERIENCE OF OUR OLDER CITIZENS.

BETWEEN THE EAST BUTTE AND WEST BUTTE AREAS LIES LAMB COTTAGE IN SKINNER'S BUTTE PARK. THIS IS A CITY FACILITY, UNDER THE CONTROL OF THE PARKS AND RECREATION DEPARTMENT. THE AREA AND THE BUILDING SPACE (A ONE-STORY WORK OR MEETING ROOM WITH AN ADJOINING KITCHEN) WOULD MAKE AN IDEAL LOCATION FOR WORKSHOPS, EDUCATIONAL MEETINGS, CANNING AND DRYING PRODUCE, WITH THE SURROUNDING PARK FOR CHILDREN

TO PLAY IN. BEING LOCATED ON THE BIKE PATH RUNNING ALONG THE RIVER ALSO MAKES IT ADVANTAGEOUS AS AN EXCHANGE LOCATION FOR EXCESS NEIGHBORHOOD FARM AND GARDEN PRODUCE, AND GLEANED FOOD.

USE OF THIS SITE, HOWEVER, HAS SOME PROBLEMS AT THIS POINT. THE COTTAGE IS USED BY THE PARKS DEPARTMENT FOR WORKSHOPS, MEETINGS AND OTHER GATHERINGS, WITH THE UNSCHEDULED TIME OPEN FOR USE BY OTHER GROUPS UPON APPROVAL BY THE PARKS DEPARTMENT. USUALLY THERE IS AN HOURLY FEE ATTACHED TO ITS USE. WE ENCOURAGE FURTHER NEGOTIATIONS BETWEEN THE NEIGHBORHOOD AND THE PARKS DEPARTMENT REGARDING THE POTENTIAL NEIGHBORHOOD USE OF LAMB COTTAGE.

IN THE WEST BUTTE AREA, ONE POTENTIAL SITE IS THE ACTIVITY CENTER OF THE SEVENTH DAY ADVENTIST CHURCH AT 90 N. LAWRENCE. THIS SITE IS JUST TWO STREETS FROM A DEVELOPING BLOCK FARM (AT CHESHIRE AND LAWRENCE), AND DAY CARE, INC. A CHILD-CARE PROGRAM WITH A GARDEN-ING PROGRAM (SEE SCHOOL FARMS SECTION, AND BLOCK FARMS SECTION.) THE ACTIVITY CENTER IS CLOSED DURING THE SUMMER AND ITS REGULAR ACTIVITIES ARE SUSPENDED, THUS MAKING IT AVAILABLE FOR FOOD USE DURING JUNE, JULY AND AUGUST ON A REGULAR BASIS, AND NEGOTIABLE AFTER THAT TIME. BESIDES HANDLING THE FOOD FROM THE BLOCK FARM (AND U-PICKS AND GLEANING), THIS SITE CAN ALSO ENCOURAGE INTER-ACTION BETWEEN RESIDENTS AND THE VERY YOUNG CHILDREN (3-5 YEARS) AT THE DAY CARE SCHOOL.

THE ADJOINING BLOCK, AT THE WEST EDGE OF SKINNER'S BUTTE PARK, WAS RECENTLY PURCHASED BY A NEIGHBORHOOD HOUSING COOPERATIVE FOR DEVEL-OPMENT AS A PROTOTYPE HOUSING/AGRICULTURAL COOPERATIVE, WITH POTENTIAL INTENSIVE FOOD PRODUCTION, WHICH CAN ALSO BENEFIT FROM THE CLOSE LOCATION OF A FOOD PRESERVATION CENTER. (SEE SECTION ON PROTOTYPE HOUSING-FOOD CLUSTERS).

A PROPOSAL IS BEING WRITTEN BY EDIBLE CITY RESOURCE CENTER STAFF AND NEIGHBORHOOD RESIDENTS TO THE SEVENTH DAY ADVENTIST BOARD OF DIRECTORS REGARDING POTENTIAL USE OF THIS FACILITY. INITIAL CON-TACTS HAVE BEEN MADE BY STAFF WITH A BOARD MEMBER. THE CHURCH'S FUTURE PLANS MAY INCLUDE CLOSING DOWN THE CHURCH'S USE OF THIS FACILITY IN FAVOR OF OPERATING OUT OF THEIR DOWNTOWN LOCATION. NEIGHBORHOOD RESIDENTS ARE ENCOURAGED TO ACTIVELY SOLICIT USE OF THIS BUILDING, WITH THE POSSIBILITY IN MIND OF SECURING THIS FACILITY FOR PERMANENT NEIGHBORHOOD USE SHOULD THE CHURCH INDEED RELOCATE THE ACTIVITIES OF ITS PRESENT CENTER.

IN THE SLADDEN AREA, TWO POTENTIAL SITES HAVE BEEN LOCATED: THE WHITEAKER COMMUNITY SCHOOL AND THE SCHOOL DISTRICT 4-J ADMINIS-TRATION BUILDING. RESIDENTS FROM THE NEIGHBORHOOD AS A WHOLE HAVE BEGUN MEETING AND ARE PURSUING THE USE OF THE WHITEAKER COMMUNITY SCHOOL CENTER FOR SUMMERTIME USE. THE IMMEDIATE GOALS OUTLINED FOR THIS FOOD PRESERVATION PROGRAM ARE: FOOD DRYING, SOLAR DEHYDRATORS, CANNING, EDUCATION AND SUPERVISION (STAFFING) OF THE CENTER. OTHER ACTIVITIES ARE PLANNED FOR LATER IN THE CENTER'S DEVELOPMENT -- ROOT STORAGE AND FREEZING, GLEANING, PRO-DUCE EXCHANGE BOARD, CHILD EDUCATION AND CHILD CARE.

A CONTRACT IS BEING FORMED BY THE NEIGHBORHOOD GROUP TO BE SUB-MITTED TO THE COMMUNITY SCHOOL FOR USE OF THE BUILDING. MEMBERS ARE LOOKING FOR EQUIPMENT, COMPILING EDUCATIONAL RESOURCES, DECIDING ON SUPERVISORY PERSONNEL, AND DEVELOPING PUBLICITY. THIS CENTER SHOULD BE READY TO OPEN IN TIME FOR THE FIRST MAJOR CROPS --STRAWBERRIES AND CHERRIES. THE BULK OF THESE FIRST CROPS WILL COME FROM U-PICK FIELDS AND WILL BE READY IN JUNE. (SEE U-PICK SECTION FOR CROP SCHEDULES AND NEIGHBORHOOD ORGANIZATION).

USING THE COMMUNITY SCHOOL CENTER IS IDEAL FOR GAINING INVOLVEMENT OF YOUNG CHILDREN OF ELEMENTARY SCHOOL AGE. THE SCHOOL'S SUMMER PROGRAM, WHICH INCLUDES GARDENING AT THE NEARBY CITY COMMUNITY GARDEN SITE, OFFERS AN OPPORTUNITY TO INTEGRATE YOUTH ACTIVITY WITH ADULT, PRESERVATION WITH PRODUCTION, AND PRACTICE WITH CUR-RICULUM. (SEE SCHOOL FARMS SECTION.)

PROPOSAL SUMMARY FORM

PROPOSED BY: ___EDIBLE CITY RESOURCE CENTER___

PROJECT TITLE: ___FOOD PRESERVATION AND GLEANING PROJECT FOR___
___WHITEAKER NEIGHBORHOOD___

TOTAL AMOUNT REQUIRED: _$13,875_

1. BRIEF DESCRIPTION OF THE PROJECT: TO PROVIDE THE INSTRUCTION, EQUIPMENT, AND FACILITIES FOR FAMILIES TO PRESERVE FOOD. TWO INITIAL SITES WOULD BE ESTABLISHED AS COMMUNITY BASES FOR SUCH COOPERATIVE EFFORTS AS CANNING, GLEANING, WORKSHOPS ON NUTRITION AND FOOD PRESERVATION.

1. RELATIONSHIP TO NEIGHBORHOOD GOALS: THIS PROJECT WILL ADDRESS THE NEED FOR LOW-COST FOOD AVAILABILITY FOR CITIZENS ON LOW FIXED INCOMES, SENIOR CITIZENS, HANDICAPPED OR HOME-BOUND PERSONS WITHIN THE NEIGHBORHOOD. THERE IS MUCH FOOD PRODUCED IN THE NEIGHBORHOOD AT PRESENT THAT IS BEING WASTED. THIS FOOD, ALONG WITH AN ANTICIPATED INCREASE IN HOME-PRODUCED FOOD, WILL BE BETTER UTILIZED, PROVIDING RESIDENTS WITH NUTRITIONALLY BETTER DIETS AT LOWER COSTS.

BUDGET INFORMATION

ITEM	AMOUNT
A. PERSONNEL (1 FTE, 10 MONTHS, MAY BE JOB-SHARED)	$ 10,500
B. FRINGE (15%)	1,575
C. CONSULTANTS	0
D. TRAVEL	50
E. TRAINING	0
F. SPACE COSTS	0
G. SUPPLIES	200
H. PRINTING	200
I. EQUIPMENT	1,350
J. OTHER DIRECT COSTS (EXPLAIN)	0
K. INDIRECT COSTS (EXPLAIN)	0
	$ 13,875

FOOD PRESERVATION AND GLEANING PROJECT FOR WHITEAKER NEIGHBORHOOD
APPLICATION FOR FUNDS

JOB DESCRIPTION

GENERAL DUTIES:

1. TO COORDINATE WITH NEIGHBORS TO DEVELOP A NEIGHBORHOOD-WIDE SYSTEM FOR THE USE AND PRESERVATION OF LOCALLY PRODUCED FOOD, WHETHER FROM URBAN OR RURAL SOURCES.

2. TO IMPLEMENT THE DEVELOPMENT OF TWO NEIGHBORHOOD FACILITIES TO SERVE AS CENTERS FOR FOOD PRESERVATION ACTIVITIES.

3. TO LOCATE AND PURCHASE NECESSARY EQUIPMENT FOR FOOD PRESERVATION, INCLUDING KITCHEN ITEMS AS WELL AS MATERIALS FOR CONSTRUCTION OF LOW-COST FOOD DEHYDRATORS.

4. TO DEVELOP A GLEANING PROGRAM TO ASSURE MAXIMUM USE OF EXCESS PRODUCE WITHIN THE NEIGHBORHOOD AND IMMEDIATE SURROUNDINGS.

5. TO DEVELOP AND OFFER PUBLIC EDUCATION WORKSHOPS ON SUCH TOPICS AS NUTRITIONAL VALUE OF LOCALLY-PRODUCED FOODS, SEASONAL CHARTS OF LOCALLY-ABUNDANT FOODS -- WHAT FOODS, WHERE THEY ARE, HOW TO GET AND HOW TO USE THEM.

6. TO DEVELOP A DISTRIBUTION SYSTEM TO ASSURE THAT UNUSED PRODUCE GOES TO PEOPLE ON LOW, FIXED INCOMES, SENIOR CITIZENS, HANDICAPPED OR HOME-BOUND PERSONS WITHIN THE NEIGHBORHOOD.

7. TO COORDINATE WITH OTHER COMMUNITY, CITY, AND COUNTY SOCIAL SERVICE PROGRAMS TO ASSURE COORDINATION OF EDUCATIONAL, SERVICE-ORIENTED, AND SELF-HELP EFFORTS.

NEIGHBORHOOD FOOD CENTER

Home Canning and Drying

Rural Truck Farms

DISTRIBUTION

to seniors & needy

solar dryers

NORTH

vegetable gardens on the contour

B.B. · R.B. · P.W.

THE SECOND SITE IDENTIFIED IN THIS SECTOR IS THE SCHOOL DISTRICT 4-J ADMINISTRATION BUILDING AT 201 N. MONROE. THIS IS ALSO THE SITE OF A POTENTIAL NEIGHBORHOOD FARM (SEE NEIGHBORHOOD FARMS SECTION.) NO NEGOTIATIONS HAVE BEEN INITIATED WITH THE SCHOOL DISTRICT FOR POTENTIAL NEIGHBORHOOD USE OF FACILITIES AT THIS TIME, BUT WE SUGGEST AN INITIAL CONTACT TO SURVEY THE FACILITIES AND TO SEE HOW THE ADMINISTRATION WOULD FEEL ABOUT THIS POSSIBILITY. AT THIS TIME, THE FUTURE PLANS OF THE ADMINISTRATION BUILDING ARE UNDETERMINED AS TO WHETHER THEY WILL EXPAND USE OF THE FACILITY BY THE SCHOOL DISTRICT OR WHETHER THEY WILL SEEK ANOTHER MORE SATISFACTORY LOCATION FOR THEIR OPERATIONS.

TYPES OF FOOD PRESERVATION

CANNING: THE SPRINGFIELD COMMUNITY CANNERY SHUT ITS DOORS IN 1978, LEAVING A VOID IN HOME CANNING BY EUGENE/SPRINGFIELD RESIDENTS. THE REMAINING OPTION WAS FOR EACH HOUSEHOLD TO BUY ITS OWN EQUIP-MENT AND PROCESS ITS FOOD WITHIN THE HOME KITCHEN.

SOMETIMES PEOPLE ARE LUCKY TO KNOW FRIENDS WITH EQUIPMENT TO LEND. IF YOU ARE NOT ONE OF THESE, DON'T GIVE UP HOPE. IN 1978 TWO NEW TOOL LIBRARIES OPENED, CONTAINING TOOLS SPECIFICALLY FOR THE PRO-DUCTION AND PRESERVATION OF FOOD. INCLUDED IN EACH LIBRARY ARE WATER BATH CANNERS, PRESSURE COOKERS, DEHYDRATORS AND MISCELLAN-EOUS UTENSILS. DURING THE 1978 AND 1979 HARVEST SEASONS ALL EQUIPMENT WAS IN CONSTANT USE, SHOWING HIGH INTEREST IN PRESERV-ING FOODS. MORE EQUIPMENT WILL BE AVAILABLE DURING THE 1980 SEASON AND AFTER. THE LIBRARY MOST AVAILABLE TO WHITEAKER RESI-DENTS IS AT PROJECT SELF-RELIANCE IN THE JEFFERSON ELEVATOR, 315 MADISON (343-2711). THE OTHER IS LOCATED AT THE URBAN FARM AT THE UNIVERSITY OF OREGON (ON THE MILLRACE BEHIND KELLEY'S RESTAURANT), 686-3647. THERE IS NO CHARGE FOR BORROWING THIS EQUIPMENT.

CANNING ALLOWS A WIDE VARIETY IN PRESERVED DISHES, AS MANY FOODS CAN BE COMBINED IN TASTY COMBINATIONS WHICH ARE BEAUTIFUL TO LOOK AT AND IMMEDIATELY AVAILABLE FOR FUTURE USE. INITIAL COSTS CAN BE HIGH FOR INDIVIDUALS FOR EQUIPMENT AND JARS, ETC., BUT BY SHARING RESOURCES THIS COST CAN BE REDUCED. ACCURATE INFORMATION IS ESSENTIAL WHEN CANNING, TO REDUCE SPOILAGE AND CONTAMINATION, ESPECIALLY WITH LOW-ACID FOODS. COMPARED TO OTHER FORMS OF PRESERVATION, CANNING CAUSES THE GREATEST NUTRIENT AND FLAVOR LOSS THROUGH THE NECESSARY HIGH-TEMPERATURES USED IN PROCESSING. FOR MORE INFORMATION ON RECIPES AND HOW TO PROCESS VARIOUS FOODS, WATCH FOR SUMMER WORKSHOP SCHEDULES BY VARIOUS LOCAL GROUPS AND AGENCIES.

DRYING FOODS IS A PRESERVATION METHOD WHICH HAS BEEN GAINING IN-CREASING POPULARITY IN THIS AREA, AND IS A PRACTICE THAT WE FEEL HAS MANY ADVANTAGES OVER THE CANNING PROCESS. MAINLY, IT'S DELICIOUS; IT'S EASY (TO PREPARE, TO STORE, TO USE); IT'S HEALTHY; IT'S INEXPENSIVE.

ALMOST ANY FRUIT, VEGETABLE OR HERB CAN BE DRIED. HERBS ARE OFTEN BEST DONE BY SIMPLY HANGING THEM IN A BUNCH UNTIL DRY, USING THEM AS NEEDED. FRUITS AND VEGETABLES DRY BETTER WITH A LITTLE MORE HEAT, SUCH AS DIRECTLY IN THE SUN OR IN AN ELECTRIC DRYER. THIS ALLOWS TEMPERATURES OF 140-150 DEGREES, WHICH IS IDEAL FOR DRYING BUT NOT HOT ENOUGH TO COOK OUT VITAMINS AND MINERALS. (COMPARE WITH TEMPERATURES OF 212+ DEGREES IN WATER BATH CANNING, AND 240 DEGREES WHEN USING A PRESSURE COOKER). WE DO NOT BLANCH FOODS BEFORE DRYING THEM, NOR DO WE USE SULPHER OR OTHER PRESERVATIVES, AS MANY PEOPLE DO. WE HAVE FOUND THE TASTE AND CONSISTENCY OF FOODS WE HAVE DRIED TO BE EXCELLENT, AND ARE NOT CONCERNED WITH POSSIBLE COLOR CHANGES IN SOME OF THE FRUIT.

SOLAR DRYERS WITH ELECTRIC BACKUP ARE BEING PLANNED AT THE PRESENT TIME FOR NEIGHBORHOOD USE, TO BE INSTALLED ON A FLAT GARAGE TOP IN THE BLAIR AREA. FRUITS AND VEGETABLES ARE BEAUTIFUL WHEN SLICED AND DRIED. THEY ARE DELICIOUS AS SNACKS JUST AS THEY ARE, OR CAN BE EASILY RECONSTITUTED IN LIQUID FOR USE IN SOUPS, SAUCES, ETC. CHECK THE BIBLIOGRAPHY AT THE END OF THIS SECTION FOR BOOKS ON DRYING VARIOUS TYPES OF FOOD, WITH RECIPES AND HELPFUL SUGGESTIONS.

FREEZING FOODS: MANY FRUITS AND VEGETABLES CAN ALSO BE FROZEN. THIS IS VERY GOOD FOR VITAMIN AND MINERAL RETENTION (ESPECIALLY WHEN THE FOOD IS VERY FRESH). SOMETIMES HOWEVER, THERE IS A TENDENCY TOWARD TEXTURE CHANGE WHICH SOME PEOPLE FIND UNPLEASANT. TRY FREEZING SOME OF YOUR FAVORITE FOODS TO SEE HOW THEY ARE AFFECTED. FREEZING RELIES ON EXPENSIVE EQUIPMENT, AND ON CON-STANT ENERGY USAGE, AND IS NOT THE MOST ECONOMICAL PRESERVATION METHOD. IF YOU DO NOT HAVE AN AVAILABLE FREEZER, CHECK WITH LOCAL GROCERY STORES AND ASK IF THEY HAVE FREEZER SPACE TO RENT OUT. THIS USUALLY COSTS ABOUT $20 A YEAR FOR ABOUT NINE CUBIC FEET OF SPACE.

ROOT STORAGE: ANOTHER EASILY-DONE METHOD OF PRESERVING ROOT CROPS IS TO STORE THEM IN A BOX, OR DIRECTLY IN THE GROUND. TURNIPS AND CARROTS AND OTHER ROOT CROPS CAN OVERWINTER VERY WELL IN THE GARDEN AND BE GREAT FOR WINTER AND EARLY SPRING TREATS. OR, THEY CAN BE HARVESTED AND STORED IN A SPECIALLY-DESIGNED PIT DUG INTO THE GROUND, OR BROUGHT INTO A COOL MOIST DARK ROOM AND STORED IN A BOX. PROJECT SELF-RELIANCE (343-2711), THE COMMUNITY GARDENS PROGRAM (687-5329), OR THE URBAN FARM (686-3647) HAVE MORE INFOR-MATION ON BUILDING PITS, OR REFER TO THE BIBLIOGRAPHY FOR READING SELECTIONS.

WORKSHOPS ON BUILDING SOLAR DRYERS, CANNING FOOD, FREEZING AND ROOT STORAGE (AND OTHER TOPICS) HAVE BEEN DEVELOPED BY VARIOUS APPROPRIATE TECHNOLOGY GROUPS AROUND EUGENE, AND WILL BE OFFERED EACH YEAR THROUGH THE COALITION FOR COMMUNITY SELF RELIANCE. MEMBER GROUPS OF THIS COALITION INCLUDE PROJECT SELF-RELIANCE, AMITY FOUNDATION, THE EDIBLE CITY RESOURCE CENTER, THE URBAN FARM, AND EUGENE COMMUNITY GARDENS. WATCH FOR FLYERS OR CALL ONE OF THE ABOVE GROUPS FOR SCHEDULE INFORMATION ON FUTURE WORK-SHOPS.

THE NATURE OF FOOD

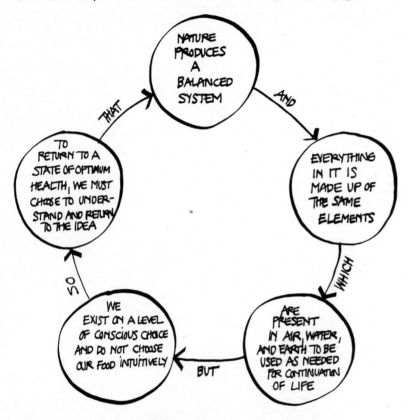

NATURE PRODUCES A BALANCED SYSTEM

AND

EVERYTHING IN IT IS MADE UP OF THE SAME ELEMENTS

WHICH

ARE PRESENT IN AIR, WATER, AND EARTH TO BE USED AS NEEDED FOR CONTINUATION OF LIFE

BUT

WE EXIST ON A LEVEL OF CONSCIOUS CHOICE AND DO NOT CHOOSE OUR FOOD INTUITIVELY

SO

TO RETURN TO A STATE OF OPTIMUM HEALTH, WE MUST CHOOSE TO UNDERSTAND AND RETURN TO THE IDEA

THAT

SO MUCH INFORMATION IS AVAILABLE ON THE SUBJECT OF NUTRITION THAT WE WILL NOT EVEN TRY TO DUPLICATE IT IN ANY IN-DEPTH MANNER. HERE WE WILL JUST MENTION THE BASIC FUNCTION OF FOOD, A SOURCE LIST OF VITAMINS AND MINERALS, AND A COMPARISON CHART OF VITAMIN/MINERAL CONTENT BETWEEN FRESH, FROZEN AND CANNED PRODUCE.

ANY CHANGE IN THE FOODS YOU PUT INTO YOUR BODY WILL, BY DEFINITION, CHANGE YOUR BODY ITSELF. REGARDLESS OF THE DESIRABILITY OF POSITIVE CHANGES IN A PATTERN OF OVERALL HEALTH, IT MUST BE DONE SLOWLY AND GRADUALLY TO PREVENT UNFAVORABLE AND DRASTIC REACTIONS. MAKE CHANGES ONLY AS YOUR BODY ALLOWS YOU TO AND REMEMBER THAT IT IS IMPORTANT NOT TO OVERWHELM YOURSELF WITH CHANGES TOO GREAT FOR YOUR BODY TO HANDLE.

WE NEED TO MAKE A CONSCIOUS EFFORT AT THIS POINT TO LOOK AT THE FOODS THAT MAKE UP OUR DAILY DIETS, WHERE THESE FOODS COME FROM, HOW THEY ARE GROWN AND PROCESSED. WE NEED TO KNOW HOW TO PROVIDE THE BODY WITH THE FOOD NECESSARY FOR ITS MAINTENANCE AND GROWTH AT ANY GIVEN TIME IN ITS DEVELOPMENT. THIS MEANS A CHANGE IN THE WAY OF THINKING OF THE ENTIRE WESTERN CULTURE; OR ON A SMALLER SCALE, OF THOSE INDIVIDUALS READY TO BEGIN THINKING IN TERMS OF OPTIMUM HEALTH AND BIOLOGICAL STABILITY.

TO BRING ABOUT THIS CHANGE WE NEED TO KNOW WHAT THE NATURAL ELEMENTS ARE AND HOW THEY ARE USED IN THE HUMAN BODY. THE BODY PRODUCES ENERGY FOR MENTAL AND PHYSICAL GROWTH BY COMBINING THESE ELEMENTS INTO ESSENTIAL NUTRIENTS.

CARBOHYDRATES (SUGARS AND STARCHES) ARE QUICKLY CHANGED INTO GLUCOSE WHICH IS THE PRINCIPLE FOOD OF THE BRAIN AND WHICH PROVIDES US WITH IMMEDIATE ENERGY. PROTEIN ALSO CONVERTS INTO GLUCOSE, BUT MORE SLOWLY THAN CARBOHYDRATES, AND THIS GLUCOSE IS STORED IN THE LIVER TO BE USED WHEN THE CARBOHYDRATE-PRODUCED GLUCOSE IS USED UP. THE BRAIN NEEDS TO MAINTAIN ITS GLUCOSE LEVEL. THIS MEANS THAT IF YOU ARE GOING TO DEMAND MORE ENERGY FROM YOUR MIND AND BODY YOU WILL NEED TO INCREASE YOUR CARBOHYDRATE LEVEL (INTAKE).

OUR BODIES CAN PRODUCE THE GLUCOSE FOR ENERGY AND GROWTH, BUT WE CANNOT PRODUCE THE VITAMINS WE NEED. THESE MUST COME FROM THE FOODS WE EAT. FOLLOWING IS A LIST OF FIVE OF THE MOST ESSENTIAL VITAMINS, WHAT THEY DO, WHERE TO GET THEM, AND HOW TO TELL IF YOU GET ENOUGH.

BASIC VITAMIN GUIDE

VITAMIN A:

SOURCES: HOT RED PEPPERS, DANDELION GREENS, CARROTS, APRICOTS, KALE, SWEET POTATOES, PARSLEY, SPINACH, TURNIP AND MUSTARD GREENS, SWISS CHARD, EGGS, MILK AND MILK PRODUCTS.

STABILITY: INSOLUABLE IN WATER, SOLUABLE IN FATS, STABLE TO LOW HEATS.

USES IN BODY: STORED IN LIVER, AIDS IN CONDITIONING OF EYES, SKIN, LUNGS, GALL BLADDER, KIDNEYS, PROMOTES HEALTHY APPETITE AND NORMAL DIGESTION, PROMOTES LONGEVITY.

SIGNS OF DEFINIENCY: NIGHT BLINDNESS, ABNORMALITIES OF MUCOUS MEMBRANES, POOR BONE AND TOOTH DEVELOPMENT, BLADDER AND KIDNEY PROBLEMS, DANDRUFF AND DRY SKIN, FINGERNAILS THAT BREAK EASILY, REPRODUCTION PROBLEMS, SLOW GROWTH, ACNE.

VITAMIN B:

SOURCES: BREWER'S YEAST, RICE POLISH AND BRAN, WHEAT GERM, SUNFLOWER AND SESAME SEEDS, HOT RED PEPPERS, KELP, NUTS, BLACKSTRAP MOLASSES, YOGURT, WHOLE GRAIN BREADS.

STABILITY: MOST B COMPLEX VITAMINS ARE SOLUABLE IN WATER BUT VARY ON STABILITY IN AIR AND HEAT.

USES IN BODY: LIMITED STORAGE IN BODY, AIDS IN RESISTANCE TO DISEASE, BALANCES STOMACH ACIDITY, FORTIFIES NERVES AGAINST STRESS, AIDS DIGESTION, PROMOTES HEALTHY MUSCLES, CONDITIONS EYES, GUMS, SKIN, HAIR, PROTECTS AGAINST ACCUMULATION OF CHOLESTEROL.

SIGNS OF DEFICIENCY: LOSS OF SELF-CONFIDENCE, MALNUTRITION, LOWERED MORALE, SORE TONGUE, FAILURE OF GROWTH, REPRODUCTION PROBLEMS, ANEMIA, LOSS OF APPETITE, CONSTIPATION AND DIARRHEA, BURNING AND DRYNESS OF EYES.

VITAMIN C:

SOURCES: CHERRIES, ROSE HIPS, SWEET RED AND GREEN PEPPERS, HOT RED PEPPERS, KALE, PARSLEY, COLLARD GREENS, RAW BROCCOLI, BRUSSEL SPROUTS, CITRUS FRUITS.

STABILITY: HIGHLY SOLUABLE IN WATER, EASILY DESTROYED BY HEAT AND AIR, ACID INHIBITS DESTRUCTION, NOT DESTROYED BY FREEZING.

USES IN BODY: LIMITED STORAGE, HELPS PREVENT ALLERGIES, AIDS IN ABSORPTION OF IRON, PRESERVES NORMAL VISION, PROMOTES SUPPLE JOINTS, LOWERS HIGH BLOOD PRESSURE, ACTS AS A DEFENSE AGAINST ASTHMA, HAY FEVER, ECZEMA, HIVES, POISON IVY AND OAK, HELPS REPAIR DAMAGED NERVE TISSUE, PROMOTES STRONG BLOOD VESSELS, AIDS IN HEALING BROKEN BONES, PROMOTES GOOD BONE AND TEETH FORMATION, DETOXIFIES SOME POISONS, PROMOTES HEALTHY GUMS.

SIGNS OF DEFICIENCY: IMPROPER USE OF CALCIUM, SHORTNESS OF BREATH, FATIGUE, APATHY, BRITTLE BONES, SUSCEPTIBILITY TO INFECTION, SLOW HEALING OF WOUNDS, ANEMIA, SCURVY, COLDS, STERILITY, BRUISES.

VITAMIN D:

SOURCES: SUNLIGHT, FISH LIVER OILS, EGGS, MILK, MILK PRODUCTS (PROVIDED THE COWS AND CHICKENS HAVE HAD PLENTY OF SUN.

STABILITY: FAT SOLUBLE, STABLE TO LOW HEAT AND OXIDATION.

USES IN BODY: STORED IN LIVER, PROMOTES NORMAL BONE AND TOOTH DEVELOPMENT, REGULATES ABSORPTION, FIXATION, UTILIZATION OF CALCIUM AND PHOSPHORUS.

SIGNS OF DEFICIENCY: RICKETS, SOFTENING OF BONES, TOOTH DECAY, PYORRHEA.

VITAMIN E:

SOURCES: WHEAT AND RICE GERM, BROWN RICE, BARLEY, RYE, PARSLEY, GREEN LEAFY VEGETABLES, KALE, NUTS, LEGUMES.

STABILITY: SOLUABLE IN OIL AND FAT, STABLE TO HEAT AND ALKALIES IN ABSENCE OF OXYGEN, STABLE IN ACID AT LOW TEMPERATURES, OXIDIZES IN RANCID FATS, PRESENCE OF IRON, LEAD, SILVER SALTS, AND ULTRA-VIOLET RAYS.

USES IN BODY: NOT STORED, MAJOR FOOD OF ADRENAL, SEX, AND PITUITARY GLANDS, PREVENTS LIBERATION OF HEMAGLOBIN FROM RED BLOOD CELLS, AIDS IN NORMAL REPRODUCTION, PROMOTES HEALTHY BLOOD PRESSURE AND HEART, AND RETARDS AGING.

SIGNS OF DEFICIENCY: POOR HEALING OF BRUISES, BURNS, WOUNDS; OXIDATION OF VITAMIN A IN BODY, POOR MUSCLE TONE, IRREGULAR MENSTRUATION AND EARLY MENOPAUSE, IMPROPER UTILIZATION OF FATTY ACIDS, SCAR TISSUE, STERILITY.

MINERALS

SUGARS, STARCHES, FATS AND PROTEINS MAKE UP ALL PLANT AND ANIMAL MATTER. BUT FOR LIFE, SOMETHING ELSE IS NEEDED. THIS OTHER SOMETHING IS MINERALS. ONLY A VERY SMALL AMOUNT OF MINERAL MATTER IS NEEDED BY THE BODY, AND THIS (WITH THE EXCEPTION OF IRON) IS USUALLY SUPPLIED IN A VITAMIN-BALANCED DIET. (NOTE: FOODS THAT ARE "ENRICHED" WITH VITAMINS BECAUSE PROCESSING HAS REMOVED MUCH OF THE ORIGINAL NUTRITION, ARE NOT ENRICHED WITH ALL THE MINERALS THAT HAVE BEEN REMOVED, AND, AS STATED, THESE ARE ALSO VITAL TO HEALTH.) THEREFORE, IT IS NOT USUALLY NECESSARY TO CONSIDER MINERAL INTAKE SEPARATELY WHEN PLANNING A DIET. IF VITAMIN INTAKE IS SUFFICIENT, MINERAL INTAKE WILL ALSO BE ADEQUATE.

HOWEVER, IF THE BODY IS IN SOME SORT OF STRESS CONDITION; I.E., BROKEN BONE, SEVERE FLESH WOUND, EXTREME MENTAL STRESS, ETC., THE INCREASE OF SOME MINERALS CAN BE HELPFUL. THESE ARE AS FOLLOWS:

CALCIUM:

NEEDED FOR: BONE AND TOOTH DEVELOPMENT, HEALING OF BROKEN BONES, MAINTENANCE OF MUSCLE TONE, AND STRENGTHENING NERVOUS SYSTEM.

FOUND IN: SOYBEANS, BLACKSTRAP MOLASSES, MILK AND MILK PRODUCTS, BROCCOLI, AND UNHULLED SESAME SEEDS (GROUND INTO MEAL OR BUTTER FOR DIGESTIBILITY).

MAGNESIUM:

NEEDED FOR: ALMOST EVERY CHEMICAL REACTION IN THE BODY (ESPECIALLY THOSE INVOLVING ENERGY PRODUCTION), ALLOWING THE BODY TO USE CALCIUM, CURING INSOMNIA, RELIEVING CONSTIPATION, MAINTAINING STEADY NERVES AND PATIENCE.

FOUND IN: NUTS, SOYBEANS, GREEN LEAFY VEGETABLES.

SODIUM:

NEEDED FOR: PART OF THE COMPLEX BLOOD FILTRATION SYSTEM IN THE LIVER.

FOUND IN: MOST GREEN VEGETABLES. (SINCE SO MANY FOODS NOW USE SODIUM IN PROCESSING -- CHECK LABELS FOR SODIUM COMPOUNDS -- PLUS WHAT WE USE IN COOKING AND AT THE TABLE, THERE IS MUCH MORE OF A PROBLEM OF GETTING TOO MUCH SODIUM RATHER THAN TOO LITTLE.)

POTASSIUM:

NEEDED FOR: MAINTAINING PROPER BOWEL FUNCTIONING, PREVENTION OF INSOMNIA AND FATIGUE, MAINTAINING NORMAL BLOOD PRESSURE AND BLOOD SUGAR.

FOUND IN: FRUITS, VEGETABLES, UNREFINED GRAINS, FISH, AND MEAT.

PHOSPHORUS:

NEEDED FOR: HEALING OF BROKEN BONES AND SKIN WOUNDS.

FOUND IN: LIVER, YEAST, WHEAT GERM, MEAT, GRAPES.

ZINC:

NEEDED FOR: CELL FORMATION AND SYNTHESIS OF BODY PROTEIN.

FOUND IN: NUTS AND GREEN LEAFY VEGETABLES, IF GROWN ON ORGANIC
SOILS.

ALTHOUGH ZINC IS ONE OF A GROUP OF MINERALS KNOWN AS
"TRACE" MINERALS (NEEDED ONLY IN MINUTE AMOUNTS), IT
IS VERY EFFECTIVE IN LARGE DOSES IN HEALING SKIN
WOUNDS.

IRON:

NEEDED FOR: PRODUCTION OF RED BLOOD CELLS.

FOUND IN: BREWER'S YEAST, ROSE HIPS, RAISINS, WHEAT GERM,
APRICOTS, BLACKSTRAP MOLASSES, FISH LIVER OILS, EGGS,
PRUNE JUICE, LIVER (ESPECIALLY PORK), AND COOKING IN
CAST IRON POTS.

UNLIKE OTHER MINERALS, IRON IS NOT SO EASILY OBTAINED
IN SUFFICIENT QUANTITIES, AND MANY PEOPLE -- ESPECIALLY
WOMEN AND CHILDREN -- SHOW SIGNS OF IRON DEFICIENCY.

NUTRITION WORKSHOPS

TO PROVIDE SOME NUTRITION INFORMATION TO NEIGHBORHOOD RESIDENTS,
THE EDIBLE CITY RESORCE CENTER AND COMMUNITY HEALTH AND EDUCATION
CENTER (CHEC) CO-SPONSORED A SERIES OF NUTRITION WORKSHOPS.
THESE WERE FACILITATED BY TWO NUTRITION COUNSELORS, AND ALL
WORKSHOPS WERE HELD AT CENTRAL NEIGHBORHOOD LOCATIONS. SURATA SOY-
FOODS AND THE COMMUNITY NATURAL FOODS AND GENERAL STORE DONATED
FOOD FOR EACH WORKSHOP'S RECIPES. THE ATTACHED FLYER DESCRIBES
EACH WORKSHOP, AND AN EVALUATION OF THE PROCESS USED IS ALSO IN-
CLUDED.

WORKSHOP EVALUATION: BY REGGI NORTON AND BARBARA WEINSTEIN

WE PLANNED, PUBLICIZED, AND PRESENTED 5 NUTRITION WORKSHOPS IN
WHITEAKER NEIGHBORHOOD IN JANUARY 1980. THESE WORKSHOPS, FUNDED
THROUGH THE NATIONAL CENTER OF APPROPRIATE TECHNOLOGY MONIES FOR
WHITEAKER, WERE FREE TO WHITEAKER RESIDENTS.

1. THE GREAT GOOD FOR YOU GOODIES WORKSHOP (FOR AGES 6-12)

WE MADE A COOKBOOK TO HAND OUT WITH SOME WHOLESOME SNACKS.
OUT OF THIS, WE CHOSE THREE SNACKS TO MAKE, PLUS TEA. THE
THREE SNACKS WERE BRAN MUFFINS, PEANUT BUTTER BALLS, AND
PEANUTTY SPREAD. WE INVOLVED THE CHILDREN IN MEASURING
AND MIXING AND CUTTING UP CELERY AND APPLES TO GO WITH THE
SPREAD. WE ALSO INVOLVED THEM IN THE CLEANUP.

GENERALLY, THE WORKSHOP WAS SUCCESSFUL, WITH TWELVE CHILD-
REN PRESENT, AND THE CHILDREN LOVED MAKING AND EATING THE
"GOODIES", ESPECIALLY THE PEANUT BUTTER BALLS SINCE THEY
COULD MAKE THEIR OWN AND ADD WHAT THEY WANTED. HOWEVER, IT
MIGHT BE BETTER TO DO SEVERAL WORKSHOPS OR A SERIES OF
CLASSES AND PREPARE ONE ITEM EACH TIME, SO THEIR ATTENTION
WON'T BE DIVIDED.

2&3. THE GREAT EDIBLES SHOPPING SPREE AND THE "LESSER" COOKING
WORKSHOP

FOR THE "SHOPPING SPREE" WE HAD A HANDOUT ON "WHAT TO BUY IN
A NATURAL FOODS STORE." WE PLANNED A PRESENTATION ON HOW
TO USE AND COOK WHOLE GRAINS, BEANS, ETC. AND HOW TO SPROUT
SEEDS, IN ADDITION TO THE NUTRITIVE VALUE OF WHOLE FOODS
AND FRESH PRODUCE. WE THEN PLANNED A TRIP TO THE COMMUNITY
STORE.

FOR THE "LESSER COOKING" WORKSHOP WE PRESENTED HANDOUTS:
"EATING FOR HEALTH: A GUIDE FOR MAKING DIETARY CHANGES",
"HOW TO ASSESS YOUR OWN NUTRITIONAL NEEDS", RECIPES FOR
CUTTING DOWN ON FATS AND SUGAR; NATURAL FOODS SUBSTITUTIONS;
"LESSER" RECIPES; AND A SELF-ASSESSMENT QUESTIONNAIRE.

WE HAD PREPARED VEGETABLE SOYBEAN STEW AS AN EXAMPLE OF PRO-
TEIN COMPLIMENTARITY; APPLE BUNDT CAKE AS AN EXAMPLE OF
BAKING WITHOUT ANY FATS OR CONCENTRATED SWEETENERS; OATMEAL
RAISIN COOKIES AS AN EXAMPLE OF SWEETS USING WHOLE GRAINS;
GOMASIO AS A SUBSTITUTE FOR SALT; LOWFAT MOCK SOUR CREAM
AS A SUBSTITUTE FOR SOUR CREAM; AND TOFU MAYONNAISE AS AN
EXAMPLE OF A LOW-FAT, HIGH-PROTEIN SUBSTITUTE FOR MAYONNAISE.

WE PLANNED PRESENTATIONS ON PROTEIN COMPLIMENTARITY; WHY AND
HOW TO CUT DOWN ON FATS, SUGAR, AND SALT; STRETCHING THE FOOD
DOLLAR WITH PLANT PROTEINS; AND INDIVIDUAL DIET ASSESSMENTS
BASED ON THE FOOD EXCHANGE SYSTEM. THE CONTENT OF THESE
WORKSHOPS WAS REQUESTED BY WHITEAKER RESIDENTS FROM THE
HEALTH SURVEY DONE IN THE NEIGHBORHOOD IN SEPTEMBER. WE

SENT OUT 500 FLYERS AND HAD AN ARTICLE IN THE MONTHLY
WHITEAKER NEWSLETTER IN JANUARY. NONETHELESS, ATTENDANCE
WAS POOR. WE QUESTION WHETHER SATURDAY IS A GOOD DAY, AND
PERHAPS MORE ONE-TO-ONE CONTACT IS NEEDED IN PUBLICITY.
ALSO, POSTERS FOR INDIVIDUAL EVENTS MIGHT HELP, RATHER THAN
ONE POSTER FOR ALL OF THE WORKSHOPS.

4. SOYFOODS: THE VERSATILE BEAN

SURATA SOYFOODS DONATED THE TOFU AND TEMPEH. HANDOUTS IN-
CLUDED "LESSER RECIPES" AND HANDOUTS ON TOFU AND TEMPEH FROM
SURATA. WE ALSO HAD A DISPLAY OF TOFU COOKBOOKS AND RECIPES.
WE PREPARED SOYMILK AND SHOWED HOW TO USE OKARA, THE SOY
PULP LEFT FROM MAKING SOYMILK. WE ALSO PREPARED A TOFU-
GARLIC DIP, FRIED TEMPEH STRIPS, AND BASIC SAUTEED TOFU, AND
HAD PREPARED A TOFU CHEESECAKE. WE SERVED RYE SOURDOUGH
BREAD MADE WITH OKARA AND VEGETABLES WITH THE DIP. WE TALKED
ABOUT THE NUTRITIVE VALUES OF THESE FOODS AND COST AND HEALTH
FACTORS AND THE MANY WAYS OF USING SOY.

THIS WORKSHOP WAS WELL ATTENDED (18 PEOPLE), AND THE RESPONSE
WAS VERY POSITIVE. THE ATTENDANCE MAY BE ATTRIBUTED TO IN-
DIVIDUAL POSTERS POSTED ALL OVER TOWN AND SPECIFIC MEDIA
AND ONE-TO-ONE CONTACT DONE. PUBLICITY WAS SO SUCCESSFUL
THAT 70% OF THE PARTICIPANTS WERE FROM OUTSIDE THE WHITEAKER
NEIGHBORHOOD.

5. SENIORS ARE SPECIAL NUTRITION WORKSHOP

FOOD WAS DONATED BY SURATA SOYFOODS, SOLSTICE BREAD, JUST
PRODUCE, THE COMMUNITY STORE, AND SPROUT CITY. WE STARTED
WITH INTRODUCTIONS AND WHY PEOPLE WERE THERE AND THEN DIS-
CUSSED THE SPECIAL NEEDS OF SENIORS, ADDRESSING THEIR
SPECIFIC NEEDS AND QUESTIONS. WE TALKED ABOUT HOW TO CUT
DOWN ON FATS (PARTICULARLY SATURATED), SALT, AND SUGAR (AND
WHY), AND ABOUT INCLUDING FIBER AND HOW TO BUY AND PREPARE
WHOLE GRAIN FOODS AND FRESH PRODUCE. HANDOUTS INCLUDED:
"SENIORS ARE SPECIAL"; "BULK FOOD BUYING"; SAMPLE MENUS
AND RECIPES; AND RECIPES FOR CUTTING DOWN ON FATS AND SUGAR;
AND "EATING FOR HEALTH: A GUIDE FOR MAKING CHANGES IN YOUR
DIET".

THE LUNCH MENU INCLUDED: SCRAMBLED TOFU, BAKED POTATO WITH
MOCK SOUR CREAM, FRESH GREEN SALAD WITH DRESSING (LOW-FAT
TOMATO OR LOW-FAT TOFU DRESSINGS), SOLSTICE WHOLEGRAIN
BREAD WITH SOY MARGARINE, AND HERB TEA.

WE DEMONSTRATED MAKING MOCK SOUR CREAM, TOFU DRESSING, AND
THE SCRAMBLED TOFU. KEZI T.V. NEWS WAS THERE AND THE WORK-
SHOP WAS SPOTLIGHTED ON THE EVENING NEWS.

THIS WORKSHOP WAS WELL ATTENDED (23 SENIOR CITIZENS), AND
INTEREST WAS HIGH. AGAIN, PUBLICITY INCLUDED SPECIFIC WORK-
SHOP POSTERS AND INDIVIDUAL CONTACT WITH EACH SENIOR CENTER,
AND, ONCE AGAIN, 60-70% OF THE PARTICIPANTS WERE FROM OUTSIDE
THE WHITEAKER AREA.

Nutrition  for all

GREAT GOOD FOR YOU GOODIES
Friday, January 18, 2:30-4 pm
This workshop is for ages 6 to 12 and will happen at the
Whiteaker Community Center. We will discuss and make
some nutritious and easy goodies.

"LESSER" COOKING
Saturday, January 19, 1-5:00 pm
This workshop will emphasize cooking with LESS —less
sugar, salt, fat, and MONEY. We will look at easy. tasty
ways of meeting one's daily nutritional needs with LESS.
Individual diet assessments will be done. Meet at the
Whiteaker Community Center, corner of Jackson and
Clark.

SENIORS ARE SPECIAL TOO
Tuesday, January 29, 9:30am to noon
This workshop will discuss the special needs of seniors
and how to meet those needs with easy, tasty foods.
Lunch will be prepared and served. The workshop will
meet at Campbell Senior Center, 155 High St.

GREAT EDIBLES SHOPPING SPREE
Saturday, January 19, 9:30am to noon
We will meet at the Whiteaker Community Center (corner
of Jackson and Clark) to discuss food values and how to
use bulk foods and then we'll go on a shopping spree.

SOYFOODS: THE VERSATILE BEAN
Sunday, January 27, 3-6:00 pm
At this workshop we will discuss the food values of a variety of soy products and demonstrate how to prepare tasty
dips, snacks, and meals from soy. Tastes included! Meet
at Whiteaker Community Center, corner of Jackson and
Clark.

These workshops, sponsored by the Community Health
and Education Center (CHEC) and The Edible City Resource Center, are partially funded through funds from
the National Council of Appropriate Technology and are
FREE to Whiteaker residents. A $1.00 fee will be charged
to non-Whiteaker residents for the "Lesser" Cooking and
the Soyfoods workshops.
The Workshops are taught by Reggi Norton and Barbara
Weinstein, nutrition counselors at CHEC. For more information call the Community Health and Education Center
(CHEC), 485-8445.

**FREE to
Whiteaker residents.**

**sponsored by
the Community Health & Education Center and Edible City
Resource Center. For information, call 485-8445.**

IF YOU WANT TO GROW UP RIGHT, EAT MORE THAN ONE KIND OF FOOD.

PLANT MORE THAN ONE KIND OF FOOD

Tomatoes, carrots, deep yellow and dark green vegetables, peas, fruits, sprouts all contain VITAMIN A (you need fresh, daily, for healthy skin, eyes, etc.) Sprouts peas, seeds, nuts, whole grains contain the VITAMIN B COMPLEX (you need for your nerves, and stamina). Citrus fruits, tomatoes, sprouts, peppers, leafy greens, cabbage family,* peas, beans, contain VITAMIN C (for defense). Sunshine will give you VITAMIN D, and you've already eaten VITAMIN E and the rest by now!

"I ONLY EAT CARROTS"

"I LIKE PEAS"

TO NAME A FEW:

Seeds, nuts, beans, whole grains, fresh sprouts, peas, potatoes contain PROTEIN (the basic material for all cells). Whole grains, ripe fruits, peas, beans, potatoes contain CARBOHYDRATES (quick fuel). Whole grains, potatoes, beans contain STARCH (slow burning fuel and stored energy). MINERALS—IRON (strong blood)—CALCIUM (strong bones)—MAGNESIUM—PHOSPHOROUS—ZINC—COPPER—BIOTIN—IODINE—POTASSIUM—SULPHUR—(ETCETERA)—are available from leafy greens, peas, beans, whole grains, prunes, turnip greens, beets, parsley and all of their relatives in varying proportions. Vegetable protein should be supplemented by animal protein (milk and milk products, eggs), fresh and whole— to complete the spectrum of EIGHT ESSENTIAL AMINO ACIDS the body uses to make all kinds of cells, organs, miracles.

Brussel Sprouts_Cauliflower_Broccoli_Curly Kale_Crinkled Savoy_Kohl Rabi_are all members of the Cabbage Family*

B "seed cases" Bell Peppers

A "legumes" Peanuts

B "seed cases" Crook Neck Squash

C "stems" Celery

D "leaves" Parsley

A "legumes" Bush Beans

D "leaves" Brussel Sprouts

D "leaves" Cauliflower

E "tubers" Turnips

E "tubers" Carrots

D "leaves" Boston Lettuce

E "tubers" Sweet Potatoe

E "tubers" Onion

E "tubers" Beets

A "legumes" Peas

B "seed cases" Tomatoes

RESOURCES:

ANDERSON, JEAN, THE GREEN THUMB PRESERVING GUIDE; WILLIAM MORROW AND CO., INC.: NEW YORK. 1976

BEYER, BEE, FOOD DRYING AT HOME THE NATURAL WAY; J.P. TARCHER, INC.: LOS ANGELES, CALIFORNIA. C 1976

BRITZ, BARBARA, AND BARBARA BESSEY, "THE NATURE OF FOOD", IN THE ADVENTURES OF GLEN AND DALE; PENTAGON PRESS: PHOENIX, ARIZONA. 1974

COSPER, NANCY, YOU CAN CAN WITH HONEY; MCKENZIE BRIDGE, OREGON. 1976-77

EATING IN AMERICA: DIETARY GOALS FOR THE UNITED STATES; REPORT OF THE SELECT COMMITTEE ON NUTRITION AND HUMAN NEEDS, U.S. SENATE, MIT PRESS. 1978

FIRST HEALTH AND NUTRITION EXAMINATION SURVEY, PRELIMINARY FINDINGS; PUBLIC HEALTH SERVICE, HEALTH RESOURCES ADMINISTRATION, HEW. 1972

FOOD FIRST RESOURCE GUIDE: DOCUMENTATION ON THE ROOTS OF WORLD HUNGER AND RURAL POVERTY, INSTITUTE FOR FOOD AND DEVELOPMENT POLICY, SAN FRANCISCO, CALIFORNIA. 1979

FULLER, JOHN G., 200,000,000 GUINEA PIGS; G.P. PUTNAM'S SONS: NEW YORK. 1972

GIBBONS, EUELL, STALKING THE WILD ASPARAGUS.

HERTZBERG, RUTH, BEATRICE VAUGHAN, AND JANET GREENS, PUTTING FOOD BY; THE STEPHEN GREENE PRESS: BRATTLEBORO, VERMONT. 1975

HIGHTOWER, JIM, EAT YOUR HEART OUT; CROWN PUBLISHERS, INC.: NEW YORK. 1975

HOME CANNING OF FRUITS & VEGETABLES, HOME AND GARDEN BULLETIN #8, U.S.D.A.: WASHINGTON, D.C. 1965

LAPPE, FRANCES M., DIET FOR A SMALL PLANET; FRIENDS OF THE EARTH/ BALLANTINE BOOKS: NEW YORK. 1972

LERZA, CATHERINE AND MICHAEL JACOBSON (EDS.), FOOD FOR FUN NOT FOR PROFIT; BALLANTINE BOOKS: NEW YORK. 1975

McCANN, ALFRED W., STARVING AMERICA; GEORGE H. DORAN CO.: NEW YORK. 1912

OREGON STATE UNIVERSITY EXTENSION SERVICE, DRYING FRUITS & VEGE-TABLES; EXTENSION CIRCULAR 889: CORVALLIS, OREGON. REPRINTED APRIL 1977

OREGON STATE UNIVERSITY EXTENSION SERVICE, HOW TO BUILD A PORTABLE ELECTRIC FOOD DEHYDRATOR; EXTENSION CIRCULAR 855: CORVALLIS, OREGON. 1979

OREGON STATE UNIVERSITY EXTENSION SERVICE, HOME FREEZING OF FRUITS AND VEGETABLES; EXTENSION CIRCULAR 864: CORVALLIS, OREGON. 1978

PARKER, RUSSELL C., AND JOHN M. CONNOR, "ESTIMATES OF CONSUMER LOSS DUE TO MONOPOLY IN THE U.S. FOOD MANUFACTURING INDUSTRY"; FOOD SYSTEMS RESEARCH GROUP OF NORTHCENTRAL RESEARCH PROJECT, NC-117. JULY 1978

RODALE AND STAFF, STOCKING UP; RODALE PRESS, INC.: EMMAUS, PA. 1977

SCHLINK, F.J., EAT, DRINK AND BE WARY; CONSUMER RESEARCH, INC.: WASHINGTON, NEW JERSEY. 1935

SCHLINK, F.J. AND ARTHUR KALLET, 100,000,000 GUINEA PIGS; THE VAN-GUARD PRESS: NEW YORK. 1933

SIMMONS, PAULA, THE GREEN TOMATO COOKBOOK.

TARR, YVONNE YOUNG, NEW YORK TIMES BREAD AND SOUP COOKBOOK; BALLAN-TINE BOOKS: NEW YORK. 1972

THE FAMILY FARM IN CALIFORNIA; EMPLOYMENT DEVELOPMENT DEPT., GOVERN-OR'S OFFICE OF PLANNING AND RESEARCH, DEPARTMENT OF FOOD AND AGRICULTURE, AND THE DEPARTMENT OF HOUSING AND COMMUNITY DE-VELOPMENT. NOVEMBER 1977

WHITEAKER REFINEMENT PLAN; CITY OF EUGENE PLANNING DEPARTMENT: EUGENE, OREGON. 1978

Chapter 2

The School Farms Program

I have helped the grade school students at Condon School till, plant, tend, and harvest a small vegetable garden for the past two years. Gardening teaches timeless truths sorely lacking among this society's young people. You reap what you sow. Everything has its season. The world must be managed carefully. Endeavors are easily destroyed.

Furthermore, in time of political upheaval the ability to grow food on a small intensive scale can be invaluable to a population. Finally, to quote Peter Chen's old Chinese proverb; "... if you wish to be happy for the rest of your life, become a gardener."

Tom Bittman

WHY SCHOOL FARMS?

AS OUR COUNTRY'S GIANT AGRIBUSINESS SYSTEM GAINS MORE AND MORE CONTROL OVER OUR FOOD SUPPLY, THE LOSS WE SUFFER SPREADS TO SEVERAL AREAS. THE QUALITY OF PRODUCE THAT WE BUY IN GROCERY STORES (ESPECIALLY FRUITS AND SOME VEGETABLES) IS DECLINING BECAUSE FOOD IS GROWN FOR APPEARANCE AND SHIPPING QUALITIES, AND MOST OFTEN IS PICKED TOO EARLY TO RIPEN WITH ITS FULLEST FLAVOR. WE HAVE LITTLE SAY ABOUT THE VARIETY OF FOODS THAT ARE AVAILABLE TO BUY, AS FOODS ARE GROWN ACCORDING TO DEMAND IN THE MARKET. MOST OF ALL, THE MORE FOOD THAT COMES TO US FROM LARGE CONGLOMERATE FARMS AND PROCESSORS, AND FROM PLACES FAR FROM OUR HOMES YEAR-ROUND, THE FURTHER WE GET FROM HAVING CONTROL OVER OUR OWN FOOD SUPPLIES AND A SENSE OF THE LIFE CYCLES AND SEASONS WHICH PRODUCE THE FOOD THAT WE EAT EVERY DAY.

TEACHING FARMING AS PART OF A SCHOOL CURRICULUM WILL GIVE CHILDREN THIS KNOWLEDGE OF CYCLES AND FOOD ORIGINS, AND OF ALTERNATIVES TO THE CURRENT BUSINESS ORIENTED AND ECOLOGICALLY DISASTROUS FOOD SYSTEM IN OUR COUNTRY. IT MAKES SENSE FOR CHILDREN TO LEARN FARMING SKILLS AT A TIME WHEN FAMILY BUDGETS MAY BE UNDER PRESSURE AND A CHILD MAY END UP INFLUENCING A WHOLE FAMILY TO PARTICIPATE IN A HOME GARDEN WHICH COULD SUBSTANTIALLY REDUCE THE FAMILY'S TOTAL FOOD BILL.

SCHOOL GARDENS PROMOTE HEALTH IN CHILDREN THROUGH RECREATION AND NUTRITION, AND THE GARDEN IS EXCELLENT AS A STRESS REDUCING TOOL FOR CHILDREN. GARDENS ARE FUN AND CREATIVE AND PROVIDE A WONDERFUL EDUCATIONAL EXPERIENCE FOR THE YOUNG AND OLD ALIKE. THE ADDITION OF SMALL PRODUCTIVE ANIMALS TRANSFORMS THE GARDEN INTO A FARM, SINCE MUTUALLY BENEFICIAL CYCLES OF LIVING THINGS SUPPORT EACH OTHER.

CHILDREN WILL GAIN A SENSE OF SATISFACTION IN GROWING THEIR OWN FOOD, AND THIS ABILITY TO PROVIDE FOR ONESELF IS A PRICELESS TOOL TO TEACH. WHEN IT MAKES SO MUCH SENSE TO BE FARMING FOR ONESELF ON A SMALL OR NEIGHBORHOOD SCALE, THE QUESTION BECOMES WHY NOT SCHOOL FARMS?! THE SOONER THE BETTER!

A SENSE OF THE LIFE CYCLES AND SEASONS WHICH PRODUCE FOOD THAT WE EAT EVERY DAY.

SCHOOL FARMS

PLANT THE SEEDS FOR AN AGRICULTURAL FUTURE

(KIDS)

WHY → 1. WHY SCHOOL FARMS?

WHAT → 2. THE SCOPE: ·SOILS (FOUNDATION OF LIFE)
·FARMING
·RECYCLING
·ENERGY
·HEALTH

HOW → 3. LEARNING MATRIX (STRUCTURE)

4. TEACHERS' GUIDE

5. LESSON PLANS -- SCHOOL FARMS ACTIVITIES
·AUTUMN
WHEN → ·WINTER
·SPRING
·SUMMER

WHO → 6. RESOURCES (BOOKS, PERIODICALS, FILMS)

7. ORGANIZATIONS LOCAL REGIONAL NATIONAL

8. HELPFUL PEOPLE

WHERE →

SOME THINGS FOR TEACHERS TO CONSIDER...

THE GARDEN IS ONE OF THE MOST TOTAL LEARNING ENVIRONMENTS FOR A CHILD. A YEAR-ROUND FARMING PROGRAM IN SCHOOLS WILL PROVIDE OPPORTUNITIES FOR GROWTH IN SEVERAL AREAS OF LEARNING. FARMING COMBINES PRACTICAL EXPERIENCE WITH CONCEPTS OF SEASONS, CHANGE, GROWTH, RESPONSIBILITY, AND AN AWARENESS OF OUR ENVIRONMENT AS THE SOURCE OF OUR FOOD. ALSO, FARMING WITH CHILDREN NATURALLY INCLUDES USING AND DEVELOPING EXPERIENCES WITH READING, SCIENCE, ART, SOCIAL SCIENCE, GEOGRAPHY, ETC., AND WILL INTEGRATE THESE SUBJECTS IN A WAY WHICH IS FUN AND EXCITING FOR THE CHILD. THE CHILD WILL BE ABLE TO SEE THE ONGOING RESULTS OF HIS/HER EFFORTS AS THE SEASONS PROGRESS, GARDEN TASKS CHANGE, AND THE GARDEN FLOURISHES, PRODUCING FOOD TO EAT AND MARVEL AT, AND BRIGHT COLORS TO ENJOY.

THE CURRICULUM ITSELF IS BASED ON A STRUCTURE OF <u>DO</u> -- <u>THINK</u> -- <u>COMMUNICATE</u>, AN IDEA WHICH WE HAVE BORROWED FROM HUMBOLDT COUNTY SCHOOL DISTRICT'S ENVIRONMENTAL EDUCATION PROGRAM GUIDE, "THE GREEN BOX". IN EACH OF OUR LESSONS, <u>DOING</u> WILL BE A PHYSICAL ACTIVITY RELATING TO ONE OF THE FIVE FOCUS AREAS OF FARM LIFE TO BE COVERED: <u>SOILS</u>, <u>FARMING</u>, <u>RECYCLING</u>, <u>ENERGY</u>, OR <u>HEALTH</u>. <u>THINKING</u> WILL BE AN EXPLANATION BY THE TEACHER OF FACTUAL MATERIAL RELATED TO THE ACTIVITY AND A TIME FOR THE CHILDREN TO FREELY EXPLORE IDEAS WHILE THE TEACHER ACTS AS FACILITATOR. <u>COMMUNICATE</u>-ING SOLIDIFIES THE PHYSICAL AND CONCEPTUAL EXPERIENCES FOR THE CHILDREN THROUGH AN ORAL OR WRITTEN REPORT, DISCUSSION, ART OR MUSIC PROJECT, OR A GAME.

A LEARNING MATRIX

THE MATRIX IN THE TOP RIGHT HAND CORNER OF THE FIRST PAGE OF EACH LESSON PLAN ILLUSTRATES WHICH OF EIGHT ACADEMIC AREAS WILL BE TOUCHED ON IN EACH LESSON. THE FOLLOWING ARE EXAMPLES OF HOW EACH AREA MAY BE COVERED THROUGHOUT THE YEAR:

<u>LANGUAGE</u> -- CHILDREN SPEAKING IN DISCUSSION GROUPS OR IN FRONT OF THE CLASS, LISTENING WHILE OTHERS SPEAK AND EXPLAIN IDEAS, WRITING REPORTS OR MAKING CHARTS, READING SEED CATALOGS AND OTHER GARDENING INFORMATION.

<u>SCIENCE</u> -- BASIC SCIENCE CONCEPTS WILL BE PRESENTED: WEATHER CYCLES, PLANT AND ANIMAL LIFE CYCLES, AND HOW THE PARTS OF OUR ENVIRONMENT WORK TOGETHER.

<u>ART</u> -- VISUAL AND PERFORMING ARTS WILL BE A PART OF THE CURRICULUM. THERE WILL BE ART PROJECTS AS WELL AS SOME DRAMA ACTIVITIES.

<u>SOCIAL SKILLS</u> -- SHARING, COOPERATION, DEVELOPING PATIENCE, AND WORKING IN GROUPS WILL HAPPEN THROUGHOUT THE YEAR IN THE LEARNING PROCESS.

THE GOALS AND SCOPE OF SCHOOL FARMS.

SCHOOL FARMS

PROVIDE OPPORTUNITIES FOR STUDENTS, TEACHERS, FACILITATORS, NEIGHBORS, AND FRIENDS TO EXPERIENCE THE SPONTANEOUS JOYS OF LIVING AND LEARNING WITH MANY ASPECTS OF FARM LIFE INTEGRATED INTO THE SCHOOL CURRICULUM. ALONG WITH THIS THEY WILL HELP SUSTAIN AN AGRICULTURAL TRADITION AND ECONOMY IN THE WILLAMETTE VALLEY.

FOR CHILDREN OF ALL AGES

OUR GOALS INCLUDE:

1. TO FACILITATE THE CHILD LEARNING TO <u>DO</u>, TO <u>THINK</u>, AND TO <u>COMMUNICATE</u>.

2. TO HELP THE CHILD SENSE THE SEASONAL ASPECTS AND OTHER CYCLES OF LIFE, AND TO DISCOVER THE INTERRELATEDNESS OF ALL THINGS.

3. TO FOSTER THE GROWTH OF SELF-RELIANCE WITHIN THE CHILD BY GROWING SOME OF OUR OWN FOOD AND CREATING SCHOOL FARMS.

4. TO REDUCE STRESS AND PROMOTE BETTER HEALTH FOR THE CHILD AND THE COMMUNITY.

THE SCHOOL FARM GROWS WITH THE CHILDREN USING THE MATERIALS, TOOLS, AND ENERGIES IN AND OUT OF THE CLASSROOM, OBSERVING AND RECORDING WHAT CAN BE SEEN, TASTED, FELT, MEASURED, AND HEARD--CARING FOR AND IMPROVING THE ENVIRONMENT.

<u>THE SCOPE OF THE SCHOOL FARM CURRICULUM</u> WILL COVER FIVE SEPARATE "FOCUS" AREAS IN AN INTERRELATED AND CYCLICAL MANNER.

1. <u>SOILS</u> -- OUR FOUNDATION OF LIFE, pH, MINERALS, TESTING, CYCLES, CONSERVATION

2. <u>FARMING</u> -- SEEDS, STARTS, PLANTS, GARDENING TECHNIQUES, INSECTS, WORMS, FISH FARMING, RABBITS, CHICKENS, CYCLES, CONSERVATION

3. <u>RECYCLING</u> -- COMPOSTING, FISHY WATER, CHICKEN AND RABBIT POOP, EARTHWORMS

4. <u>ENERGY</u> -- SUN, RAIN, COLD FRAMES AND GREENHOUSES, REAL COST OF AGRIBUSINESS FOOD COMPARED TO SCHOOL FARM FOOD

5. <u>HEALTH</u> -- NUTRITION, FAST FOODS COMPARED TO FRESH, CANNING AND DRYING, HERBS, RECIPES

MATH -- MEASURING GARDEN PLOTS, CONSTRUCTING COLD FRAMES, PLANTING AND SPACING SEEDS AND PLANTS UTILIZES MATH SKILLS WITHIN THE PROJECTS.

MUSIC -- THERE WILL BE SINGING, RHYTHM EXERCISES, AND SOME LISTENING TO MUSIC IN THE CURRICULUM.

PHYSICAL DEVELOPMENT -- GROSS MOTOR SKILLS WILL BE DEVELOPED THROUGH SHOVELING AND CONSTRUCTING CONTAINER BOXES FOR PLANTING, WHILE FINER HAND-EYE COORDINATION IS PART OF PLANTING SEEDS AND OTHER DELICATE GARDEN TASKS.

SENSORY DISCRIMINATION -- THIS INVOLVES EXPLORING THE SENSES, COLORS, TEXTURES, SHAPES, TASTES, SMELLS, ETC.

THERE ARE ENOUGH LESSONS PRESENTED HERE FOR A YEAR-ROUND, ONCE-A-WEEK, PROGRAM HAVING TO DO WITH FARMING. THE CURRICULUM PROGRESSES IN A SEQUENCE WHICH BUILDS UPON KNOWLEDGE FROM THE PREVIOUS LESSONS, ALTHOUGH USING A PORTION OF LESSONS THROUGHOUT THE YEAR AS TIME ALLOWS WOULD WORK JUST AS WELL. IT IS MAINLY UP TO YOU, THE TEACHER, AND THE STUDENTS TO DECIDE HOW MUCH YOU CAN FIT INTO YOUR CURRICULUM, AND WHICH LESSONS ARE BEST FOR YOUR CLASS, AS WELL AS TO PROVIDE THE CONTINUITY BETWEEN THE LESSONS AS THE SEASONS AND YEAR PROGRESS. THE YEAR CURRICULUM IS DIVIDED INTO FOUR SEASONAL SECTIONS AND EACH IS A COMPLETE CYCLE OF ACTIVITIES IN ITSELF, SO THAT A SCHOOL FARM COULD BEGIN WITH ANY SEASON DURING THE YEAR.

THE SUMMER LESSONS ARE INCLUDED IN HOPES THAT THE PROGRAM WILL CONTINUE YEAR-ROUND. IF THIS CANNOT HAPPEN, PERHAPS YOU CAN SET UP A PROGRAM OR SCHEDULE WITH CHILDREN AND INTERESTED PARENTS TO MAINTAIN THE GARDEN FOR THE SUMMER MONTHS. THERE SHOULD BE LITTLE PROBLEM FINDING INTERESTED KIDS SINCE SUMMER BRINGS THE BEGINNINGS OF HARVEST AND BLOOMS.

THE LESSONS HERE ARE OPEN-ENDED. THERE ARE INFINITE POSSIBILITIES FOR EXPANSION WITHIN THEM. BE CREATIVE, CHANGE THEM ANY WAY YOU WOULD LIKE TO MAKE THEM FIT YOUR SPECIAL INTERESTS OR CLASS NEEDS. THERE ARE MANY IDEAS FOR ALTERNATIVE OR FURTHER PROJECTS IN THE APPENDIX, AND ALSO, USE THE CHILDREN'S NATURAL CURIOSITY TO EXPAND ON THE LESSONS. THEIR QUESTIONS AND IDEAS CAN BE USED TO LEARN MORE ABOUT WHAT THEY ARE DRAWN TO IN THE COURSE OF THE YEAR.

THE LESSONS ARE WRITTEN FOR A CLASS LEVEL OF THE 4TH GRADE, SO TEACHERS OF OTHER GRADES WILL NEED TO GEAR THE MATERIAL UP OR DOWN ACCORDING TO GRADE LEVEL. EACH LESSON CAN TAKE ANYWHERE FROM 45 MINUTES TO AN HOUR TO COMPLETE, BUT THEY COULD EASILY BE EXPANDED TO TWO SESSIONS. WHEN A PROJECT INVOLVES PUTTING SOMETHING TOGETHER THAT WILL BE CHECKED ON PERIODICALLY, FOR INSTANCE GROWING SPROUTS INSIDE, TIME MUST BE SET ASIDE TO INCLUDE THIS LATER IN THE WEEK. THERE IS MUCH INFORMATION TO EXPAND ON, AND IF THIS IS DONE, ALL THE MORE LEARNING WILL TAKE PLACE.

THE ACTIVITIES IN THE SCHOOL GARDEN CURRICULUM ARE AS DIVERSE AS A WHOLE SCHOOL ACADEMIC PROGRAM. IT IS NATURAL FOR SOME CHILDREN TO ENJOY OR HAVE MORE CONFIDENCE IN THEIR ABILITIES IN ONE ACTIVITY OVER ANOTHER. STRESS THROUGHOUT THE PROGRAM THAT THERE IS NO FAILURE IN THE GARDEN, ONLY GROWTH AND LEARNING, AND THAT ALL COMPONENTS OF THE GARDENING PROCESS ARE INTERDEPENDENT AND NECESSARY IN TERMS OF THE CYCLES IN LIFE. WITH PLENTY OF PATIENT, CARING FACILITATING FROM A TEACHER, A CHILD WILL EXPERIENCE FULLY THE POSSIBLY-NEW EXPERIENCE OF FARMING, EXCELL IN MANY SKILLS AND HAVE THE SATISFACTION OF CARRYING THROUGH FROM BEGINNING TO END A PROJECT WHICH GIVES TANGIBLE RESULTS, FOOD TO EAT AND TO SHARE.

WE HOPE THAT THROUGH A SCHOOL FARM PROGRAM, A CHILD WILL BEGIN TO SENSE HIM/HERSELF AS AN ACTIVE PART OF THIS CYCLE, AND TO SEE THE CHOICES THAT HE/SHE HAS FOR LIVING A WHOLE AND HEALTHFUL LIFE. THE

CURRICULUM HERE IS A BEGINNING, AND WE NEED THE CONCERN AND COMMITMENT FROM YOU TO MAKE THIS HAPPEN AND FLOURISH WITHIN THE CHILDREN AND THE SCHOOLS.

SUGGESTIONS FOR ADDITIONS TO THE CURRICULUM: "THE GREEN BOX" IS A TOTAL COMPREHENSIVE ENVIRONMENTAL EDUCATION TOOL IN ITSELF WHICH CAN BE USED FOR IDEAS OR IN ITS ENTIRETY IN THE CLASSROOM. IT HAS ACTIVITIES FOR A VARIETY OF CLASS LEVELS, AND IS A VALUABLE TOOL FOR ANY SCHOOL. FOR INFORMATION AND ORDERING, WRITE TO:

HUMBOLDT COUNTY SCHOOLS
OFFICE OF ENVIRONMENTAL EDUCATION
COURT HOUSE
EUREKA, CALIFORNIA 95501
(707) 445-7207

GARDEN SPACE, TOOLS AND SUPPLIES

LACK OF A POTENTIAL GARDEN SPOT AT ANY SCHOOL NEED NOT PREVENT YOU FROM HAVING A SCHOOL GARDEN. IF THERE IS NO SUITABLE, SUNNY SPOT AVAILABLE FOR TILLING AND DIGGING, HAVE A CONTAINER GARDEN. THE DIRECTIONS ARE INCLUDED HERE. WINDOW LEDGE GARDENS WOULD BE EASY TO MAKE AND ARE EASILY ACCESSIBLE TO THE CHILDREN. PLANT SEEDS IN ANY PATCH OF GROUND YOU CAN FIND ALONG THE BUILDING. PLANT PEAS OUTSIDE IN LOW CONTAINERS BELOW THE CLASSROOM WINDOWS, WITH A STRING TRELLIS FOR THEM TO CLIMB. THERE ARE MANY GARDEN ACTIVITIES TO DO EVEN WITHOUT AN ACTUAL GARDEN. BE CREATIVE!

THE QUANTITY OF TOOLS AND SUPPLIES YOU'LL NEED WILL DEPEND ON THE NUMBER OF CHILDREN TO BE WORKING AT ONCE AND ON YOUR AVAILABLE RESOURCES.

ITEMS	FREE SOURCES
·LONG-HANDLED SHOVELS	·WHITEAKER NEIGHBORHOOD
·HOES	PROJECT SELF-RELIANCE TOOL LIBRARY
·RAKES	315 MADISON (JEFFERSON ELEVATOR)
·TROWELS (HAND SHOVELS)	EUGENE, OREGON 343-2711
·CHILDREN'S SHOVELS	
·HAMMER	·URBAN FARM TOOL LIBRARY
·PITCHFORKS	UNIVERSITY OF OREGON
·WHEELBARROW	EUGENE, OREGON 686-3647
·GRASS SHEARS	
·SPRINKLERS	·DONATIONS
·HOSE	
·PLANTS AND HERBS	
·STAKES	
·STRING	
·BUCKETS	·FAST FOOD RESTAURANTS
·SEEDS	·GARDEN STORES OFTEN DESTROY LAST SEASON'S STOCK--ASK FOR DONATIONS
·LEAF MOLD	·CITY WILL DELIVER LEAVES IN THE FALL
·MANURE	·LOCAL FARMERS LOVE TO HAVE THEIR BARNS CLEANED OUT
	·LOCAL WORM FARMS
·WORM CASTINGS	·SCHOOL CAFETERIA; LOCAL MARKETS
·ORGANIC GARBAGE	·WOOD STOVES OR FURNACES
·WOOD ASHES	·CITY OR COUNTY ROAD AND PARK MAINTENANCE DEPTS.
·GRASS CLIPPINGS	
·A TRUCK	·WILLING PARENT; SCHOOL FLEET SVCS.
·INFORMATION AND SPEAKERS	·LOCAL GROUPS; COUNTY EXTENSION SVC.

THERE ARE SOME SCHOOLS IN OUR AREA WHICH HAVE GARDENS, TO VISIT,
TO GET IDEAS FROM, AND TO MODEL AFTER. THESE INCLUDE:

CRESWELL HEADSTART FARM
33795 HARVEY ROAD
CRESWELL, OREGON
(503) 895-3106   RICHARD SNELL

CHILD CARE, INCORPORATED
169 N. WASHINGTON
EUGENE, OREGON
(503) 344-1165   SHERRY CISNEY

   THERE WILL BE A MODEL SCHOOL GARDEN HERE BEGINNING IN
   SPRING 1980, UTILIZING THIS WRITTEN CURRICULUM AND A
   PLAN DESIGN DEVELOPED BY MARY TRUAX, A LANDSCAPE ARCHI-
   TECTURE STUDENT AT THE UNIVERSITY OF OREGON.

WHITEAKER COMMUNITY SCHOOL, LATCH KEY PROGRAM
(503) 687-3505   YONA ASH, SANDY SCHUSTER AND KIM HARPER

LINCOLN COMMUNITY SCHOOL
(503) 687-3468

RIVER ROAD SCHOOL
(503) 687-3307

LAUREL HILL SCHOOL (EACH CLASSROOM HAS A PLOT; EXTRAS ARE
(503) 687-3288   MIKE LEWINE   RENTED OUT)

NEW BEGINNING DAY CARE (GARDEN AREAS AND FRUIT TREES)
(503) 484-2735   KATHY NEWFIELD

IN YOUR SCHOOL GARDEN PROGRAM, ENCOURAGE PARENTS, ADMINISTRATORS,
NEIGHBORS AND COMMUNITY GROUPS TO GET INVOLVED WITH THE CHILDREN
IN ORGANIZING AND DOING THE GARDEN. THE POSSIBILITIES WILL BE
ENDLESS WITH THE INTEREST, EXPERTISE, AND ACTUAL LABOR OF THE
COMMUNITY MEMBERS AS A WHOLE. THE PROJECT COULD EXPAND INTO A
NEIGHBORHOOD FARM WITH RESPONSIBILITIES SHARED BY ALL. BUT TO
BEGIN, START SMALL, USE WHAT RESOURCES ARE AVAILABLE, AND EXPAND
SLOWLY AS INTEREST GROWS.

A GOOD EXAMPLE IS:

IVORENA CARE CENTER

WASHINGTON JEFFERSON OVERPASS

UNDERPASS OPENED

FRUIT & NUT ORCHARD

GREEN-HOUSE

CHICKENS

COMPOST

VEGETABLES

PLAY YARD

FLOWER GARDEN

CHILD CARE, INC.

A SCHOOL FARM

WHITEAKER URBAN
INTEGRATED COMMUNITY
MARY TRUAX LA DESIGN 489
WINTER 1980

N

CHILD CARE INC.
A SCHOOL FARM

WHITEAKER URBAN
INTEGRATED COMMUNITY
MARY TRUAX LA DESIGN 489
WINTER 1980

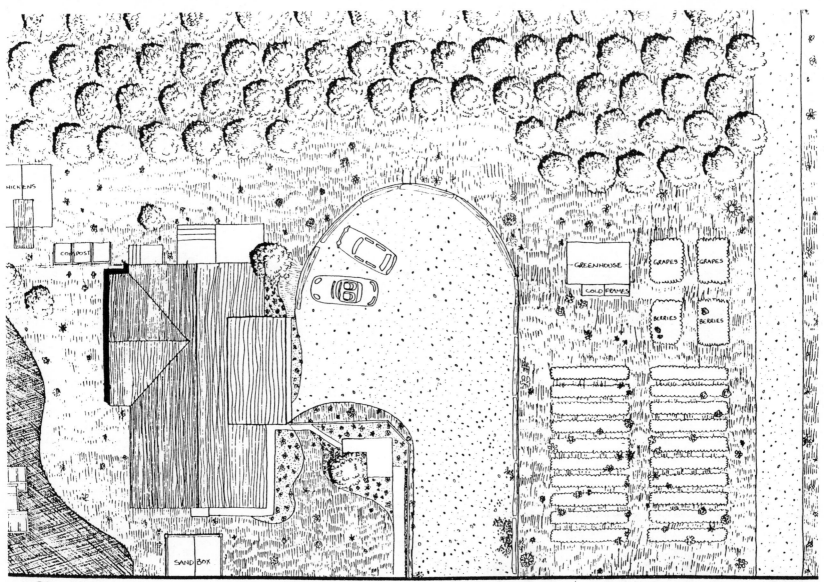

CHILD CARE INC.
A SCHOOL FARM

WHITEAKER URBAN
INTEGRATED COMMUNITY
MARY TRUAX LA DESIGN 489
WINTER 1980

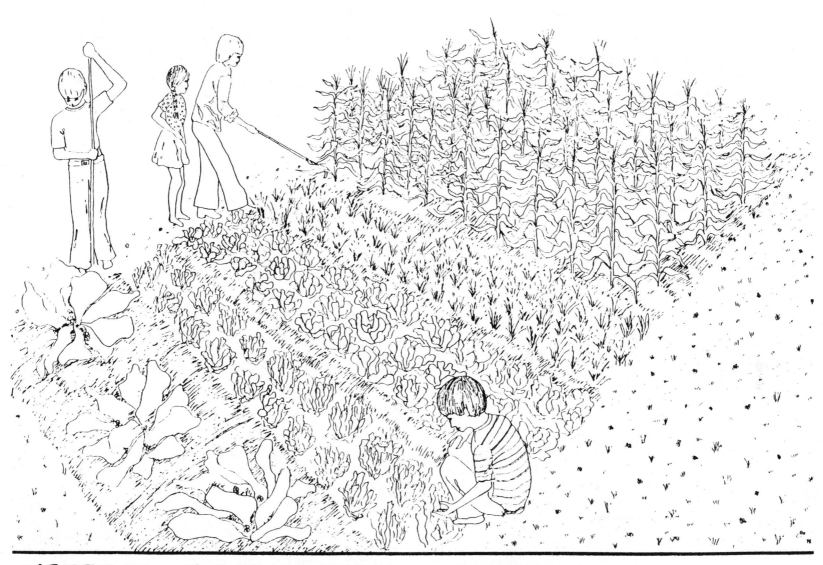

CHILD CARE INC.
A SCHOOL FARM

WHITEAKER URBAN
INTEGRATED COMMUNITY
MARY TRUAX LA DESIGN 489
WINTER 1980

USING STAKES AND TWINE, LOCATE THE
POSITIONS OF EACH PLOT

SHAPE EACH BED INTO A MOUND, FLAT ON
TOP WITH GENTLY SLOPING SIDES. ALLOW
A NARROW FURROW BETWEEN PATH
& BED FOR DRAINAGE

SOIL CAN BE ROTOTILLED OR HAND
SPADED, REMOVING STONES & BREAKING
UP CLODS

SOD SHOULD BE REMOVED WITH A
SHARP SHOVEL AND ADDED TO COMPOST

RAKE THE SURFACE OF BEDS CAREFULLY
TO FORM A SMOOTH SEEDBED

CHILD CARE INC.
A SCHOOL FARM

WHITEAKER URBAN
INTEGRATED COMMUNITY
MARY TRUAX LA DESIGN 489
WINTER 1980

REMOVE VINES & PLANT DEBRIS FROM
GARDEN TO COMPOST BIN

COMPACT COMPOST BIN & COVER WITH
SOD, PLASTIC, MANURE, ETC.

DIG ROOT CROPS & STORE FOR THE
WINTER BETWEEN LAYERS OF
STRAW IN COLD FRAMES

STORE POLES & TRELLISES IN A
DRY PLACE FOR THE WINTER

SPREAD MANURE, LEAF MOLD OR
COMPOST OVER BEDS

CHILD CARE INC.
A SCHOOL FARM

WHITEAKER URBAN
INTEGRATED COMMUNITY
MARY TRUAX LA DESIGN 489
WINTER 1980

FALL

FOCUS: **ENERGY**

LESSON: **WE NEED THE SUN**

MATERIALS: ORANGE TAGBOARD, CREPE PAPER, PAPER FOR CHARTS, MAGA-
ZINES FOR PICTURES, SCISSORS, PENCILS (COLORED OR REGULAR)

TEACHER'S BACKGROUND: ANCIENT MAN WORSHIPPED THE SUN, CALLING IT
"THE GIVER OF LIFE". THE TEMPERATURE AT THE SURFACE OF THE
SUN IS ABOUT $12,000°$ FAHRENHEIT. GREAT AMOUNTS OF RADIANT
ENERGY SPOUT FROM THE SUN YEAR AFTER YEAR, GOING OUT IN ALL
DIRECTIONS. ONLY A TINY PORTION OF IT STRIKES THE EARTH.

ENERGY TRAVELS FROM THE SUN WITH THE SPEED OF LIGHT:
186,000 MILES PER SECOND. IT TAKES 8 MINUTES TO REACH THE
EARTH, BUT THE CLOUDS REFLECT ABOUT 50% OF IT BACK INTO
SPACE.

DO: USING A LARGE SHEET OF ORANGE
TAGBOARD, CONSTRUCT A SUN.
ON THE RAYS, PRINT SOME OF
THE DIFFERENT WAYS WE GET
ENERGY FROM THE SUN.

HAVE STUDENTS IDENTIFY AS
MANY WAYS AS POSSIBLE THAT
THE SUN PROVIDES ENERGY.
BUILD A CHART FOR EACH OF
THESE MAIN DIVISIONS:
FOOD--HEAT--WEATHER--WIND--
WATER POWER--FUELS.

THINK: FOOD CAN BE DIVIDED INTO
THOSE WE GET DIRECTLY FROM
PLANTS (FRUITS, VEGETABLES,
GRAINS), AND THOSE WE GET FROM
ANIMALS WHICH FEED ON PLANTS
(MILK-EGGS-MEAT-FISH-CHEESE).

HEAT (SOLAR HEATING): CON-
STRUCT A CHART SHOWING THE
USES OF SOLAR POWER FOR HEAT.
HAVE STUDENTS TAKE TEMPERA-
TURE READINGS IN THE SUN AND
IN THE SHADE.

WEATHER (WIND, WATER POWER): THE HEAT OF THE SUN WARMS THE
AIR WHICH RISES AND KEEPS IT ON THE MOVE AND CAUSES WIND.
THE SAME HEAT MAKES THE OCEAN CURRENTS. WEATHER CONTAINS
TWO ENERGY SOURCES:

WIND POWER WHICH MOVES SAILBOATS, SAILS GLIDERS, LIFTS
HOT AIR BALLOONS, TURNS WINDMILLS

WATER POWER WHICH GRINDS GRAIN, POWERS SAW MILLS, PUMPS
WATER, AND TURNS TURBINES.

FUELS (OIL, GAS, WOOD, COAL)

OIL AND GAS ARE CREATED BY THE DECAYING REMAINS OF PLANTS
AND ANIMALS. THIS TAKES PLACE IN THE SEA.

WOOD AS FUEL: THE TREE STORES THE SUN'S ENERGY UNTIL IT
IS RELEASED AS HEAT WHEN WE BURN THE TREE.

COAL IS THE END RESULT OF A PROCESS THAT TOOK MILLIONS
OF YEARS. VEGETATION DIED AND FELL INTO SWAMPS WHERE
IT WAS COVERED WITH MUD AND WATER. IT DECAYED VERY
SLOWLY. A HIGH PERCENTAGE OF CARBON REMAINED--COAL.

COMMUNICATE: HAVE STUDENTS USE MAGAZINES AND SCISSORS TO SUPPLY
PICTURES FOR THE CHART UNDER FOOD.

HAVE STUDENTS KEEP A DAILY CALENDAR OF TEMPERATURES, UNDER
HEAT, AND DECIDE AT WHICH TEMPERATURE RANGE THEY ARE MOST
COMFORTABLE.

CONSTRUCT A CHART SHOWING THE DIFFERENT USES OF WIND AND
WATER POWER.

HAVE THE STUDENTS CONSTRUCT A CHART SHOWING DIFFERENT USES
OF FUELS. HAVE CHILDREN IDENTIFY WAYS THEY USE EACH OF THE
FUELS.

USE A KEROSENE LAMP.

HAVE A MARSHMALLOW ROAST USING WOOD IN A HIBACHI.

FOCUS: HEALTH

LESSON: BLACKBERRY JAMMING!

MATERIALS: STERILIZED CANNING JARS AND LIDS, 5 CUPS OF WASHED BLACK-BERRIES, A STOVE, LARGE SAUCEPAN AND LID, 3 CUPS OF HONEY, DOUBLE BOILER, PARAFFIN, HOT PADS

TEACHER'S BACKGROUND: ARRANGE TO USE THE SCHOOL'S EQUIPMENT AND FACILITIES, OR DO YOUR JAMMING AT SOMEONE'S HOME OR AT A LOCAL COMMUNITY CENTER.(THE COUNTY EXTENSION SERVICE KITCHEN MAY ALSO BE AVAILABLE). THIS IS THE TIME THAT BLACKBERRIES ARE RIPE, SO PICK ALL YOUR BERRIES FROM THE ABUNDANT NEIGH-BORHOOD BERRY BUSHES.

DO: GATHER THE MATERIALS AND SOME ENTHUSIASTIC STUDENTS AND TRAIPSE INTO AN AVAILABLE KITCHEN. BE SURE TO MENTION THE NEED FOR CAUTION WITH HOT STUFF. DECIDE HOW SWEET YOU WANT THE JAM TO BE, AND ADD ENOUGH HONEY TO THE BLACKBERRIES. IN A LARGE SAUCEPAN BRING THE BLACK-LERRIES AND HONEY TO A QUICK BOIL, AND CONTINUE FOR 8-10 MINUTES, STIRRING OCCASION-ALLY. LET COOL FOR SEVERAL HOURS, AND ONE HOUR BEFORE GOING HOME, STERILIZE THE JARS WITH BOILING WATER, AND ADD ENOUGH OF THE JAM TO THE JAR SO THAT IT IS ABOUT AN INCH AWAY FROM BEING FULL. MELT A BLOCK OF PARAFFIN IN THE DOUBLE BOILER, AND POUR A 1/2 INCH LAYER OF PARAFFIN ON THE TOP OF THE JAM. LABEL THE JAR WITH THE NAME AND DATE, AND PLACE IN STORAGE.

THINK: DISCUSS WHY IT IS IMPORTANT TO WASH THE FRUIT. WHY DO WE ADD HONEY TO THE JAM? WHY DO WE NEED TO STERILIZE THE JARS? WHY IS STORE-BOUGHT JAM SO EXPENSIVE? WHY DOES STORE-BOUGHT JAM OFTEN INCLUDE PRESERVATIVES, ARTIFICIAL FLAVORS, AND ARTIFICIAL COLORS? HOW MUCH ENERGY MIGHT WE SAVE IF WE MAKE THE JAM OURSELVES?

COMMUNICATE: TAKE SOME OF THE SCHOOL-MADE JAM TO OTHER CLASSES AND EXPLAIN HOW IT WAS MADE. INVITE INTERESTED STUDENTS TO YOUR KITCHEN WORKSHOP TO LEARN HOW TO DO IT. DRAW A STEP-BY-STEP PICTURE DIAGRAM TO HELP INTERESTED FOLKS UNDERSTAND THE PROCESS. BE CAREFUL TO NOTE THE PROCESS AND RECIPES IN THE SCHOOL FARM JOURNAL AND OTHER APPROPRIATE FILES OR BOOKS.

FOCUS: ENERGY

LESSON: WINDS AND WINDMILLS

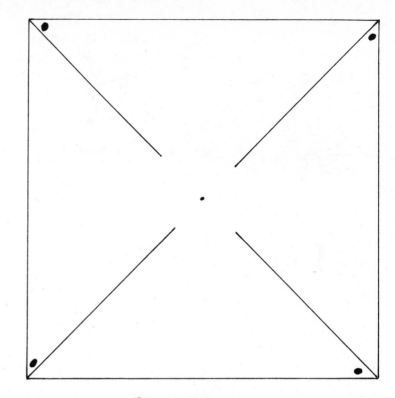

MATERIALS: PINWHEEL PATTERN, STRAWS AND STRAIGHT PINS (ONE PER CHILD), SCISSORS

TEACHER'S BACKGROUND: SOLAR ENERGY MANIFESTS ITSELF IN MANY USABLE FORMS: LIGHT, HEAT, PLANTS AND <u>WIND</u>. THE SUN'S RAYS HEAT THE EARTH AND THE WARM AIR RISES CAUSING WIND CURRENTS. THIS WIND CAN BE USED TO DO MANY THINGS -- PUSH A SAILBOAT, TURN A WINDMILL TO PUMP WATER, AIR, AND TO PRODUCE ELECTRICITY. WIND IS NOT CONSTANT AND ITS VARIABILITY CAUSES PROBLEMS FOR USERS.

DO: PASS OUT SUPPLIES. INSTRUCT CHILDREN TO CUT ON PATTERN'S LINES AND STOP WHEN LINE STOPS. AID THEM IN PLACING PIN THROUGH DOTS AND INTO STRAW. HAVE THEM SWING THE PINWHEELS. HAVE THEM BLOW SOFTLY ON PINWHEELS -- HAVE THEM BLOW HARD ON PINWHEELS.

THINK: DOES IT MOVE FASTER WHEN YOU BLOW HARDER? WHAT TURNS THE PINWHEEL? WHAT WERE WINDMILLS USED FOR ON FARMS? CAN YOU SEE PROBLEMS IN THE USE OF WIND AS AN ENERGY SOURCE? (WHEN WIND STOPS, THE WORK STOPS UNLESS THERE IS A WAY TO STORE THE WIND'S ENERGY.)

COMMUNICATE: DISCUSS AS A GROUP HOW WIND AS ENERGY IS USED. WHAT DOES WIND DO TO THE ENVIRONMENT? TALK ABOUT HOW WIND COULD BE USED MORE FOR OUR ENERGY NEEDS. HAVE EACH CHILD TALK ABOUT HIS/HER OWN ENERGY NEEDS.

FOLLOW-UP POSSIBILITIES: TOUR AMITY FOUNDATION'S GREENHOUSE/ AQUACULTURE/WINDMILL SYSTEM.
TOUR THE URBAN FARM WINDPUMP/IRRIGATION SYSTEM.

PINWHEEL PATTERN

FOCUS: RECYCLING

LESSON: WHERE HAS OUR WATER GONE?

MATERIALS: A 10-GAL. CONTAINER, 6 ONE-CUP MEASURING CUPS, AN EYE DROPPER, PLATE, COLORED PAPER, BRADS, PASTE, SCISSORS, POINTER FOR A SPINNING WHEEL, SIPHON

TEACHER'S BACKGROUND: SEE REFERENCE SHEET ENTITLED "USABLE FRESH WATER: IS THERE A SHORTAGE?" IT IS ALSO USEFUL TO KNOW THAT WITH POPULATION OF ALMOST FOUR BILLION PERSONS, THERE IS AN AVERAGE OF ABOUT 100 BILLION GALLONS OF WATER PER PERSON. HOWEVER, MOST OF THIS WATER IS NOT USABLE BY HUMANS. WATER IS A RETRIEVABLE RESOURCE. OUR PLANET HAS A VAST, FIXED SUPPLY OF WATER. ONLY A SMALL PORTION OF THIS WATER, HOWEVER, IS READILY AVAILABLE FOR USE BY PEOPLE. THE REST IS IN OCEANS, GLACIERS, AND UNDERGROUND.

THIS LESSON IS TO 1) DEMONSTRATE THE SMALL PERCENTAGE OF THE WORLD'S USABLE WATER, 2) IMPROVE MATH SKILLS, 3) STRESS CONSERVATION OF WATER.

DO: HAVE ONE PERSON FROM THE CLASS DEMONSTRATE THE DIAGRAM WHICH REPRESENTS THE TOTAL WORLD WATER SUPPLY (FILL THE 10-GALLON JAR), THE TOTAL FRESH WATER SUPPLY (SIPHON WATER FROM THE JUG TO 5 OF THE ONE-CUP MEASURERS), THE TOTAL FRESH WATER THAT IS NOT ICE (POUR ONE OF THE 5 CUPS INTO THE 6TH CUP, THE TOTAL USABLE FRESH WATER SUPPLY (DRAW 9 DROPS WORTH OF WATER FROM THE 6TH CUP AND SQUEEZE THESE ONE-BY-ONE ONTO THE PLATE).

THINK: REFERRING TO THE LIST OF HOW AMERICANS USE WATER, ASK THE STUDENTS TO DISTINGUISH BETWEEN DIRECT AND INDIRECT USE OF WATER. GIVE SAMPLES OF EACH. FOR EXAMPLE, WHEN WE EAT ONE EGG WE INDIRECTLY CONSUME THE 40 GALLONS OF WATER WHICH HAVE GONE INTO THE FEEDING OF THE CHICKEN, THE CLEANING AND COOLING OF THE EGG, THE GENERATION OF ELECTRICITY (NW) FOR STORAGE, ETC. AN EXAMPLE OF THE DIRECT USE OF WATER INCLUDES STANDING IN THE SHOWER USING 5 GALLONS PER MINUTE. IN THE CASE OF HOT WATER, 72 GALLONS OF WATER HAVE BEEN USED INDIRECTLY TO PRODUCE THE ELECTRICITY TO HEAT EACH GALLON OF YOUR HOT SHOWER WATER.

COMMUNICATE: 1. HAVE THE STUDENTS WORK IN GROUPS TO DETERMINE WAYS THAT FRESH USABLE WATER CAN BE SAVED BOTH INDIRECTLY AND DIRECTLY. HAVE EACH GROUP PRESENT THEIR FINDINGS TO THE CLASS OR PRODUCE AN ARTICLE FOR A SCHOOL PUBLICATION LISTING THEIR SUGGESTIONS.

FOR EXAMPLE: IF ONE POUND OF BEEF REPRESENTS THE INDIRECT USE OF 2,500 GALLONS OF WATER, THEN EACH OUNCE REPRESENTS $16\sqrt{2,500}$ OR 156 GALLONS USED. THEREFORE, 156 GALLONS LESS WATER WOULD BE USED FOR EVERY OUNCE OF BEEF NOT PRODUCED. ANOTHER EXAMPLE: IF A NEW CAR REPRESENTS 100,000

GALLONS OF WATER, THEN THIS AMOUNT COULD BE SAVED BY KEEPING AN OLDER CAR.

2. DEPENDING UPON THE GRADE LEVEL OF THE CLASS, MAKE AND PLAY THE FOLLOWING GAME:

DIRECTIONS FOR MAKING THE SPIN WHEEL--

    A. CUTOUT A SPINNING WHEEL.

    B. COLOR AND LABEL THE DIFFERENT SECTIONS.

    C. MAKE A POINTER OUT OF CARDBOARD AND BRADS.

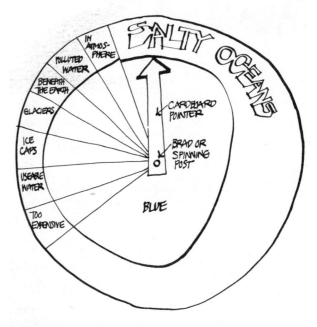

DIRECTIONS FOR PLAYING THE GAME:

    A. HAVE THE CLASS DECIDE WHICH CLASSIFICATIONS OF WATER ARE USABLE AND FOR WHAT PURPOSE; E.G., SEE CHART.

CLASSIFICATION	USEABLE	NOT USEABLE	WHY
SALTY	X		TRANSPORTATION, FISH, CLIMATE
TOO EXPENSIVE		X	DIFFICULT TO OBTAIN, REMOTE
USEABLE WATER	X		FRESH WATER LAKES, RIVERS DRINKING WATER, ETC.

    B. ASSIGN EACH USABLE CLASSIFICATION A NUMBER VALUE; E.G. 1 = SALTY; 4 = USABLE; 1/2 = GLACIER; ETC.

    C. EACH STUDENT GETS A CHANCE TO SPIN THE POINTER.

D. WHEN THE POINTER STOPS ON "USABLE" AREAS, THE STUDENTS
   ADD THE ALLOTED NUMBER OF POINTS. IF THE POINTER STOPS
   ON AN "UNUSABLE" AREA, THE STUDENT RECEIVES NO POINTS
   (EXCEPTION: IF THE POINTER STOPS ON "POLLUTED", HE OR
   SHE MUST DEDUCT ONE POINT).

E. THE FIRST PERSON TO GAIN 5 POINTS WINS.

F. DISCUSS THE DIFFERENCE BETWEEN "USABLE" AND "DRINKING"
   WATER AND POINT OUT WHY THE "USABLE" WATER IS WORTH 4
   POINTS WHEN THE OTHER TWO ARE ONLY WORTH 1 OR 1/2
   POINT.

   TEACHER'S NOTE: FROM THE REFERENCE PAGE "USABLE FRESH
   WATER: IS THERE A SHORTAGE", YOU WILL NOTE THAT THE
   RELATIVE PERCENTAGES OF WATER NOT IN THE OCEANS HAS
   BEEN CHANGED FOR PURPOSES OF THE GAME.

G. SHOW FILM: "WATER: OUR MOST VITAL RESOURCE", AVAILABLE
   FROM THE UNITED NATIONS ENVIRONMENT PROGRAM, NEW YORK,
   NEW YORK.

USABLE FRESH WATER; IS THERE A SHORTAGE?

OUR ~~PLANT~~ PLANET HAS A VAST, FIXED SUPPLY OF WATER BUT ONLY A TINY FRAC-
TION IS READILY AVAILABLE FOR USE BY MAN.

| TOTAL WORLD | TOTAL FRESH | FRESH WATER | USABLE |
| WATER SUPPLY | WATER SUPPLY | THAT'S NOT ICE | FRESH WATER |

DOES THE WORLD HAVE A WATER SHORTAGE?

THE REAL QUESTION IS: IS THERE OR WILL THERE BE A SHORTAGE OF
USABLE WATER? IN ORDER TO ANSWER THE QUESTION ONE NEEDS TO KNOW
WHERE THE EARTH'S WATER IS LOCATED AND HOW IT IS USED.

LOCATION:

97.20%	OF THE WORLD'S WATER IS IN THE SALTY OCEANS.
2.80%	REMAINS; OF THAT
2.48%	IS TIED UP IN ICECAPS AND GLACIERS, LIES TOO DEEP UNDER THE SURFACE, OR IS IN THE ATMOSPHERE OR TOPSOIL.
0.32%	FRESH WATER REMAINS (THE WATER FOUND IN LAKES AND RIVERS); OF THAT
99.00%	IS TOO EXPENSIVE TO GET, IS NOT READILY AVAILABLE (REMOTE SIBERIAN RIVERS), OR IS POLLUTED.
0.003%	IS THE TOTAL AMOUNT OF USABLE WATER OF THE TOTAL SUPPLY ON EARTH.

HOW WATER IS USED BY AMERICANS:

DIRECT PERSONAL USE: 8% = 160 GALLONS PER DAY

30-40 GAL	= BATH
5 GAL/PER MIN.	= SHOWER
3 GAL	= SHAVING (WATER ON)
20-30 GAL	= WASH CLOTHES
10 GAL	= WASH DISHES
8 GAL	= COOKING
3-5 GAL	= FLUSH TOILET
35 GAL/PER DAY	= LEAK IN TOILET
80 GAL	= SPRINKLE 8,000 SQ. FT. OF LAWN
8 GAL	= SCRUB AND CLEAN HOUSE
200,000 GAL	= FAUCET AND TOILET LEAKS IN N.Y. CITY

INDIRECT AGRICULTURAL USE: 33% OF TOTAL = 600 GALLONS PER DAY.

    40 GAL = ONE EGG
    80 GAL = ONE EAR OF CORN
   150 GAL = ONE LOAF OF BREAD
   230 GAL = ONE GALLON OF WHISKEY
   375 GAL = 5 LBS. OF FLOUR
2,500 GAL = 1 LB. OF BEEF

INDIRECT INDUSTRIAL USE: 59% OF TOTAL = 1,040 GALLONS PER DAY.

   720 GAL/PER PERSON/PER DAY = COOLING WATER FOR ELECTRICAL
                                     POWER PLANTS
   320 GAL/PER PERSON/PER DAY = INDUSTRIAL PRODUCTION THINGS.
   280 GAL   = SUNDAY PAPER
   300 GAL   = 1 LB. SYNTHETIC RUBBER
 1,000 GAL   = 1 LB. ALUMINUM
    35 GAL   = 1 LB. STEEL
  7-25 GAL   = 1 GALLON GASOLINE
100,000 GAL   = 1 CAR

COMPARE THE AVERAGE AMOUNT OF WATER USED BY AN AMERICAN WITH THE
AMOUNT USED BY A PERSON OF AN UNDERDEVELOPED COUNTRY:

1,800 GAL/PER DAY   = AMERICAN PERSON
   12 GAL/PER DAY   = UNDERDEVELOPED COUNTRY PERSON

NOW MAKE A LIST OF WAYS YOU CAN CONSERVE USABLE FRESH WATER. LIST
BOTH DIRECT AND INDIRECT WAYS.

SOURCE: MILLER, G. TYLER, JR., LIVING IN THE ENVIRONMENT: CON-
    CEPTS, PROBLEMS AND ALTERNATIVES. BELMONT, CA: WADSWORTH
    PUBLISHING COMPANY, INC., 1975. EXCELLENT TEACHER REFERENCE
    (PAGE 253)

    FILM: "WATER: OUR MOST VITAL RESOURCE", AVAILABLE FROM THE
    UNITED NATIONS ENVIRONMENT PROGRAM; NEW YORK, N.Y., 10017

FOCUS **HEALTH**

LESSON **DRYING FRUIT AND VEGETABLES**

(wheel diagram with segments: SCIENCE, PHYSICAL, SENSORY)

MATERIALS: AN ELECTRIC FOOD DRYER (FROM THE TOOL LIBRARY), VARIOUS FRUITS, VEGETABLES, AND HERBS (APPLES, CARROTS, ZUCCHINI, GREEN BEANS, SAGE, BASIL, PARSLEY, ETC.), KNIVES, CUTTING BOARDS. (AN OPTION MIGHT BE TO BUILD A SOLAR DRYER!)

TEACHER'S BACKGROUND: YOU MAY WANT TO SIMPLIFY THIS LESSON BY DRYING ONLY ONE OR TWO FOODS IN THE LESSON. MAKE SURE YOU POINT OUT ALL THE POSSIBILITIES OF VEGETABLES AND FRUITS WHICH CAN BE DRIED, THOUGH. DRYING WILL TAKE 24 HOURS OR MORE TO COMPLETE, DEPENDING ON WHAT FOOD YOU USE AND WHAT SIZE THE PIECES OR SLICES ARE. APPLES SHOULD BE LEATHERY; CARROTS AND SQUASH SHRIVELLED AND PAPERY; HERBS DRY ENOUGH TO CRUMBLE; TOMATOES AND JUICY FRUITS TOUGH AND LEATHERY.

DO: HAVE EACH CHILD PREPARE SOME KIND OF FOOD FOR DRYING. APPLES NEED TO BE SLICED VERY THIN, VEGETABLES IN THIN SLICES OR SMALL CHUNKS, HERBS IN SMALL SPRIGS. HAVE THE CHILDREN ARRANGE THE SLICED FOOD CLOSE TOGETHER BUT NOT TOUCHING ON THE SCREENS OR TRAYS OF THE FOOD DRYER. PUT THEM INSIDE AND TURN ON THE DRYER.

THINK: DISCUSS BRIEFLY THE SEASONAL ASPECT OF FOOD. MANY FOODS COME TO BE RIPE ALL AT ONCE, OR WITHIN A FEW WEEKS, AND SOMETIMES IT IS IMPOSSIBLE TO EAT IT ALL UP. TO HAVE FOOD YEAR-ROUND WE HAVE TO PRESERVE SOME OF IT FOR USE LATER. COVER BRIEFLY THE METHODS OF PRESERVING FOOD--FREEZING, CANNING, DRYING, OR ROOT STORAGE--AND TALK ABOUT THE PROCESSES OF EACH. FREEZING MEANS THAT ALL YOU NEED IS A PLASTIC CONTAINER OR BAG, BUT THAT YOU NEED TO HAVE A BIG FREEZER TO HAVE ENOUGH ROOM FOR A LOT OF FOOD. FREEZERS USE A LARGE AMOUNT OF ENERGY AS WELL. WITH CANNING, YOU HAVE MORE POSSIBILITIES FOR VARIETY IN PRESERVED FOODS. PICKLES, JAM, TOMATO SAUCE, AS WELL AS PLAIN FRUITS AND VEGETABLES. CANNING TAKES SOME ENERGY TOO, AND SUBSTANTIAL NUTRIENT LOSS OCCURS DURING THE COOKING. IT IS HARDER TO DO --WORKING WITH BIG POTS OF HOT WATER AND HOT JARS. DRYING IS THE EASIEST WAY TO PRESERVE FOOD. STORAGE IS EASY. ALL YOU NEED ARE JARS OR BAGS TO STORE THE FOOD IN. YOU CAN SNACK ON DRIED FRUITS AND VEGETABLES JUST AS THEY ARE, OR USE THEM IN SOUP, SAUCES OR CASSEROLES WHERE THEY ABSORB LIQUID AGAIN AND LOOK ALMOST LIKE THEY DID BEFORE.

EXPLAIN THAT DRYING WILL DESTROY VERY LITTLE OF THE NUTRIENTS IN THE FRUITS AND VEGETABLES. EXPLAIN THAT THE ONLY DIFFERENCE BETWEEN THE RAW FOOD AND THE DRIED FOOD IS THAT YOU HAVE TAKEN THE WATER OUT. THE CHILDREN MAY BE SURPRISED TO SEE THAT THE FOOD SHRINKS SO MUCH IN THE DRYER, AND THAT IT MAY CHANGE COLOR. THE FOOD WE EAT IS MADE MOSTLY OF WATER, ESPECIALLY JUICY FOODS LIKE PEACHES, ZUCCHINI AND TOMATOES. WHEN YOU BRING OUT THE DRIED FOODS,

REMIND THE CHILDREN THAT WHEN THEY PUT THEM INTO THE DRYER THEY WERE SO CLOSE TOGETHER THAT THEY WERE ALMOST TOUCHING.

COMMUNICATE: IN A DAY OR TWO, THE FOOD SHOULD BE DRY (HERBS TAKE LESS TIME) AND READY TO TAKE OUT OF THE DRYER. HAVE EVERYONE IN THE CLASS DESCRIBE HOW THE FOOD HAS CHANGED THROUGH THE DRYING PROCESS. WHAT DOES IT LOOK LIKE NOW? DOES IT LOOK LIKE THE FRUIT THAT IT IS? TASTE EACH FOOD AND DISCUSS HOW IT TASTES IN COMPARISON TO THE ORIGINAL FOOD. IF IT HAS A DIFFERENT KIND OF FLAVOR, IS IT STILL GOOD TO EAT? HAVE EVERYONE SHARE THEIR THOUGHTS. STORE ANY EXTRA FOOD IN PLASTIC BAGS OR JARS, WITH LABEL AND DATE.

FOLLOW-UP POSSIBILITIES: YOU CAN HAVE CONTINUOUS BATCHES OF FOOD DRYING IN THE DRYER AS LONG AS THERE IS PRODUCE FROM THE GARDEN OR SCHOOL FARM FRUIT TREES.

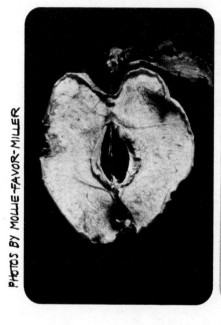

PHOTOS BY MOLLIE-FAVOR-MILLER

FOCUS: RECYCLING AND ENERGY

LESSON: WHAT IS A FOOD CHAIN?

MATERIALS: SEE <u>FOOD CHAIN GAME</u>.

TEACHER'S BACKGROUND: SEE <u>FOOD CHAIN GAME</u>. THE SUN PROVIDES ENERGY FOR ALL LIFE. THE SUN'S ENERGY IS CAPTURED BY INDIVIDUAL PLANTS AND TRANSFERRED TO ANIMALS THROUGH FOOD CHAINS. THIS LESSON HAS THE FOLLOWING OBJECTIVES: 1) TO LEARN TO PREDICT A LIKELY FOOD CHAIN FOR A GIVEN HABITAT, 2) TO UNDERSTAND THAT ENERGY IS LOST THROUGH BREATHING, HEATING AND MOVING, 3) TO UNDERSTAND THAT ENERGY IS LOST THIS WAY EACH TIME IT PASSES FROM ONE ORGANISM TO ANOTHER.

<u>FOOD CHAIN GAME</u> (FROM OUTDOOR BIOLOGY INSTRUCTIONAL STRATEGIES (OBIS), LAWRENCE HALL OF SCIENCE, UNIVERSITY OF CALIFORNIA, BERKELEY, CALIFORNIA 94720.

BACKGROUND: FEEDING RELATIONSHIPS ARE OFTEN DIFFICULT TO OBSERVE. IN THIS ACTIVITY, YOUNGSTERS GAIN SOME UNDERSTANDING OF THESE RELATIONSHIPS BY ASSUMING THE ROLES OF ANIMALS, PLAYING TAG, AND SIMULATING FEEDING RELATIONSHIPS. POPCORN IS SPREAD OVER A LAWN AREA. THE KERNELS REPRESENT PLANTS, WHICH ARE FOOD SOURCES FOR THE PLANT EATERS. SOME YOUNGSTERS PLAY GRASSHOPPERS (PLANT EATERS); SOME PLAY FROGS (WHICH EAT GRASSHOPPERS); AND SOME PLAY HAWKS (WHICH EAT FROGS). THE OBJECT OF THE GAME TO FOR EACH ANIMAL TO GET SOMETHING TO EAT WITHOUT BEING EATEN BEFORE THE "DAY" (FIVE MINUTES) IS OVER. IN NATURE, THE POPULATIONS OF PLANTS AND ANIMALS ARE USUALLY LARGE ENOUGH TO INSURE CONTINUATION OF THE SPECIES IF SOME ARE LOST. IN THIS GAME, POPULATIONS (POPCORN PLANTS, HOPPERS, FROGS, HAWKS) ARE SO SMALL, THAT THE SURVIVAL OF <u>EVEN ONE</u> OF EACH KIND WILL BE CONSIDERED AN INDICATION OF A "BALANCED", ONGOING COMMUNITY. YOU CAN REPEAT THIS GAME MANY TIMES DURING ONE ACTIVITY SESSION. WITH EACH REPEAT, ENCOURAGE THE YOUNGSTERS TO CHANGE RULES OF BEHAVIOR AND NUMBERS OF EACH KIND OF ANIMAL UNTIL A "BALANCE" IS ACHIEVED IN YOUR CORN-HOPPER-FROG-HAWK FOOD CHAIN.

OBJECTIVE: SURVIVE AS AN ANIMAL IN A MAKE-BELIEVE FOOD CHAIN BY GETTING ENOUGH TO EAT WHILE AVOIDING BEING EATEN YOURSELF.

MATERIALS: FOR EACH ANIMAL: SASHES ABOUT 7" X 40" (SEE PREPARATION SECTION)

ONE PLASTIC BAG "STOMACH" (SANDWICH BAG)

FOR THE GROUP: 4-5 QUARTS OF POPPED CORN
1 DATA BOARD, MARKING PEN
KITCHEN TIMER WITH BELL
ROLL OF 1-INCH MASKING TAPE

PREPARATION: AT LEAST TEN YOUNGSTERS ARE NEEDED FOR BEST RESULTS.

SASHES: MAKE THE SASHES FROM STRIPS OF CLOTH IN THREE DIFFERENT COLORS. HAVE ENOUGH SASHES FOR 3/4 OF THE GROUP TO BE GRASSHOPPERS, 1/3 TO BE FROGS, AND 1/3 TO BE HAWKS. THE UNBALANCED RATIO PROVIDES THE OPPORTUNITY TO CHANGE THE POPULATION NUMBERS IN THE GAME. MAKE THE SASHES ABOUT 7 INCHES WIDE AND 40 INCHES LONG. PREPARE "STOMACH" BAGS. PLACE A STRIP OF MASKING TAPE ACROSS THE SANDWICH BAG SO THE BOTTOM EDGE OF THE TAPE IS 1-1/2 INCH FROM THE BOTTOM OF THE BAG.

SITE SELECTION: A SECTION OF LAWN 50 FEET ON A SIDE IS SUFFICIENT. THE GROUP MAY DECIDE TO DESIGNATE POTENTIAL "HOME BASES" SUCH AS TREES, A WALKWAY, ETC., WHERE HOPPERS AND FROGS CAN HIDE, OR BE "SAFE".

DO: <u>FOOD CHAIN GAME</u>

1. DESCRIBE THE LIMITS OF THE GAMING AREA. SPREAD POPCORN OVER THE AREA. (SAVE A LITTLE FOR LATER.) TELL THE GROUP THAT YOU ARE DISTRIBUTING PLANTS THAT GRASSHOPPERS EAT.

2. HAND OUT A PLASTIC BAG AND A GRASSHOPPER SASH (ALL ONE COLOR) TO 1/3 OF YOUR GROUP. TELL THE KIDS TO PUT THEIR "FOOD" (POPCORN) INTO THEIR "STOMACHS" (BAGS) WHEN THE GAME STARTS.

3. HAND OUT A BAG AND A FROG SASH TO A SECOND 1/3 OF THE GROUP AND HAWK SASHES TO THE LAST 1/3. WHEN THE GAME STARTS, FROGS WILL TRY TO CAPTURE (TAG) HOPPERS, AND THE HAWKS WILL PURSUE FROGS. WHEN A FROG CAPTURES A HOPPER, THE HOPPER'S STOMACH CONTENTS ARE TRANSFERRED TO THE STOMACH OF THE FROG. WHEN THE HAWK CAPTURES A FROG, HE TAKES THE FROG'S WHOLE STOMACH. HAWKS DO NOT EAT HOPPERS IN THIS GAME.

4. STATE THE CHALLENGE. SET THE TIMER FOR FIVE MINUTES AND HOLLER "GO!". THE FIRST GAME USUALLY LASTS ONLY A FEW SECONDS WITH ONE OF TWO THINGS HAPPENING. HOPPERS ARE GOBBLED UP BEFORE THEY HAVE A CHANCE TO FORAGE, OR THE FROGS ARE GOBBLED UP AND HOPPERS CONTINUE TO EAT POPCORN AND GET FAT.

5. ANALYSIS: HOW MANY ANIMALS SURVIVE? FOR A HOPPER TO SURVIVE, POPCORN MUST FILL THE STOMACH BAG TO THE BOTTOM OF THE TAPE (1-1/2 INCH). FOR A FROG TO SURVIVE, POPCORN MUST FILL THE STOMACH BAG TO THE TOP OF THE TAPE (2-1/2 INCHES). HAWKS MUST HAVE THE EQUIVALENT OF ONE FROG WITH SUFFICIENT FOOD TO SURVIVE. IF AT LEAST ONE OF EACH KIND OF ANIMAL SURVIVES, YOU HAVE AN ONGOING FOOD CHAIN. RETURN THE POPCORN TO THE ACTIVITY AREA AFTER EACH GAME.

6. INSTANT REPLAY: LEARNING BY MAKING RULE VARIATIONS. ASK FOR SUGGESTIONS ON RULE CHANGES THAT MIGHT RESULT IN MORE OF A BALANCE AFTER THE FIVE-MINUTE DAY. USUALLY <u>ONE</u> RULE IS CHANGED FOR EACH REPLAY. WHEN YOU HAVE SETTLED ON YOUR NEW RULES, PLAY AGAIN. SUGGEST THESE CHANGES IF THE KIDS CAN'T OFFER ANY:

A. CHANGE THE NUMBER OF HOPPERS AND/OR FROGS AND/OR HAWKS.

B. LET EACH HOPPER COME BACK AS ANOTHER HOPPER ONCE AFTER BEING CAPTURED AND TRANSFERRING "STOMACH" CONTENTS.

C. PROVIDE A "SAFETY ZONE" FOR FROGS AND/OR HOPPERS WHERE THEY CAN BE SAFE.

D. TIMED RELEASES. LET HOPPERS GO FIRST TO FORAGE UNMOLESTED. ONE MINUTE LATER RELEASE THE FROGS, AND LATER THE HAWKS.

E. SPREAD OUT MORE POPCORN.

NOTE: YOU MAY WANT TO ELIMINATE BICKERING OVER WHO WILL BE WHICH ORGANISM BY DRAWING MARKERS FROM A HAT TO ASSIGN ROLES FOR REPLAYS.

THINK: INTRODUCING FOOD CHAINS: ASK THE PARTICIPANTS IF THEY KNOW WHAT MICE EAT AND WHAT EATS MICE. "MICE EAT SEEDS AND SNAKES EAT MICE", THEY MAY RESPOND. DIAGRAM THE RELATIONSHIP THEY DESCRIBE AND INTRODUCE IT AS A FOOD CHAIN. ARROWS POINT IN THE DIRECTION THAT THE FOOD GOES. FOR EXAMPLE: SEEDS ⟷ MICE ⟶ SNAKES.

ASK THE KIDS IF THEY CAN THINK OF OTHER FOOD CHAINS, INCLUDING A FOOD CHAIN THAT CONTAINS MAN.

COMMUNICATE: AFTER EACH GAME, ANALYZE THE RESULTS. HOW MANY HOPPERS GOT A FULL STOMACH? HOW MANY FROGS? THE HAWKS?

ENCOURAGE YOUNGSTERS TO COMPARE GAME RESULTS AFTER EACH RULE CHANGE, AND TO COMMENT ON HOW THE GAME "BALANCE" COMPARES WITH BALANCE IN THE REAL WORLD. IN NATURE'S BALANCE, THERE ARE MORE PLANTS THAN PLANT EATERS AND MORE PLANT EATERS THEN ANIMAL EATERS. YOU MIGHT WISH TO GRAPHICALLY REPRESENT THE RESULTS ON YOUR DATA BOARD.

--WHAT WOULD HAPPEN IF THERE WERE ONLY HALF AS MANY POPCORN PLANTS? WHAT WOULD HAPPEN TO THE ANIMAL THAT DEPENDS ON THOSE PLANTS?

--IF THERE WERE NO FROGS, WHAT WOULD HAPPEN TO THE PLANT POPULATION? THE HOPPER POPULATION? THE HAWK POPULATION?

--DO HAWKS NEED PLANTS TO SURVIVE? EXPLAIN!

--CAN YOU DESCRIBE SOME FOOD CHAINS THAT YOU ARE PART OF?

--ARE THERE ANY PLANTS OR ANIMALS THAT ARE NOT PART OF ANY FOOD CHAINS?

FOCUS: RECYCLING

LESSON: NATURE'S RECYCLING PLAN

MATERIALS: PICTURES OF PLANTS AND ANIMALS (INCLUDING DECOMPOSERS), PAPER (LARGE PIECE), BLACKBOARD, ERASER, COLORED CHALKS, COLORED PAPER, SCISSORS, GLUE, CRAYONS, YARN

TEACHER'S BACKGROUND: REFER TO "WHAT IS A FOOD CHAIN" LESSON PLAN. THIS LESSON PLAN IS TO HELP THE STUDENT IDENTIFY THE ROLE OF PLANTS, HERBIVORES, CARNIVORES AND DECOMPOSERS IN THE FOOD CHAIN. TO UNDERSTAND THAT THE ACTION OF DECOMPOSERS COMPLETES THE NUTRIENT CYCLE LINKING PLANTS, HERBIVORES AND CARNIVORES FOUND IN A VARIETY OF HABITATS.

HUMAN AND ANIMAL RECYCLING OF ORGANIC WASTES COMPLETES THE NATURAL NUTRIENT CYCLES AND IMPROVES SOIL STRUCTURE AND FERTILITY.

DO: AFTER PLAYING THE FOOD CHAIN GAME, HAVE STUDENTS CHOOSE A HABITAT AND WITH PICTURES, PASTE AND CONSTRUCTION PAPER, DEPICT SOME OF THAT HABITAT'S POSSIBLE FOOD CHAINS. AFTER THESE DIAGRAMS ARE MADE, ASK STUDENTS TO IDENTIFY THE PLANTS (PRODUCERS), PLANT EATERS (HERBIVORES), ANIMAL EATERS (CARNIVORES) AND DECOMPOSERS. ASK STUDENTS TO DESCRIBE WHAT THEY THINK THE HABITAT WOULD BE LIKE WITHOUT DECOMPOSERS (FEWER NUTRIENTS WOULD BE RECYCLED).

THINK: LOOK AT A FARM AS A HABITAT. IDENTIFY THE PLANTS, HERBIVORES, CARNIVORES, AND DECOMPOSERS THAT MIGHT BE FOUND IN A FARMER'S FIELD.

1. WHICH WAYS CAN A FARMER USE THE CARNIVORES TO REGULATE HERBIVORE POPULATIONS? WHAT TECHNIQUES DO FARMERS COMMONLY USE TO RID THEIR FIELDS OF HERBIVORES?

2. HOW CAN A FARMER USE DECOMPOSERS TO ENRICH HIS/HER SOIL? WHICH OTHER TECHNIQUES DO FARMERS COMMONLY USE TO IMPROVE THEIR SOIL'S FERTILITY AND TILTH (TEXTURE)?

3. COMPARE THE NUTRIENT CYCLES IN A FARMER'S FIELD USING THE CYCLES YOU IDENTIFIED IN THE OTHER HABITATS.

COMMUNICATE: SEE DO ABOVE.

FOLLOW-UP: TOUR OF AMITY'S GREENHOUSE AND GARDEN.

FOCUS: ENERGY

LESSON: OILING THE HUMAN FOOD CHAIN

TEACHER'S BACKGROUND: HUMANS ARE UNIQUE IN THEIR ABILITY TO TAP INTO A SOURCE OF ENERGY (STORED FOSSIL FUELS) UNAVAILABLE TO ALL OTHER CREATURES. THE OBJECTIVES OF THIS LESSON ARE: 1) TO COMPARE THE ENERGY INVOLVED IN NON-HUMAN FOOD CHAINS WITH THAT INVOLVED IN HUMAN FOOD CHAINS, AND 2) TO UNDERSTAND HOW HUMAN FOOD CHAIN DEPENDS UPON FOSSIL FUELS.

DO: AFTER PLAYING THE FOOD CHAIN GAME, HAVE STUDENTS MAKE A MENU FOR A MEAL THEY WOULD LIKE TO PREPARE. HAVE GROUPS OF STUDENTS CHOOSE THE INGREDIENTS NEEDED TO MAKE THE MEAL.

THINK: 1. IN THE CLASSROOM: A) ON A MAP LOCATE WHERE ALL THE INGREDIENTS WERE PRODUCED, B) LIST THE EQUIPMENT USED IN THE PRODUCTION OF THE FOODS, C) LIST ALL THE FUELS USED TO OPERATE THE EQUIPMENT, AND D) LIST ALL THE RESOURCES (METAL, PAPER, ETC.) USED TO PRODUCE THE FOOD.

2. COMPARE THESE LISTS WITH LISTS STUDENTS DREW UP FOR AN ANIMAL'S FOOD CHAIN. HOW ARE THESE LISTS THE SAME? HOW DO THEY DIFFER?

3. SHOW THE FILM "TOAST" WHICH DISCUSSES WHAT ENERGY WOULD BE NEEDED TO PRODUCE A PIECE OF TOAST. FILM "TOAST" BY DAN HOFFMAN IS AVAILABLE FROM EARTH CHRONICLES, 1714 N.W. OVERTON, PORTLAND, OREGON; OR CALL EDUCATIONAL SERVICE DISTRICT 121, 242-9400. (14 MINUTES)

COMMUNICATE: HAVE STUDENTS DO THE FOLLOWING AT HOME OR IN THE CLASSROOM.

1. IN THE CENTER OF THE PAGE, DRAW A PICTURE OF ONE OF THE INGREDIENTS.

2. ASK THEIR PARENTS WHERE THE INGREDIENT CAME FROM OR FIND OUT FROM THE ENCYCLOPEDIA.

3. AROUND THE EDGE OF THE PAGE, HAVE THEM DRAW SOME OF THE EQUIPMENT NECESSARY TO GET THE INGREDIENT TO THEIR HOUSE. SOME OF THE THINGS TO THINK ABOUT ARE:

--HOW WERE THE PLANTS GROWN?

--WHAT FOOD WAS FED TO ANIMALS THAT YOU EAT?

--HOW WAS THE FOOD PREPARED?

--WAS IT PACKAGED? WHERE DID THE PACKAGES COME FROM?

--WAS THE FOOD TRANSPORTED? HOW?

4. HAVE THE STUDENTS SHOW THEIR PICTURE TO THEIR PARENTS. CAN THEY THINK OF SOME OTHER STEPS INVOLVED IN BRINGING THE FOOD TO THE HOME? HAVE STUDENTS SHARE THEIR DRAWINGS WITH THE CLASS.

PRODUCTION OF FOODS BY FIRST, SECOND, AND THIRD LARGEST PRODUCERS

	FIRST	SECOND	THIRD
WHEAT	U.S.S.R.	U.S.A.	CHINA (PRC)
CORN	U.S.A.	U.S.S.R.	BRAZIL
BARLEY	U.S.S.R.	U.S.A.	UNITED KINGDOM
RICE	CHINA (PRC)	INDIA	PAKISTAN
POTATOES	U.S.S.R.	POLAND	EAST GERMANY
TOBACCO	U.S.A.	CHINA (PRC)	INDIA
COCOA	GHANA	NIGERIA	BRAZIL
TEA	INDIA	CEYLON	CHINA (PRC)
COFFEE	BRAZIL	COLUMBIA	IVORY COAST
APPLES	FRANCE	U.S.A.	ITALY
BANANAS	BRAZIL	ECUADOR	INDIA
CITRUS FRUIT (ORANGES, ETC.)	U.S.A.	BRAZIL	SPAIN
GRAPES	ITALY	FRANCE	SPAIN
SOYA BEANS	U.S.A.	CHINA (PRC)	BRAZIL
SUNFLOWER SEED	U.S.S.R.	ARGENTINA	ROMANIA
PEANUTS	INDIA	CHINA (PRC)	NIGERIA
MILK	U.S.S.R.	U.S.A.	EAST GERMANY
CHEESE	U.S.A.	FRANCE	U.S.S.R.
BEEF	U.S.A.	U.S.S.R.	ARGENTINA
MUTTON/LAMB	U.S.S.R.	AUSTRALIA	CHINA (PRC)
PORK	CHINA (PRC)	U.S.A.	U.S.S.R.
SUGAR (CANE)	CUBA	BRAZIL	INDIA

FOCUS: **ENERGY**

LESSON: **HOW FAR DID YOUR BREAKFAST TRAVEL?**

MATERIALS: WORLD MAP OR BLACKBOARD, YARN, MAGAZINES WITH PHOTOS OF FOOD, SCISSORS, TACKS.

TEACHER'S BACKGROUND: DEMAND FOR OUT-OF-SEASON CROPS, A WIDE VARIETY OF FOODS, AND THE INCREASING CENTRALIZATION OF DISTRIBUTION SYSTEMS INTENSIFY THE USE OF ENERGY IN THE FOOD SYSTEM.

THIS LESSON WILL BE FOCUSING ON LOCATING THE ORIGINS OF THE FOODS STUDENTS EAT; SHOWING WHICH FOODS ARE GROWN LOCALLY AND WHICH ARE NOT; IDENTIFYING AND PREPARING A MEAL OF FOODS GROWN OR RAISED WITHIN OUR GEOGRAPHIC AREA (I.E., LOCAL, OR STATE, OR U.S.)

DO: ASK THE CHILDREN TO MAKE A COLLAGE OF THE FOODS THEY ATE FOR BREAKFAST.

THINK: 1. USING A LARGE WORLD MAP (OR BLACKBOARD) FIND OUT WHERE THESE FOODS ORIGINATED. HAVE A LARGE CUT-OUT OF OREGON STATE. SELECT FOODS THAT ARE NOT HIGHLY PROCESSED BUT COMPOSED OF ONE MAIN INGREDIENT; I.E. BANANAS--SOUTH AMERICA; GRAPEFRUIT--TEXAS, CALIFORNIA, FLORIDA; ORANGE JUICE--TEXAS, CALIFORNIA, FLORIDA. USING YARN, MAKE CONTACT BETWEEN OUR TOWN AND THE ORIGIN OF THE FOOD.

2. DISCUSS WITH THE CLASS HOW FAR SOME FOODS TRAVEL TO GET TO US, AND HOW SOME FOODS CAN BE FOUND CLOSER TO HOME. WHAT FOODS WE MIGHT HAVE TO DO WITHOUT IF FUELS TO RUN TRAINS, PLANES, TRUCKS BECOME LESS AVAILABLE.

COMMUNICATE: 1. ASK CHILDREN TO BRING IN PICTURES OF FAVORITE FOODS. PUT ON DISPLAY BOARD. DISCUSS WHAT FOODS WE DO NOT HAVE REGIONALLY. WHAT FOODS ARE WE USED TO OR DEPEND UPON THAT ARE NOT LOCALLY AVAILABLE? ARE THEY NECESSARY TO OUR DIET OR ARE THEY A "LUXURY"?

2. INVITE A FARMER (OR SPEAKER FROM LANE COUNTY EXTENSION OFFICE) TO TALK ABOUT WHAT KINDS OF FOOD ARE AVAILABLE IN OUR REGION AND WHEN THEY ARE IN SEASON.

3. HAVE THE CHILDREN MAKE SMALL BOOKLETS USING PICTURES (EITHER FROM MAGAZINES OR SELF-DRAWN) OF LOCAL FOODS.

4. PLAN AND PREPARE A MEAL USING LOCAL FOODS.

FOCUS: **ENERGY AND RECYCLING**

LESSON: **VISIT TO AMITY**
SOLAR GREENHOUSE, WINDPUMP, AND GARDENS, (FISH TOO!!)

MATERIALS: BRING A COTTAGE CHEESE CONTAINER TO CARRY EARTHWORMS.

TEACHER'S BACKGROUND: THIS FIELD TRIP WILL TIE TOGETHER THE PREVIOUS LESSON PLANS ON ENERGY, FOOD CHAINS, AND FOOD PRODUCTION. PREPARE STUDENTS BY HIGHLIGHTING PAST LESSONS. TELL THEM THEY WILL SEE HOW RENEWABLE RESOURCES--SUN, WIND, WATER--ARE USED TO RAISE PLANTS AND ANIMALS (FISH, EARTHWORMS). PLEASE CALL AMITY FOUNDATION (484-7171) TO ARRANGE THE VISIT AND GET FURTHER BACKGROUND. CALL TWO WEEKS AHEAD OF TIME.

DO: VISIT AMITY'S GREENHOUSE.

--SOLAR GREENHOUSE

--MICROSCOPE: LOOK AT TINY ALGAE AND AQUATIC ANIMALS

--WINDMILL: USED TO PROVIDE OXYGEN TO THE FISH

--AQUACULTURE: WATER USED IN GREENHOUSE AS HEAT STORAGE AND TO RAISE FISH

--GARDENS: INSIDE GREENHOUSE, AND OUTSIDE RAISED BEDS

--EARTHWORMS: NATURE'S SOIL TILLER

THINK: THE TOUR LEADERS WILL EXPLAIN HOW EACH SYSTEM WORKS AND WHY IT IS NECESSARY.

COMMUNICATE: HAVE THE STUDENTS DRAW ONE OF THE SYSTEMS THAT THEY SAW AT AMITY, INCLUDING ALL OF THE PARTS OF THE CYCLE INVOLVED IN EACH. TALK ABOUT THESE IN CLASS AS A GROUP. STRESS THE NECESSITY OF EACH PART OF THE CYCLE. IF ONE PART WERE LEFT OUT, THE CYCLE COULDN'T WORK.

FOLLOW-UP LESSON: THE LITTLE SOIL TILLER.

FOCUS: RECYCLING

LESSON: THE LITTLE SOIL TILLERS

MATERIALS: TWO TERRARIUMS OR GLASS CONTAINERS, CLAY, SAND, SOIL, LEAVES, EARTHWORMS (FROM AMITY'S WORM BINS)

TEACHER'S BACKGROUND: THE EARTHWORM CONSUMES ITS OWN WEIGHT IN LEAVES AND GRASS EVERY 24 HOURS AND IS A HUMUS FACTORY RETURNING OTHERWISE UNUSABLE NUTRIENTS TO THE SOIL. WORM CASTINGS CONTAIN MINERALS. WORM TUNNELS OPEN SPACES FOR ROOTS AND WATER. WORMS CANNOT HEAR OR SEE BUT THEY ARE SENSITIVE TO LIGHT AND VIBRATIONS. THEY WILL EAT DEAD LEAVES, TWIGS AND INSECTS AND IN TURN, ARE EATEN BY SOME BIRDS AND BURROWING ANIMALS.

HUMAN AND ANIMAL RECYCLING OF ORGANIC WASTES COMPLETES THE NATURAL NUTRIENT CYCLES AND IMPROVES SOIL STRUCTURE AND FERTILITY.

DO: SET UP TWO CONTAINERS AS SHOWN IN THE DIAGRAM.

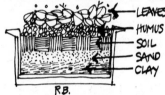

LEAVES
HUMUS
SOIL
SAND
CLAY

R.B.

PLACE WORMS COLLECTED AT AMITY IN ONLY <u>ONE</u> OF THE CONTAINERS. KEEP THE SOIL OF THIS CONTAINER SLIGHTLY MOIST.

COMMUNICATE: DESCRIBE AND DRAW PICTURES OF THE SOILS OF BOTH CONTAINERS. ASK STUDENTS TO PREDICT WHAT WILL HAPPEN TO THE SOILS.

COVER THE WORM CONTAINER WITH BLACK CLOTH OR PAPER. ALLOW AIR HOLES FOR BREATHING.

AFTER 3-4 DAYS, REMOVE THE PAPER AND OBSERVE THE WORMS AND SOIL LAYERS. COMPARE THE SOIL IN THE TWO CONTAINERS.

THINK: HOW IS THE WORM LIKE A TILLER? HOW IS IT ABLE TO RECYCLE MATERIALS IN THE ENVIRONMENT?

CONTINUE TO OBSERVE AND ENJOY THE ACTIVITIES OF THE WORMS. PLACE WORMS IN THE SECOND CONTAINER AND WATCH THEM PLOW THE SOIL.

FOCUS: NUTRITION

LESSON: GRINDING WHEAT BERRIES INTO FLOUR

MATERIALS: STALKS OF WHEAT, 3-4 CUPS OF WHEAT BERRIES, TWO MEDIUM MIXING BOWLS, PICTURES OF WHEATFIELDS, CASSETTE TAPE RECORDERS, LOTS OF SMALL JARS WITH LIDS

TEACHER'S BACKGROUND: CHILDREN LOVE TO TURN THE HANDLE OF A MILL AND WATCH HOW FLOUR IS MADE. BORROW A MILL OR ASK THE LOCAL TOOL STORES IF THEY WILL POSSIBLY RENT ONE FOR AN AFTERNOON.

DO: FIND A SUITABLE PLACE FOR A HAND TURNING GRAIN MILL AND SOME CURIOUS FRIENDS TO HELP PUT THE PIECES TOGETHER. MENTION THAT YOU NEED SOME FLOUR AND CAN GRIND YOUR OWN. ONE AT A TIME AND WITH YOUR FRIENDS, EXAMINE WITH THE SENSES, STALKS OF WHEAT, A HANDFUL OF WHEAT BERRIES AND A LOAF OF BREAD. EACH PERSON CAN HAVE THE OPTION OF TURNING THE HANDLE OF THE MILL, FEELING THE TENSION OF THE MACHINE, HEARING THE SOUND OF GRINDING, SMELLING, AND TASTING THE TEXTURE OF THE FLOUR. COUNT THE TURNS IN CADENCE.

THINK: THINK OF THE GRADUAL TRANSFORMATION OF WHEAT BERRIES FROM THERE TO HERE TO THERE--POSSIBLY WITH TAPE RECORDERS LISTENING, THE CHILDREN CAN DISCUSS WAYS TO MIX THE WHEAT INTO BREAD--A TIME FOR RECIPES. IF POSSIBLE, SHOW THOSE INTERESTED A GARDEN OF WHEAT GRASS GROWING, OR A BOWL OF WHEAT PASTE ALL SET FOR PAPIER MACHE. COUNT THE NUMBER OF PARTICIPANTS WHO WANT A STASH OF FLOUR AND DIVIDE THE FLOUR INTO HALVES, AND HALVES AGAIN, UNTIL EVERYONE HAS HIS SHARE.

COMMUNICATE: WHILE TAKING THE MILL APART, AND WITH TAPE RECORDER LISTENING, OR A LITERATE FRIEND RECORDING, DISCUSS WHAT WE MIGHT DO WITH THE FRESH FLOUR AND WHEN. DO WE WANT TO BAKE A LOAF OF BREAD, FINGER-PAINT, FLOUR ON GLUE PICTURES, PAPIER MACHE...TODAY? SOME GROUPS MAY CHOOSE TO COMBINE THEIR FLOUR TO MAKE A BUNCH OF BISCUITS.

FOLLOW-UP POSSIBILITIES: INCLUDE GRINDING PEANUTS, SOYBEANS, RICE. TAKE THE TIME TO MIX AND KNEAD A BATCH OF BREAD DOUGH. SCULPT THE DOUGH OR BAKE A LOAF. TREAT THE CLASS NEXT DOOR TO HOMEMADE BREAD AND PEANUT BUTTER...GROW WHEAT GRASS ON A RHYTHMIC BASIS AND CHART THE GROWTH...DRAW PICTURES OF FIELDS OF WHEAT, A MACHINE TO GRIND FOOD, A LOAF OF BREAD YOU WOULD BAKE FOR A FRIEND...FIND THE WORD "WHEAT" IN THE DICTIONARY, IN THE ENCYCLOPEDIA...TRY TO GROW SOME WHEAT.

WINTER

FOCUS: FARMING · SEEDS

LESSON: ALFALFA SPROUTING

FOLLOW-UP POSSIBILITIES: HARVEST THE SPROUTS AND SERVE A MID-
MORNING SNACK TO YOUR FRIENDS. TRY SPROUTING RADISH SEEDS,
WHEAT BERRIES, LENTILS, BEANS AND SUNFLOWER SEEDS. PLANT
THE YOUNG SEEDLINGS IN A WIDE-MOUTH JUG WITH PLASTIC OVER
THE TOP, AND YOU'LL CREATE AN INSTANT GREENHOUSE. LOOK AT
THE ALFALFA SEEDS THAT SOAKED OVERNIGHT AND NOTICE THE
CHANGES IN THE SEEDS AND THE WATER. LEARN AND SING "DOWN
WHERE THE SUNFLOWERS GROW". NOTE: SPROUTS ARE GREAT ON
CRACKERS, WITH CREAM CHEESE.

MATERIALS: ALFALFA SEEDS, PICTURES OF GROWING ALFALFA, ALFALFA
SPROUTS, SMALL BOWL, 4 JARS, CHEESE CLOTH, RUBBER BANDS,
MEASURING CUP, WATER, PAPER, PENCILS, TAPE, TAPE RECORDERS,
RADISH SEEDS, WHEAT BERRIES, LENTILS, BEANS, SUNFLOWER SEEDS.

TEACHER'S BACKGROUND: ANOTHER POSSIBILITY HERE IS TO GROW LARGE
BEANS (KIDNEY, PINTO) IN JARS WITH A MOISTENED PAPER TOWEL.
THE ROOTS AND STEMS SHOW UP BEAUTIFULLY.

DO: FIND A COMFORTABLE WORK SPACE AND SOME WILLING FRIENDS AND
LOOK AT ALFALFA SEEDS, AND SOME PICTURES OF ALFALFA GROWING
IN PASTURES. ALSO EXAMINE SOME SPROUTS AND TASTE THE TEX-
TURES AND FLAVORS. WASH OUT 4
JARS AND ADD 1/3 CUP OF ALFALFA
SEEDS TO 3 JARS. WITH RUBBER
BANDS, ATTACH SOME CHEESE CLOTH
TO THE OPEN END. POUR WATER INTO
2 OF THE JARS. ALLOW TO SIT FOR
10-15 MINUTES, AND CAREFULLY
RINSE. PUT ONE OF THE RINSED
SEED JARS ON ITS SIDE ON A
SUNNY INDOOR WINDOW LEDGE. PUT
THE OTHER JAR WITH WET SEEDS IN A
DARK, WARM LOCATION. INTO THE
THIRD JAR ADD SOME ALFALFA SEEDS
AND NO WATER AND PLACE THE JAR
NEXT TO THE ONE IN THE DARK WARM
SPACE. LABEL EACH JAR AND MAKE
A SPROUT CHART FOR 6 DAYS. IN
THE FOURTH JAR, ADD 1/3 CUP
ALFALFA SEEDS AND POUR IN A CUP
OF WATER AND ALLOW TO SIT IN A
SUNNY INDOOR SPACE.

THINK: THINK ABOUT WHAT WILL HAPPEN TO EACH JAR AND RECORD, WITH
TAPE RECORDER, OR PENCIL AND PAPER. WHAT DO PLANTS NEED TO
GROW WELL? WHAT OTHER KINDS OF SEEDS COULD SPROUT? WHAT IS
SO GREAT ABOUT SPROUTS? DO OTHER SEEDS FOLLOW SIMILAR
SPROUTING PATTERNS? HOW LONG WILL IT TAKE FOR THE SEEDS TO
SPROUT?

COMMUNICATE: SHOW YOUR FRIENDS THE JARS, THE CHARTS, AND THE PRE-
DICTIONS. EXPLAIN TO THEM WHAT PROCESS HAS BEEN USED SO FAR,
AND FIND PEOPLE WHO WOULD LIKE TO HELP IN THE DAYS TO COME.
DRAW PICTURES OF SPROUTING SEEDS AND LABEL THE VARIOUS PARTS.
SPECULATE ON WHAT WILL HAPPEN TO SPROUTS THAT NEVER SEE THE
SUN.

FOCUS: FARMING

LESSON: 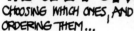VEGETABLES WE CAN GROW

CHOOSING WHICH ONES, AND ORDERING THEM...

FOUND FREE SEEDS THROUGH A GARDEN STORE OR ANOTHER SOURCE. IF SO, YOU MAY WANT TO LET THE CLASS SORT THROUGH THE SEEDS TO FIND THE ONES YOU WILL BE NEEDING FOR THE GARDEN.

MATERIALS: LOCAL SEED CATALOGS, VARIOUS VEGETABLES (TOMATO, CARROT, HEAD OR LEAF LETTUCE, GREEN PEPPER)

TEACHER'S BACKGROUND: SOME VEGETABLES WILL BE EASIER AND MORE PRACTICAL TO GROW IN THE SCHOOL GARDEN THAN OTHERS, THOUGH IT IS GOOD FOR CHILDREN TO KNOW ALL OF THE VEGETABLES POSSIBLE TO PLANT AND HOW THEY GROW. MAKE SURE THAT THE CHILDREN CHOOSE A WIDE VARIETY OF VEGETABLES FOR THEIR GARDEN.

DO: ASK THE CHILDREN TO IDENTIFY THE VEGETABLES YOU HAVE ON HAND. IDENTIFY THE PARTS OF THE VEGETABLE THAT WE EAT. THROUGH DISCUSSION, DEVELOP THE UNDERSTANDING THAT THE CARROT IS A ROOT OF A PLANT, THE LETTUCE CONSISTS OF LEAVES AND SOME STEM, WHILE THE PEPPER OR TOMATO IS A FRUIT OF A PLANT. HAVE THE CHILDREN MAKE A LIST OF ALL THE VEGETABLES THAT THEY KNOW OF. CLASSIFY THE VEGETABLES ON THE LIST INTO ROOT, LEAF, OR FRUIT TYPES.

LOOK AT SEED CATALOGS AND SEE IF THERE ARE ANY VEGETABLES THAT THEY HAVE NEVER EATEN. STRESS THAT THOUGH THERE ARE MANY VEGETABLES PICTURED IN THE CATALOG, THERE ARE ONLY SO MANY THAT WE CAN FIT OR GROW IN OUR SCHOOL GARDEN.

THINK: REFER AGAIN TO THE VEGETABLES. ASK THE CHILDREN IF THEY CAN TELL WHICH VEGETABLES TOOK THE LONGEST TO GROW. DIS-CUSS THE REASONS FOR THEIR CHOICES. EXPLAIN WITH DISCUSSION THAT SOME CROPS CAN BE GROWN IN 25-30 DAYS, AND OTHER CROPS MAY TAKE AS LONG AS 100-150 DAYS. THE CHILDREN SHOULD KNOW THAT RADISHES ARE ONE OF THE FASTEST GROWING CROPS, PRO-DUCING EDIBLE PARTS IN 25-30 DAYS AFTER THEY ARE SOWN IN THE GARDEN. GREEN BEANS WILL TAKE FROM 50-60 DAYS, SEED TO HARVEST. TOMATOES WHICH ARE NOT SEEDED DIRECTLY OUTDOORS, BUT ARE SET OUT AS TRANSPLANTS, TAKE FROM 70-90 DAYS AFTER BEING SET OUT IN THE GARDEN BEFORE THEY PRODUCE FRUIT.

COMMUNICATE: AS A GROUP, DISCUSS WHICH VEGETABLES YOU WOULD LIKE TO GROW IN THE SCHOOL GARDEN. MAKE A LIST OF EVERYONE'S CHOICES AND DISCUSS WHY EACH VEGETABLE IS A GOOD CHOICE IN TERMS OF TIME OF PLANTING UNTIL HARVEST, NUTRITION, TYPE OF PLANT, ETC. TRIM OR ADD TO THIS LIST UNTIL MOST EVERYONE IS SATISFIED.

FOLLOW-UP POSSIBILITIES: IF TIME ALLOWS, HAVE THE CHILDREN FILL OUT AN ORDER BLANK TO THE NURSERY FOR SEEDS. YOU MAY HAVE

FOCUS:

LESSON **FEELING SOIL TEXTURES**

MATERIALS: SAMPLES OF CLAY, SILT, AND SANDY SOILS. ASK THE COUNTY EXTENSION AGENCY FOR THESE OR FOR INFORMATION ABOUT GETTING THEM. A SPRAY BOTTLE OF WATER, PAPER TOWELS, POSTER BOARD, FELT PENS

TEACHER'S BACKGROUND: SOILS ARE MADE UP OF MANY THINGS, SOME OF THEM BEING PARTICLES OF MINERALS THAT ARE OF DIFFERENT TEXTURES. THE TYPES OF PARTICLES ARE SAND, SILT AND CLAY. WITH BATCHES OF SOIL, THE DIFFERENCE SHOULD BE EASY TO SPOT. THE PARTICLE SIZE AFFECTS DRAINAGE OF THE SOIL AND ITS ABILITY TO HOLD NUTRIENTS.

DO: IN EACH STUDENT'S HAND, PUT A SMALL MOUND (1-2 TABLESPOONS) OF ONE OF THE THREE SOILS. DO THESE ONE AT A TIME TO AVOID CONFUSION. SPRAY A SMALL AMOUNT OF WATER INTO EACH CHILD'S HAND--ENOUGH TO MAKE A PASTE FROM THE SOIL. HAVE THE CHILDREN FEEL THIS AND SEE WHAT THEY CAN MAKE OUT OF IT. THE CLAY SOIL WILL BE VERY MUCKY AND WHEN DRIED A LITTLE, WILL FORM A LONG SNAKE IF ROLLED BETWEEN THE PALMS. TRY TO DO THIS WITH EACH SOIL. THE CLAY WILL DO THIS BEST. THE SILT SOIL WILL FEEL SLIPPERY--ALMOST SOAPY IN THE PALM. IT WILL NOT MAKE AS NICE OF A SNAKE, BUT CAN BE MADE TO PRODUCE ONE. THE SANDY SOIL WILL FEEL GRAINY AND ROUGH, AND WILL BARELY HOLD TOGETHER TO MAKE A SNAKE. ENCOURAGE THE CHILDREN TO BE ABLE TO TELL THESE SOILS APART.

THINK: DRAW ON A POSTER DURING CLASS OR BEFORE, THE PARTICLE SHAPES OF SAND, SILT AND CLAY. SAND PARTICLES ARE LARGE AND IRREGULARLY SHAPED. WHEN THEY ARE BUNCHED TOGETHER, THERE IS MUCH SPACE IN BETWEEN THEM. WE ARE ABLE TO FEEL INDIVIDUAL SAND PARTICLES.

SILT HAS THE TEXTURE OF TALCUM POWDER. THE PARTICLES ARE VERY FINE, AND CANNOT BE FELT INDIVIDUALLY. SILT WILL MIX WITH SOIL TO MAKE A NICE, LOOSE LOAM.

CLAY PARTICLES ARE MICROSCOPIC. WE CANNOT SEE THEM INDIVIDUALLY. THEY ARE FLAT SHAPED LIKE PAPER PAGES AND HAVE A LARGE SURFACE AREA OF WATER. WHEN WET, THEY WILL STICK TOGETHER AND WATER HAS DIFFICULTY SIFTING THROUGH THEM. CLAY PARTICLES ALSO HOLD ONTO CHEMICAL NUTRIENTS SO TIGHTLY THAT THEY ARE UNAVAILABLE TO PLANTS. CLAY IS NOT BAD TO HAVE IN THE SOIL, BUT IT SHOULD BE IN BALANCE WITH THE OTHER PARTICLES.

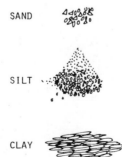

COMMUNICATE: HAVE THE CHILDREN BRING A SAMPLE OF SOIL FROM HOME AND, ONE AT A TIME, FEEL EACH CHILD'S SAMPLE AS A GROUP AND TALK ABOUT ITS TEXTURE AND WHAT SOIL PARTICLES YOU THINK ARE IN EACH SAMPLE.

FOCUS  SOILS

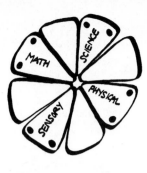

LESSON SOIL TYPES AND PLANTING

MATERIALS: THREE OR SO POTS OR PLANTING CONTAINERS OF EACH OF THE FOLLOWING: CLAY SOIL, SILT SOIL, SANDY SOIL, AND A BALANCED LOAM OF ALL THREE. ASK THE COUNTRY EXTENSION OFFICE ABOUT GETTING SOIL SAMPLES. MASKING TAPE, PENS, RADISH, BEAN OR SUNFLOWER SEEDS, WATERING CAN, PLANT GROWING LIGHTS OR AN OUTSIDE SUNNY SPOT, A WOODEN FLAT OR SHALLOW BOX.

TEACHER'S BACKGROUND: TO MAKE THIS A PARTLY MATH ORIENTED LESSON, FIGURE OUT THE AMOUNT OF SOIL TO GO INTO EACH POT SO THAT THIS CAN BE DONE BY THE CHILDREN IN MEASURED AMOUNTS. THIS LESSON WILL BE A CONTINUOUS ONE FOR A WEEK OR SO, UNTIL THE SEEDS GERMINATE. CHECKING ON THE SEEDS AND THE FINAL DIS- CUSSION CAN BE OF SHORTER TIMES.

DO: HAVE THE CHILDREN MEASURE INTO EACH POT OR CONTAINER, ENOUGH SOIL TO FILL IT ½ -1 INCH BELOW THE TOP. HAVE THE CHILDREN LABEL THESE POTS WITH THE SOIL THAT IS IN THEM. HAVE THEM PLANT 2 OR 3 SEEDS IN EACH POT. MAKE SURE THAT THE SEEDS AREN'T PLANTED TOO DEEP IN THE POTS. PLACE THE POTS OF SOIL AND SEEDS OUT IN THE SUN OR UNDER GROW LIGHTS AND WATER THEM WELL, ALLOWING THEM TO DRAIN. THESE WILL HAVE TO REMAIN WHERE THEY ARE UNTIL THE SEEDS BEGIN TO GERMINATE AND GROW. THE PROJECT WILL BE ONGOING UNTIL IT BECOMES OBVIOUS WHICH SEED- LINGS ARE SURVIVING AND WHICH ARE NOT.

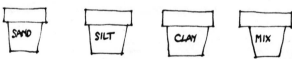

THINK: ASK THE CHILDREN TO PREDICT WHICH SEEDS WILL GROW THE BEST IN WHICH SOIL. ASK THEM TO SAY WHY THEY THINK SO. TALK ABOUT THE TEXTURES OF EACH SOIL--THE SAND IS LOOSE, THE WATER FELL RIGHT THROUGH IT, THE CLAY TOOK A LONG TIME TO DRAIN, AND SO DID THE SILT. THE MIXTURE WAS CRUMBLY AND DRAINED QUITE WELL. THE CLAY MIGHT HAVE A SHINY SURFACE AFTER WATERING AND DRAIN- AGE, AND MIGHT DRY WITH A HARD CRUST. HAVE THE CLASS KEEP AN EYE ON THE SOIL'S BEHAVIOUR AND THE GERMINATION OF THE SEEDS. WATER THESE POTS EVERY DAY.

COMMUNICATE: WHEN THE SEEDS GERMINATE AND IT BECOMES CLEAR WHICH ONES ARE DOING THE BEST, DISCUSS AS A GROUP WHY THIS MIGHT BE SO. THE SANDY AND SOIL MIXTURE POTS SHOULD HAVE THE BEST LOOKING SEEDLINGS. THIS IS BECAUSE THE ROOTS HAD ENOUGH ROOM TO GROW EASILY AND THERE IS ENOUGH DRAINAGE TO KEEP THE SEEDS FROM WASHING AROUND OR DROWNING. THE CLAY AND SILT SOILS WILL HAVE SMALLER SEEDLINGS, LESS, OR NONE COMPARED TO THE OTHERS. FEEL THE SOIL HERE. IT WILL PROB- ABLY BE DENSE AND COMPACTED, WITH LITTLE ROOM FOR THE SEED- LINGS TO GROW AND ROOT, AND PERHAPS A CRUST ON THE TOP PRE- VENTS THE FIRST LEAVES FROM EMERGING FROM THE SOIL. DISCUSS THE IMPORTANCE OF SOIL STRUCTURE FOR PLANTS AND ASK THE CHILDREN WHERE THEY WOULD FEEL MOST COMFORTABLE GROWING IF THEY WERE A SEED. ASK THEM TO DESCRIBE OR ACT OUT WHAT IT WOULD FEEL LIKE TO BE A SEED IN EACH TYPE OF SOIL.

FOLLOW-UP POSSIBILITIES: LOOK IN THE GARDEN AND SEE HOW THE SOIL THERE COMPARES TO YOUR TEST SOILS INSIDE. THERE MAY BE DIF- FERENCES IN THE TYPES OF SOIL IN YOUR GARDEN IF IT IS A LARGE ONE.

FOCUS:  FARMING

LESSON:  STARTING STARTS

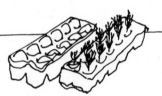

MATERIALS: POTTING SOIL, SEVEN EGG CARTON PLANTERS (SEE ACCOMPANY-
ING DIAGRAM), PACKETS OF RADISH, LETTUCE, SPINACH, AND CABBAGE
SEEDS, PAPER, PENCILS, TAPE, SPRAY BOTTLE, CASETTE TAPE RE-
CORDERS, RULERS.

TEACHER'S BACKGROUND" IF YOU HAVE A COLD FRAME, PLACE YOUR PLANTERS
IN IT TO SPROUT.

DO: GATHER TOGETHER THE MATERIALS AND THE CHILDREN IN SMALL GROUPS
OF THREE. PUT THE PLANTERS TOGETHER AND ADD SOIL TO EACH SEC-
TION UNTIL 2/3 FULL. EACH CHILD
SHOULD HAVE THREE SECTIONS OF EGG
CARTON PLANTER TO WORK WITH. IN
EACH SECTION, A CHILD CAN PUSH A
FINGERTIP ONE FINGERNAIL'S DEPTH
INTO THE SOIL. NOT TOO FAR NOW..
IN ONE OF THE SECTIONS, A CHILD
CAN PLACE ONE RADISH SEED. BE
SURE TO DRAW A DIAGRAM OF YOUR
PLANTER TO KEEP TRACK OF THE LO-
CATION OF SEEDS. IN EACH SECT-
ION, PLANT ONE SEED AT A TIME OF
LETTUCE, SPINACE AND CABBAGE.
REMEMBER TO MARK THESE ON THE
DIAGRAM TOO. MIST WATER OVER
THE SOIL IN THE PLANTERS UNTIL
IT IS WELL SOAKED. PLACE IN A
SUNNY SPACE, OR IN YOUR COLD FRAME
OUTSIDE.

THINK: DOES ANYONE WANT TO PREDICT WHICH SEEDS WILL EMERGE FIRST?
KEEP TRACK OF THE PREDICTIONS ON A CHART. CONSIDER GROWING
SOME STARTS WITHOUT ANY SUNLIGHT AT ALL. WHAT DO CHILDREN
THINK WILL HAPPEN?

COMMUNICATE: ENCOURAGE THE PARTICIPATING CHILDREN TO MAKE A REPORT
TO OTHERS IN THEIR CLASS OR NEARBY CLASSROOMS, DESCRIBING
WHAT'S BEEN HAPPENING WITH THE SEEDS AND MATERIALS SO FAR.
IF POSSIBLE, USE A CASETTE TAPE RECORDER TO KEEP TRACK OF
THE CHILDREN'S CONVERSATIONS. AN ARTICLE CAN BE TRANSCRIBED
FROM THE TAPES AND POSSIBLY BE PUBLISHED IN A SCHOOL FARMS
NEWSLETTER CREATED BY THE STUDENTS, OR IN THE SCHOOL'S NEWS-
PAPER. SHOW THE DIAGRAMS AND PREDICTIONS TO YOUR FRIENDS.
HELP EACH OF THE CHILDREN PUT TOGETHER GROWTH CHARTS TO
KEEP TRACK OF EACH PLANT'S GROWTH PATTERN. ENCOURAGE THE
USE OF RULERS AND GRAPH PAPER WHEN MAKING THE GROWTH CHARTS.

FOLLOW-UP POSSIBILITIES: BE SURE TO PLAN AHEAD WITH STARTS. IT
IS IMPORTANT TO BE READY TO TRANSPLANT THE SEEDLINGS TO
LARGER CONTAINERS SO THAT THE PLANTS DO NOT BECOME ROOT-
BOUND. BE SURE TO "HARDEN OFF" THE STARTS IN THE GARDEN
BEFORE FINALLY TRANSPLANTING THE STARTS INTO THE PREPARED
SOIL OF THE RAISED BEDS. REDUCE THE CHANCES OF SHOCK TO
THE PLANT BY USING COLD FRAMES OR CLOCHES AROUND THE STARTS
DURING COLD NIGHTS.

WITH ENOUGH STARTS, IT WOULD BE POSSIBLE TO SELL SOME OF THE
STARTS THAT YOU GROW, AT THE SATURDAY MARKET. CALL THE
SATURDAY MARKET OFFICE, 686-8885, TO SET THIS UP IN ADVANCE.

FOCUS: **SOILS**

LESSON: **CLAY IN SURFACE AND SUB-SURFACE SOILS**

MATERIALS: TWO MEDIUM-SIZED JARS, SHOVEL, 5-6 SPOONS, PAPER, PEN-CILS, TAPE, WATER, NON-SUDSING DETERGENT.

TEACHER'S BACKGROUND: PLAN AHEAD--THIS IS AN OVERNIGHT LESSON. FIX THE SOIL JARS ONE DAY AND EXAMINE THEM THE NEXT.

DO: GATHER SOME FRIENDS TOGETHER WITH TWO MEDIUM-SIZED JARS, A SHOVEL, AND SEVERAL SPOONS AND STEP OUTSIDE TO THE LOCA-TION OF A POSSIBLE GARDEN PLOT. DIG ENOUGH SURFACE SOIL TO FILL ONE JAR 1/3 FULL AND LABEL THE JAR "SURFACE SOIL". NEXT, USE THE SHOVEL TO DIG DOWN AT LEAST 6-8 INCHES. PUT ENOUGH OF THIS SUBSOIL INTO A JAR LABELED "SUBSOIL" TO FILL IT 1/3 FULL. BACK IN THE CLASSROOM, BE SURE TO BREAK UP ANY LUMPS, AND THEN FILL EACH JAR 2/3 FULL OF WATER. ADD 3 TABLESPOONS OF NON-SUDSING DETERGENT TO EACH JAR, PLACE THE LID TIGHTLY IN PLACE AND SHAKE THE JARS FOR AT LEAST 5-10 MINUTES, OR UNTIL ALL THE SOIL PARTICLES ARE BROKEN APART. PUT THE JARS IN A PLACE WHERE THEY WILL NOT BE DISTURBED FOR 24 HOURS.

VS.

THINK: WHY IS IT THAT THE COARSE SOIL PARTICLES ARE SETTLING QUICKLY? AFTER 24 HOURS, THE LAYERS OF SOIL WILL BE DIVIDED INTO COARSE SAND ON THE BOTTOM, FINE SAND ABOVE THE COARSE GRAINS, SILT ABOVE THE FINE SAND, AND CLAY ON TOP. WHY? HOW DO THE LAYERS OF SOIL IN EACH JAR COMPARE? WHY IS THERE AN APPARENT DIFFERENCE IN AMOUNTS OF CLAY, SILT, AND SAND IN THE TWO SAMPLES OF SOILS?

COMMUNICATE: PLACE A PIECE OF CARDBOARD OR STIFF PAPER BEHIND EACH JAR AND MARK THE LINES OF THE SEPARATE LAYERS. LABEL THESE TWO GRAPHS AND MEASURE THE THICKNESSES OF EACH LAYER. TRY TO EXPLAIN TO A SMALL GROUP WHY THE TWO SAMPLES OF SOIL HAVE DIFFERENT RELATIVE QUANTITIES OF SAND, SILT, AND CLAY.

FOLLOW-UP POSSIBILITIES: INTO A CLAY POT MARKED "SURFACE SOIL" ADD ENOUGH SURFACE SOIL TO PLANT A BEAN SEED. IN ANOTHER CLAY POT LABELED "SUBSOIL" ADD ENOUGH SUBSOIL TO PLANT ANOTHER BEAN SEED. WATER BOTH PLANTS ADEQUATELY AND ALSO PROVIDE GOOD AMOUNTS OF SUNLIGHT. HOW DO THE PLANTS COMPARE AFTER A FEW WEEKS? HOW CAN ANY APPARENT DIFFERENCES BE EXPLAINED?

FOCUS: **NUTRITION**

LESSON: **GOOD SOURCES FOR MAIN MINERALS**

MATERIALS: FOUR CARDBOARD BOXES, LOTS OF APPROPRIATE FOOD, FELT PENS, PAPER, PENCILS, TAPE RECORDER.

DO: LABEL FOUR MEDIUM-SIZED CARDBOARD BOXES WITH THE WORDS CALCIUM, PHOSPHORUS, IRON, IODINE. TAKE ONE PIECE OF FOOD FROM A LARGE BOWL, IDENTIFY IT AND USING THE CHART ACCOM-PANYING THIS LESSON, PLACE THE PIECE IN A BOX WITH THE AP-PROPRIATELY-LABELED MINERAL. EXAMPLE: MILK, WHICH IS A GOOD SOURCE OF PHOSPHORUS, WOULD BE PUT INTO THE PHOSPHORUS BOX. THIS WOULD CONTINUE WITH INTEREST UNTIL HOPEFULLY EACH BOX CONTAINED THREE OR FOUR ITEMS OF FOOD. CONSIDER COM-BINING ONE FOOD ITEM FROM EACH BOX INTO A RAW FOODS DELIGHT. DISCUSS THE VARIETY OF COMBINATIONS ONE COULD HAVE.

THINK: WHAT MIGHT BE THE OPTIMUM MENU THAT WOULD INCLUDE AT LEAST ONE ITEM FROM EACH BOX? WILL IT CAUSE GASTRIC DISTRESS? WILL THE RIGHT COMBINATION OF AMINO ACIDS INTERACT WITH VARIOUS MENUS? WHAT COULD TASTE GOOD? WHAT WILL LOOK DE-LICIOUS? HOW MIGHT THE TEXTURES CHANGE IF THE RAW FOODS WERE ~~SUDDENLY~~ SUDDENLY FRIED, BOILED, OR BAKED?

COMMUNICATE: WRITE DOWN FAVORITE POSSIBLE MENUS AND GRADUALLY EAT THEM. SHARE WITH YOUR CLASSMATES YOUR FAVORITE DISH AT AN AFTERNOON-AFTER SCHOOL-COMMUNITY SCHOOL POTLUCK. TEACH YOUR FRIENDS HOW TO PLAY THE RELAY. TEACH AN AFTERNOON FOOD COMBINING CLASS, AND PLAY THE TAPE RECORDINGS ON THE SCHOOL FARMS RADIO SHOW.

FOLLOW-UP POSSIBILITIES: CREATE A NEW DANCE CALLED THE "OTHER MINERALS EIGHT STEP". LABEL EIGHT BOXES WITH THE NAMES CHLORINE, FLUORINE, MAGNESIUM, POTASSIUM, SILICON, SODIUM AND SULPHER, AND PROVIDE APPROPRIATE FOODS WITH WHICH TO FILL THEM.

FOCUS: **ENERGY**

LESSON: **COLD FRAMES**
ADVANTAGES AND DESIGNS

MATERIALS: PLANS AND PICTURES OF SEVERAL DIFFERENT SIMPLE COLD FRAME DESIGNS.

TEACHER'S BACKGROUND: COLDFRAMES HELP YOU SAVE MONEY BY GROWING YOUR OWN STARTS AND ARE A NECESSITY FOR A SELF-SUFFICIENT GARDEN. A PLAN FOR A COLDFRAME IS INCLUDED HERE.

DO: LOOK AT PICTURES OF COLDFRAMES (FROM GARDENING BOOKS AND THE COLDFRAME DESIGN INCLUDED). EXPLAIN WHAT A COLDFRAME IS AND WHAT IT IS MADE OF. ASK IF THE CHILDREN KNOW HOW THE AIR INSIDE A COLDFRAME WILL FEEL COMPARED TO THE AIR OUTSIDE ON A COLD DAY. TALK ABOUT HOW THE GLASS WILL HELP COLLECT THE HEAT AND THE BOX STRUCTURE WILL HOLD IT IN.

THINK: EXPLAIN IN A DISCUSSION THE USES AND ADVANTAGES OF COLD-FRAMES IN THE GARDEN. THEY ALLOW YOU TO GROW SEEDLINGS OUTSIDE WAY BEFORE IT IS WARM ENOUGH FOR PLANTS. THEY ARE EASY TO MAKE AND ARE LIKE A MINIATURE GREENHOUSE. EXPLAIN THAT THE COLDFRAME MUST BE FACING TOWARDS THE SOUTH TO CATCH THE LOW WINTER SUN AS IT PASSES ACROSS THE SKY. GO INTO THE SCHOOL GARDEN AREA AND ASK THE GROUP TO PICK AREAS WHICH WOULD BE GOOD FOR A COLDFRAME, KEEPING IN MIND THE LIGHTING, DIRECTION AND AVAILABILITY TO THE CLASSROOM.

COMMUNICATE: FORM GROUPS OUT OF THE WHOLE CLASS WHICH WILL DO THE DIFFERENT STEPS OF BUILDING THE COLD FRAMES. OUTLINE TO EACH GROUP WHAT THEY WILL BE DOING AND HOW. THE TASKS CAN BE DIVIDED UP INTO:

--MARKING AND SAWING BOARDS

--PUTTING BOARDS IN PLACE AND HAMMERING THE NAILS IN

--CUTTING PLASTIC

--TACKING PLASTIC ONTO THE FRAME

--MARKING AND SCREWING IN THE COLD FRAME HINGES

--DIGGING THE SOIL FOR THE COLDFRAME AND ADDING ORGANIC FERTILIZER MATERIALS TO IT

FOLLOW-UP POSSIBILITIES: BUILD THE COLDFRAMES IN THE NEXT LESSON. THIS WILL TAKE SOME PLANNING AS THE TEACHER WILL HAVE TO DIRECT THE ORDER OF TASKS DONE AND MAKE SURE THAT THE CHILD-REN'S WORK IS ACCURATE.

FOCUS: **ENERGY**

LESSON: **COLDFRAMES - BEGIN CONSTRUCTION**

MATERIALS: THESE ARE ACCORDING TO THE COLD FRAME PLANS PICKED BY YOU AND YOUR CLASS. INSTRUCTIONS ARE INCLUDED HERE. YOU WILL NEED SAWHORSES OR TABLES, HAND SAWS, NAILS, HAMMER, LUMBER, TAPE MEASURE, ETC. FOR CONSTRUCTION.

TEACHER'S BACKGROUND: LUMBER FOR THE COLD FRAME CAN MOST LIKELY BE SALVAGED OR DONATED TO THE SCHOOL GARDEN. THE HARDWARE -- HINGES AND NAILS -- MAY HAVE TO BE BOUGHT. IN THE LONG RUN, THE COLDFRAME IS NOT EXPENSIVE.

DO: HAVE EACH TASK GROUP GET TO-GETHER AND GO OVER WHAT THEIR PART OF THE COLD-FRAME CON-STRUCTION WILL BE. BEGIN CON-STRUCTION BY HAVING THE MARKING AND SAWING GROUP GET STARTED ON THE LUMBER. THE CHILDREN MAY NEED HELP WITH THESE SKILLS IF THEY ARE UNFAMILIAR WITH THEM. EXPLAIN TO THE REST OF THE CLASS HOW AND WHY EACH STEP IS DONE AS IT IS BEING DONE.

THINK: WHEN IT COMES TO ASSEMBLY, ARRANGE THE BOARDS SO THAT THEY ARE IN THE ARRANGEMENT TO BE NAILED TOGETHER. THIS MAKES IT LESS CONFUSING. FOLLOW THE PLAN DIRECTIONS AND NAIL THE COLD FRAME BODY TOGETHER. CONTINUE WITH CONSTRUCTION, INCLUDING THE WHOLE CLASS WHENEVER POSSIBLE IN THE CONSTRUCTION AND UNDERSTANDING PROCESS.

COMMUNICATE: THE ACTUAL CONSTRUCTION MAY TAKE UP ALL OF YOUR CLASS TIME UNLESS YOU SPREAD IT OUT IN THE WEEK. AT SOME POINT DISCUSS AS A GROUP WHAT IT FELT LIKE TO BUILD THE COLD FRAME WAS IT HARD? HAD THE CHILDREN DONE ANYTHING LIKE THIS BEFORE? LET THE CHILDREN KNOW THAT THE COLD FRAME IS AN IMPORTANT PART OF ANY GARDEN, AND LET THEM FEEL PROUD OF THEIR ACCOM-PLISHMENT.

FOLLOW-UP POSSIBILITIES: PERHAPS VISIT A GARDEN OR FARM WHERE COLD FRAMES ARE IN OPERATION (I.E. THE URBAN FARM, THE COMMUNITY GARDENS IN WHITEAKER).

FOCUS: SOILS

LESSON: SOIL NUTRIENTS AND SOURCES

MATERIALS: SHOVELS, PITCHFORKS, LARGE PLASTIC GARBAGE BAGS, TRANSPORTATION TO A SOURCE OF FERTILIZER, POSTER PAPER, CRAYONS OR COLORED PENCILS, DRAWING PAPER, SCISSORS

TEACHER'S BACKGROUND: SINCE SOIL NUTRIENTS CAN BE A BORING SUBJECT, A FIELD TRIP WITH SOME DISCUSSION WILL JAZZ THIS UP. FIND A SOURCE OF LEAF MOLD, MANURE (A STABLE--THEY WILL USUALLY BE HAPPY TO DONATE FOR THE ASKING), WOOD ASH (MESSY), AND PLAN AN EXCURSION TO PICK SOME UP. IT CAN BE CARRIED BY THE CHILDREN IN PLASTIC GARBAGE BAGS AND INTO A VAN, BUS, TRUCK OR STATION WAGON.

DO: HAVE A FIELD TRIP WITH THE CHILDREN TO GATHER SOME SORT OF FERTILIZER FOR THE GARDEN. LOADING THE MATERIAL INTO THE PLASTIC BAGS WILL BE EASY IF YOU HAVE THE CHILDREN WORK IN PAIRS--ONE HOLDING THE BAG AND ONE LOADING. THE JOBS CAN BE TRADED OFF. BACK AT THE SCHOOL GARDEN, FIND A SPOT TO EMPTY THE BAGS OF FERTILIZER FOR GRADUAL USE.

THINK: HAVE A TALK ABOUT WHAT THE SOIL NEEDS IN ORDER TO GROW HEALTHY VEGETABLES. THE ELEMENTS CARBON, HYDROGEN AND OXYGEN ARE ABSORBED INTO PLANTS FROM THE AIR. WE DON'T NEED TO WORRY ABOUT PROVIDING THESE FOR OUR PLANTS. NITRO-GEN, POTASSIUM AND PHOSPHORUS ARE THE THREE THAT WE MOSTLY NEED TO PROVIDE OUR PLANTS WITH. THERE ARE MANY OTHER ELE-MENTS WHICH ARE NEEDED THAT WILL PROBABLY BE IN THE DIFFERENT SORTS OF FERTILIZERS THAT WE GATHER FOR OUR GARDEN. TALK ABOUT THE ORGANIC FERTILIZERS COMPARED TO CHEMICAL ONES. THE ORGANIC FERTILIZER GIVES US BULK MATERIAL FOR THE SOIL STRUCTURE AS WELL AS THE ELEMENTS WHICH THE SOIL NEEDS IN ORDER TO GROW VEGETABLES. THE CHEMICAL FERTILIZER IS PURE ELEMENTS AND ONLY HELPS IN ONE WAY, PROVIDING THESE ELEMENTS. ORGANIC FERTILIZER IS BULKIER AND REQUIRES MORE WORK, BUT IT GIVES THE SOIL MUCH MORE THAN CHEMICAL FERTILIZER DOES. MAKE SURE TO MENTION THE SCHOOL COMPOST PILE AND THAT ALL OF THE GARBAGE THAT YOU HAVE THROWN INTO IT WILL EVENTUALLY HELP TO NOURISH THE PLANTS THAT YOU GROW.

COMMUNICATE: MAKE A LARGE CHART AS A GROUP FOR EACH OF THE THREE MOST IMPORTANT NUTRIENTS, NITROGEN, POTASSIUM AND PHOSPHORUS. FROM THE LIST BELOW, HAVE THE CHILDREN DRAW SMALLER PICTURES OF THE SOURCES OF THESE ELEMENTS AND POST THEM ON THE APPRO-PRIATE POSTER. FOR MANURE, THEY MAY WANT TO DRAW THE ANIMAL, NOT THE MANURE.

SOURCES OF NITROGEN	SOURCES OF POTASSIUM	SOURCES OF PHOSPHORUS
ALFALFA HAY	SEAWEED	BONE MEAL
BLOOD MEAL	COMPOST	COMPOST
CHICKEN MANURE	GRANITE DUST (FROM STATUE MAKERS)	GRANITE DUST
RABBIT MANURE	WOOD ASHES	ROCK PHOSPHATE (POWDERED STONE, COMES IN SACKS. IT IS MINED.)
COW MANURE		
HORSE MANURE		
FISH SCRAPS		
COMPOST		
LAWN CLIPPINGS		

TALK ABOUT HOW ALL THE PARTS OF ANIMALS ARE USED IN SLAUGHT-ERING AND THAT IT IS GOOD TO USE THESE TO MAKE OUR GARDENS BETTER. IT IS A RECYCLING PROCESS AND AVOIDS WASTE.

FOLLOW-UP POSSIBILITIES: GET MORE FERTILIZERS AS TIME ALLOWS. TRY EXPERIMENTING WITH THEM IN THE GARDEN. SEE THE DIFFERENT EFFECTS. MAKE MANURE OR COMPOST TEAS FROM THEM AND TRY THESE OUT ON THE GARDEN--THE PLANTS LOVE THEM!

SPRING

FOCUS: RECYCLING

LESSON: MAKING A COMPOST PILE

MATERIALS: SHOVELS, MANURE, WEEDS, GRASS CLIPPINGS, LEAVES, SOIL, KITCHEN COMPOST (FROM THE CAFETERIA OR HOME), ASHES, BONE OR BLOOD MEAL.

TEACHER'S BACKGROUND: FOR A COMPOST PILE, USE AS MANY MATERIALS AS YOU HAVE AROUND YOU TO BUILD IT. YOU CAN BUILD YOUR COMPOST HEAP INSIDE AN ENCLOSURE OF CINDER BLOCKS OR WOODEN BOARDS OR YOU CAN SIMPLY MAKE A PILE AND COVER IT WITH BLACK PLASTIC TO HOLD IT DOWN AND HOLD IN THE HEAT. SEE WHAT MATERIALS YOU CAN COME UP WITH. COMPOST CAN BE A REAL SCIENCE IF YOU WANT TO GET VERY INVOLVED.

DO: TO MAKE THE COMPOST HEAP, HAVE THE CHILDREN CLEAR A SPOT IN THE GARDEN, 4x5 FEET OR SO, DEPENDING ON HOW MANY COMPOSTING MATERIALS YOU HAVE. START BY PILING A LAYER (6-8") OF KITCHEN SCRAPS OR OTHER ORGANIC PLANT MATERIAL YOU HAVE ONTO YOUR COMPOST SPOT. ON TOP OF THIS LAYER SPREAD A 2-5" LAYER OF MANURE WHICH YOU HAVE GATHERED ON YOUR FERTILIZER FIELD TRIP. THEN, PUT A 2-3" LAYER OF REGULAR SOIL FROM THE GARDEN PLUS ANY POWDERED SOIL ADDITIVES WHICH YOU HAVE ON HAND. REPEAT THIS LAYERING IN

YOUR COMPOST UNTIL IT IS ABOUT 3-4 FEET HIGH OR YOU RUN OUT OF MATERIALS. WATER THE PILE UNTIL IT IS MOIST AND COVER IT WITH BLACK PLASTIC. IN A FEW DAYS, IT SHOULD START TO HEAT UP AND STEAM ON ITS OWN. CHECK IT FOR STEAM WITH THE CLASS. THIS IS HOW YOU KNOW IT IS WORKING. IF IT IS DRY, WATER IT A LITTLE BIT.

THINK: EXPLAIN TO THE CLASS HOW COMPOST WORKS. THE DECOMPOSERS IN THE SOIL HAVE A WONDERFUL ENVIRONMENT IN WHICH TO THRIVE IN THE COMPOST HEAP. THERE IS PLENTY OF FOOD FOR THEM TO EAT, AND THE ACTIVITY OF THESE MICROORGANISMS IS WHAT HEATS UP THE PILE. WHEN THE DECOMPOSITION IS OVER, THE PILE WILL COOL OFF AND THEN THE COMPOST WILL BE A CRUMBLY MASS OF RICH BLACK FERTILIZER FOR THE GARDEN. COMPLETE DECOMPOSITION CAN TAKE FROM TWO WEEKS TO TWO YEARS, SO TO HELP THE COMPOST HEAP ALONG, YOU CAN TURN IT OVER TO GET ALL OF THE MATERIALS TO THE CENTER AND INVOLVED IN THE DECOMPOSITION PROCESS. TURN THE HEAP ABOUT TWO WEEKS AFTER IT HAS BEEN BUILT AND CONTINUE TURNING IT PERIODICALLY UNTIL THE COMPOST IS FINISHED.

COMMUNICATE: HAVE THE CHILDREN DRAW PICTURES OF WHAT HAPPENS IN THE CENTER OF THE COMPOST HEAP. THE MICROBES ARE ACTIVE AND ARE BREAKING DOWN THE MATERIALS. IT IS A BIG PRODUCTION FACTORY WHERE EVERYONE IS AT WORK AND BUSY MAKING SOMETHING TO HELP THE SOIL AND THE GARDEN.

FOLLOW-UP POSSIBILITIES: IF YOU HAVE ENOUGH MATERIALS FOR SEVERAL COMPOST HEAPS, YOU MAY WANT TO HAVE TWO GOING AT THE SAME TIME. ONE OPERATING AND ONE BEING BUILT GRADUALLY WITH AVAILABLE MATERIALS. THIS WAY, YOU WILL HAVE A CONSTANT SUPPLY OF COMPOST.

FOCUS: **ENERGY**

LESSON: **CONSTRUCTING A CLOCHE**

MATERIALS: FOURTEEN 24" STAKES, STRING, EIGHT FEET OF 36"-WIDE PLASTIC SHEETING, TAPE MEASURE, 2-3 HAMMERS, PAPER, PENCILS, PICTURES OF COLD FRAMES, GREENHOUSES, CLOTHES PINS, CAMERAS, TAPE RECORDER, THERMOMETER, SEVEN 36"-SECTIONS OF PVC PIPE, PHONE BOOK, TELEPHONE.

TEACHER'S BACKGROUND: THE CLOCHE IS EASY TO BUILD AND USEFUL. COMPARE THE TWO WHEN YOU GET TO "BUILDING THE COLD FRAME".

DO: STEP INTO YOUR BEAUTIFUL GARDEN, WHEREVER YOU FIND IT, AND PUSH 24" WOODEN STAKES ABOUT SIX INCHES INTO THE SOIL IN SUCH A WAY TO CREATE A LONG ROW OF A-FRAME STAKES ATTACHED BY STRING IN TRICKY KNOTS. LIKE THIS... PERHAPS 12" APART. BEFORE PROCEEDING FURTHER, MEASURE THE LENGTH OF THE BED, THE LENGTH OF STAKE FRAMEWORK, AND THE LENGTH OF PLASTIC. IS EVERYTHING LINING UP OKAY? PROCEED LOGICALLY AND WHIMSICALLY. MAYBE WHISTLING.

THINK: WILL THERE BE ANY HASSLE WHEN PUTTING THE PLASTIC OVER THE STAKES? WHY ARE WE PUTTING THE PLASTIC OVER THE STAKES? A GOOD TIME TO MENTION COLD FRAMES AND GREENHOUSES...MAYBE SHOW A FEW PICTURES. FIGURE OUT THE BEST WAY TO STRETCH THE PLASTIC OVER THE STAKES AND DO SO. WHAT ABOUT THE ENDS? HOW CAN WE CLOSE OFF THE COLD NIGHT WINDS? MAYBE CLOTHES PINS WILL WORK.

COMMUNICATE: WHILE COVERING THE EDGES OF THE PLASTIC, SOME OF THE FRIENDS MIGHT WANT TO TAKE PHOTOGRAPHS OF THE STRUCTURE, WITH FRIENDS AT WORK AND PLAYING AS COLD LITTLE PLANTS FIRST BLOWN BY FREEZING NORTHERN WINDS AND THEN WARMLY SNUG INSIDE THEIR SHELTER. A TAPE RECORDER COULD BE USED FOR INTERVIEWS WITH THOSE ON THE JOB. OTHERS MIGHT WISH TO DRAW A PICTURE OF THE STRUCTURE AND TO LABEL THE PARTS.

FOLLOW-UP POSSIBILITIES: SOME PEOPLE MIGHT WANT TO PLACE ONE THERMOMETER INSIDE THE STRUCTURE WITH ANOTHER OUTSIDE, TO GATHER TEMPERATURE INFORMATION, AND TO CHART THE FINDINGS -- WHAT'S THE VALUE OF A CLOCHE? THINK OF OTHER STRUCTURAL POSSIBILITIES FOR THE CLOCHE FRAMEWORK.

AFTER A HEAVY RAIN, OR A COLD NIGHT, COMPARE PLANTS INSIDE AND OUTSIDE OF THE CLOCHE.

FOCUS: **RECYCLING**

LESSON: **PLANTERS FROM RECYCLED MATERIALS**

MATERIALS: YOGURT CONTAINERS, WOOD FLATS, EGG CARTONS, MILK CARTONS, (QUART AND HALF-GALLON SIZES), SCISSORS, TAPE, STAPLER, PEAT CUPS, GRAVEL, TIN CANS, POTTING SOIL (EQUAL PARTS PEAT MOSS, SAND, SOIL), PENCILS AND PAPER, WATER, JUG, WATER MISTING BOTTLE, TAPE RECORDERS, CAMERAS, CRAYONS, SMALL SEEDS.

TEACHER'S BACKGROUND: TRY TO SCAVENGE SOME MATERIALS OR GET THEM DONATED SOMEHOW. THE COUNTY EXTENSION OFFICE, HORTICULTURE SOCIETY, OR GARDEN SUPPLY STORES MAY PROVIDE SOME HELP.

DO: FIND SOME FRIENDLY VOLUNTEERS, A GOOD PLACE TO WORK AND PLAY, AND THE NEEDED MATERIALS, AND TREAD A FEW MORE STEPS INTO YOUR GARDEN. BE SURE TO ALLOW FOR GOOD DRAINAGE BY EITHER CAREFULLY PUNCHING A FEW HOLES IN THE BOTTOMS OF CONTAINERS, WHEN POSSIBLE, OR BY ADDING A SLIGHT LAYER OF GRAVEL ON THE BOTTOM OF SOLID-BOTTOMED CONTAINERS. ADD SOIL TO THE CONTAINERS UNTIL THEY ARE NEARLY FULL.

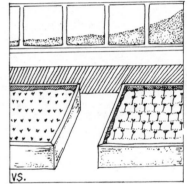

THINK: IS THERE ANY CHANCE THAT THE CONTAINER IS DISEASED OR COATED WITH POISONS? IF SO, CONSIDER NOT USING THE CONTAINER, OR STERILIZING THE CONTAINER WITH BOILING WATER. ALSO, IS THE DEPTH OF THE SOIL DEEP ENOUGH TO ALLOW THE PLANT STARTS ENOUGH ROOM FOR THEIR ROOTS TO SPREAD OUT BEFORE TRANSPLANTING IS NECESSARY? WILL THE LAYER OF GRAVEL INTERFERE WITH ROOT GROWTH? HOW QUICKLY WILL THE WATER PASS THROUGH THE NEWLY-ADDED SOIL? TRY IT. WATCH HOW POURED WATER ERODES THE SOIL...WHAT WOULD HAPPEN TO A SPROUTING SEED NESTLED JUST BENEATH THE SURFACE, IF WE POURED WATER INTO THE PLANTER? WHAT OTHER WATERING METHOD COULD WE USE. TRY IT.

COMMUNICATE: HOPEFULLY THE STUDENTS ARE INTO DOCUMENTING THEIR PROCESS AND PROGRESS WITH SKETCHES, NOTES, PHOTOS, AND CASETTE TAPE RECORDINGS. IF SO, ENCOURAGE THE SCHOOL FARMERS TO ENTER REPORTS IN THE SCHOOL FARM JOURNAL. MAYBE THEY WOULD WANT TO VISIT OTHER CLASSROOMS AND NEIGHBORHOOD CENTERS TO HELP PEOPLE GET PLANT STARTS HAPPENING.

FOLLOW-UP POSSIBILITIES: SPROUT AND GROW WHEATBERRIES INDOORS. EAT THE GRASS AND SHARE A SWEET TREAT WITH A FRIEND.

FOCUS: ENERGY AND FARMING

LESSON: USING GARDEN TOOLS

MATERIALS: SHOVELS, GARDEN RAKES, A HOSE, TROWELS, A BUCKET, A PATCH OF SOIL OUTSIDE, TAPE RECORDER

TEACHER'S BACKGROUND: YOU WILL NEED TO MAKE AN OVERHEAD TRANS-PARENCY OR A LARGE POSTER TO ILLUSTRATE THE COMMON USES OF TOOLS. THERE IS A DRAWING INCLUDED HERE TO USE. THIS LESSON CAN BE DONE OUTSIDE AS WEATHER PERMITS, OR INSIDE SIMPLY PRACTICING THE TOOL TECHNIQUES. STRESS TOOL SAFETY, CLEANLINESS AND CARE THROUGHOUT THIS LESSON.

DO: EACH CHILD SHOULD BE ABLE TO NAME THE PIECES OF EQUIPMENT HE OR SHE WILL MOSTLY ENCOUNTER IN REGULAR USE IN THE GARDEN. THESE ARE THE SHOVEL, GARDEN RAKE, HOE, TROWEL AND BUCKET. DISCUSS EACH ITEM AND LIST ITS POSSIBLE USES. USE THE OVERHEAD TRANSPARENCIES OR POSTER AND DEMONSTRATE HOW TO HOLD AND USE A RAKE. THE RAKE IS USED TO SMOOTH THE SURFACE OF THE SOIL, TO KILL WEEDS, OR TO MAKE FURROWS FOR SEEDS OR PLANTS TO BE PLANTED INTO. IT SHOULD BE USED WITH A SWEEPING MOTION, LIKE A BROOM. HAVE THE CHILDREN PRACTICE USING A RAKE, OUTSIDE SMOOTHING THE SOIL, OR INSIDE PRACTICING THE MOTION.

PRACTICE USING A HOE TO CHOP WEEDS OUTSIDE. HAVE THE CHILD-REN HOLD THE HOE AND STAND WITH THEIR BACKS STRAIGHT SO THAT THE HOE BLADE IS FAIRLY PARALLEL TO THE GROUND. SLICE THE GROUND UNDERNEATH THE WEED TO SIMPLY CHOP IT OFF. KEEP THE CHILDREN FROM DIGGING DEEP HOLES WITH THE HOE. KEEPING THE BLADE PARALLEL TO THE GROUND TO BE CHOPPED HELPS.

DIG HOLES IN THE EARTH WITH THE TROWELS TO GET A FEEL FOR THE SOIL AND THE TOOL. HAVE THE CHILDREN BE GENTLE WITH THE SOIL AND TO PLACE IT BACK INTO THE HOLES CAREFULLY. THIS IS PRACTICE FOR WHEN YOU WILL BE PLANTING SEEDS AND PLANTS INTO THE GARDEN.

SHOW THE CHILDREN HOW TO COIL AND UNCOIL THE GARDEN HOSE. A HOSE WILL LAST LONGEST AND BE OF BEST USE WHEN STORED AWAY WHEN YOU ARE FINISHED WITH IT.

HAVE THE CHILDREN DIG WITH THE SHOVEL BRIEFLY, STEPPING ON THE EDGE OF IT FIRMLY TO PUSH IT INTO THE SOIL DEEPLY, AND TAKING UP SCOOPS OF SOIL. DO THIS JUST ENOUGH TO LET THE CHILDREN CATCH ON, AS THEY WILL BE DOING PLENTY MORE WHEN THEY BEGIN BUILDING THEIR RAISED BEDS.

HAVE THE CHILDREN COMPARE CARRYING ONE BUCKET WITH THE BAL-ANCE OF CARRYING ONE IN EACH HAND, ESPECIALLY WHEN THEY ARE HEAVY AND FULL.

THINK: AS A GROUP, DISCUSS THE IMPORTANCE OF USING TOOLS IN THE GARDEN. TALK ABOUT HOW TOOLS SAVE US ENERGY BY HELPING US DO THINGS WHICH WE COULDN'T DO WITH OUR BARE HANDS. DIS-CUSS LARGE MECHANICAL TOOLS, WHICH ONES ARE THE CHILDREN FAMILIAR WITH? WHAT DO THEY DO? EXPLAIN THAT THESE MACHINES DO MUCH MORE WORK THAN PEOPLE DO AND THAT THEY USE GASOLINE TO OPERATE. DISCUSS WITH THE CHILDREN THE ENERGY DIFFER-ENCE BETWEEN MANY PEOPLE PLANTING OR HARVESTING AND A MACHINE DOING THE SAME WORK. WHICH WAY DO THEY LIKE BETTER AND WHY? TALK ABOUT DIMINISHING FOSSIL FUELS AND HOW MANY PEOPLE LOSE JOBS WHEN MACHINES DO WORK THAT PEOPLE ONCE DID. ASK THE CHILDREN WHAT SORT OF SOLUTIONS THEY SEE.

COMMUNICATE: HAVE THE CHILDREN PRODUCE A TAPE RECORDING OF TOOL DESCRIPTION, USES, AND SAFETY PRECAUTIONS TO BE USED IN OTHER CLASSROOMS AS A PRELIMINARY FOR THE SCHOOL GARDEN. HAVE DIFFERENT GROUPS DEVELOP A SHORT TALK ON EACH TOOL. EACH GROUP CAN WRITE OUT THE TALK FROM A LIST OF QUESTIONS GIVEN TO THEM SUCH AS:

--DESCRIBE A HOE (SHOVEL, TROWEL, BUCKET). "A HOE IS A TOOL USED IN THE GARDEN. IT HAS A LONG WOODEN HANDLE AND A METAL BLADE SHAPED SORT OF LIKE A FIN...."

--WHAT IS IT USED FOR?

--HOW DO YOU USE IT?

--HOW DO YOU TAKE CARE OF IT AND AVOID GETTING HURT WITH A HOE?

FOLLOW-UP POSSIBILITIES: HAVE THE CHILDREN PRESENT TOOL DEMON-STRATIONS IN OTHER CLASSROOMS. DO SOME PRACTICE HOEING OR SHOVELLING IN AREAS OF THE SCHOOL THAT NEED WEEDING BEFORE THE GARDEN DEMANDS ALL OF THE ATTENTION.

FOCUS: HEALTH

LESSON: WHERE DOES FOOD COME FROM?....

MATERIALS: MAGAZINES AND SEED CATALOGS FOR PICTURES. POSTER BOARD, TAPE OR GLUE, FELT PENS.

TEACHER'S BACKGROUND: THERE ARE NO SET RULES ABOUT KNOWING WHICH FOOD COMES FROM WHICH PLANT. KNOWING THIS IS A MATTER OF EXPERIENCE, ESPECIALLY IN THE CASE OF FRUITS, NUTS AND VEGETABLES. THIS LESSON IS TO EXPAND AWARENESS OF FOOD SOURCES AND IF THE CHILDREN MAKE MISTAKES ABOUT FOOD SOURCES, IT IS PROBABLY BECAUSE THEY HAVEN'T SEEN THESE GROW BEFORE AND ARE MAKING AN EDUCATED GUESS. IN THIS LESSON, LEAVE ROOM FOR DISCUSSION.

DO: DISCUSS WITH THE CLASS THAT THE FOODS WE EAT ARE EITHER THE FRUIT OF A TREE, THE FRUIT OF A BUSH OR SMALL PLANT, A ROOT OF A PLANT, SEED FROM A PLANT, OR A PRODUCT OF THE OCEAN OR AN ANIMAL. PASS OUT MAGAZINES AND SEED CATALOGS AND HAVE THE CHILDREN CUT OUT PICTURES OF DIFFERENT KINDS OF FOODS. AS A GROUP LOOK AT THESE PICTURES AND DETERMINE THE SOURCE OF EACH VERBALLY. IF THERE IS A PICTURE OF A LOAF OF BREAD, ASK WHAT THIS IS MADE OF -- WHEAT -- AND ASK THE CHILDREN IF THEY KNOW WHAT PLANT IT COMES FROM. WHEAT IS THE SEED OF A GRASS, ADAPTED TO BE LARGER THAN THE KIND OF GRASS OUT-SIDE IN THE SCHOOL YARD, BUT STILL A GRASS. ANY PASTA FOOD IS THE SAME. THE VEGETABLE OR FRUIT PICTURES ARE EASIER TO TRACE TO THEIR ORIGINS. DO THIS WITH ALL OF THE DIFFERENT FOODS THAT THE CHILDREN CUT OUT OF THE MAGAZINES.

THINK: DISCUSS THE DIVERSITY OF FOODS AND OF WHERE FOODS COME FROM. DIFFERENT KINDS OF FOODS CAN ALL COME FROM TREES. APPLES, PEARS, CITRUS FRUITS, AVOCADOES, PLUMS, PEACHES, MANY NUTS, AND BANANAS COME FROM TREES. THERE ARE ROOT PLANTS THAT, IN ORDER TO EAT, WE MUST PULL UP THE WHOLE PLANT: PO-TATOES, YAMS, CARROTS, BEETS, RADISHES, PEANUTS, ONIONS. SOME FOOD FROM BUSH-TYPE OR SMALL PLANTS ARE BLUEBERRIES, RASPBERRIES, BLACKBERRIES, TOMATOES, SQUASH, PEPPERS, EGG-PLANT, GREEN BEANS, PEAS, BRUSSEL SPROUTS, CAULIFLOWER, BROCCOLI, CABBAGE. THE GRASSES PRODUCE ALL OF OUR STAPLE GRAINS -- WHEAT, RICE, RYE, CORN, BARLEY, ETC. THE OCEAN GIVES US FISH, SHELLFISH, AND SOME PEOPLE EAT KELP. ANIMALS PRODUCE MEAT AND ANIMAL PRODUCTS -- MILK AND EGGS.

COMMUNICATE: LIST ON A POSTER THE DIFFERENT SOURCES OF OUR FOOD: GRASSES, TREES, BUSHES, ROOTS, THE OCEAN, ANIMALS. ILLUS-TRATE EACH SOMEHOW WITH AN EXAMPLE. THEN HAVE THE CHILDREN POST THEIR MAGAZINE PICTURES UNDERNEATH THE RIGHT SOURCE. IF THERE ARE MISTAKES, DISCUSS WHERE THE FOOD COMES FROM AND CLARIFY ANY CONTROVERSY.

GRASS	TREE	ROOT	"BUSH"	OCEAN	ANIMAL

FOLLOW-UP POSSIBILITIES: ASK THE CHILDREN TO PAY ATTENTION AT HOME TO THE FOODS THEY EAT AND WHERE THEY COME FROM.

FOCUS:

LESSON: **Ph SOIL TEST**

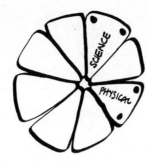

MATERIALS: SIX SPOONS, SIX PLASTIC CONTAINERS, PENCIL, PAPER, TAPE, SIX pH STRIPS, WATER, SPONGES, CLOTH TOWELS.

TEACHER'S BACKGROUND: TESTING SOIL CAN TELL YOU A LOT ABOUT WHAT IS IN STORE FOR YOUR GARDEN. KNOWING DEFICIENCIES AND STRENGTHS OF NUTRIENTS IN YOUR SOIL WILL LET YOU KNOW WHAT YOU WILL NEED TO ADD.

DO: MENTION TO THE CHILDREN THAT IT IS POSSIBLE TO HAVE A SCHOOL GARDEN. FIND A GROUP OF ENTHUSIASTIC VOLUNTEERS AND GO OUTSIDE WITH THE SPOONS AND CONTAINERS. FIND A SUNNY SPOT, WITH GOOD DRAINAGE, THAT SEEMS LIKE A FRIENDLY PLACE FOR A GARDEN AND DRAW A MAP OF THE DREAM GARDEN'S OUTLINE. WHO KNOWS THE WAY NORTH? EAST? IN SMALL GROUPS, SPOON SMALL AMOUNTS OF SOIL FROM SIX SEPARATE GARDEN LOCATIONS, INTO THE SIX CONTAINERS. LABEL EACH CONTAINER AND INDICATE ON THE MAP WHERE EACH SAMPLE WAS GATHERED. ADD A LITTLE WATER TO EACH CONTAINER MAKING A MUDDY SOUP. MENTION THAT SOILS HAVE CHARACTERISTICS OF BASICITY (ALKALINITY) AND ACIDITY THAT CAN BE MEASURED WITH pH STRIPS DIPPED INTO THE MUDDY SOUPS. INTO EACH OF THE SIX CONTAINERS, DIP A SINGLE pH STRIP AND REMOVE IT TO SEE WHAT COLORS APPEAR. IF THE PAPER TURNS PINK OR RED, THE SOIL IS LIKELY TO BE ACIDIC; AND IF THE PAPER TURNS BLUISH, THE SOIL IS PROBABLY MORE ALKALINE. PLACE EACH pH STRIP ON THE PAPER MAP OF THE GARDEN.

THINK: ARE THE STRIPS THE SAME COLOR? IS THE GARDEN MORE ACID? OR BASIC? MENTION THAT MOST PLANTS PREFER A SLIGHTLY ACID SOIL (BE SURE TO STUDY REASONS WHY BEFOREHAND). ASK THE CHILDREN IF THEY WANT TO FEEL THE TEXTURE OF THE SOIL IN THEIR HANDS. DOES THE SOIL FEEL GRITTY, SILKY, STICKY? DO THE SAMPLES FEEL THE SAME? WHY OR WHY NOT? EXAMINE THE COLOR OF THE VARIOUS SAMPLES. ARE SOME DARKER THAN OTHERS? WHAT ARE SOME OF THE REASONS THE CHILDREN MIGHT OFFER FOR COLOR VARIATIONS?

COMMUNICATE: PLACE THE MAP OF THE GARDEN ON THE WALL AND ASK FOR SOMEONE TO LABEL IT ONTO PAPER WITH AN APPROPRIATE PHRASE OR WORD. ASK IF ANY OF THE PARTICIPANTS MIGHT WANT TO DESCRIBE IN WORDS AND ACTIONS HOW A WORM MIGHT FEEL TRAVELING THROUGH GRITTY, SILKY, OR STICKY SOIL. FEEL FREE TO GIVE YOUR OWN INTERPRETATION. CONSIDER USING A CASETTE TAPE

RECORDER IF YOU WANT TO CAPTURE THE SPONTANEOUS CREATIVITY. THIS PROCESS CAN INVOLVE DIVIDING THE CHILDREN INTO TWO GROUPS -- ONE GROUP ACTING FIRST AS THE AUDIENCE WHILE THE OTHER GROUP PLANS AND PERFORMS THE INTERPRETATION. THEN SWITCH.... CAN PEOPLE DISCOVER WHICH WORM IS IN GRITTY SOIL?

FOLLOW-UP POSSIBILITIES: ADD ENOUGH WATER TO THE SOIL IN THE GLASS JAR SO THAT THE SOIL SWIRLS ABOUT WHEN STIRRED. LET THE SOIL SETTLE OUT OVERNIGHT, AND OBSERVE WITH THE CHILDREN THE LAYERS OF CLAY, SILT AND SAND.

WITH THE CHILDREN, GATHER SOME SOIL FROM THE BEST GARDEN SITE AND MAIL IT TO A SOIL TESTING LABORATORY TO SEE WHAT NUTRIENTS NEED TO BE ADDED TO THE PROPOSED GARDEN PLOT. THE EXTENSION SERVICE OR THE SOIL CONSERVATION SERVICE WILL DO THIS FOR YOU, AT A COST OF $5.00.

FOCUS: FARMING

LESSON: A LOOK AT DIFFERENT SEEDS

MATERIALS: BEAN SEEDS FOR EACH CHILD, AN APPLE, ORANGE (WITH
   SEEDS), WALNUT, CHESTNUTS, ACORNS, ANY EDIBLE OR NON-EDIBLE
   SEED TO EXAMINE.

TEACHER'S BACKGROUND: THE NIGHT BEFORE THIS LESSON, SOAK ONE BEAN
   SEED FOR EACH CHILD IN A SAUCER OF WARM WATER AND ALLOW THEM
   TO SOAK OVERNIGHT UNTIL YOU BEGIN.

DO: GIVE A SOAKED BEAN SEED TO EACH CHILD. ASK EACH CHILD TO
      CAREFULLY REMOVE THE SEED COAT. IT SHOULD SLIP OFF RATHER
      EASILY. THE SEED COAT IS A
      PROTECTIVE MEMBRANE FOR THE
      SEED. THEN ASK EACH CHILD
      TO SEPARATE THE TWO HALVES
      OF THE SEED. THESE SEED
      HALVES ARE CALLED COTYLEDONS.
      THE COTYLEDONS CONTAIN A
      FOOD SUPPLY FOR THE TINY
      PLANT OR EMBRYO. THERE IS
      ENOUGH FOOD IN THESE COTY-
      LEDONS TO KEEP THE PLANT
      ALIVE AND GROWING UNTIL IT
      DEVELOPS ITS FIRST TRUE GREEN
      LEAVES. AT THAT TIME, THE
      PLANT STARTS TO MANUFACTURE
      ITS OWN FOOD. HAVE EACH
      CHILD CAREFULLY LOOK AT
      EACH END OF EACH COTYLEDON.
      ATTACHED AT THE END OF ONE
      OF THE COTYLEDONS WILL BE A
      VERY TINY STRUCTURE THAT LOOKS LIKE A MINIATURE PLANT. DE-
      SCRIBE HOW A PLANT BEGINS TO GROW FROM THIS EMBRYO BY SEND-
      ING OUT A LONG SIMPLE ROOT. WHEN THE ROOT GROWS TO A CERTAIN
      LENGTH, THE STEM OF THE EMBRYO GROWS UP OUT OF THE SOIL.
      SOME PLANTS WILL PUSH THE COTYLEDONS ABOVE THE GROUND WHEN
      SPROUTING, AND OTHERS LEAVE THEM DOWN BELOW THE SURFACE.

      CUT OPEN THE FRUITS AND NUTS AND EXAMINE THE SEEDS INSIDE.

THINK: EXPLAIN HOW DIFFERENT FRUITS HAVE DIFFERENT SORTS OF SEEDS.
   NOTICE THE DIFFERENT STRUCTURE OF HOW SEEDS ARE IN THE FRUITS.
   THE WALNUT -- THE PART WE EAT -- ARE COTYLEDONS, JUST LIKE
   THE BEAN. THEY HAVE THAT SHRIVELLED LOOK BECAUSE THEY ARE
   VERY DRY. EXPLAIN THAT ALL PLANTS, EVEN GIANT TREES START
   FROM SMALL SEEDS. SEEDS USUALLY HAVE A SOFT OR HARD OUT-
   SIDE COAT, AND FOOD STORAGE ORGANS FOR THE SEEDLING TO LIVE
   ON UNTIL THE PLANT GETS BIG ENOUGH TO MAKE ITS OWN FOOD FROM
   THE SUNSHINE AND SOIL.

COMMUNICATE: HAVE THE CHILDREN PICK ONE KIND OF SEED WHICH THEY
   PARTICULARLY LIKE AND PRETEND THEY ARE THAT SEED WAITING IN
   THE GROUND TO SPROUT. HAVE THEM GET THEIR BODIES SOMEHOW
   INTO THE SHAPE THAT THEIR SEED IS. THEN ANNOUNCE THAT THERE
   IS NOW BRIGHT SUNLIGHT, OR PERHAPS TURN ON THE LIGHTS, AND
   HAVE THEM PUT OUT ROOTS AND SEND THEIR FIRST GREEN LEAVES
   UP TO THE LIGHT.

FOLLOW-UP POSSIBILITIES: ASK THE CHILDREN TO BRING UNUSUAL SEEDS
   FROM HOME IN TO SHOW THE CLASS.

FOCUS: GARDENING

LESSON: MAKING CHINESE RAISED BEDS

MATERIALS: GARDEN STRING, STAKES, SHOVELS, RAKES, MANURE, COMPOST, LEAF MULCH OR SOME OTHER BULKY FERTILIZER.

TEACHER'S BACKGROUND: AN EXTREMELY USEFUL BOOK WITH STEP-BY-STEP INSTRUCTIONS FOR MAKING RAISED BEDS AND REASONS WHY WE USE RAISED BEDS IS PETER CHAN'S, <u>BETTER VEGETABLE GARDENS THE CHINESE WAY</u>. RAISED BEDS REQUIRE SOME EFFORT, BUT ARE BEST FOR THIS AREA BECAUSE OUR SOIL STAYS WET FOR SO LONG THAT WE NEED SOME EXTRA HELP IN DRAINAGE. IN A RAISED BED, THE SOIL IS UP OFF THE REGULAR GROUND LEVEL ABOUT FOUR INCHES, AND WE CAN GET GARDENS IN SOONER BY NOT HAVING TO WAIT UNTIL THE RAINS STOP. MAKING THE RAISED BEDS MAY TAKE SEVERAL DAYS OF WORK WITH THE CLASS. THE CHILDREN CAN PAIR UP ON BEDS, HAVE ONE BED EACH, OR WORK IN GROUPS, DEPENDING ON HOW MUCH GARDEN SPACE YOU HAVE.

DO: IN THE GARDEN, HAVE THE CHILDREN HELP WITH MEASURING AND MARK-ING THE FOUR CORNERS OF EACH BED WITH STAKES, AND THEN OUT-LINING EACH BED WITH STRING. THE BEDS CAN BE AS LONG AS YOU LIKE (25 FEET OR MORE), SO ADJUST THIS LENGTH TO YOUR GARDEN AREA. MAKE THE BEDS THREE FEET ACROSS AT THE BASE (THE SLOPING SIDES OF A RAISED BED MAKE THE TOP ONLY 2 FEET ACROSS). THIS IS NARROWER THAN REGULAR RAISED BEDS, AND YOU WON'T GET A MAXIMUM YIELD ON THESE BEDS, BUT THEY ARE SMALL ENOUGH THAT THE CHILDREN CAN REACH TO THE CENTER FROM EITHER SIDE WITHOUT HAVING TO STEP ON THE BED.

AFTER THE BEDS ARE MARKED OFF, THEY SHOULD BE DUG ONE-SHOVEL-FUL DEEP THROUGHOUT. THIS PUTS AIR IN THE SOIL, AND LETS YOU MORE EASILY SIFT OUT ROCKS AND WEED CLUMPS. AT THE EDGES OF THE BED, DIG STRAIGHT DOWN TO GET ALL OF THE BED SOIL INTO THE MIDDLE. PILE THE SOIL ALONG THE CENTER OF THE BED, FORM-ING A RIDGE OF SOIL ALONG THE ENTIRE LENGTH OF YOUR BEDS.

TO MAKE THE RAISED BED, SHAPE THE SOIL INTO A MOUND WITH A FLAT TOP. IT WILL BE FROM 4-6 INCHES HIGHER THAN THE GROUND. WHEN THE GROUND LEVEL WIDTH OF THE BED IS 3 FEET, MAKE THE TOP LEVEL 2 FEET, OR ONE FOOT LESS THAN THE WIDTH OF THE BED. RAKE IT UNTIL SMOOTH. THE CHILDREN WILL HAVE HAD PRACTICE WITH THEIR TOOLS FROM THE PREVIOUS LESSON ON TOOL USE. WHEN THE BED IS SMOOTH AND NEAT, REMOVE THE STRING. THE STAKES CAN REMAIN AS MARKERS IF YOU LIKE.

DURING THE DIGGING OF THE BED, YOU MAY WANT TO ADD SOME OR-GANIC MATTER, ESPECIALLY IF THE AREA HAS NOT BEEN USED AS A GARDEN BEFORE. AFTER MARKING THE BED WITH STAKES AND STRING, HAVE THE CHILDREN PILE WHATEVER ORGANIC MATERIAL YOU HAVE OVER THE BED AREA. THEN CONTINUE DIGGING THE RAISED BED.

THINK: TALK TO THE CLASS ABOUT WHY WE ARE USING RAISED BEDS. DIS-CUSS HOW THEY HELP WITH SOIL DRAINAGE IN OUR CLIMATE BECAUSE OF THE RAIN THAT WE HAVE. EXPLAIN THAT IF WE HAD TO WAIT UNTIL THE SOIL DRIED OUT BEFORE BEGINNING TO GARDEN, THEN WE WOULD BE WAITING UNTIL JUNE OR LATER, AND WE WOULDN'T BE ABLE TO HAVE SO MANY VEGETABLES SO EARLY. ASK THE CHILDREN IF THEY HAVE IDEAS ABOUT WHY RAISED BEDS ARE A GOOD IDEA. VEGETABLES CAN BE GROWN MORE DENSELY IN A RAISED BED BECAUSE THE ROOTS CAN MOVE DEEPER INTO THE SOIL WHICH YOU DUG AND PUT AIR SPACE INTO. THE RAISED BEDS LOOK NEAT AND ARE EASY TO WORK WITH. AND BECAUSE OF THE LOOSE SOIL, WEEDS ARE EASY TO PULL OUT. TALK ABOUT SOIL COMPACTION AND ASK THE CHILDREN WHAT WOULD HAPPEN IF THEY WALKED ON THE RAISED BED.

COMMUNICATE: HAVE THE CHILDREN GET INTO PAIRS OR GROUPS TO BRAIN-STORM ABOUT WHAT THEY WANT TO SEE GROWING IN THEIR RAISED BEDS. HAVE THEM MAKE LISTS IF THEY WANT TO. HAVE THEM DISCUSS AND SHARE ABOUT THE WORK THEY DID IN THE GARDEN THAT DAY. WAS IT TOO MUCH WORK? DID THEY ENJOY IT? EMPHASIZE THAT IN ORDER TO GET FOOD AND OTHER THINGS FROM THE GARDEN, WE MUST PUT EFFORT INTO IT AS AN EXCHANGE. THAT'S HOW THE GARDEN WORKS. THAT'S HOW ENERGY WORKS AS WELL.

FOLLOW-UP POSSIBILITIES: MAKING ENOUGH BEDS FOR A WHOLE SCHOOL GARDEN SHOULD BE ENOUGH FOLLOW-UP WORK BY ITSELF!

FOCUS  FARMING

LESSON COMPANION PLANTING

MATERIALS: PAPER, PENCILS, CASETTE TAPE RECORDERS, COLORED PENCILS

TEACHER'S BACKGROUND: NEIGHBORS ARE USUALLY MORE THAN HAPPY TO SHOW
    OFF THEIR FLOURISHING GARDENS. MAKE SURE THAT THE CHILDREN
    KNOW THAT THEY ARE NOT TO WALK ON THE BEDS OR PICK THESE
    PLANTS WHICH ARE NOT THEIRS.

DO: WITH A GROUP OF FRIENDLY STUDENTS, VISIT A FLOURISHING GAR-
    DEN IN THE NEIGHBORHOOD AND DISCUSS COMPANION PLANTING.
    EMPHASIZE THAT PLANTS GROWING TOGETHER CLOSELY IN A SMALL
    SPACE OF A GARDEN COMPETE FOR SPACE, LIGHT, NUTRITION AND
    WATER. PLANTS WILL COMPLEMENT AND NOURISH EACH OTHER IF
    THEY ARE COMPATIBLE (SUCH AS LEAFY VEGETABLES LIKE SPINACH,
    LETTUCE AND CABBAGE) GROWING NEAR LIGHT FEEDERS SUCH AS
    POLE BEANS AND PEAS. ALSO TRY TO FIND PLANT NEIGHBORS WHOSE
    ROOTS OCCUPY DIFFERENT LEVELS IN THE SOIL OR FIND THE LIGHT
    REQUIREMENTS THAT SUIT THEIR NEEDS.

THINK: OF COMBINATIONS OF PLANTS THAT WOULD HELP OR HINDER EACH
    OTHER. WHY MIGHT IT BE A MISTAKE TO PLANT A LOT OF LETTUCE,
    CUCUMBERS, SPINACH AND CORN TOGETHER IN A RAISED BED? RE-
    MEMBER THE CONSIDERATIONS OF SPACE, LIGHT, NUTRIENTS AND
    WATER. TRY TO DETERMINE HOW DEEPLY VARIOUS COMPANION PLANTS'
    ROOTS REACH INTO THE SOIL. WHAT MIGHT BE THE RESULT OF
    PLANTING SUNFLOWERS AND BEANS IN THE SAME BED? ENCOURAGE
    THE GATHERING OF NOTES ABOUT THESE QUESTIONS AND POSSIBLE
    ANSWERS WITH PAPER, PENCILS, TAPE RECORDERS...SKETCHES...

COMMUNICATE: ENCOURAGE THE STUDENTS TO DRAW "SIDE-VIEW" SKETCHES
    OF POSSIBLE COMPANION PLANT COMBINATIONS THAT MIGHT HAVE A
    GOOD CHANCE OF SUCCESS. PERHAPS LIGHT FEEDING PLANTS COULD
    BE OF ONE COLOR. BE SURE TO INCLUDE THE RELATIVE SIZE AND
    DISTRIBUTION OF THE PLANT'S ROOT STRUCTURES AS WELL. SHOW
    THESE SKETCHES TO NEARBY COMPANIONS AND TELL THEM WHAT HAS
    BEEN LEARNED. ADD NOTES TO THE SCHOOL FARM'S JOURNAL UNDER
    THE HEADING "COMPANION PLANTING".

FOLLOW-UP POSSIBILITIES: TRAVELING IN SMALL GROUPS WITH THE AID
    OF A TAPE RECORDER, INTERVIEW FARMERS AND GARDENERS IN THE
    NEIGHBORHOOD ON THE SUBJECT OF COMPANION PLANTING. DO THE
    PEOPLE BELIEVE SOME PLANTS REPEL INSECTS? WHY? DO SOME OF

THE PLANTS TEND TO LURE OTHER INSECTS? WHY? WITH A BROAD-
ER UNDERSTANDING OF COMPANION PLANTING DYNAMICS, PLANT THREE
OR FOUR COMPANION PLANTS IN A RAISED BED AND RECORD WHAT
HAPPENS OVER AN EXTENDED PERIOD OF TIME. IF ENOUGH SOIL AND
SPACE IS AVAILABLE, PLANT A RAISED BED OF PLANTS THAT ARE NOT
COMPANION PLANTS (SUCH AS FOUR HEAVY FEEDERS AND NOTHING ELSE)
TRY TO PLANT THESE IN A BED IN THE SAME VICINITY AS THE OTHER
RAISED BED (WITH COMPANION PLANTS) TO SEE HOW THE RESULTS
COMPARE...MAKE COLLAGES OF COMPANION PLANTS CLUSTERED TO-
GETHER.

FOCUS  FARMING

LESSON SOWING SEEDS IN THE GARDEN

MATERIALS: BEAN, RADISH, CARROT, LETTUCE, PEA OR ANY SEEDS TO BE PLANTED IN THE GARDEN. HOES, STAKES, STRING, WATERING CANS.

TEACHERS' BACKGROUND: SOME OF THE RAISED BEDS WILL NEED TO BE RAKED FLAT AND MADE READY FOR PLANTING IN ORDER TO DO THIS LESSON. IF THIS IS AN EARLY SPRING PLANTING, PEAS, SPINACH, RADISH, CARROTS, AND BEANS WILL BE HARDY ENOUGH TO PLANT.

DO: HAVE THE CHILDREN COMPARE RADISH SEEDS AND BEAN SEEDS FOR SIZE AND SHAPE. POINT OUT THAT WE AS GARDENERS CONSIDER THE SIZE OF THE SEED WHEN DETERMINING HOW DEEP TO PLANT THEM. A RULE OF THUMB, SEEDS ARE PLANTED ABOUT THREE TO FIVE TIMES AS DEEP AS THEIR DIAMETER.
IN A GARDEN BED, HAVE TWO CHILDREN DEMONSTRATE TYING STRING TIGHT BETWEEN TWO STAKES AT EACH END OF THE RAISED BED. THE STRING SHOULD BE RIGHT AT GROUND LEVEL. THEN, HAVE THEM MAKE A FURROW FOR WHATEVER TYPE OF SEED THEY ARE PLANNING TO PLANT THERE. DEEPER FOR BEANS, SHALLOW FOR RADISH OR LETTUCE SEEDS. FOR SPACING, REFER TO THE PLANTING GUIDE BEHIND THE "PLANTING PLANTS IN THE GARDEN" LESSON. HAVE EACH CHILD MAKE A FURROW THIS WAY IN THE GARDEN BEDS. CHILDREN TEND TO PLANT ALL OF THEIR SEEDS WITHIN A SHORT DISTANCE IN THEIR ROW. TO REMEDY THIS, HAVE THE CHILDREN POUR ABOUT HALF OF THE SEEDS INTO THE PALM OF THEIR HAND, AND PICKING UP ONE OR TWO SEEDS WITH THE OTHER HAND, HAVE THEM DROP THE SEEDS 4 TO 6 INCHES APART IN THE ENTIRE ROW. THEN GO BACK OVER THE ROW AGAIN WITH ANOTHER BATCH OF SEEDS UNTIL THE SEEDS ARE ALL IN. ONCE THE SEEDS ARE IN ALL OF THE CHILDREN'S FURROWS, DEMONSTRATE HOW TO COVER THE SEEDS. USE THE THUMB AND FOREFINGERS TO PULL OR PINCH THE SOIL FROM EACH SIDE OF THE FURROW TO COVER THE SEEDS. MONITOR THE AMOUNT OF SOIL THAT THE CHILDREN PUT OVER THE SEEDS. MAKE SURE THAT TOO MUCH SOIL IS NOT PUT OVER THE SEEDS. FINALLY, GENTLY WATER THE ROWS WITH WATERING CANS.

THINK: EXPLAIN TO THE CHILDREN THAT TOO MUCH SOIL OVER THE SEEDS MEANS THAT THE SEEDLINGS WILL NOT BE ABLE TO WORK THEIR WAY THROUGH TO THE SUNNY AIR WHICH THEY LOVE AND LIVE ON. TALK ABOUT WHY GARDENS ARE PUT INTO ROWS, FOR EASE IN WEEDING AND TAKING CARE OF THE GARDEN, AND LOOKING NEAT. DISCUSS THE APPEARANCE OF A GARDEN WHICH WAS PLANTED IN ROWS COMPARED TO ONE IN WHICH THE SEEDS ARE SIMPLY SCATTERED AROUND. ASK THE CHILDREN WHICH GARDEN WOULD BE EASIER TO TAKE CARE OF. DISCUSS THE IMPORTANCE OF BEING GENTLE WITH WATERING SO AS NOT TO WASH THE SEEDS FROM THE SOIL.

COMMUNICATE: HAVE THE CHILDREN TALK ABOUT OTHER PLANTING EXPERIENCES AND WHAT THEY LEARNED FROM THE PLANTING THAT DAY. SOME CHILDREN MAY HAVE DISCOVERED THEIR OWN TECHNIQUES FOR FURROW MAKING OR COVERING SEED THAT THEY WOULD LIKE TO HAVE SHARED WITH THE CLASS.

FOLLOW-UP POSSIBILITIES: THESE NEW ROWS WILL NEED TO BE WATERED IF RAIN DOES NOT TAKE CARE OF THIS. EVERYDAY WATERING IS BEST WHEN THE SEEDS ARE GERMINATING. A LIGHT MULCH SUCH AS DRY GRASS, FINE COMPOST, WILL HELP RETAIN MOISTURE IF THE WEATHER IS DRY.

FOCUS **FARMING**

LESSON **PLANTING IN THE GARDEN**

MATERIALS: PLANTS FOR TRANSPLANTING: CABBAGE, BRUSSEL SPROUTS,
CAULIFLOWER, ETC. PEPPERS, TOMATOES. RAKES, STAKES
STRING, TROWELS, WATERING CANS

DO: IN A DISCUSSION DEVELOP AN UNDERSTANDING OF HOW MUCH SMALL
PLANTS WILL GROW BEFORE THEY MATURE , AND THAT EVERY PLANT
NEEDS ENOUGH ROOM TO GROW FULLY. HAVE THE CHILDREN ESTI-
MATE THE PROPER SPACING. FOLLOW THE PLANTING GUIDE IN THE
BACK OF THIS LESSON FOR SPACING OF PLANTS. IN THE GARDEN,
MARK A ROW WITH STAKES AND STRING AS IN THE SEED PLANTING
LESSON. RAKE THE SOIL SMOOTH UNDER THE STRING, MEASURE
OUT AND MARK WHERE THE TRANSPLANTS WILL GO. THE HOLES DUG
FOR THE PLANTS SHOULD BE LARGER THAN THE ROOT BALL OF THE
PLANTS. THE ROOTS MUSTN'T BE JAMMED INTO A SMALL HOLE,
THEY NEED ROOM TO GROW EASILY. THE PLANTS SHOULD BE PLACED
IN THE GROUND AT THE SAME LEVEL AS IN THE POT OR FLAT. THE
EXCEPTION IS THE TOMATO PLANT WHICH CAN BE
PLANTED DEEPER SO THAT ADDITIONAL ROOTS
DEVELOP ON THE STEM. HAVE EACH CHILD
DIG A HOLE FOR THE PLANT, THEN REMOVE
A PLANT FROM A POT BY GENTLY KNOCKING
IT ON ITS SIDE UNTIL IT IS LOOSENED
ENOUGH TO PULL OUT BY HOLDING ON TO
THE LEAVES AND STEM. WHEN THE PLANT
IS PLACED INTO THE HOLE, SHOW THE
CHILDREN HOW TO PUT LOOSE SOIL GENT-
LY AROUND THE ROOT BALL. HAVE THEM
PAT THE SOIL VERY GENTLY AROUND THE
PLANTS ONLY HARD ENOUGH TO KEEP THE
PLANT UPRIGHT. MAKE SURE THEY DO NOT
PAT THE SOIL TOO HARD AND COMPACT THE
SOIL AROUND THE PLANT. AFTER PLANTING
HAVE THE CHILDREN WATER THEIR PLANTS
GENTLY, YET THOROUGHLY. THEN LET THE
CHILDREN PLANT THE REST OF THE PLANTS
ON THEIR OWN, PAYING ATTENTION TO THE
TECHNIQUES THAT THEY HAVE LEARNED.

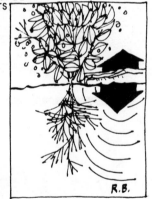
R.B.

THINK: DISCUSS THE TIME IT TAKES TO GROW DIF-
FERENT CROPS TO MATURITY. COMPARE RADISH
AND TOMATO. TALK ABOUT THE CROPS WHICH HAVE
A LONG GROWING PERIOD AND THAT MAY NOT COME TO
MATURITY FAST ENOUGH IF THE SEEDS ARE PLANTED DIRECTLY INTO
THE GROUND. THOSE CROPS ARE USUALLY STARTED INDOORS BEFORE
IT IS WARM ENOUGH OUTDOORS FOR THEM TO SURVIVE, AND THEN THEY
ARE PLACED OUTSIDE WHEN THEY ARE BIGGER AND IT HAS WARMED UP
ENOUGH OUTSIDE. DISCUSS THE OUTSIDE TEMPERATURE, AND THOUGH
IT MAY BE COLD TO THE CHILDREN, IT IS NOT TOO COLD FOR PLANTS.
MENTION FROST, HOW IT CAN FREEZE PLANTS IF THEY ARE OUTSIDE,
AND WHEN IN THE SPRING, THE FROST STOPS IN THE MORNINGS.

COMMUNICATE: A WATERING SCHEDULE WILL BE NEEDED TO WATER THE NEW
STARTS AND THE WHOLE GARDEN. TALK TO THE CLASS ABOUT THE
IMPORTANCE OF BEING RESPONSIBLE FOR THE GROWING PLANTS IN
THE GARDEN. THEY NEED SPECIAL CARE. AS A GROUP, WORK OUT
A SCHEDULE OF WATERING AFTER SCHOOL, WHERE ONE OR TWO CHILD-
REN TAKE RESPONSIBILITY FOR CHECKING ON THE GARDEN AND
SEEING IF IT NEEDS WATERING, AND WATERING IF NECESSARY. KEEP
THE SCHEDULE POSTED IN THE CLASSROOM.

FOCUS:

LESSON:

MATERIALS: PICTURES OF SOME BENEFICIAL AND HARMFUL INSECTS, JARS
FOR COLLECTING INSECTS, PAPER, COLORED PENCILS. THIS LESSON
WILL TAKE PLACE OUT IN THE GARDEN.

TEACHER'S BACKGROUND: SOME OF THE MOST COMMON HARMFUL GARDEN IN-
SECTS IN THIS AREA INCLUDE CABBAGE WORMS, CUTWORMS, CUCUMBER
BEETLES (GREEN WITH BLACK SPOTS), APHIDS, FLEA BEETLES,
SLUGS, ETC. IT WOULD BE GOOD TO FAMILIARIZE YOURSELF WITH
THESE IF YOU AREN'T ALREADY, BY GETTING AN INSECT PAMPHLET
AVAILABLE AT THE COUNTY EXTENSION OFFICE. LADYBUGS AND BEES
ARE AMONG THE HELPFUL INSECTS IN THE GARDEN. STRESS THE
AVAILABILITY OF ALTERNATIVES TO PESTICIDES IN THE GARDEN FOR
INSECT CONTROL.

DO: GO OUT INTO THE GARDEN AND LOOK FOR INSECTS OR EVIDENCE OF
THEM. LOOK FOR DAMAGED PLANTS AND TRY TO SEE WHETHER THIS
DANGER WAS DONE BY INSECTS. SOME
INSECT DAMAGE WOULD INCLUDE
PLANTS WHICH HAVE HOLES CHEWED
IN THEM, OR SECTIONS OF LEAVES
CHEWED AWAY OR CERTAIN LEAVES
THAT ARE TIGHTLY CURLED OR
TWISTED OR HAVE BUMPS ON THEM.
POINT OUT THAT SOME INSECTS
CHEW PLANT PARTS, WHILE OTHER
INSECTS HAVE BEAKS LIKE MOS-
QUITOES AND SUCK PLANT JUICES
CAUSING THE LEAVES TO CURL OR
TWIST OR CAUSE BUMPS ON THE
LEAVES. WHERE INSECT-DAMAGED
PLANTS ARE FOUND, HAVE THE
CHILDREN CAREFULLY LOOK UNDER
AND AROUND THE LEAVES AND STEM
OF THE PLANT TO SEE IF THEY CAN
FIND THE INSECT THAT CAUSED THE
DAMAGE. ALSO, SEE IF THE
CHILDREN CAN FIND SOME HELPFUL INSECTS PREYING ON THE HARMFUL
ONES.

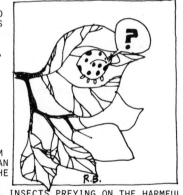

THINK: DISCUSS AS A GROUP THE INSECTS SEEN IN THE GARDEN. LIST THEM
AND OTHERS AND CLASSIFY THEM AS HARMFUL OR HELPFUL INSECTS.
EXPLAIN WHAT MAKES INSECTS HARMFUL OR HELPFUL.

TALK ABOUT BEES. THE BEES ARE IMPORTANT BECAUSE THEY CARRY
POLLEN FROM FLOWER TO FLOWER WHICH IS NECESSARY TO MAKE THE
FRUITS GROW ON OUR VEGETABLES. LADYBUG BEETLES AND THEIR YOUNG
EAT APHIDS AND ARE IMPORTANT IN HELPING TO KEEP APHIDS UNDER
CONTROL.

SHOW THE CHILDREN PICTURES OF BENEFICIAL INSECTS AND HELP THEM
UNDERSTAND THAT NOT ALL INSECTS ARE HARMFUL. TALK ABOUT HOW
MANY PEOPLE HAVE BEES LIVING IN HIVES ON THEIR FARMS AND THAT
THEY TAKE VERY GOOD CARE OF THEM SO THAT THE BEES WILL POLLIN-
ATE THEIR GARDENS, ORCHARDS OR FIELDS, AND WILL GIVE THEM GOOD
HONEY IN RETURN.

ASK THE CHILDREN HOW WE CAN CONTROL HARMFUL INSECTS IN OUR
GARDEN. THEIR ANSWERS MAY BE SPRAYING, AND DUSTING WITH CHEM-
ICALS. STRESS THAT THERE ARE ALTERNATIVES AND OTHER APPROACHES
IN INSECT CONTROL. ONE WAY IS TO TAKE GOOD CARE OF OUR GAR-
DENS. AVOID INJURING PLANTS BY STEPPING ON THEM OR BUMPING
THEM SEVERELY BECAUSE PLANTS WHICH ARE INJURED ARE MORE SUS-
CEPTIBLE TO INSECT INFESTATION. HEALTHY PLANTS TOLERATE INSECT
DAMAGE BETTER. WEEDS SHOULD BE PICKED BECAUSE SOME INSECTS
HIDE IN WEEDS UNTIL THE VEGETABLE PLANTS ARE READY FOR THEM
TO EAT. GARLIC, ONION OR MILD SOAPY WATER SOLUTIONS WILL
DETER INSECTS FROM PLANTS.

COMMUNICATE: HAVE THE CHILDREN DRAW AN INSECT THAT THEY SAW IN THE
GARDEN OR A PICTURE INCLUDING WHAT THE INSECT LIVES ON AND
WHAT IT DOES. THESE PICTURES COULD BE DISCUSSED IN CLASS.

HAVE THE CHILDREN BE AN INSECT WHICH THEY SAW IN THE GARDEN
AND ACT OUT WHAT THEY LEARNED ABOUT THE INSECT'S HABITS.
HAVE THE REST OF THE CLASS GUESS WHAT INSECT THEY ARE.

FOLLOW-UP POSSIBILITIES: YOU MAY WANT TO HAVE SOMEONE FROM A LOCAL
BEE CLUB TALK TO THE CLASS ABOUT THE IMPORTANCE OF BEES, HOW
THEY LIVE, AND HOW THEY ARE CARED FOR. YOU MAY HAVE THE
CHILDREN CATCH SOME INSECTS AT HOME WHICH WEREN'T TALKED ABOUT
OR SEEN DURING THE LESSON, AND DISCUSS THEM WITH THE CLASS.

FOCUS **FARMING**

LESSON **THINNING GARDEN PLANTS**

MATERIALS: APPROXIMATELY THREE FEET OF A ROW OF SEEDLINGS WHICH
ARE ABOUT THREE INCHES HIGH FOR A DEMONSTRATION SITE. EACH
CHILD SHOULD HAVE A ROW OR MORE TO THIN ALSO. RULERS OR
YARDSTICKS, SCISSORS.

TEACHER'S BACKGROUND: CHECK WITH THE PLANTING GUIDE (WITH THE
"PLANTING PLANTS IN THE GARDEN" LESSON TO KNOW THE SPACING
OF THE PLANTS TO BE THINNED.) THE PLANTS WILL MOST LIKELY
BE YOUR RADISHES, CARROTS, LETTUCE, SPINACH, CHARD, BUSH
BEANS, BEETS.

DO: IN A ROW OF PLANTS WHICH NEED THINNING, SELECT A LARGE,
HEALTHY PLANT NEAR THE END OF THE ROW TO SAVE. PLACE A
FINGER FIRMLY ON THE SOIL ON EITHER SIDE OF THIS PLANT AND
GENTLY PULL AWAY ONE OR TWO STARTS FROM AROUND THE SELECTED
START. ANOTHER METHOD IS TO TAKE A SHARP SCISSORS AND SNIP
OFF THE STARTS SURROUNDING THE ONE YOU ARE GOING TO SAVE.
THEN PLACE A RULER ALONG THE ROW WITH THE ZERO MARK AT THAT
FIRST PLANT THAT YOU SAVED. MEASURE THE SPACING FOR THE
PARTICULAR PLANTS THAT YOU ARE THINNING (REFER TO THE
PLANTING GUIDE IN THE "PLANTING PLANTS IN THE GARDEN" LESSON)
AND SELECT THE SECOND PLANT WHICH IS TO BE GROWN TO MATURITY.
THIN AROUND THAT PLANT, AND ALL THE OTHER PLANTS IN BETWEEN
THE FIRST TWO. CONTINUE ON TO THE NEXT PLANT IN THE ROW
UNTIL THE CHILDREN UNDERSTAND. IF THERE ARE NOT ENOUGH RULERS
TO GO AROUND, HAVE THE CHILDREN DETERMINE HOW MANY FINGERS
APART THE PLANTS WILL NEED TO VE. THEY CAN DETERMINE THIS
BY PLACING THEIR FINGERS BETWEEN THE PLANTS THAT THEY HAVE
THINNED. THE CHILDREN CAN THEN GO AND THIN THEIR OWN ROWS.
SUPERVISE CAREFULLY, AS THIS IS NOT EASY TO DO RIGHT AWAY
AND DAMAGE CAN EASILY BE DONE TO THE PLANTS THAT ARE TO STAY.

THINK: REVIEW THE VEGETABLES THAT YOU WILL BE GROWING IN YOUR GAR-
DEN. TALK ABOUT THE SIZE OF EACH ONE AND THE KIND OF SPACE
THAT IT WILL NEED. MENTION THAT THOUGH THINNING CAN SEEM
LIKE A MEAN THING TO DO, IT HELPS THE PLANTS THAT REMAIN
AND THEY PRODUCE MORE THAN A ROW OF CROWDED PLANTS WOULD.
TALK ABOUT THE CYCLE OF ORGANIC MATTER BREAKING DOWN TO BE-
COME SOIL AND HOW THE PULLED PLANTS WILL BECOME PART OF THE
REMAINING PLANTS EVENTUALLY ANYWAY.

COMMUNICATE: MARK OFF ON THE GROUND A 4 x 5' AREA. HAVE ALL OF
THE CHILDREN STAND WITHIN THAT AREA FOR A FEW MINUTES WHILE
YOU DISCUSS THE EFFECTS OF CROWDING OF PLANTS. EXPLAIN HOW
PLANTS NEED THE SAME KINDS OF THINGS AS HUMAN BEINGS; FOOD,
AIR, WATER, LIGHT, AND SPACE TO GROW. PLANTS THAT REMAIN
CROWDED WILL NOT GROW WELL. THEN HAVE THE CHILDREN BREAK
APART AND STAND IN A GROUP ABOUT AN ARM'S LENGTH FROM EVERY-
ONE ELSE. DISCUSS HOW IT FEELS TO HAVE MORE ROOM AND HOW
MUCH MORE COMFORTABLE IT IS FOR THEM. EXPLAIN THAT PLANTS
MUST FEEL THIS WAY TOO.

FOLLOW-UP POSSIBILITIES: PUT THE PULLED PLANTS INTO THE COMPOST
PILE OR USE THEM AS MULCH FOR THE REMAINING PLANTS. YOU MAY
WANT TO HAVE A SEPARATE PRACTICE ROW OF PLANTS FOR THINNING
PRACTICE. ALSO, TRY LEAVING ONE SECTION OF A ROW UNTHINNED
SO THAT THE CHILDREN CAN SEE HOW OVERCROWDING STUNTS PLANT
GROWTH.

FOCUS: URBAN FARMING

LESSON: FIELD TRIP TO THE URBAN FARM

MATERIALS: PHONE, PAPER, PENCILS, CRAYONS, CAMERAS, CASSETTE TAPE RECORDERS, BUTCHER PAPER, MASKING TAPE, WATER COLORS.

TEACHER'S BACKGROUND: TO SET UP THE FIELD TRIP TO THE UNIVERSITY OF OREGON URBAN FARM, CALL 686-3647.

DO: INVITE STUDENT PARTICIPATION IN THE PROCESS OF ARRANGING FOR YOUR VISIT TO THE URBAN FARM IN EUGENE, OREGON, INCLUDING INITIAL CONTACT WITH REPRESENTATIVES OF THE FARM, DECISIONS ON VISITING TIME, AND TRANSPORTATION DETAILS. WHILE VISITING THE FARM WITH THE STUDENTS, BE SURE TO ENCOURAGE EVERYONE TO PAY CLOSE ATTENTION TO THE SYSTEMS WITH ALL THEIR SENSES, KEEPING NOTE OF QUESTIONS, POINTS OF INTEREST, SUGGESTIONS FOR FOLLOW-UP. IF POSSIBLE, FACILITATE THE USE OF CAMERAS, TAPE RECORDERS, PAPER, PENCILS, CRAYONS, TO RECORD THE EXPERIENCES.

THINK: ABOUT WHAT'S SO GREAT ABOUT AN URBAN FARM? WHAT ARE THE PROBLEMS, IF ANY? WHAT MIGHT BE SOME OF THE BENEFITS? BUT, ISN'T IT A HASSLE TO WORK IN A GARDEN: BUT ISN'T IT A HASSLE TO DEAL WITH CANCER YOU MIGHT GET FROM THE PESTICIDE-SPRAYED CARROT YOU GOBBLED TWENTY YEARS AGO? WOULD WE WANT TO HAVE A FARM AT OUR SCHOOL? WHAT WOULD WE NEED TO DO TO HAVE ONE THERE? HMMMM....

COMMUNICATE: SHOW THE SKETCHES YOU DREW AND THE NOTES YOU TOOK AT THE URBAN FARM TO YOUR CLASSMATES BACK AT SCHOOL. ARRANGE A TIME WHEN VOLUNTEERS CAN DESCRIBE WHAT THE VISIT WAS LIKE FOR THEM. PERHAPS THE VISITORS' EXCITEMENT MAY RUB OFF ON THE OTHERS AND A BRAINSTORM SESSION (WITH A TAPE RECORDER ROLLING, OF COURSE) MIGHT FOCUS ON A SEQUENTIAL PROCESS TO FACILITATE THE DEVELOPMENT OF ANY OR ALL ASPECTS OF A SCHOOL FARM. SPECIFIC ASPECTS LIKE COMPOST PILES CAN BE DISCUSSED AND COMPARED TO SIMILAR SYSTEMS IN THE SCHOOL OR THE SCHOOL'S NEIGHBORHOOD. TRY TO DOCUMENT THE STEP-BY-STEP PROCESS WITH NOTES, SKETCHES, AND PHOTOGRAPHS IN A SCHOOL FARMS JOURNAL-- AND KEEP THE CASSETTE TAPES AS A RESOURCE TOO. A MURAL ON A

BUTCHER-PAPERED CLASSROOM WALL CAN SLOWLY EVOLVE TO EVOKE THE CHILDREN'S VISION OF A SCHOOLYARD TRANSFORMING INTO FUNCTIONAL AND JOYFUL GARDENS OF LEARNING.

FOLLOW-UP POSSIBILITIES: IF THERE IS A STRONG FEELING TO PARTICIPATE IN SUCH A TRANSFORMATION AT SCHOOL, THEN BEGIN BY SPEAKING WITH EVERYONE INVOLVED TO SEE WHAT THE RULES ARE, WHO MAKES THE RULES, AND HOW ARE THE RULES AMENDED TO FACILITATE THE GROWTH OF SCHOOL FARMS. IT WILL SOON BE TIME FOR DECISIONS WHETHER TO PURSUE THE GOAL AND TO ACTIVELY BECOME INVOLVED IN THE PROCESSES.

SUMMER

FOCUS: SOIL

LESSON: MULCHING FOR SUMMER

MATERIALS: SOME COMPLETELY PLANTED BEDS, STRAW, COMPOST, GRASS CLIPPINGS, LEAF MOLD, OR OTHER MULCH MATERIALS. TWO LARGE FLOWER POTS OR WOODEN PLANTERS, FILLED WITH WATER.

TEACHER'S BACKGROUND: MULCHING WILL SAVE TIME AND ENERGY IN THE GARDEN BY PREVENTING EXCESSIVE WEED GROWTH AND CUTTING DOWN ON THE WATERING NECESSARY TO KEEP THE GARDEN CONTINUOUSLY MOIST. USE MULCHING MATERIAL WHICH IS THE MOST AVAILABLE TO YOU, UNLESS YOU WANT A VARIETY OF MULCH MATERIAL TO EXPERIMENT WITH. STRAW CAN BE BOUGHT BY THE BALE OR DONATED BY A HORSE STABLE. NEWSPAPERS ARE EASILY FOUND. PLASTIC WOULD MOST LIKELY HAVE TO BE BOUGHT.

DO: ON THE RAISED BEDS IN THE GARDEN, HAVE THE CHILDREN PLACE LARGE HANDFULS OF MULCH MATERIAL CLOSELY AROUND THE STEMS OF THE PLANTS. MAKE SURE THAT THE CHILDREN DON'T COVER UP THE PLANTS WITH MULCH. HAVE THEM COVER THE WHOLE BED WITH MULCH ABOUT 1-2 INCHES DEEP. THEN HAVE THEM THOROUGHLY WATER EACH MULCHED BED.

PLACE THE TWO CONTAINERS OF SOIL IN A SUNNY PLACE IN THE GARDEN. EXPLAIN TO THE CHILDREN THAT ONE WILL HAVE A MULCH LAYER ON IT AND ONE WILL NOT. AFTER A FEW DAYS OF SUNNY WEATHER, YOU WILL ALL CHECK TO SEE HOW THE SOIL IN EACH FEELS AND LOOKS.

THINK: DISCUSS AS A GROUP SOME REASONS FOR MULCHING THE GARDEN. ASK THE CHILDREN IF THEY CAN GUESS WHY MULCHING HELPS THE GARDEN. EXPLAIN HOW RAISED BEDS TEND TO DRY OUT QUICKLY BECAUSE THEY ARE UP ABOVE THE REST OF THE GROUND. THIS HELPED FOR DRAINAGE WHEN THERE WAS LOTS OF RAIN, BUT WHEN HOT WEATHER COMES, THERE IS LESS RAIN AND MORE NEED FOR RETAINING WATER IN THE SOIL. TALK ABOUT HOW WEED SEEDS ARE KEPT FROM GROWING WITH A LAYER OF THICK MULCH BETWEEN THEM AND THE SUN.

COMMUNICATE: AFTER SEVERAL DAYS OF HOPEFULLY DRY IF NOT SUNNY WEATHER, BRING THE CONTAINERS OF SOIL, ONE MULCHED AND ONE NOT, INTO THE CLASSROOM TO LOOK AT. FEEL THE SOILS AND DISCUSS AS A GROUP HOW THEY FEEL DIFFERENT AND ASK THE CHILDREN WHICH SOIL THEY WOULD LIKE TO GROW IN IF THEY WERE A PLANT. RELATE THIS EXPERIENCE TO THE PLANTS AND ENVIRONMENT IN THE GARDEN. NOTICE HOW MOIST, COOL, AND LUSH THE SOIL FEELS UNDER THE MULCH IN THE GARDEN. THIS RICHNESS IS ONE THING THAT BRINGS LIFE TO THE GARDEN.

FOLLOW-UP POSSIBILITIES: IF THERE IS SPACE IN THE GARDEN, EXPERIMENT WITH DIFFERENT MULCH MATERIALS SUCH AS NEWSPAPER, BLACK PLASTIC, CLEAR PLASTIC, BARK MULCH, ETC. SEE IF THE GARDEN

RESPONDS DIFFERENTLY WITH THESE MATERIALS. THE CLEAR PLASTIC WILL ACT AS A GREENHOUSE FOR WEEDS, SO IF YOU USE THIS, IT WILL BE AN EXAMPLE OF WHAT NOT TO USE. BLACK PLASTIC ABSORBS SO MUCH HEAT WITHOUT LIGHT THAT IT KILLS OFF YOUNG WEED SEEDLINGS SOON AFTER GERMINATION.

FOCUS: RECYCLING

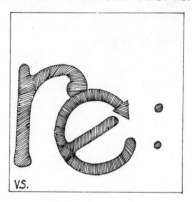

LESSON: NEIGHBORHOOD RECYCLING WALK

MATERIALS: SIX BURLAP BAGS, STIFF PAPER, SCISSORS, FELT PENS, SAFETY PINS, BROOM/DUST PAN, SHOVEL, CARDBOARD BOXES, PHONE, PHONE BOOK, STRING, PAPER, PENCILS.

TEACHER'S BACKGROUND: IF YOU LIVE IN AN ESPECIALLY CLEAN NEIGHBOR-HOOD, YOU MAY HAVE TO TRAVEL TO FIND TRASH. TRY A CITY PARK WHERE PICNICS MAY HAVE BEEN GOING ON.

DO: GATHER SOME GUNNY SACKS AND SOME OF YOUR LITTLE FRIENDS. ASK IF ANY OF THE CHILDREN MIGHT LIKE TO MAKE SOME PAPER LABELS THAT SAY OTHER PAPER, GLASS, METAL, NEWSPAPER, PLASTIC, AND COMPOST. WHEN THE LABELS ARE FINISHED, MAYBE SOME OF THE CHILDREN CAN USE SAFETY PINS TO ATTACH THE LABELS TO THE BAGS. WHEN THE CHILDREN ARE DRESSED WARMLY ENOUGH FOR THE OUTSIDE, STEP INTO THE DAY, SEARCHING FOR LITTER, TRYING TO PUT THE MATERIAL INTO THE APPROPRIATE BAGS. BE SURE TO USE A BROOM AND DUST PAN TO GATHER BROKEN GLASS.

THINK: ABOUT WHERE ALL THE LITTER CAME FROM.... WHY WOULD PEOPLE THROW THE STUFF ONTO THE GROUND? WHERE COULD PEOPLE HAVE PUT THE MATERIAL INSTEAD? CONSIDER WHO MADE THE GLASS, THE TIN CANS, THE PAPER WRAPPINGS.... WHERE DID THEY MAKE IT, AND WHY? THINK OF PACKAGING DYNAMICS. THINK OF OPTIONS TO STANDARD PACKAGING TECHNIQUES. FIGURE OUT THE BEST RESTING PLACE FOR THE COM-POST MATERIALS. CONSIDER THE REASONS FOR WASHING HANDS BE-FORE EATING WITH THEM.

COMMUNICATE: SHOW THE SEPARATED MATERIAL TO YOUR FRIENDS AT SCHOOL AND EXPLAIN WHERE THE GROUP FOUND THE STUFF. DISCUSS WHAT PEOPLE CAN DO WITH IT INSTEAD.... WHAT WILL THE STUDENTS DO WITH THE BAGS NOW? CONSIDER BEGINNING A RECYCLING CENTER IN THE CLASSROOM OR ON THE SCHOOL GROUNDS. MAYBE SOME OF THE STUDENTS WOULD LIKE TO LABEL SOME CARDBOARD BOXES FOR THE SEPARATED RECYCLABLES, SO THAT THE BAGS ARE AVAILABLE FOR LATER WALKS. ONE OR TWO OF THE STUDENTS MAY WANT TO CALL GARBAGIO'S OR ANOTHER RECYCLING GROUP TO SEE HOW MUCH THEIR SERVICES COST. INTERESTED FRIENDS CAN MAKE REPORTS ON WHAT THEY HAVE LEARNED, AND MAYBE THE GROUP WILL DISCUSS WAYS TO EARN THE NEEDED MONEY TO GAIN THE RECYCLING SERVICES.

FOLLOW-UP POSSIBILITIES: MIGHT INCLUDE MAKING A MOBILE OR SCULPTURE WITH SOME OF THE RECYCLED MATERIALS. A SMALL GROUP MIGHT WANT TO CREATE A RECYCLING QUESTIONNAIRE, TO USE DOOR-TO-DOOR IN THE SCHOOL'S NEIGHBORHOOD TO DETERMINE HOW MANY RESIDENTS ARE CONSCIOUSLY RECYCLING. STUDENTS MIGHT WANT TO KEEP CHARTS INDICATING HOW MUCH OF EACH MATERIAL IS FOUND OVER AN EXTENDED PERIOD OF TIME. PERHAPS SCHOOL FARM PARTICIPANTS COULD DEVELOP A LONG-TERM RECYCLING CENTER FOR THE SCHOOL AND THE NEIGHBORHOOD -- A PROCESS INVOLVING MANY PEOPLE, AND TUGGING AT THE IMAGINATION. HOWDY NEIGHBORS.

FOCUS:  NUTRITION

LESSON: MAKING BUTTER

MATERIALS: FRESH CREAM, SALT, STERILIZED JARS AND LIDS, PAPER, PENCILS, SCOTCH TAPE, SCISSORS

TEACHER'S BACKGROUND: HOMEMADE BUTTER IS THE BEST. THE CHILDREN MAY NOT LIKE THE TASTE OF FRESH UNSALTED BUTTER AT FIRST, BUT ENCOURAGE THEM TO TRY IT.

DO: ASK EVERYONE TO WASH THEIR HANDS AND THE JARS IN WARM, SOAPY WATER. RINSE THEM WELL. ADD FRESH CREAM TO EACH JAR--SO THAT THE CONTAINERS ARE ABOUT HALF FULL. TAKING TURNS, CAREFULLY, BUT VIGOROUSLY SHAKE THE JARS FOR AS LONG AS IT TAKES FOR SOLIDS IN THE CREAM TO LUMP TOGETHER INTO BUTTER. SING YOURSELF A SONG. ADD SALT IF YOU LIKE, AND SPREAD ON CRACKERS, BREAD, OR FRESH COOKED CORN ON THE COB FOR A TASTE DELIGHT.

THINK: DOES IT TASTE LIKE THE $2.79 SPREAD? HOW DOES OUR BUTTER MAKING PROCESS COMPARE TO A CREAMERY BUTTER PRODUCTION OP-ERATION? HOW DO THE TWO COMPARE IN TERMS OF ENERGY USED TO SUPPLY THE BY PRODUCT, BUTTER? WHAT WOULD HAPPEN TO BUTTER IF LEFT IN A WARM PLACE FOR ONE HOUR, FOUR DAYS, THREE WEEKS, FOUR FULL MOONS, TWO BIRTHDAYS, AND A CENTURY? RE-CORD THE PREDICTIONS IN THE SCHOOL FARMS JOURNAL, PERHAPS IN THE FORM OF SPECULATIVE FICTION. OBSERVE THREE WEEK OLD, WARM BUTTER AND RECORD YOUR FINDINGS IN THE JOURNAL. TRY TASTING SOME BUTTER WITHOUT SALT. WHAT IS THAT WATERY STUFF AT THE BOTTOM OF THE JAR? EXPLAIN TO THE CLASS HOW THE CREAM CHANGES SO THAT THE FAT GLOBS TOGETHER. THIS IS THE BUTTER. IF YOU WHIP CREAM LONG ENOUGH, IT TURNS TO BUTTER.

COMMUNICATE: CAN YOU DESCRIBE UNSALTED BUTTER IN ONE WORD? TWO WORDS? WHAT COLOR ARE YOU REMINDED OF WHEN YOU THINK OF BUTTERED, SALTED HOT CORN ON THE COB? TAKE TURNS TELLING YOUR FRIENDS (WITH A TAPE RECORDER LISTENING IN) ABOUT YOUR FAVORITE MEAL WITH BUTTER, AND MABYE A BRIEF DESCRIPTION OF HOW TO CREATE THE CONCOCTION. ADD THE BUTTER MAKING PROCESS TO THE SCHOOL FARM JOURNAL. CONSIDER PACKAGING YOUR BUTTER IN STERILIZED JARS WITH LIDS, AND TRADING IT FOR VALUABLE ITEMS FROM OTHER FARMS IN THE AREA. BE SURE TO INCLUDE THE BUTTER SHAKERS' NAMES AND THE LIST OF INGREDIENTS.

FOLLOW UP POSSIBILITIES: ORGANIZE A SCHOOL FARMS STORE TO SELL BUTTER, EGGS, FRESH VEGETABLES, FRUIT, PEANUT BUTTER, WHOLE WHEAT BREAD. THE STUDENTS CAN DO THE WHOLE THING, INCLUDING SHARING THE PROFIT, WITH CREATIVE FRIENDLY SHARED POWERS FROM TEACHERS ACTING AS FACILITATORS.

FOCUS:  SOILS

LESSON:  POTTING SOIL SHAKERS

MATERIALS: CARDBOARD BOXES, DIRT, GRAVEL, SAND, PEAT MOSS, NEWS-
PAPER, COFFEE CANS, YOGURT CONTAINERS, LIDS, STYROFOAM EGG
CARTONS, TAPE, PENCIL, PAPER, TAPE RECORDERS, SEEDS.

TEACHER'S BACKGROUND: IMPROVISE WITH CONTAINERS AND MATERIALS FOR
INSIDE OF THE SHAKERS. ANYTHING AVAILABLE TO YOU CAN WORK.

DO: WITH A BUNCH OF TOE-TAPPING FRIENDS, GATHER TOGETHER BOXES OF
GRAVEL, DIRT, SAND, AND PEAT MOSS AND ADD VARIOUS COMBINA-
TIONS OF THE MATERIAL INTO EMPTY COFFEE CANS, AND PLASTIC
YOGURT CONTAINERS WITH MATCHING PLASTIC LIDS. STYROFOAM
EGG CARTONS SERVE AS USEFUL CONTAINERS ALSO. ADD VARYING
AMOUNTS OF THE COMBINATIONS AND USE TAPE TO SEAL THE LIDS
TO THE CONTAINERS.

THINK: LISTEN CLOSELY TO EACH SHAKER
AND THEN CLOSE YOUR EYES AND
LISTEN AGAIN. CAN YOU PICK OUT
A PARTICULAR SOUND IN A ROOM-
FULL OF SHAKERS? TRY COUNTING
WITH THE SHAKERS PROVIDING A
BACKGROUND RHYTHM.... THINK
OF OTHER CONTAINERS WE COULD
USE FOR SHAKERS. THINK/PREDICT
HOW PLANTS COULD GROW IN VARIOUS
GRADES OF SOIL, SAND, GRAVEL,
PEAT MOSS AND COMBINATIONS. RE-
CORD THE PREDICTIONS.

COMMUNICATE: WITH A TAPE RECORDER
LISTENING, CREATE A RHYTHM EN-
SEMBLE. INVITE THE CLASS NEXT
DOOR TO HEAR YOUR TAPE AS YOU
DANCE AS VEGETABLES GROWING IN
THE BALANCED SOIL.

FOLLOW-UP POSSIBILITIES: GATHER OTHER KINDS OF CONTAINERS TO MAKE
MORE SHAKERS. USE SEEDS INSTEAD OF SOILS. PLANT A FEW RADISH
SEEDS IN VARIOUS SOIL/SAND/GRAVEL/PEAT MOSS COMBINATIONS AND ALONE.
WATER AND PAY ATTENTION OVER A PERIOD OF DAYS TO SEE WHAT HAPPENS
TO THE SEEDS. KEEP TRACK OF THE RESULTS AND CHECK THE PREDICTIONS.

FOCUS:  SOIL

LESSON: MANURE AND COMPOST "TEA"

MATERIALS: BURLAP FABRIC OR BAGS, BUCKETS (5-GALLON), WATERING CANS
OR POURING CONTAINERS, MANURE OR COMPOST.

TEACHER'S BACKGROUND: MANURE AND COMPOST TEAS ARE DYNAMITE FOR
PLANTS AND ARE EASY TO MAKE.

DO: HAVE THE CHILDREN MEASURE 4-5
CUPS OF MANURE INTO BURLAP BAGS
OR BURLAP FABRIC AND TIE IT
CLOSED SOMEHOW SO THAT NO MANURE
WILL ESCAPE. PLACE ONE OF THESE
INTO EACH BUCKET (HOWEVER MANY
YOU HAVE), AND FILL THEM CLOSE
TO THE TOP WITH WATER. LET
THESE SIT OVERNIGHT, AND THE
NEXT DAY PULL OUT THE BAG OF
MANURE AND LET IT DRAIN. THE
"TEA BAGS" ARE REUSALBE FOR 2-3
MORE BATCHES OF MANURE TEA. PUT
THE BROWN MANURE TEA INTO WATER-
ING CANS OR JUST CARRY THE
BUCKET AROUND BY HAND, AND HAVE
THE CHILDREN WATER INDIVIDUAL
PLANTS WITH THE TEA. THE PLANTS
WILL RESPOND! IF THE "TEA" LOOKS
VERY DARK OR VERY STRONG, DILUTE
IT WITH WATER, SO YOU DON'T BURN
THE PLANT'S ROOTS.

THINK: EXPLAIN TO THE CLASS BRIEFLY HOW PLANTS USE NUTRIENTS FOR
GROWTH, AND REMIND THEM THAT NITROGEN IS THE ONE ELEMENT
PLANTS USE IN THE MOST ABUNDANCE. THE WATER THAT THE MANURE
"TEA BAG" SITS IN GETS MUCH OF THE NITROGEN AND OTHER NUTRIENTS
FROM THE MANURE INTO IT. THE PLANTS LOVE IT BECAUSE IT IS
SUCH A CONCENTRATED DOSAGE OF THE NUTRIENTS THAT THEY LOVE.

COMMUNICATE: CHOOSE SEVERAL PLANTS OUTSIDE TO EXPERIMENT ON IN THE
GARDEN. WATER SOME WITH MANURE TEA FOR SEVERAL DAYS AND DO
NOTHING TO THE OTHERS EXCEPT TO WATER THEM WITH PLAIN WATER.
HAVE THE CHILDREN PREDICT HOW THE PLANTS WILL DIFFER AFTER A
WEEK OR TWO. MAKE GUESSES AS TO HOW MUCH DIFFERENCE IN HEIGHT
THE PLANTS WILL BE AFTER TWO WEEKS. WRITE THESE DOWN AND THEN
CHECK THEM WITH WHAT ACTUALLY HAPPENS.

FOLLOW-UP POSSIBILITIES: USE MANURE TEA PERIODICALLY THROUGHOUT THE
SEASON ON THE GARDEN. IT IS EASY AND REALLY GIVES PLANTS A
BOOST.

FOCUS NUTRITION

LESSON MAKING CARROT JUICE

MATERIALS: TEN POUNDS OF CARROTS, CUTTING BOARDS, MEDIUM SHARP
KNIVES (CAREFUL PLEASE) AN ELECTRIC CARROT JUICER, BOWLS,
CUPS, COMPOST BUCKET, TAPE RECORDER, PAPER, PENCILS.

TEACHER'S BACKGROUND: YOU MAY HAVE TO BORROW A JUICER FROM SOME-
ONE YOU KNOW. IF YOU CAN'T FIND A JUICER, DO THE SAME LES-
SON WITH ORANGES AND A SIMPLE GLASS CITRUS JUICER.

DO: AFTER TURNING ON THE RECORDER, WASHING THE CARROTS AND LOTS OF
READY HANDS, PUT ALL THE JUICER'S PARTS TOGETHER IN AN AP-
PROPRIATE PLACE. CAREFULLY CUT THE CARROTS IN PIECES THAT
WILL FIT EASILY INTO THE JUICER, AND LOOK AT THE GRAIN, FEEL
THE TEXTURES, SNIFF THE ESSENCE, AND ENJOY THE TASTE. CUT
A CARROT IN HALF AND COUNT THE CIRCLES.

THINK: WHY ARE CARROTS YELLOW? DO WE EAT THE LEAVES? WHY ARE
THERE RINGS INSIDE THE ROOT? LOOK AT AN UNCUT CARROT PLANT
AND WONDER HOW IT GROWS FROM THE SEED. WHAT OTHER FOODS ARE
ROOTS LIKE CARROTS? BEFORE PUTTING THE CARROT SECTIONS INTO
THE JUICER, MENTION THAT WE OUGHT TO BE CONSCIOUS OF CARE IN
USING THE MACHINE. ASK WHAT MIGHT BE LEFT AFTER THE CARROTS
GO INTO THE MACHINE. HOW MUCH ENERGY DOES IT TAKE TO GET
CARROT JUICE BY SLOWLY CHEWING A CARROT? COMPARE THE COST
OF MOUTH-JUICED CARROTS TO CARROTS JUICED BY MACHINE. WHY DO
PEOPLE MAKE CARROT JUICE? IS THE PULP USEFUL? HOW MUCH
DOES AN ELECTRIC JUICER COST?

COMMUNICATE: KEEP A RECORD OF HOW MANY CARROTS IT TAKES TO GET A
QUART OF JUICE. HOW MUCH PULP IS LEFT WHEN WE MAKE ONE QUART
OF CARROT JUICE? PERHAPS A SHORT-LIVED DISPLAY OF JUICER,
CARROTS, JUICE AND PULP (ALL NEATLY LABELED) WILL BE USEFUL
FOR LEARNING ABOUT VOLUME AND WEIGHT. WEIGH FIVE CARROTS
BEFORE JUICING AND RECORD THE WEIGHT. AFTER JUICING THE
CARROTS, WEIGH THE JUICE AND PULP AND SEE HOW THE WIEGHTS
COMPARE (BE SURE TO WEIGH THE JUICE CONTAINER BEFORE FILLING)
ENCOURAGE THE NOTES ON CARROT JUICING TO BE ENTERED INTO THE
SCHOOL FARM JOURNAL.

FOLLOW-UP POSSIBILITIES: IF THE CHILDREN AGREE, MAKE ENOUGH CARROT
JUICE TO SHARE WITH CLASSMATES AND HAVE A MID-MORNING SNACK
TOGETHER. KEEP THE TAPE RECORDINGS AVAILABLE FOR INTERESTED
LISTENERS. CUT A CARROT TOP FROM THE ROOT SECTION (LEAVING A

SECTION OF ORANGE ROOT ATTACHED TO THE LEAVES) AND PLACE IN
A SHALLOW DISH WITH A SMALL AMOUNT OF WATER. PLACE THE DISH
IN A WELL LIGHTED LOCATION AND WATCH THE CHANGES OVER THE
NEXT SEVERAL WEEKS. MAKE VEGETABLE PRINTS USING CUT CARROTS
AND FRIENDS DIPPED IN WATER SOLUABLE INK.

GROUP MADE THE RULES. DISCOVER THE RULE MAKING PROCESS. DE-
CIDE IF YOU HAVE THE ENERGY TO INTERVENE TO TRY TO CHANGE THE
RULES SO THAT YOU CAN ENJOY THE RABBITS AND CHICKENS AT
SCHOOL. CONSIDER THE INITIATIVE PETITION PROCESS, SUPPORT
CANDIDATES WHO SUPPORT YOUR GOALS. RUN FOR OFFICE. SHARE
WHAT YOU'VE LEARNED WITH YOUR CLASSMATES AND NEIGHBORS.

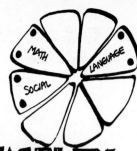

FOCUS: FARMING

LESSON: RESEARCHING FEASIBILITY OF ANIMALS AT YOUR SCHOOL FARM....

MATERIALS: PHONE, PHONE BOOK, PAPER, PENCILS, ENVELOPES, STAMPS,
    LOG BOOK, CASETTE TAPE RECORDER.

TEACHER'S BACKGROUND: THIS IS A FINE LESSON FOR A RAINY DAY. EVEN
    IF YOUR SCHOOL GARDEN HAS NO SPACE FOR ANIMALS, CHECK INTO
    THE LEGALITY OF IT ALL.

DO: SUGGEST TO A SMALL GROUP OF
        INTERESTED STUDENTS THAT IT
        MIGHT BE FUN AND HEALTHY TO
        HAVE A BUNCH OF CHICKENS AND
        RABBITS AT THE SCHOOL FARM
        --AFTER ALL, WHAT'S A FARM
        WITHOUT CHICKENS AND RABBITS?
        NOW, IT MIGHT BE AGAINST THE
        RULES TO HAVE THE ANIMALS
        AT SCHOOL.

THINK: HMMM? I WONDER WHOSE RULES
        WE'D NEED TO DEAL WITH, AND
        HOW WE MIGHT BE ABLE TO GET
        THE RULES CHANGED TO A
        HAPPIER PLACE, SO THAT WE
        COULD ENJOY, FEED AND CARE
        FOR OUR ANIMALS AT SCHOOL?
        MAYBE WE COULD TALK TO THE
        PRINCIPAL, THE SCHOOL DISTRICT
        ADMINISTRATION, THE SCHOOL BOARD, THE CITY PLANNERS, THE
        ZONING DEPARTMENT, THE CITY COUNCIL, THE COUNTY HEALTH DEPART-
        MENT. MAYBE WE COULD EITHER WRITE, VISIT, OR CALL THE APPRO-
        PRIATE PEOPLE TO FIND OUT ABOUT EXISTING RULES THAT MIGHT
        PROHIBIT CHICKENS AND RABBITS LIVING AT THE SCHOOL.

COMMUNICATE: FACILITATE THE DISCOVERY OF SPECIFIC CONTACT PEOPLE
        WITH THE VARIOUS OFFICES OR GROUPS MENTIONED ABOVE, VARIOUS
        ADDRESSES, PHONE NUMBERS AND OTHER PERTINENT INFORMATION.
        BEGIN MAKING CONTACT WITH THE VARIOUS RULE MAKERS/ENFORCERS
        VIA PHONE CALLS AND LETTERS. MAKE APPOINTMENTS WITH THE
        CONTACT PEOPLE. BEGIN KEEPING A LOG OF INFORMATION GATHERED.
        VISIT WITH SOME CHICKENS AND TELL THEM WHAT YOU ARE DOING.
        ASK FOR THEIR ADVICE.

FOLLOW-UP POSSIBILITIES: BE PROMPT AND COURTEOUS AT YOUR APPOINT-
        MENTS WITH VARIOUS OFFICIALS. KEEP A CLEAR RECORD OF ALL
        THAT IS SAID -- USE A CASETTE TAPE RECORDER WHENEVER POSSIBLE.
        AN INTERVIEW...FIND OUT JUST WHAT ARE THE RULES AND WHO/WHICH

FOCUS: **HEALTH**

LESSON: **TASTING RAW AND COOKED VEGETABLES AND FRUIT**

FOOD	RAW	COOKED
🥕		
🍎		
🥬		
🍐		

MATERIALS: VARIOUS VEGETABLES AND FRUIT WHICH ARE GOOD RAW OR
COOKED -- APPLES, PEARS, CARROTS, CELERY, CAULIFLOWER, PEAS,
BROCCOLI, ONIONS, BEANS, ETC.; A HOT PLATE OR STOVE OF SOME
KIND; TWO OR MORE SAUCEPANS, STEAMERS; KNIFE, PLATES OR
BOWLS FOR THE CLASS, FORKS FOR HOT FOOD, POSTER BOARD, FELT
PENS.

TEACHER'S BACKGROUND: YOU MAY WANT TO KEEP THIS VERY SIMPLE, WITH
JUST FOUR FOODS. IF THERE IS NO WAY TO COOK IN THE CLASS-
ROOM, PERHAPS A DIFFERENT ROOM WOULD BE AVAILABLE. MAKE
SURE THAT THE CHILDREN FEEL THAT WHATEVER THEY LIKE OR DIS-
LIKE IS FINE AND NOT TO BE JUDGED. THIS LESSON IS TO BECOME
AWARE OF OPTIONS IN NUTRITION AND TO LEARN ABOUT WHAT IS
GOOD FOR OUR BODIES.

DO: TAKE THE DIFFERENT FRUITS AND VEGETABLES, CUT THEM INTO BITE-
SIZED PIECES, AND MAKE TWO EQUAL PILES OF EACH FOOD, MAKING
SURE THERE IS ONE PIECE ON EACH PILE FOR EACH CHILD. IN
TWO SAUCE PANS WITH STEAMERS, HAVE THE CHILDREN PUT ABOUT
1/2-3/4 OF A CUP OF WATER FOR STEAMING IN EACH ONE. STEAM
THE VEGETABLES AND FRUIT UNTIL THEY ARE TENDER, BUT NOT TOO
WELL DONE. WHEN THEY ARE SLIGHTLY COOLED OFF, ARRANGE THEM
ON PLATES AND PLACE NEXT TO THEIR RAW COUNTERPARTS. GO
THROUGH EACH VEGETABLE AND FRUIT ONE AT A TIME AND HAVE EACH
CHILD TASTE A RAW FOOD AND THEN THE COOKED FOOD, AND DECIDE
FOR HIM/HERSELF WHAT THEY LIKE OR DISLIKE ABOUT EACH.

THINK: HAVE A DISCUSSION ABOUT THE DIFFERENCE BETWEEN RAW AND
COOKED FOODS. GET THE CHILDREN TO UNDERSTAND THAT COOKING
IN ITSELF (HEAT) WILL DESTROY SOME VITAMINS IN FOODS AND
THAT BOILING VEGETABLES (WATER) WILL DESTROY SOME AS WELL.
STEAMING IS A GOOD COOKING METHOD FOR PRESERVING SOME OF
THE NUTRIENTS IN FOODS. EXPLAIN THAT RAW VEGETABLES ARE
GOOD FOR DIGESTION BECAUSE THEY HAVE "FIBER", ALSO CALLED
"ROUGHAGE", WHICH CLEANS OUT THE INTESTINES AS IT PASSES
THROUGH OUR BODIES. TALK ABOUT HOW ANIMALS CAN EAT MUCH
ROUGHER FOODS THAN WE CAN BECAUSE THEY ARE ABLE TO DIGEST
THINGS LIKE GRASS AND HAY, BUT WE CAN'T.

COMMUNICATE: MAKE A CHART ON THE WALL OF THE DIFFERENT FOODS WHICH
YOU TASTED RAW AND COOKED, OR SIMPLY DISCUSS THEM ONE AT A
TIME. ASK EACH CHILD WHAT HE/SHE LIKED OR DISLIKED ABOUT
EACH FOOD, AND TALK ABOUT IT OR LIST THEIR RESPONSES ON THE
CHART. LIST THE HEALTH ASPECTS OF EACH COOKED AND RAW FOOD
AS WELL.

FOLLOW-UP POSSIBILITIES: ASK THE CHILDREN TO WRITE DOWN HOW MANY
DIFFERENT FOODS THEY EAT AT HOME WHICH ARE COOKED OR RAW.
BRING THESE LISTS TO SCHOOL AND TALK ABOUT THEM.

FOCUS: RECYCLING

LESSON: PHOTOSYNTHESIS OXIDATION FOOD CHAIN DANCE!

MATERIALS: THIRTY FEET OF ROPE, LOTS OF CARDBOARD PIECES LABELED, APPROPRIATELY, PHOTOSYNTHESIS, OXIDATION FLOW DIAGRAM.

TEACHER'S BACKGROUND: PHOTOSYNTHESIS DIAGRAMS CAN BE FOUND IN MOST PLANT BIOLOGY TEXTBOOKS. IT MAY HAVE TO BE SIMPLIFIED, OR DONE IN PICTURES. PICTURES ARE EASIEST AND ARE SIMPLE FOR THE CHILDREN TO UNDERSTAND.

DO: HELP THE CHILDREN GATHER THE MATERIALS AND MAKE THE VARIOUS LABELS. ON A LARGE FLOOR OR SCHOOL GROUND, LAY ON THE GROUND A THIRTY-FOOT LENGTH OF ROPE IN THE SHAPE OF A STEM AND LEAF. CHILDREN LABELED "CHLOROPHYLL" SHOULD BE BUBBLING AROUND INSIDE THE "LEAF", AND FROM THE SURROUNDING AREAS, CHILDREN LABELED CO_2 (CARBON DIOXIDE), H_2O (WATER) AND E (ENERGY FROM THE SUN), GRADUALLY FILTER INTO THE INSIDE OF THE "LEAF" TO BUBBLE AND BUMP AROUND TOGETHER FOR A WHILE. AFTER A LITTLE LAPSE OF TIME, SOME OF THE CHILDREN WILL TURN THEIR CO_2, E AND H_2O LABELS AROUND TO SHOW A SIDE WHICH READS O_2 (OXYGEN). THE CHILDREN LABELED O_2 SHOULD BUBBLE ON OUT OF THE "LEAF" AND BE SURROUNDED BY FOUR OR FIVE CHILDREN HOLDING HANDS, LABELED ANIMAL LUNGS. OTHER CHILDREN INSIDE THE "LEAF" CAN TURN THEIR LABELS OF CO_2 H_2O & E TO THE OTHER SIDE WHICH SAYS SUGAR. AFTER A LITTLE MORE BUBBLING AROUND, THE CHILDREN INSIDE THE "LEAF" LABELED SUGAR FOLD UP THE LABELS, PUT THEM IN A POCKET, AND UNFOLD AND DISPLAY ONE OF TWO NEW LABELS SAYING CARBOHYDRATES AND PROTEINS. THESE NEWLY-LABELED CHILDREN CONTINUE TO BUBBLE AROUND FOR AWHILE UNTIL FINALLY A LARGE GROUP OF CHILDREN, HOLDING HAND, AND LABELED ANIMAL MOUTH, SURROUNDS THE WHOLE "LEAF", ROPE, LABELED CHILDREN, AND ALL. SOME O_2-LABELED CHILDREN OCCASIONALLY GET INSIDE THE "MOUTH", AND EVENTUALLY ALL THE CHILDREN WHO ARE INSIDE THE MOUTH SWITCH THEIR LABELS FROM CARBOHYDRATES, O_2 AND PROTEIN, TO CO_2, H_2O, AND E. OCCASIONALLY THESE CHILDREN CAN BUBBLE OUT OF THE MOUTH. FORGET TO BURP THE BABY, AND WATER THE PLANTS.

THINK: WHILE LOOKING AT THE TWO DIAGRAMS (SUGAR + OXYGEN CARBON DIOXIDE + WATER + ENERGY = PHOTOSYNTHESIS, AND CARBON DIOXIDE + WATER + ENERGY   SUGAR + OXYGEN = OXIDATION), ASK THE CHILDREN IF THEY UNDERSTAND WHY THE TWO

SYSTEMS ARE SO IMPORTANT FOR PLANTS AND ANIMALS. FOR ALL LIVING THINGS TO STAY ALIVE, THEY NEED A CONSTANT SUPPLY OF ENERGY AND FOOD. WHAT HAPPENS IF PLANTS NO LONGER EXIST? CAN ANIMALS PHOTOSYNTHESIZE? DO PLANTS NEED ANIMALS?

COMMUNICATE: ASK A GROUP OF CHILDREN IF THEY WOULD LIKE TO MAKE A LARGE CHART OF THE PHOTOSYNTHESIS AND OXIDATION CYCLES. PERFORM THE "DANCE" EXPLAINED ABOVE TO THE SCHOOLMATES DOWN THE CORRIDOR. HAVE A NARRATOR EXPLAIN WHAT IS HAPPENING. REPORT WHAT'S BEEN LEARNED IN THE SCHOOL FARM JOURNAL.

FOLLOW-UP POSSIBILITIES: CREATE AND TAPE RECORD SOME BACKGROUND MUSIC FOR THE "PHOTOSYNTHESIS/OXIDATION FOOD CHAIN DANCE", DRAW PICTURES OF SOME TYPICAL FOOD CHAINS -- WHO'S EATING WHO.

FOCUS: RECYCLING

LESSON: FARMERS' SCAVENGER HUNT

MATERIALS: PAPER, PENCILS, WAGONS, BIKE CARTS, BAGS, BOXES, MASKING TAPE, 3x5 CARDS, FILE BOX.

TEACHER'S BACKGROUND: THIS LESSON IS WONDERFUL FOR SPREADING THE NEWS ABOUT SCHOOL GARDENS AND EVOKING NEIGHBORHOOD PARTICI-PATION. TELL NEIGHBORS ALL ABOUT YOUR GARDEN PROJECT.

DO: WITH A GROUP OF FRIENDS, MAKE A LIST OF ITEMS THAT MIGHT BE USEFUL AT A SCHOOL FARM, SUCH AS SEEDS, STRING, STAKES, EGG CARTONS, CANS, OLD TOOLS, BIG OLD BLANKETS, GARDENING BOOKS, SAND, COMPOST,.... GATHER TOGETHER SOME MORE FRIENDS, DIVIDE INTO GROUPS OF 4 OR 5 PER BIG PERSON, AND TAKE LISTS, CONTAINERS (BOXES, PLASTIC BUCKETS, CANS, ETC.) AND WAGONS AROUND THE NEIGHBORHOOD. SMALLER GROUPS OF STUDENTS CAN WALK DOOR-TO-DOOR SAYING HELLO, MENTIONING THE GROWING SCHOOL FARMS, AND THE NEED FOR NEIGHBORHOOD SUPPORT AND MATERIALS IF POSSIBLE.

THINK: WHAT POSSIBLE ITEMS OR ENER-GIES MIGHT THE NEIGHBORS HAVE THAT COULD BE USEFUL AT THE SCHOOL FARM? HOW CAN THESE PEOPLE BE INTEGRATED INTO THE EVOLVING STRUCTURE OF THE FARM? DO THE NEIGHBORS HAVE IDEAS FOR THE FARM? BE SURE TO KEEP TRACK OF ALL SUGGESTIONS, AND NAMES, AND CONTACT NUMBERS, OF PEOPLE WILLING TO HELP CREATE THE SCHOOL FARM.

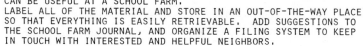

COMMUNICATE: RETURN TO THE SCHOOL WITH YOUR TREASURES, AND SHOW YOUR AMAZED FRIENDS. EXPLAIN TO ANY FOLKS INTERESTED JUST HOW AN OLD CAN OR AN OLD SHIRT CAN BE USEFUL AT A SCHOOL FARM. LABEL ALL OF THE MATERIAL AND STORE IN AN OUT-OF-THE-WAY PLACE SO THAT EVERYTHING IS EASILY RETRIEVABLE. ADD SUGGESTIONS TO THE SCHOOL FARM JOURNAL, AND ORGANIZE A FILING SYSTEM TO KEEP IN TOUCH WITH INTERESTED AND HELPFUL NEIGHBORS.

FOLLOW-UP POSSIBILITIES: INVITE THE NEIGHBORS TO LUNCH TO SEE ALL OF YOUR TOOLS, MATERIALS, AND FRIENDS AND DECIDE ON HOW TO PROCEED WITH THE SCHOOL FARM. PERHAPS EVERYONE CAN MIX SOME POTTING SOIL TO GET SOME SEEDS STARTED ON THEIR WAY.

FOCUS ENERGY

LESSON THE REAL COST OF A CARROT

MATERIALS: ONE ORGANIC CARROT, AND ONE INORGANIC, AGRA-BUSINESS CARROT, PAPER, PENCILS, ACCESS TO LIBRARIES, CASETTE TAPE RECORDERS, RAILROAD BOARD, FELT PENS

TEACHER'S BACKGROUND: ORGANIC CARROTS CAN BE PRUCHASED AT THE COMMUNITY STORE AT 444 LINCOLN IN EUGENE, OREGON, THE COM-MERCIALLY GROWN CARROTS AT ANY SUPERMARKET.

DO: SITTING WITH A CLOSE KNIT GROUP OF STUDENTS CLOSELY EXAMINE TWO CARROTS. ONE IS FROM A LOCALLY OWNED-OPERATED NEIGHBOR-HOOD STORE, GROWN ORGANICALLY CLOSE TO TOWN. THE OTHER CAR-ROT WAS PURCHASED AT A BIG CHAIN GROCERY STORE, POSSIBLY GROWN THOUSANDS OF MILES AWAY, WITH LOTS OF INORGANIC FER-TILIZERS, AND POTENTIALLY DANGEROUS PESTICIDES, AND SHIPPED VIA TRUCK OR TRAIN USING ENORMOUS AMOUNTS OF ENERGY.

THINK: WHAT ARE THE REAL COSTS OF THE TWO CARROTS. HOW DO THE CARROTS COMPARE IN TERMS OF FUEL, ENERGY, TIME, HEALTH RISKS AND ENVIORNMENTAL DAMAGE. TRY TO FIND THIS INFORMATION USING SOME RESOURCES OF THE SCHOOL, CITY OR NEARBY COLLEGE LIBRARY. USING PAPER, PENCILS, TAPE RECORDERS AND PHOTOCOPY MACHINES, GATHER AS MUCH DOCUMENTATION TO BRING BACK TO CLASS.

COMMUNICATE: THE STUDENTS CAN CONSTRUCT A MATRIX TO COMPARE THE TWO CARROTS ACCORDING TO THE POSITIVE AND NEGATIVE CRITERIA MEN-TIONED ABOVE, FILLING IN THE BLANKS WITH THE INFORMATION RE-SEARCHED BY THE CHILDREN. AT SOME POINT, SOME IN THE GROUP MAY VENTURE THEIR OPINIONS ABOUT THE "REAL" COST OF THE CARROTS. ENCOURAGE THE CHILDREN TO SUPPORT THEIR OPINIONS WITH WELL DOCUMENTED INFORMATION. PERHAPS SMALL GROUPS CAN EXAMINE EACH OF THE TWO CARROTS ACCORDING TO THE SPECIFIC CRITERIA MENTIONED BEFORE. RECORD THE FINDINGS IN THE SCHOOL FARM JOURNAL.

FOLLOW-UP POSSIBILITIES: WHEN THE FINDINGS ARE "COMPLETED", ARRANGE THE TIME FOR YOUNG RESEARCHERS TO ADDRESS THEIR CLASS MATES ON THE VARIOUS COSTS FOR CARROTS, BOTH ORGANIC AND INORGANIC. PUBLISH THE FINDINGS. PUT 20 ORGANIC CARROTS IN ONE BOWL AND 20 INORGANIC CARROTS IN ANOTHER BOWL. LABEL BOTH BOWLS AP-PROPRIATELY AND SUGGEST THAT ONE CHILD CHOOSE ONE CARROT TO EAT. WHAT ARE THE RESULTS?

FOCUS:  FARMING

LESSON: FIELD TRIP TO CRESWELL HEAD START FARM

MATERIALS: TRANSPORT VEHICLES, FUEL, PENCILS, PAPER, CASETTE
TAPE RECORDERS

TEACHER'S BACKGROUND: TO SET UP THIS TRIP, CALL RICHARD SNELL
AT 1-895-3106 TWO WEEKS IN ADVANCE PREFERABLY.

DO: AFTER MAKING NECESSARY ARRANGEMENTS AND RESERVATIONS, THE
CHILDREN CAN VISIT RICHARD SNELL AND HIS FRIENDS AT THE
HEAD START FARM NEAR CRESWELL, OREGON. BE SURE TO RECORD
AS MANY IMPRESSIONS AS POSSIBLE WITH TAPE RECORDED INTER-
VIEWS, HAND-WRITTEN NOTES, SKETCHES, OR PHOTOGRAPHS. ASK
IF ANY HANDOUT INFORMATION SHEETS ARE AVAILABLE. HANDLE
THE TOOLS, SMELL THE HAY, TASTE SOME VEGETABLES.

THINK: SUGGEST THAT THE CHILDREN DISCOVER HOW THE SCHOOL FARM
CAME TO LIFE. KEEP TRACK OF THE PROCESS FOR LATER USE.
FIND OUT HOW THE FARM'S CYCLES OF PLANTS AND ANIMALS (IN-
CLUDING THE HUMAN CARETAKERS) INTERRELATE AND ENHANCE THE
FARM'S HEALTHY GLOW. HOW COULD SOME OR ALL OF THE HEAD
START FARM'S LIVING SYSTEMS BE INTEGRATED INTO AN URBAN
ELEMENTARY SCHOOL SYSTEM? WHAT ABOUT RULES, MONEY, TIME,
ENERGY? WHAT ARE THE POSSIBLE NEGATIVE CONSEQUENCES OF
WORKING TOWARD A SCHOOL FARM IN OUR OWN CITY SCHOOL?
WHAT ARE SOME OF THE BENEFITS OF SCHOOL FARMS FOR THE
STUDENTS, TEACHERS, SCHOOL STAFF, NEIGHBORS AND THE CITY
AS A WHOLE?

COMMUNICATE: FORM SMALL GROUPS OF THREE TO FIVE STUDENTS TO
BRAINSTORM ANSWERS TO SOME OF THE QUESTIONS. RECORD THE
PROCESS AND CONTENT, USING RECORDERS AND/OR JOURNALS. IN
LARGER GROUPS, ASK SMALL GROUP REPRESENTATIVES TO REPORT
THEIR FINDINGS TO THEIR FRIENDS. SHARE THE INFORMATION
WITH OTHER CLASSES, TEACHERS AND SCHOOL STAFF AND SET TIMES
FOR ALL INTERESTED PARTICIPANTS TO SHARE THEIR EXCITEMENT
AND TO PLAN FURTHER ACTION. KEEP MINUTES OF MEETINGS AND
PROVIDE CASETTE TAPES OF DISCUSSIONS AND INTERVIEWS WHEN-
EVER POSSIBLE. PROVIDE A SUGGESTION BOX FOR PEOPLE TO
ENHANCE COMMUNICATION.

FOLLOW-UP POSSIBILITIES: PLANT THE SEEDS OF THE SCHOOL FARM. IF
POSSIBLE, GAIN PERMISSION TO CREATE AN APPROPRIATELY-SITED
RAISED BED. TURN AND NOURISH THE SOIL. GET SOME WELL-TIMED
STARTS GROWING, THEN TRANSFER INTO THE RAISED BED, BEING CARE-
FUL TO ALLOW FOR A GRADUAL TRANSITION TO THE OUTSIDE WORLD.
KEEP TRACK OF ALL PROCESSES, INFORMATION, SUGGESTIONS AND RE-
SOURCES IN WELL-ORGANIZED FILES AND JOURNALS. ENCOURAGE YOUR
CLASSMATES IN CROSSTOWN SCHOOLS TO DO THE SAME.

FOCUS: HEALTH

LESSON: HARVESTING A SALAD FROM YOUR GARDEN

MATERIALS: BASKETS OR BAGS FOR HARVESTING, CUTTING BOARDS, KNIVES
FOR CHOPPING, A LARGE BOWL FOR SALAD, PAPER PLATES, FORKS,
SALAD DRESSING (MADE IN THE CLASSROOM WITH A BLENDER, OR
BROUGHT IN).

TEACHER'S BACKGROUND: HARVESTING AND EATING FROM THE GARDEN IS ONE
OF THE MOST EXCITING EVENTS IN THE GARDEN CYCLE. STRESS THE
APPRECIATION OF WHAT YOU HAVE DONE FOR THE PLANTS AND HOW
THEY WILL NOW HELP YOU.

DO: TALK ABOUT WHEN TO HARVEST VEGETABLES, THE SIZE THEY SHOULD
BE AND WHAT COLOR. WITH SOME VEGETABLES, YOU PICK THE ENTIRE
PLANT, SUCH AS CARROTS, RADISHES,
AND BEETS. WITH OTHERS, YOU
TAKE ONLY THE FRUIT OR USABLE
PART, LIKE TOMATOES, BEANS, OR
SQUASH. GO INTO THE GARDEN AND
HAVE THE CHILDREN PICK SOME
VEGETABLES WHICH ARE READY.
SINCE YOU ARE MAKING A SALAD,
PICK SALAD SORTS OF THINGS,
ALTHOUGH ALMOST EVERY VEGETABLE
WILL BE GOOD RAW IN THE SALAD.
BRING THEM INSIDE, WASH OFF ANY
DIRT, AND CHOP THE CARROTS, TO-
MATOES, SQUASH, GREEN BEANS,
CUCUMBERS, BEETS AND RADISHES
INTO BITE-SIZED PIECES. TEAR
LETTUCE INTO SMALL PIECES AND
PLACE EVERYTHING IN THE LARGE
BOWL. IF YOU MAKE DRESSING,
SIMPLY BUZZ CUCUMBER, HERBS,
OIL AND A LITTLE VINEGAR, ONION AND SALT IN THE BLENDER. IT
IS SIMPLE, AND EXTREMELY TASTY.

THINK: DISCUSS THE HARVESTING OF VEGETABLES, THE GIVE-AND-TAKE OF
THE GARDEN AND THE HEALTH ASPECTS OF THE FRESH VEGETABLES
YOU HAVE GROWN YOURSELF. GO OVER THE VEGETABLES THAT YOU
HARVESTED AND TALK ABOUT THE KIND OF VEGETABLES THEY ARE, IF
YOU PICKED THE WHOLE PLANT, WHETHER YOU PICKED SOME AND MORE
WILL COME LATER, ETC. STRESS BEING CAREFUL WHILE HARVESTING
VEGETABLES NOT TO DISTURB THE REST OF THE FRUITS ON THE PLANT
OR THE SURROUNDING PLANTS AS YOU PULL A ROOT VEGETABLE OUT
OF THE GROUND.

COMMUNICATE: PASS OUT SALAD AND BEFORE EATING, HAVE EACH CHILD SAY
OUT LOUD HOW HE/SHE FEELS ABOUT THE GARDEN RIGHT THEN. THEN
DIG IN AND ENJOY TOGETHER.

FOLLOW-UP POSSIBILITIES: CONTINUE HARVESTING VEGETABLES THAT ARE
READY AND USE THEM IN OTHER CLASSROOM ACTIVITIES -- MAKE
SALADS FOR OTHER CLASSES OR GIVE EXCESS VEGETABLES AWAY TO
NEIGHBORS AND SENIOR CITIZENS.

SOME RESOURCES:

AG WORLD: INSIGHT INTO THE FORCES AFFECTING AGRICULTURE, 1186 WEST SUMMER ST., ST. PAUL, MINNESOTA, 55113.

ALLEN, FLOYD, ED. WESTERN ORGANIC GARDENING (EMMAUS, PENNSYLVANIA: RODALE PRESS, 1972.)

ALLEN, GERTRUDE. EVERYDAY INSECTS. (HOUGHTON, 1962.)

ALLISON, LINDA. THE REASON FOR SEASONS: THE GREAT COSMIC MEGAGALACTIC TRIP WITHOUT MOVING FROM YOUR CHAIR (TORONTO:LITTLE, BROWN AND CO., 1975)

ALLISON, LINDA. THE SIERRA CLUB SUMMER BOOK (NEW YORK:SCRIBNER'S AND SONS, 1977.)

BELL, GWEN, ED. STRATEGIES FOR HUMAN SETTLEMENTS: HABITAT AND ENVIORNMENT (HONOLULU: THE UNIVERSITY PRESS OF HAWAII, L976)

BENDER, TOM. ENVIORNMENTAL DESIGN PRIMER (NEW YORK: SCHOCKEY BOOKS, 1973)

BLISHEN, EDWARD, ED. THE SCHOOL THAT I'D LIKE (MIDDLESEX, ENGLAND: PENGUIN BOOKS, LTD. 1971)

BLOUGH, GLENN D. THE INSECT PARADE (ROW, 1953)

CALLENBACH, ERNEST, LIVING POOR WITH STYLE (NEW YORK: BANTUM BOOKS 1972.)

CELLARIUS, DORIS. SCHOOL GARDENS (INFORMATION SERVICES, SIERRA CLUB, 530 BUSH, SAN FRANCISCO, SINGLE COPY, 15¢)

CHAN, PETER. BETTER VEGETABLE GARDENS THE CHINESE WAY (PORTLAND OREGON: GRAPHIC ARTS CENTER PUBLISHING CO., 1977)

COBB, VICKI. SCIENCE EXPERIMENTS YOU CAN EAT (NEW YORK: LIPPENCOTT 1972)

COOKE, EMOGENE. FUN TIME WINDOW GARDEN (CHICAGO: CHILDREN'S PRESS 1957)

CRAIG, GERALD S. SCIENCE FOR THE ELEMENTARY SCHOOL TEACHER (BOSTON: GINN AND CO. L940)

DARROW, KEN AND RICK PAM. APPROPRIATE TECHNOLOGY SOURCE BOOK (STANFORD CALIFORNIA: VOLUNTEERS IN ASIA, INC. 1976)

DENSMORE, FRANCES. HOW INDIANS USE WILD PLANTS FOR FOOD, MEDICINE AND CRAFTS (NEW YORK: DOVER PUBLICATIONS, INC. 1974)

DORF, RICHARD AND YVONNE HUNTER. APPROPRIATE VISIONS (SAN FRANCISCO: BOYD AND FRASER PUBLISHING COMPANY, 1978)

DOTY, WALTER, ALL ABOUT VEGETABLES (SAN FRANCISCO: CHEVRON CHEMICAL CO. 1973)

DUNKS, THOM AND PATTY GARDENING WITH CHILDREN: A GUIDE FOR PARENTS AND CHILDREN

EKHOLM, ERIK P. LOSING GROUND (NEW YORK: W.W. NORTON AND CO. 1976)

FENTEN, DX. GREENHOUSING FOR PURPLE THUMBS (SAN FRANCISCO: 101 PUBLICATIONS, 1976)

FOSTER, WILLENE K AND QUAREE, PEARL. SEEDS ARE WONDERFUL (CHICAGO:MELMOST PUBLISHERS INC. 1960)

FRANZ, MAURICE, ED. THE CALENDAR OF ORGANIC GARDENING (NEW YORK: RANDOM HOUSE, 1960)

GOODMAN, PAUL AND PERCIVAL. COMMUNITAS (NEW YORK: RANDOM HOUSE, 1960)

GOODMAN, PAUL. LIVING THE GOOD LIFE (NEW YORK: SCHOCKEN BOOKS 1970)

GOODWIN, MARY AND POLLEN, GERRY. CREATIVE FOOD EXPERIENCES FOR CHILDREN (CENTER FOR SCIENCE IN THE PUBLIC INTEREST, 1757 "S" STREET, N.W. WASHINGTON DC, 200009, 1974)

HESS, KARL. COMMUNITY TECHNOLOGY (NEW YORK: HARPER AND ROW. 1979)

HOME, FARM, AND GERDEN RESEARCH ASSOCIATES. LET AN EARTHWORM BE YOUR GARBAGE MAN (NORTON, CONN:THE COUNTRY BOOKSTORE 1954)

HUCKABY, GLORIA AND SKELSEY, ALICE. GROWING UP GREEN

HUNGER ACTION CENTER. FARMERS MARKET ORGANIZERS' HANDBOOK (OLYMPIA WASHINGTON COMMUNITY SERVICES ADMINISTRATION 1978)

JEAVONS, JOHN. HOW TO GROW MORE VEGETABLES (PALO ALTO, CALIF: ECOLOGY ACTION OF THE MID-PENNINSULA. 1974)

JOBB, JAMIE MY GARDEN COMPANION

JAMES, SANDY LEARNING FOR LITTLE KIDS (BOSTON: HOUGHTON MIFFLIN CO. 1979)

KIRK, DONALD. WILD EDIBLE PLANTS OF THE WESTERN UNITED STATES (HEALDSBURP CALIFORNIA: NATUREGRAPH PUBLISHERS. 1970)

KLOSS, JETHRO. BACK TO EDEN (COALMONT, TENN: LONGVIEW PUBLISHING HOUSE, 1971)

KULVINSKAS, VIKTORAS. SURVIVAL INTO THE 21ST CENTURY (WEATHERSFIELD CONN: OMANGOD PRESS, 1975)

LANE CO. DEPT. OF HOUSING AND COMMUNITY DEVELOPMENT. SCHOOL GARDENS EUGENE, OREGON.

LAPPE', FRANCIS MOORE. DIET FOR A SMALL PLANET (BALLENTINE, REV. ED. 1975)

LAUREL, ALICIA BAY. LIVING ON THE EARTH (BERKELY: BOOKWORKS, 1970)

LEE, ALLYSON, MALOOL, SUSAN AND MCPHAIL, BARBARA. GROWING WITH CHILDREN: OUR EXPERIENCE WITH SCHOOL GARDENS IN EUGENE (TILTH; BIOLOGICAL AGRICULTURE IN THE NORTHWEST, VOL. 5 NO 1, SPRING, 1979)

MABE, REX. BACKYARD VEGETABLE GARDENING (GREENSBORO: POTPOURRI PRESS 1974.

MALONEY, MARK. GARDENING NOTEBOOK (WALNUT CREEK, CAL:ECO HOUSE 1978)

MASEFIELD, G.B.,ET.AL. THE OXFORD BOOK OF FOOD PLANTS (OLD TRAF- FORD, ENGLAND: OXFORD UNIVERSITY PRESS, 1969)

MC CRUMMEN, J.B. ED. FOOD PRESERVATION (HUNGER ACTION CENTER 1978)

MCHALE, JIM. FUTURE OF RURAL TECHNOLOGY (COUNTRYSIDE MAGAZINE, JUNE, 1978)

MILLER, G.TYLER JR. REPLENISH THE EARTH (BELMONT, CA.: WADSWORTH PUBLISHING CO. 1972)

MORRIS, DAVID AND KARL HESS. NEIGHBORHOOD POWER (BOSTON:BEACON PRESS, 1975)

NASH, HUGH. PROGRESS AS IF SURVIVAL MATTERED (SAN FRANCISCO: FRIENDS OF THE EARTH, 1977.)

NATIONAL ASSOCIATION FOR THE EDUCATION OF YOUNG CHILDREN: SCIENCE EXPERIENCES FOR NURSURY SCHOOL CHILDREN (MERIL-PALMER INST.)

OLKOWSKI, HELGA AND WILLIAM. THE CITY PEOPLE'S BOOK OF RAISING FOOD. (EMMAUS, PENN: RODALE PRESS, INC.1975.)

ORGANIC GARDENING AND FAMILY MAGAZINE STAFF. TEACHING ORGANIC GARDENING (EMMAUS PENN: RODALE PRESS, INC.)

PARKER, BERTHA M. FLOWERS, FRUITS, SEEDS (ROW, 1953.)

_____, GARDEN INDOORS (ROW, 1953)

_____, LEAVES (ROW, 1954)

_____, SEEDS AND SEED TRAVELS (ROW, 1952)

PAUL, AILEEN. KIDS GARDENING (NEW YORK: DOUBLEDAY, 1972)

PETERSON, CHRISTINA AND AUGELL, TONY. ENERGY, FOOD AND YOU (OFFICE OF ENVIORNMENTAL EDUCATION, WASHINGTON D.C. C/O SHORELINE SCHOOL, DIST ADMINISTRATION BLDG. N.E. 158TH AND 20TH AVE N.E., SEATTLE, WASHINGTON 98155)

PETERSON, DIANNE. ABC'S OF SCHOOL GARDENING (COOPERATIVE EXTENSION UNIVERSITY; 960 E. STREET, P.O. BOX 351, PITTSBURG, CAL.94565)

PETRICH, PATRICIA AND KALTON, ROSEMARY. THE KIDS' GARDEN BOOK (CONCORD, CALIF: NITTY GRITTY PRODUCERS. 1974)

REYNOLDS, CHRISTOPHER. SMALL CREATURES IN MY GARDEN (S. AND G. 1966.)

RIVERS, PATRICK. THE SURVIVALISTS (NEW YORK: UNIVERSITY BOOKS, 1975)

RODALE, ROBERT, ED. THE BASIC BOOK OF ORGANIC GARDENING (NEW YORK: BALLENTINE, RODALE PRESS. 1971)

SCHUMACHER, E.F. SMALL IS BEAUTIFUL (NEW YORK: HARPER AND ROW 1973)

SELSAN, MILLICENT E. SEEDS AND MORE SEEDS (NEW YORK: HARPER AND ROW 1959)

SHAW, A.C., LAZELL AND FOSTER. PHOTOMICROGRAPHS OF THE FLOWERING PLANT (LONDON: WILLIAM CLOWES AND SONS, LTD. 1971)

SIMMONS, ROBIN RECYCLOPEDIA (BOSTON: HOUGHTON MIFFLIN CO. 1976)

SWAN, LESTER A. BENEFICIAL INSECTS (NEW YORK:HARPER AND ROW 1964)

SWENSON, ALLAN A. MY OWN HERB GARDEN (EMMAUS, PENN: RODALE PRESS 1976)

THEOBALD, ROBERT AND J.M. SCOTT. TEGS 1884 (CHICAGO: SWALLOW PRESS 1972)

WEBER, IRMA E. UP ABOVE AND DOWN BELOW (SCOTT, W.R. 1943)

WENT, FRITZ W. THE PLANTS LIFE NATURE LIBRARY (NEW YORK: TIME, INC 1963)

WIGGINTON, ELIOT, ED. THE FOXFIRE BOOK (NEW YORK: DOUBLEDAY 1972)

WOTAWIEC, PETER. HOW TO START A SCHOOL GARDENING PROGRAM (NORWALK: GARDENS FOR ALL, BOX 2302, NORWALK, CONN. 06852, 1975) A GARDEN BOOK WITH GRADED LESSON PLANS.

WOTAWIEC, PETER. "TIPS FOR A SUCCESSFUL GARDENING PROGRAM" ENVIORN- MENTAL ACTION BULLETIN RODALE PRESS, EMAUS, PA. 1975, EMMAUS REPRINTS AVAILABLE)

ALBANY FOOD COOP
P.O. BOX 1483
OR
229 E. 4TH
ALBANY, OREGON
97321
PH. (503) 926-8926

AMITY FOUNDATION
P.O. BOX 7066
EUGENE, OREGON 97401

CITY OF EUGENE
PARKS AND RECREATION DEPT.
COMMUNITY GARDENS PROGRAM
301 N. ADAMS ST.
EUGENE OREGON 97402
(503) 687-5329

COMMUNITY HEALTH AND
EDUCATION CENTER (CHEC)
433 W. 10TH AVE.
EUGENE, OREGON 97401
(503) 485-8445

GROWERS MARKET FOOD COOP
454 WILLAMETTE
EUGENE, OREGON 97401
(503) 686-1145

LANE CO. COMMUNITY FOOD BANK
135 E. 6TH
EUGENE, OREGON 97401
(503) 687-4038

LANE CO. EXTENSION SERVICE
950 W. 13TH
EUGENE, OREGON 97402
AGRICULTURE 687-4243
YOUTH NUTRITION
134 E. 13TH
EUGENE, OREGON
(503) 687-4281

SOIL TESTING LAB
DEPT. OF SOIL SCIENCE
OREGON STATE UNIVERSITY
CORVALLIS, OR. 97331
(503) 754-2441

ORGANICALLY GROWN INC.
A PRODUCTION COOPERATION
1640 EAST BEACON
EUGENE, OREGON 97404
(503) 688-3176

OREGON FOOD ACTION COALITION
1414 KINCAID
EUGENE, OREGON 97401
(503) 484-1707

SOLAR ENERGY COMMUNITY
BOX 1055
CORVALLIS, OR. 97330
(503) 976-6527

UNIVERSITY OF OREGON FOOD-OP
1535 AGATE ST. (ON THE ALLEY)
UNIVERSITY OF OREGON CAMPUS
EUGENE, OREGON 97403
(503) 686-4911

UNIVERSITY OF OREGON URBAN FARM
DEPT. OF LANDSCAPE ARCHITECTURE
UNIVERSITY OF OREGON
EUGENE, OREGON 97405
(503) 686-3647

WHITEAKER COMMUNITY COUNCIL
21 N. GRAND
EUGENE, OREGON 97402
(503) 687-3556
OR 343-7713

WHITEAKER ENERGY PROJECT
341 VAN BUREN
EUGENE, OREGON 97402
(503) 343-7713

WHITEAKER PROJECT SELF RELIANCE
(503) 343-2711

WHITEAKER RECYCLING PROJECT
341 VAN BUREN
EUGENE, OREGON 97402
(503) 343-7713

WILLAMETTE PEOPLES FOOD COOP GROCERY
1391 E. 22ND
EUGENE, OREGON 97403

FILMS & FILMSTRIPS

ECOLOGY CHECKS AND BALANCES

ABOUT THE PREDATORY RELATIONSHIP BETWEEN THE LADYBUGS AND APHIDS.
CALL NUMBER MB 73011-2   AVAILABLE THROUGH LANE E & D

FARM ANIMALS

A GOOD FILM FOR DIFFERTIATING SIZE, COLOR, BODY PARTS AND MARKINGS.
ENCYCLOPEDIA BRITANNICA FILMS, INC.
1150 WILLAMETTE AVE.,
WILLAMETTE, ILLINOIS 60091

GROWING, GROWING

A BEAUTIFULLY PRODUCED FILM WITH A NICE SONG. NARRATED ENTIRELY
BY THE CHILDREN.
AVAILABLE THROUGH LANE E & D, CALL NUMBER MA 720448

FINDING OUT ABOUT INSECTS (INSECTS AROUND US SERIES)

ABOUT 21 FRAMES.
HOW TO TELL INSECTS APART FROM EACH OTHER AND FROM OTHER ANIMALS.
JIM HANDY ORGANIZATION

INSECTS THAT HELP US

GOOD PHOTOGRAPHY SHOWING DIFFERENT INSECTS SUCH AS BEES, ANTS,
LADYBUGS, ETC., IN THE GARDEN AND HOW THEY HELP US.
CALL NUMBER MA 672300 AVAILABLE THROUGH LANE E & D.

LEARNING ABOUT FLOWERS

IT ILLUSTRATES THAT PLANTS HAVE MANY DIFFERENT KINDS OF FLOWERS.
IT SHOWS THE MAIN FUNCTIONS OF A FLOWER, THAT IS, TO PRODUCE SEEDS.
ENCYCLOPEDIA BRITANICA FILMS, INC.

MEET THE PLANT FAMILY

ENCYCLOPEDIA BRITANICA FILMS.

PLANTS AND THE THINGS WE USE

PLANTS AND ANIMALS SERIES, 37 FRAMES.
DIFFERENCES IN THE EDIBLE PARTS OF PLANTS ARE ILLUSTRATED.
YOUNG AMERICAN FILMS TEXT--FILM DEPT.
MCGRAW HILL BOOK CO.
330 W. 42ND ST.
NEW YORK, NEWYORK, 10036.

PLANTS AROUND US

JIM HANDY ORGANIZATION

SEED GERMINATION

WONDERFUL TIME LAPSE PHOTOGRAPHY SHOWING SEEDS SPROUT INTO PLANTS
IN SECONDS.
CALL NUMBER MB 620079-2 AVAILABLE THROUGH LANE E & D

SEEDS TRAVEL

(AUTUMN IS HERE SERIES)
THERE ARE MANY TYPES OF SEEDS AND SEED COVERINGS, AND MANY WAYS OF
SEED DISPERSAL
JAM HANDY ORGANIZATION

THE FARMER'S ANIMAL FRIENDS

SERIES, ABOUT 6 FILMSTRIPS OF ABOUT 22 FRAMES EACH
ABOUT FARM ANIMALS AND THEIR DIFFERENCES
JAM HANDY ORGANIZATION
2821 E. GRAND BLVD.
DETROIT, MICH. 48211

WATCH OUT FOR MY PLANT!

AN EXCELLENT FILM ABOUT A BOY'S EXPERIENCES GROWING HIS OWN PLANT
IN THE CITY.
CALL NUMBER MB 740098 AVAILABLE THROUGH LANE E & D.

Section II

Developing a Strong
Urban Agricultural Network

Chapter 3

Neighborhood Transformation Principles

This project is a result of
continuing work spanning a
period of more than six years.
Contributors to this work
include design students at
the University of Oregon
School of Architecture and
Allied Arts, Urban Farmers
at the University of Oregon
Urban Farm (especially Linda
Smiley, Jeff Wilson, Noel
Prchal and Chris Gum), Amity
Foundation of Eugene, Oregon
and Larry Parker of Whiteaker
Project Self-Reliance; but
the original development of
this work must remain my
primary responsibility.

NEIGHBORHOOD TRANSFORMATION PRINCIPLES

BLOCK FARMS
NEIGHBORHOOD FARMS
PROGRAM

Richard Britz,   Architect
Eugene, Oregon   June 1980

Funded by a grant from THE NATIONAL CENTER FOR APPROPRIATE TECHNOLOGY
Butte, Montana and by THE EDIBLE CITY RESOURCE CENTER, EUGENE, OREGON

Whiteaker Neighborhood is the lowest income group within the City of Eugene. The physical characteristics of the neighborhood as a whole vary from sector to sector, yet overall the physical environment reflects the landscape of high resident transient rates and instability common to most low income neighborhoods. The population demographics indicate an extremely skewed population on either end of the age scale, being comprised of low income seniors (on fixed and declining incomes) and younger adults seeking ways to stabilize their situation. Simultaneously, whether home owner, renter, or apartment dweller, Whiteaker Neighborhood residents have seen continued deterioration of their environment due to land and property speculation, non-controlled industrial misuse of the land, zoning simplification of entire sections of the neighborhood, massive land purchases of roadway rights-of-way for future freeways, and general inconsideration of traffic and parking impacts to neighborhood cohesiveness by city agencies.

Whiteaker Neighborhood is divided into five separate sub-neighborhoods, each having geographic autonomy, due primarily to massive slicing of vehicular ways. Within each sector or sub-neighborhood secondary traffic ways were discovered that further demolish the personal cohesiveness and sense of calm found in higher income level neighborhoods. Levels of noise and stress are obviously high and within this swirling milieu, this project attempted to generate models which were intended to stimulate neighbors to develop cohesive approaches toward neighborhood food self-reliance.

It became apparent during this planning period that certain neighbors were willing to "stay put" and attempt to instigate change in their neighborhood by utilizing cooperative strategies to finance their homes and build permanent residence in this sea of instability. It was those people this work is intended to support.

The stabilization of a population "in place" requires a dynamic strategy which enables the cooperating residents to have PHASED DEVELOPMENT PLANS which are implemented part by part as their resources allow. Pressures from land taxes, falsely escalated property values by outside-the-neighborhood investors, and other pressures can be dealt with by having DYNAMIC MODELING PROCESSES in mind which enable neighbors to capitalize themselves into participants rather than remaining victims of the urban real estate circumstance. It is toward these goals that this work has been accomplished.

This work consists of a general-system-based interdisciplinary approach to democratic decentralization of the land planning and design process in Eugene. It encourages organized networking of separate demonstration projects equitably distributed throughout the whole neighborhood, so that highly motivated neighbors living on Class I agricultural soils might extend their impact on the land to make it personally economically useful to them.

It is premised on the hypothesis that individuals, families, and neighbors, in direct personal contact, working together for mutual improvement in their environmental and economic circumstances, can reverse the trend toward environmental entropy and social fragmentation.

It is part of a longer tradition in planning, landscape architecture and architecture, which attempts to improve urban fitness and urban ecological well-being for the lower income urbanite. Work such as that of Ebeneezer Howard's Garden City, Frank Lloyd Wright's Broadacre City, Clarence Stein and Henry Wright's Radburn, and Paul and Percival Goodman's Communitas and Double E community ideas may be updated and made more useful by bridging the gap to a real neighborood. The essence of this work is as practical as I could make it and moves in similar directions as the work of Malcolm Wells, the Institute for Neighborood Self-Reliance, and the Street Farm group in Great Britain. In the best traditions of Stearns and Montag's Urban Ecosystem, Spence Havlick's The Urban Organism, and Hunter S. Thompson's Fat City, I have called this **THE EDIBLE CITY**

This work involves principles developed especially for our immediate bioregional ecosystem that are applicable to the neighborhoods of Eugene, Oregon, but may also be adapted to other locations as well. It is an extension of our evolving political and economic atmosphere best illustrated by the Housing and Community Development Act of 1974, and other programs intending to re-involve and re-direct personal participation in the cyclic functioning of environmental harmony.

It has illustrated these principles with physical demonstrations which involved small scale demonstrations of appropriate (or necessary) technology, integrated the University (faculty and student) community with neighbors planning their own futures, and developed future contingency plans for further actualization.

The two main physical concepts involved are the **BLOCK FARM PROGRAM** and the **NEIGHBORHOOD FARM PROGRAM** which are extensively described in the following pages, along with a prototype urban farm which has functioned for five years or more.

The basic premise of the establishment of **LOCAL FOOD SELF·RELIANCE** involves strategies which require small-scale projects to grow into a larger and mutually supportive network similar to the early sketch notes on the following page.

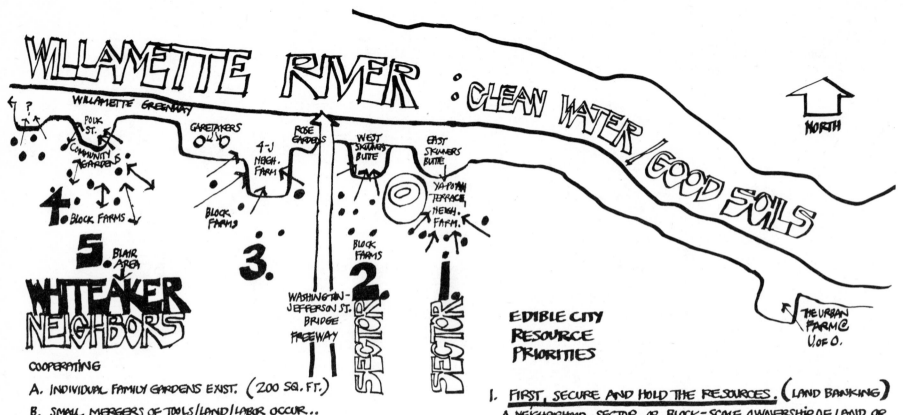

WILLAMETTE RIVER : CLEAN WATER / GOOD SOILS

NORTH

WILLAMETTE GREENWAY

POLK ST.
COMMUNITY GARDENS
CARETAKERS O
ROSE GARDENS
WEST SKINNERS BUTTE
EAST SKINNERS BUTTE
4-J NEIGH. FARM
YA-PO-AH TERRACE NEIGH. FARM

4. BLOCK FARMS
BLOCK FARMS
BLOCK FARMS

5 BLAIR AREA
WASHINGTON-JEFFERSON ST. BRIDGE FREEWAY

WHITEAKER NEIGHBORS

SECTOR 3.
SECTOR 2.
SECTOR 1.

THE URBAN FARM @ U. OF O.

COOPERATING

A. INDIVIDUAL FAMILY GARDENS EXIST. (200 SQ. FT.)

B. SMALL MERGERS OF TOOLS/LAND/LABOR OCCUR...

C. REMOVAL OF PHYSICAL BARRIERS TO COOPERATION ACCOMPANIED WITH TAX INCENTIVE PROGRAM ...

D. BLOCK GARDENS/BLOCK SECURITY STABILIZED... ANIMALS INVITED... (1 ACRE)

E. NETWORKING/ORGANIZATION OF BLOCK FARMS & SCHOOL FARMS

F. INCOME LOSS TO HIGH COST TRANSPORT SYSTEMS MINIMIZED.

G. LOCALISM ENCOURAGED WITH INTERDEPENDENCIES OF FOOD PRODUCTION DISTRIBUTION PRESERVATION

H. AGGREGATION OF BLOCK-SCALE ORGANIZATION TO LARGER SCALE (5 ACRE) NEIGHBORHOOD FARMS.

[NOTE: LOCAL BY SECTOR (1,2,3,4,5) "CARETAKERS" NECESSARY TO COORDINATE ON A CONTINUING BASIS INTER AND INTRA-SECTOR OPERATIONS — HOPEFULLY PERMANENT RESIDENTS!]

EDIBLE CITY RESOURCE PRIORITIES

1. FIRST, SECURE AND HOLD THE RESOURCES. (LAND BANKING)
   A. NEIGHBORHOOD, SECTOR, OR BLOCK-SCALE OWNERSHIP OF LAND OR
   B. CITY OWNED LAND (ACCESS TO IT)...

2. SECOND, INSURE EQUITABLE DISTRIBUTION OF RESOURCES.
   A. AMONG GEOGRAPHIC SECTORS 1,2,3,4,5 (PERMANENT
   B. AMONG AGE GROUPS; DIVERSE UNITY. (AVAILABILITY)

3. THIRD, ASSURE AND DEVELOP A PERSONAL ORGANIZATIONAL SYSTEM THAT WORKS TO SUSTAIN 1 & 2 ABOVE.
   A. MORE PERMANENT LAND-PEOPLE RELATIONSHIPS.
   B. AN EVOLVING SOCIAL/ORGANIZATIONAL/EDUCATIONAL DYNAMIC.

   [1. PROCEDURES
   2. PEOPLE
   3. PLACE]
   (CONTINUING STRUCTURE)

This supportive network for neighborhood food self-reliance is based on the establishment of block farms which grow from small scale gardens merging into larger economic units not exceeding 1 acre, the amount of land commonly held within the interiors of the blocks themselves. These block farms (so-called because they also house productive animals such as chickens and rabbits) also voluntarily increase their housing density on the north side as time goes on to contribute to the compact urban growth form which supports our urban services boundary and contributes to the maintenance of Class I agricultural soils on the perimeter of our expanding city.

Small-scale, low-cost solar retrofits enable heat trapping and solar greenhouse season extenders for year 'round food production. Continuous renewal of nutrients comes from organic waste recycling stations placed at 1-1/2 block intervals throughout the neighborhood where composting of street tree leaves, grass clippings, garbage wastes and animal manures occur. Excess organics and non-organic recyclables can be transferred the short distance to the neighborhood recycling center (indicated on the following pages by an asterisk (*). One neighborhood farm adjacent to the river (gardens with animals) of approximately five acres and several supporting block farms of approximately 1-1/2 acre each can provide each neighborhood sector with a significant step toward neighborhood self-sufficiency in food. If each sector of the neighborhood utilized the Class I (Soil Conservation Service classification) soils along the river for food production and coordinated production of growing year 'round, the majority of food needs for Whiteaker Neighborhood could be met. These five acre "neighborhood farms" could be coordinated with the Willamette Greenway provisions which allow agricultural land uses and with the Parks Department, who presently manages these lands. So where do we go from here?

Well, we could begin in our own neighborhoods that exist or make new ones of affinity groups concerned with each other and move from ways of coping to an overall strategy. We could, if we worked hard at it, transcend self-reliance (in a self-oriented culture) and emphasize community-reliance based on affinity groups with rotating spokespersons, dedicated to personal and environmental improvement in the blocks where we now live.

NEIGHBORHOOD TRANSFORMATION PRINCIPLES: CHANGE IN TIMES OF CHANGE

An upward economic spiral can be generated by carefully assessing the locally available resources at hand. The world's population is envious of our living river and unfailing water supply, our superb loam soils deposited by that river -- all of which are convertible throughout the year to opportunities for food production right where we live, thanks to our overall mild and well-watered climate. Solar conditions, while overcast, provide the basic ingredient for warmth, and insulation is right at hand in cellulose fibers from renewable plant materials. Conscious organization to stay together generates leadership and synergistic group work sessions, which, while tiring, fill us with pride in the visible changes we point to proudly in our own neighborhoods. Physical symbols, places we have transformed ourselves, serve as our best indicators that health, not decay, is present here in Eugene. Rejection of nuclear power must be accompanied by a personal commitment to changing our modes of transportation from automobile to foot and bicycle travel, which in turn, enables us to see more clearly the land around us. The private and selfish city deteriorates its "common-ground" and our reversal of these paradigms renews and reclaims the commons for both utilitarian-economic benefit, and for the culture-building vision of beauty, apparent even in our own back yards.

First of all, we have tried to work with people where they live and attempt to regenerate, with them, concepts and actions of the commonalities that we all share. It seems important to recollect smaller groups within the urban organism -- locally defined cellular units that share locations commonly and are built upon the family unit. It seems important that neighborhood (sector) availability of basic necessities should remain within walking distance for kids and senior people and that the diversity could be expanded beyond access to water, food, physical and psychological shelter and at least some kind of health care, to other more diverse economics later. As these cells "miniaturize" and we adjust our zoning and taxation structures into neighborhood control and legitimize the neighborhood with some real economic power and the responsibility this decision-making entails, we will move to genuine neighborhood government.

There are lots of things we could try, such as:

· <u>Closing or re-routing streets</u> to assure that the "slicing" of neighborhood sectors be eliminated. We could garden where the streets now are if we organize <u>and</u> own the land on either side of the street. We must keep cars away from the river.

· <u>Selling our automobiles and pooling the capital</u> -- especially in the sub-sectors that are right adjacent to downtown. We could then buy one vehicle for community use (a truck, a 15-passenger van, several small cars, and lots of bicycles for example) of each type (maybe even a limousine!) and plan ahead to schedule their usage. We must find ways to combat Nissan and General Motors and re-design our lives and our neighborhoods accordingly.

· <u>Turning around</u> (and taking the time) <u>to meet our neighbors</u> -- in turn to develop notions of commonalities --

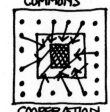

initially maybe it's just to pool our "garbage" into block-scale (50 people) composting systems filled with worms to reduce garbage hauling bills; then maybe later to use that compost on clustered gardens; and this might necessitate getting the cars out on the street where they belong, where we can monitor their oil drippings; we could share tools in block-scale workshops; playplaces for kids is a universal necessity which needs space and "hiding places"; and we could store our wood commonly; or do our laundry commonly or generally find a scale for cooperation which we are comfortable with...and this will progress through time.

· <u>Developing symbolic rallying points in each sector of the neighborhood</u>, such as the Ya-Po-Ah Terraces, benches, swings and fountains and orchards which serve as more than just places for seniors to grow their own food; and street closures which reclaim the neighborhood's historic character; plant trees along the noisy and polluting freeways in huge teams of people to provide for <u>future</u> health; build new and experimental low-cost homes to see if it can't be done without lots of federal or state financing and rules on how the money is to be spent (and suggest that they don't know much about living cheaply either!); put a wind turbine on top of the Jefferson Elevator and see if it works; save the Gum Barn as a remnant of our historic link to people of the soil and join them through our efforts in community gardening; support Helios Development in their quest for lower-cost solar housing; and acquire land in the neighborhood's behalf. Symbols of positive change mean a lot (and don't have to cost much!).

The generation of models of appropriate technology will vary from place to place, circumstance to circumstance, sector to sector, person to person, but the starting points are the commonalities of people and appreciation of our local resource base (why not exploit blackberries as a cash crop, or grow walnut trees for the nuts and the lumber, or big leaf maples for furniture, or poplar for firewood, or....).

The needs for the future require long-range vision, beyond our individual needs or even our temporary neighborhood residents' needs, but the needs of the land, the soils, the airs we breathe, the trees that shelter us and the economies of renewable resources that can sustain a healthy culture long after we are gone.

RECYCLING STATION

SUB-SECTORS

FOR RECYCLING OF ORGANIC AND OTHER "WASTES" PROVIDE VALUABLE AND CONTINUING RESOURCES FOR FOOD PRODUCTION. FALL LEAF DROP COMPOSTING AT LOCAL SITES MINIMIZES CITY-SCALE VEHICULAR NEEDS/COSTS, CONSERVES URBAN BIOMASS, AND UTILIZES IT IN MANAGEABLE UNITS.

MANAGING ORGANIC WASTE AND REGENERATING THE URBAN SOIL

○ 1½ BLOCK RAD.
✳ RECYCLING CENTER
⬢ PROP. STATION
▬ SINGLE FAM. RES.

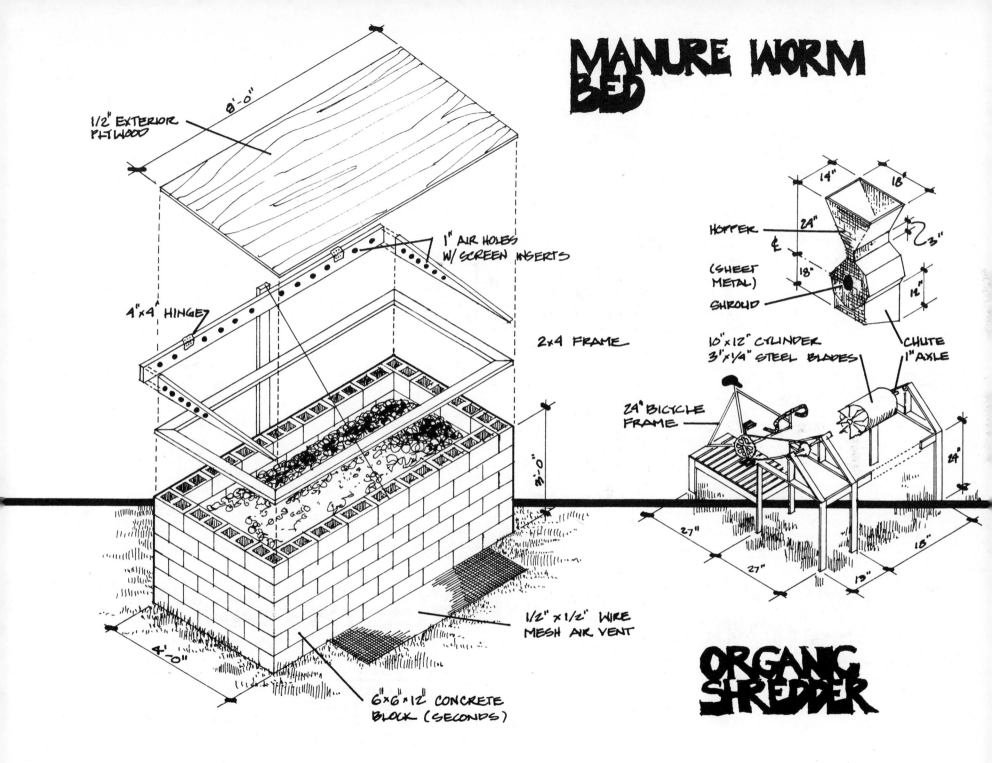

MANURE WORM BED

1/2" EXTERIOR PLYWOOD

8'-0"

1" AIR HOLES W/ SCREEN INSERTS

4"x4" HINGES

2x4 FRAME

3'-0"

1/2" x 1/2" WIRE MESH AIR VENT

4'-0"

6"x6"x12" CONCRETE BLOCK (SECONDS)

ORGANIC SHREDDER

HOPPER

(SHEET METAL)

SHROUD

14"

18"

24"

3"

18"

12"

10"x12" CYLINDER
3"x1/4" STEEL BLADES

CHUTE
1" AXLE

24" BICYCLE FRAME

24"

27"

27"

13"

18"

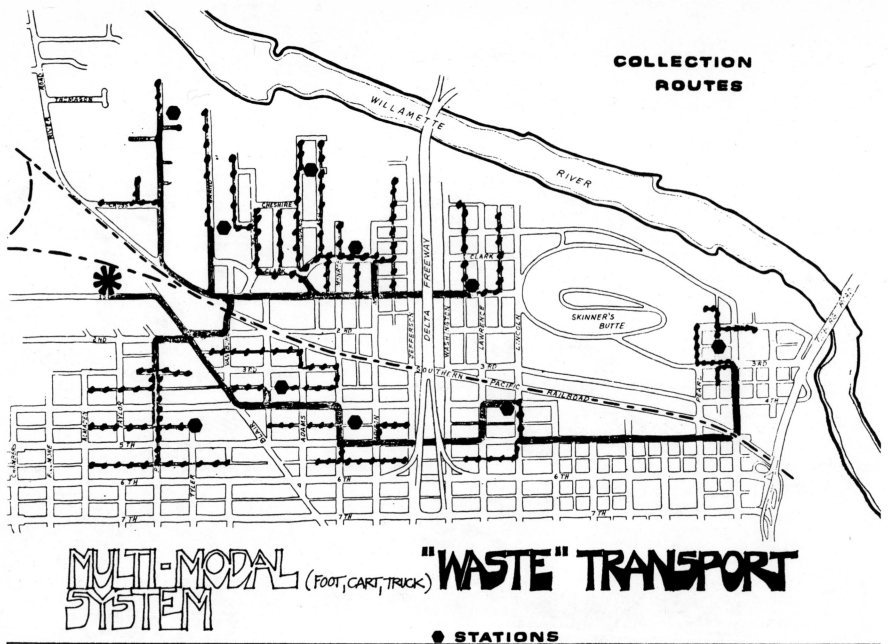

COLLECTION ROUTES

MULTI-MODAL SYSTEM (FOOT, CART, TRUCK) "WASTE" TRANSPORT

⬢ STATIONS
▬ CART/PED. ROUTE
▬ SERVICE ROUTE
✳ RECYCLING CENTER

Ecoscope / Dean Baker

Land Squeeze Could Create Newgene

Eugene is rapidly running out of vacant private land suitable for traditional new housing. That need not be bad news.

The suspicion that old kinds of housing just may not work anymore may, in fact, be one of the keys to solving a variety of economic and social problems—inflation, unemployment, interpersonal alienation and crime, for example.

A growing number of visionaries and

practical architects, including some here in Eugene, are suggesting new ideas in community-building that could help us begin to weed out all these ugly problems at once.

How so? Briefly, by beginning to create a new community-based lifestyle where people share housing, gardens, work space and tools. Certainly, not everyone is ready for this; nor will everyone ever be. But the logic is so compelling, it's hard to imagine such an idea—Richard Britz's idea, for example, about "the edible city"—failing to catch on.

Before going into that, though, a few facts.

Lifestyle changes are bound to come to Eugene no matter what is done as long as the population is allowed to grow in the metropolitan area, as projected, from the current 193,000 to 294,000 by the year 2000. Either rural lands will become city lands or housing patterns will become denser.

In Eugene alone, the present shortage of "developable land" is critical to builders and buyers wanting new traditional housing. Each year, Eugene—which covers some 19,000 acres—is consuming for housing 740 new acres of "developable land" (that is land zoned for low-density residential, in private ownership, inside the city, and in parcels of half an acre or larger). There are only about 1400 acres of such land currently available in the city. That creates pressure for annexations on the fringes, and it drives up the prices on developable land both within the city and on the edges, creating financial pressures that ruin farms while contributing to inflation.

The land shortage means that Eugene has four options for growth. It can annex and sprawl indefinitely—absorbing, for example, Coburg, Goshen, Creswell, Junction City. It can attempt to limit population growth. It can increase housing density and change building styles. Or it can combine the above. In any case, the future will bring conversion of rural lands to urban, an increase in multi-family housing, conflicts over growth issues, and further rises in housing and land prices.

These options no doubt will be debated for months now that the new metro area general plan—the proposed successor to the 1990 Plan—is off the press and set to come up for public hearings sometime this winter.

The point is that change must come with population growth.

All that visionaries like Britz are telling those who wish to hear is that change need not involve the rape of rural lands or the Los Angelization of Eugene.

Britz, a University of Oregon landscape architecture professor, points out in his article in **Seriatim** (Number 9. Winter 1978) ex-

actly how Eugene folks can change their lifestyles not only to accommodate more people but also to make neighborhoods self-reliant and interdependent in terms of food, energy and social interchange.

He advocates a roll-up-your-sleeves, unplug-your-TV, help-your-neighbor approach—a sharing of property, tools and muscle.

"We can learn to increase our humanity rather than decrease it by increasing intensity of land use," Britz writes. "As we bind ourselves to economic realities generated by increased density, our street trees can be placed into direct economic productivity through careful and personal management of their nuts and fruits, and even the streets themselves can be turned over, reclaimed as earth blankets and recycled as we garden the north-south passageways most exposed to the sunlight. . . ."

The Urban Farm, in which Britz also is involved, fits this vision.

Britz: "Other factors enter the urban farm notions, such as block interiors passing into common ownership through restructuring of private property boundaries to reduce private household taxes while pooling the block interiors under non-profit community, neighborhood, or block corporations. As such, these lands could provide a great percentage of your food year-round including rabbit, chicken, turkey, eggs, vegetables of all locally supportable varieties, cherries, berries, nuts, space for re-composting organic recyclables such as leaves, grass clippings, kitchen scraps, and hot frames, cold frames, solar greenhouses and the like. . . .

"These could even evolve into community corporation status incorporating small commercial services such as pottery pro-

duction from abundant locally available clay, woodworking shops or furniture making places. . . ."

The point is whole new concepts of housing can be developed to supplant continued sprawl. Certainly the sense of community in Britz's vision seems less alienating than the suburban tract house; would there be less crime, fewer wasted lives in the edible city?

Jobs, too, can be gained within the existing city limits.

Of Eugene's 19,000 acres, 3273 are so-called "unencumbered private vacant land," according to the metro area general plan. What if a large share of that land could be leased by the city, donated by owners or purchased and land-banked for urban farming? How many jobs could be created? How much of Eugene's food could be produced for Eugene by the unemployed, by jail inmates, by the elderly in need of useful work, by children learning about soil, plants and community? How much could be done with the 414 acres of vacant public land—not parks—that lies idle in Eugene?

There are 3409 acres of road and parking lots in Eugene—17 per cent of the total area of the city, too. Is all that needed for cars?

From January, 1977, to March, 1979, 968 acres of single-family residential subdivisions were created within the Eugene-Springfield metropolitan area. It was, according to planner Steve Gordon of Lane Council of Governments, "the largest building boom in the metro-area's history."

The boom provided jobs and houses. What else did it do? A day will come when the countryside is used up and the traditional sprawl will stop. Maybe the day will come soon. Maybe not.

Chapter 4

Urban Block Farms

STRUCTURING THE LANDSCAPE...

OUR CITY CONSISTS OF BLOCKS

SOME VISIBLE AND SOME INVISIBLE. THE VISIBLE ONES ARE THOSE WE EXPERIENCE DAILY IN OUR WALKS, OUR TRAVELS, AND OUR RETURN HOME. THEY ARE THE BUILDING BLOCKS OF AN EGALITARIAN SOCIETY THEORETICALLY, BUT HAVE PRIMARILY SERVED AS THE UNIT OF LAND DIVISION WHICH HAS DIVIDED OUR CITIES MOST THOROUGHLY FROM GENUINE HUMAN INTERACTION. IT HELPS TO SEE OUR CITY FROM THE AIR AND TO REALIZE THAT OUR BLOCKS DO NOT REFLECT PEOPLE-TO-LAND OR PEOPLE-TO-PEOPLE RELATIONS, BUT ISLANDS IN THE STREAM OF AUTOMOBILE TRAFFIC AND VEHICULAR MOVEMENT. OLD CITY PLANNING THEORIES OF THE STREET BEING THE SOCIAL INTEGRATOR ARE NOT BORNE OUT BY REALITY. STREETS ARE DIVIDERS OF PEOPLE AND ISOLATORS OF HUMAN CONTACT WITH THE SOIL. MOBILITY AND SPEED DETERMINE THE MAJORITY OF URBAN PATTERNS RATHER THAN FIXITY AND PERMANENT CARING; AND THE BLOCK ITSELF COMPRISED OF INVISIBLE PATTERNS OF ZONING, TAX LOTS AND PROPERTY LINES.

EACH HOUSE HAS ITS LOT, AND EACH LOT IS A FAIRLY CLOSED WORLD, WITH DUPLICATE "EVERYTHINGS" IN EACH UNUSED GARAGE, BASEMENT OR ATTIC. WE CAN RESTRUCTURE OUR USE OF THE LAND TO PROVIDE FOR THE SHARING AND MAINTENANCE OF SOILS, PLANTS, SOLAR RIGHTS AND SOCIAL SPACE, BUT IT REQUIRES A RATE OF CHANGE WHICH IS COMPATIBLE WITH PEOPLE'S CHANGING VALUES. SHARING AND EXCHANGE OF RESOURCES MAY BE THE MAIN CONCERN FOR OUR WORLD CITIZENRY FOR OUR IMMEDIATE FUTURE, AND MAYBE EXERCIZING THESE OPPORTUNITIES AT THE NEIGHBORHOOD SCALE WILL GIVE US THESE EXPERIENCES...

WHY NOT ORIENT TO THE GARDEN INSTEAD OF TOWARD THE STREET??

URBAN BLOCK FARMS

ORIENTING HOUSES AND OUR OUTDOOR ACTIVITIES OF GARDENING, PLAYING, WALKING, SWINGING AND OTHER USES CAN BEGIN THE PROCESS OF GETTING US OUT OF OUR AUTOMOBILES AND THEIR LINEAL ENVIRONMENTS AND AID US IN UNDERSTANDING THE CYCLES AND CONSTANT CHANGES IN OUR LANDSCAPE.

NORMALLY, CITIES ARE ZONED BY SEPARATION OF FUNCTIONS, SEPARATING LAND USES BY DISTANCES AND ENCOURAGING LONG TRAVEL DISTANCES BY AUTOMOBILE. THESE HIGH ENERGY COSTS ARE FELT NOT ONLY IN OUR PERSONAL INCOMES, BUT IN THE QUALITY OF AIR IN OUR AIRSHED. INTEGRATION OF FUNCTIONS, THAT IS, THE MULTIPLE USE OF LAND AND THE OVERLAPPING OF FUNCTIONS IS ONE WAY OF UTILIZING OUR RESOURCES MORE EFFICIENTLY. EUGENE AND OTHER CITIES COMPRISED OF BLOCKS AND ZONING CAN EXPAND MIXED-USE ZONING (NOW IN USE AS SPECIAL DISTRICT ZONING) ALLOWING CERTAIN COMMERCIAL AND COTTAGE INDUSTRY USES IN RESIDENTIAL AREAS. THESE CODES, BASED ON QUALITATIVE CRITERIA, COULD EXPAND FURTHER TO "EXPERIMENTAL" DISTRICTS TO ENCOURAGE DIVERSITY AND VARIETY IN THE LANDSCAPE OF URBAN PLACES.

TRANSFORMING ZONING

IN THIS MANNER OFFERS OPPORTUNITIES FOR GREATER NUMBERS OF PEOPLE TO PARTICIPATE IN THE BUILDING OF URBAN NEIGHBORHOODS AND TO SHARE IN CITY-WIDE AND COUNTY-WIDE CONCERNS.

MEASURES OF THIS TYPE ARE IMPERATIVE FOR PRESERVING A DECENT RATIO BETWEEN PAVED AND UNPAVED AREAS WITHIN URBAN AREAS AND PRESERVING OUR RELATIONSHIPS WITH THE LAND. THE CONTINUED CONVERSION OF SMALL SCALE PARCELS OF LAND AND THE BIOMASS THEY SUPPORT INTO PAVED OVER AREAS DECREASES THE LIFE SUPPORT CAPABILITY OF A NEIGHBORHOOD, AS WELL AS INCREASES THE NEED FOR LARGER STORM DRAINS AND HIGHER TAX COSTS. INTENSE PAVING CONCENTRATIONS ALSO INCREASE HEAT REFLECTIVITY OR URBAN AREAS (AND RESULTANT AIR INVERSIONS) AND CONTRIBUTE HEAVILY TO INCREASES IN NOISE THROUGH REVERBERATION.

SINCE BLOCKS ARE WHAT WE HAVE TO WORK WITH AND ARE WHAT WE CAN ALTER WITHOUT LOTS OF FEDERAL AID, WE HAVE WORKED ON

TRANSFORMING URBAN BLOCKS

AMONG OUR ASSUMPTIONS FOR THIS WORK HAVE BEEN THAT:

1. TRADITIONAL ZONING IS TOO GROSS A MECHANISM FOR DEVELOPING A HEALTHY URBAN ENVIRONMENT. IT ISOLATES SECTORS OF THE ECONOMY FROM ONE ANOTHER, DIVIDES ECONOMIC AND SOCIAL CLASSES OF PEOPLE INTO SEGREGATED GHETTOS, GENERATES ENORMOUS DEPENDENCE ON THE AUTOMOBILE AND IS GENERALLY ENERGY INEFFICIENT.

2. ENERGY ORIENTATION AND PRESERVATION OF SOLAR RIGHTS IS IMPERATIVE IN COMBATTING THE DEVELOPMENT OF "PIECEMEAL" PLANNING RESULTING IN A FULLY PAVED URBAN ENVIRONMENT.

BLOCK INTERIORS ARE ECONOMIC AND ENVIRONMENTAL ASSETS

THEY FORM A PATCHWORK PATTERN OF OPPORTUNITY FOR US TO KNIT OURSELVES TOGETHER, SHELTERED FROM THE DOMINANCE OF THE AUTOMOBILE. MANY GARDENS ALREADY EXIST BUT ARE LATER CONVERTED TO APARTMENTS AND PARKING LOTS, OPENING UP OUR BLOCK INTERIORS TO 100% PAVING, LEAVING NO PLACE TO GO TO BE OUTSIDE IN CONNECTION WITH THE SEASONS IN OUR

VALLEY. RESIDENT INSTABILITY IS HIGH IN THESE AREAS AND ITS COUSINS,
ABSENTEE LANDLORDS, SPECULATION AND DISPLACEMENT. GIVEN THESE CIR-
CUMSTANCES WE RECOMMEND:

1. PERMISSION--GET IT FROM YOUR LANDLORD, YOU WILL ONLY IM-
   PROVE HIS PROPERTY VALUES.
2. TEMPORARY USE--MOBILITY
3. CONVERSION TO OWNERSHIP--LAND STABILITY

(EVEN WITH A HIGH TURNOVER RATE, ORDER CAN BE CREATED AND FOOD PRO-
DUCTION CAN BE HIGH.)

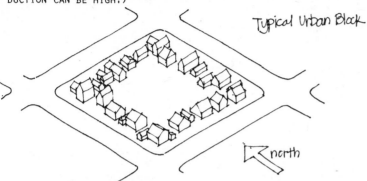

Typical Urban Block

north

MAXIMIZING SOLAR GAIN
AND STORAGE IN URBAN BLOCKS

ENTAILS SOME GUIDING PRINCIPLES. SINCE MOST OF US LIVE IN BLOCKS
AND LIVE IN RELATIVELY ORDINARY HOUSES, WE MIGHT WANT TO TAKE A
BRIEF LOOK AT WHAT WE CAN DO TO USE THE SUN AND OUR CLIMATE TO
MAXIMUM ADVANTAGE. THE FOLLOWING DRAWINGS ARE INTENDED TO SHOW A
PROGRESSION FROM A SINGLE GARDEN TO A COOPERATIVE SHARING OF SPACE
(SUCH AS AN URBAN FARM); TO THE APPLICATION OF THESE PRINCIPLES
WITHIN A TYPICAL URBAN BLOCK; TO THE DEVELOPMENT OF THAT BLOCK IN
BOTH DENSITY AND INTENSITY OF OPEN SPACE USE (CALLED AN ADVANCED
SOLAR BLOCK OR BLOCK FARM); TO THE APPLICATION OF THESE PRINCIPLES
TO A SUB-NEIGHBORHOOD PLACE (OR SECTOR); AND FINALLY TO URBAN BLOCKS
AS TIME AND ECONOMIC CIRCUMSTANCES DICTATE. WHAT IS INTENDED IN
THIS NOTION IS THAT WE CAN:

INCREASE HOUSING DENSITY IN OUR NEIGHBORHOODS

WHILE SIMULTANEOUSLY

GROWING LOTS OF OUR OWN FOOD ON THE OPEN INTERIORS OF OUR BLOCKS AND

MAINTAIN SOLAR ACCESSIBILITY FOR THE MAJORITY OF OUR HOUSES AND GARDENS AND THUS

SUBSTANTIALLY CONTRIBUTE TO THE PRESERVATION OF FARMING VALUES AND THE MAINTENANCE OF AGRICULTURAL LANDS IN THE SOUTHERN WILLAMETTE VALLEY

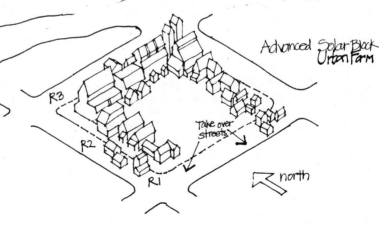

Advanced Solar Block Urban Farm

SUGGESTIONS FOR NEW CONSTRUCTION THAT MAY REPLACE THE TRADITIONAL URBAN BLOCKS

R3

R2

R1

Take over streets

north

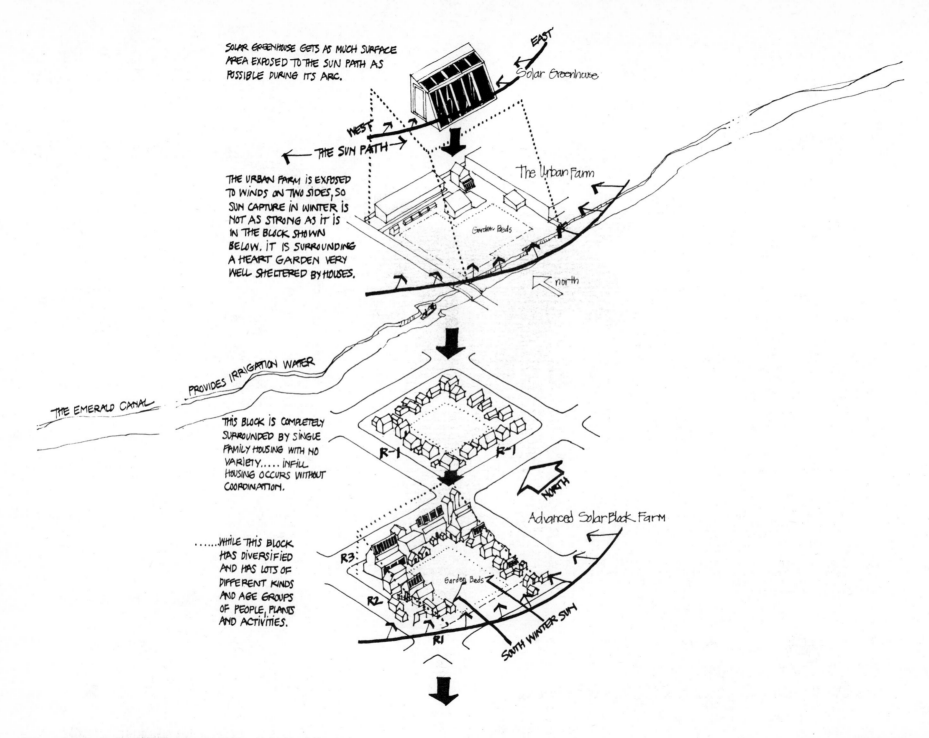

SOLAR GREENHOUSE GETS AS MUCH SURFACE AREA EXPOSED TO THE SUN PATH AS POSSIBLE DURING ITS ARC.

EAST

Solar Greenhouse

WEST

THE SUN PATH

The Urban Farm

THE URBAN FARM IS EXPOSED TO WINDS ON TWO SIDES, SO SUN CAPTURE IN WINTER IS NOT AS STRONG AS IT IS IN THE BLOCK SHOWN BELOW. IT IS SURROUNDING A HEART GARDEN VERY WELL SHELTERED BY HOUSES.

Garden Beds

north

PROVIDES IRRIGATION WATER

THE EMERALD CANAL

THIS BLOCK IS COMPLETELY SURROUNDED BY SINGLE FAMILY HOUSING WITH NO VARIETY..... INFILL HOUSING OCCURS WITHOUT COORDINATION.

R-1

R-1

NORTH

Advanced Solar Block Farm

R3

......WHILE THIS BLOCK HAS DIVERSIFIED AND HAS LOTS OF DIFFERENT KINDS AND AGE GROUPS OF PEOPLE, PLANTS AND ACTIVITIES.

R2

Garden Beds

R1

SOUTH WINTER SUN

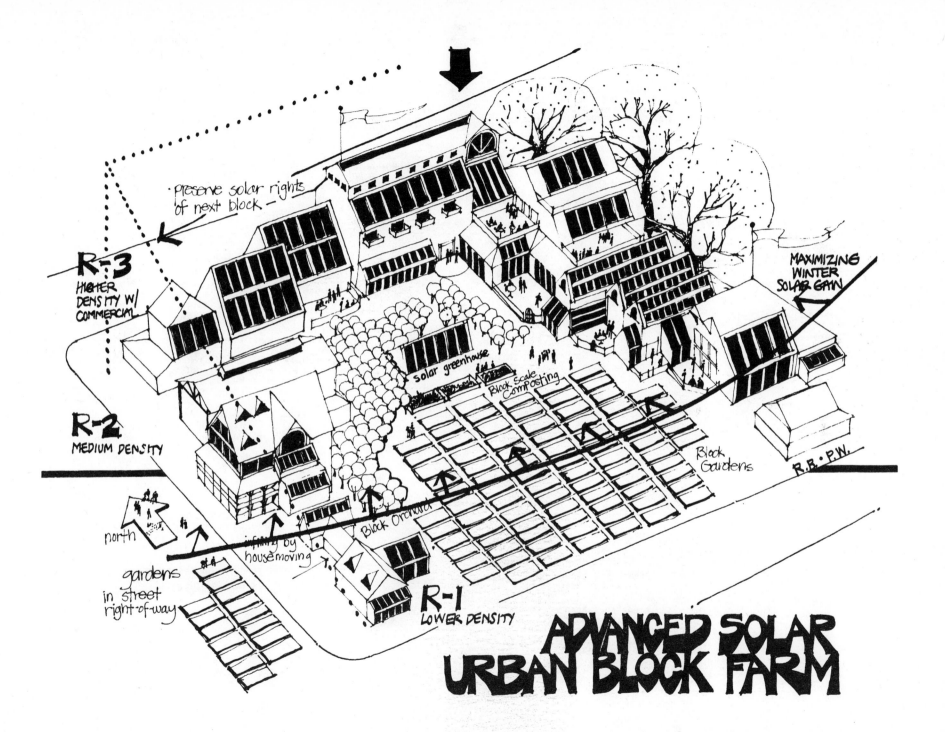

preserve solar rights
of next block

R-3
HIGHER
DENSITY W/
COMMERCIAL

MAXIMIZING
WINTER
SOLAR GAIN

solar greenhouse

Block Scale
Composting

R-2
MEDIUM DENSITY

Block
Gardens

R.B. • P.W.

north

gardens
in street
right-of-way

infilling by
housemoving

Block Orchard

R-1
LOWER DENSITY

ADVANCED SOLAR URBAN BLOCK FARM

BLOCK FARMS CAN ORGANIZE AND EVOLVE TO:

PRODUCE FOOD

"MAKE PROVISIONS FOR GROWING YR OWN!"

1. INCREASE IMMEDIATELY-ADJACENT FOOD PRODUCTION AND MAINTAIN OPEN SPACE ALLOTMENTS WITHIN THE BLOCK. INTENSIVE USE OF THIS LAND IS HIGHLY RECOMMENDED.

   A. STRUCTURE OPEN SPACE INTO RAISED BEDS AND PATHWAYS.

   B. PRESERVE SMALLER PRIVATE BACK YARD SPACE.

RECYCLE MATERIALS

"SAVE IT AND STOCKPILE IT!"

2. PROVIDE HOUSEHOLD "WASTE" RECYCLING WITHIN THE BLOCK AND CONSERVE RESOURCES NORMALLY ASSOCIATED WITH THE "WASTE STREAM".

   A. STORE AND INCREASE BIOMASS.

   B. DEVELOP BLOCK SCALE WORM BEDS.

   C. UTILIZE FENCES AND WASTE WOOD FOR WORM BEDS, SMALL ANIMAL SHELTERS, ETC.

INCREASE HOUSING DENSITY

"MINIMIZE CAPITAL LEAKAGE!"

3. ABSORB THEIR PART OF URBAN DENSITY INCREASE VOLUNTARILY AND ENTHUSIASTICALLY.

   A. DEVELOP HOUSING INFILL STRATEGIES WHICH RESPECT SOLAR RIGHTS.

   B. MOBILIZE TO MOVE HOUSES (THREATENED ELSEWHERE) ONTO AND BETWEEN EXISTING BUILDINGS.

   C. CONVERT STORAGE BUILDINGS AND UNUSED GARAGES TO RENTAL UNITS.

   D. INFILTRATE SMALL SCALE COMMERCIAL AND "COTTAGE" INDUSTRIAL LAND USES WITHIN THE BLOCK.

UTILIZE MORE LOCAL ENERGY RESOURCES

"POOL IT AND ORDER IT!"

"CAPITAL LABOR HEAT!!"

4. DEVELOP SOUND RESPECT FOR SOLAR RIGHTS AND IMPLEMENT HOUSING INFILL AND FOOD PRODUCTION SIMULTANEOUSLY.

   A. UTILIZE THE ENERGY EFFICIENCIES OF PHOTOSYNTHESIS AND GREEN PLANTS.

   B. TRAP HEAT WITH CHEAP ATTACHED SOUTH-FACING SOLAR GREENHOUSES AND/OR AIR LOCKS. EXTEND THE WARM SEASONS.

   C. RETROFIT SLOWLY (USING THE ENERGY SERVICES COMPANY) BY DIRECT GAIN SOLAR WATER HEATERS.

   D. MAINTAIN A CAREFUL BALANCE OF OPEN SPACE AND HOUSING DENSITY WITHIN THE BLOCK SUITABLE FOR A SHIFT FROM A CAPITAL ECONOMY TO A BARTER ECONOMY. CONSERVE AND COOPERATIVELY CONSOLIDATE TOOLS AND LABOR.

INCREASE HEALTH

"AN OUNCE OF PREVENTION IS WORTH A POUND OF CURE!"

"DO IT BEFORE YOU NEED IT!"

(SEE ALSO FOOD AND NUTRITION SECTION.) FOOD PRESERVATION CENTERS WHERE NEIGHBORS CAN GATHER TO PREPARE FOR FALL, WINTER, SPRING AND SUMMER, -- AND APPROPRIATELY DISTRIBUTED DEMONSTRATION PROJECTS; APPROX. ONE PER 1200 PEOPLE (6000 --5) IN EACH OF THE DISTINCTLY UNIQUE SECTORS OF THE NEIGHBORHOOD.

E. ALTER ENERGY COSTS BY REDUCING AUTO-MOBILE DEPENDENCE BY PHASING OUT CARS, PHASING IN BIKES AND BUSES AND CLOSING STREETS.

5. INCREASE TRUST AND COMMUNITY SHARING BY DEVELOPING NEIGHBORLINESS AND REJECTING INCREASING SOCIAL ISOLATION.

   A. REMOVE PHYSICAL BARRIERS SUCH AS HIGH FENCES, HEDGES, ETC.

   B. PROVIDE INTRA-BLOCK SAFETY AND VISUAL SECURITY THROUGH DIRECT OBSERVATION AND VISUAL ACCESS.

   C. ENFORCE PERSONAL RESPONSIBILITY FOR THE BLOCK LANDSCAPE.

6. PROVIDE THE BASIS FOR LARGER SCALE ORGANIZATION FOR NEIGHBORHOOD FARMS, AND OTHER NEIGHBORHOOD VENTURES.

   A. BLOCK FARMS SERVE AS DEMONSTRATION PLACES WITHIN THE NEIGHBORHOOD...

   B. INITIALLY BY SERVING AS PLACES WITHIN THE NEIGHBORHOOD TO GO TO ORGANIZE BULK BUYING, GLEANING AND NEIGHBORHOOD DELIVERIES.

   C. SECOND, BY ENCOURAGING, BY EXAMPLE, INTERDEPENDENT FOOD NETWORK SYSTEMS SUCH AS CANNING, DRYING AND STORAGE OF BULK OR EXCESS GRAINS, VEGETABLES, FRUITS, NUTS, EGGS, MEAT, AND OTHER FOODS.

   D. THIRD, BY COORDINATING EFFORTS BETWEEN BLOCK FARMS, ONCE EXPERIENCE HAS BEEN GAINED, FOR INTER- AND INTRA-NEIGHBORHOOD FOOD EXCHANGES.

   E. AND FOURTH, BY PROVIDING ORGANIZATION FOR LARGER SCALE URBAN AGRICULTURE EFFORTS DEVELOPED ON UNDERUTILIZED OR UNDERMAINTAINED PUBLIC OPEN LAND.

      (POTENTIAL NEIGHBORHOOD COMPUTER SYSTEMICS)

7. CONTRIBUTE DIRECTLY TO UNDERSTANDINGS BETWEEN "URBAN" AND "RURAL" PEOPLE ON SUCH ISSUES AS: LAND USE, TAXATION, LIFE STYLE, RESOURCES, ETC.

8. CONTRIBUTE IN GENERATING A SELF-SUPPORTING AND PERMANENT FOOD SUPPLY IN CONJUNCTION WITH EXISTING SMALL FARMERS AT THE CITY EDGE.

IN SHORTHAND SUMMARY...

1. INCREASE BIOLOGICAL DENSITY EFFICIENCY IN URBAN AREAS

- REVERSE THE DEPRECIATION OF URBAN BIOMASS BY MAXIMIZING VEGETATIVE GROWTH WITHIN THE CITY.

- RAISE FOOD ON AS MUCH OF THE OPEN LAND AS POSSIBLE AND ALTER OR ELIMINATE DELETERIOUS SUBSTANCES IN THE WATER, THE SOIL, THE AIR AND THE GENERAL URBAN HABITAT.

- CLOSELY PACK PRODUCTIVE LANDS AND INTERSPERSE THEM WITH HIGHER DENSITIES OF PEOPLE AND OTHER ANIMALS IN A WORKING, INTERDEPENDENT RELATIONSHIP.

2. TIGHTEN FEEDBACK LOOPS

- CLOSE UP RESPONSE TIMES TO INJURIOUS ENVIRONMENTAL FACTORS. (TAKE CARE OF YOUR OWN GARBAGE INSTEAD OF HAVING SOMEONE ELSE TAKE IT TO THE LANDFILL SITES WHICH ARE OVERLOADED.)

- REDUCE DEPENDENCY ON LONG-DISTANCE SUPPLIES OF FOOD, ENERGY AND MATERIALS.

- CELEBRATE LOCALISM AND INDIGENOUS PLANT MATERIALS AND ASSURE THEIR CONTINUATION THROUGH TIME.

CLOSING UP CAN MEAN COMING TOGETHER

DEVELOPING A TYPICAL LOW-DENSITY BLOCK FARM MIGHT BE LIKE THIS...

IF YOU LIVE IN A TYPICAL BLOCK, LIKE THIS EXAMPLE IN WESTSIDE NEIGHBORHOOD.

THIS EXAMPLE SHOWS THE SIMPLEST CASE OF OWNER-ACCURED VALUE IN THE NEIGHBOR-DEVELOPED BLOCK INTERIOR. THROUGH THREE PHASES OF DEVELOPMENT, MAXIMUM STRESS HAS BEEN LAID ON FOOD PRODUCTION AND ENERGY CONSERVATION WITH SOME HOUSING INFILL BETWEEN AND AMONG EXISTING HOUSES. THIS HAS BEEN ACCOMPLISHED BY (INITIALLY) RENTING UNUSED ROOMS TO NEW RESIDENTS IN EXCHANGE FOR CASH OR LABOR, WHICH IS USED AS SEED MONEY FOR BUILDING A BLOCK-SCALE ECONOMY. THROUGH GRADUAL CHANGE, THE FEW INITIAL HOMEOWNERS (WHO HAVE FORMED AN ASSOCIATION) EXPAND AND THE COOPERATIVE NATURE OF TOOL SHARING AND GARDEN SHARING BECOMES MORE FORMALIZED AND CAPABLE OF LARGER VENTURES. POOLING OF LAND FOR "COMMUNITY" VENTURES BEGINS THE FIRST STEP TOWARD DEVELOPMENT OF URBAN TRUST.

BEGINS A PROCESS OF SHARING AND CONNECTION OF PROPERTIES INTO A LARGER AND MORE SUSTAINABLE ECONOMIC UNIT. BLOCK SCALE (RATHER THAN INDIVUAL HOUSEHOLD) ORGANIC WASTE RECYCLING INTO COMPOSTING WORM BEDS BRINGS PEOPLE TOGETHER ON COMMON PATHWAYS AND ENABLES LARGE NUMBERS OF HOUSES TO DISCONTINUE GARBAGE PICK-UP, AND ELIMINATES ITS EXPENSE. ALTERNATIVES TO AUTOMOBILE TRANSPORT BECOME MORE POPULAR, ENABLING SOME AUTOS TO BE SOLD AND MORE OF THE LAND BECOMES AVAILABLE FOR FOOD PRODUCTION (OR GARAGES CONVERT TO RENTAL UNITS). LATER ON, LAND IS RECLAIMED AND THE STREET BLANKET IS ROLLED BACK AND RE-FERTILIZED WITH LEAVES AND COMPOST. FRUIT AND NUT TREES ARE PLANTED AND SOME OPEN LAND IS ALSO USED FOR RECREATION FOR CHILDREN AND ADULTS.

BUT THE BLOCK ESSENTIALLY REMAINS LOW DENSITY, WITH 24 LIVING UNITS, BUT AT THE END OF PHASE THREE, ALL OF THE UNITS ARE OWNED BY THE BLOCK AND RENTED OR LEASED TO THE OWNERS, SIMILAR TO A CONDOMINIUM OR PLANNED UNIT DEVELOPMENT.

THE DESIGN AND MODEL OF THIS BLOCK WAS DONE BY PAUL WILBERT, JIM KLEIN, SUE SCHICK, JOE MILLON, STEVEN O'CONNELL, AND NAVA NOVO-PLANSKI, STUDENTS IN THE DEPARTMENTS OF ARCHITECTURE AND LANDSCAPE ARCHITECTURE AT THE UNIVERSITY OF OREGON IN A CLASS JOINTLY TAUGHT BY G.Z. "CHARLIE" BROWN AND RICHARD BRITZ.

BY CLUSTERING THE NEW HOUSING UNITS ON THE NORTH SIDE OF THE BLOCK, THIS TECHNIQUE WILL ENABLE THE BLOCK UNIT AS A WHOLE TO CONSERVE MAXIMUM ENERGY IN ITS SOUTH FACING "U" SHAPE. HERE, PATHWAYS INTO THE BLOCK CENTER JOIN BLOCK-TO-BLOCK IN PEDESTRIAN "WAYS," AND STREETS ARE CLOSED TO VEHICULAR TRAFFIC AND PLANTED IN FRUIT AND NUT TREES.

phase one

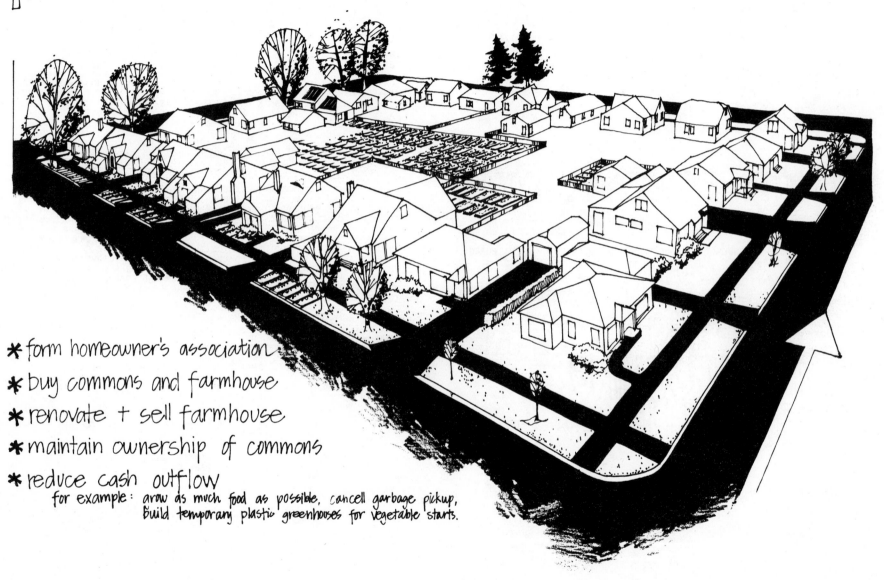

* form homeowner's association
* buy commons and farmhouse
* renovate + sell farmhouse
* maintain ownership of commons
* reduce cash outflow
  for example: grow as much food as possible, cancell garbage pickup,
  build temporary plastic greenhouses for vegetable starts.

phase one

	physical changes	behind the scenes
land ownership	RENOVATION OF FARMHOUSE REMOVING FENCES & BARRIERS & JUNK ON COMMONS	FARMHOUSE & LOT PURCHASED BY ASSOCIATION, COMMONLY OWNED !! # _____ PURCHASE PRICE LOAN PAYMENTS : # _____ PER MONTH, PER MEMBER.

financial organization	ROOMS RENTED OUT FOR EXTRA INCOME
	PART-TIME JOBS
		FORM THE ASSOCIATION BY POOLING $ _____ IN LAND
		PAY SALARIES FOR: # _____ IN ASSETS
		  PRESIDENT @ # _____ 10 HOUSES CONTRIBUTE # _____
		  SECRETARY @ # _____ EACH MONTH FOR TWO YEARS

food	PLANT STARTS IN GREENHOUSES (WINTER-SPRING)
	TRANSPLANT ALL STARTS (IN SPRING) ONTO ALL AVAILABLE
	  LAND - INCLUDING STREETS & FRONT YARDS & COMMONS
	SOIL REPLENISHED W/ ORGANICS, COMPOST
	BUY
		MAJOR TIME/LABOR COMMITMENT OF ASSOCIATION MEMBERS
		SAVE MONEY ON GROCERY EXPENSES : # _____ FOR
		  10 HOUSES FOR 2 YEARS
		SEED & START COSTS : # _____
		BUY FOOD IN BULK

construction	RENOVATE FARMHOUSE QUICKLY — SELF HELP LABOR BUILD TEMPORARY GREENHOUSES (LOW COST) FOR PLANT STARTS AND ENERGY CONSERVATION BUILD BEAUTIFUL GATES IN THE FENCES ONTO COMMON LAND. WEATHERIZE & REPAIR & RE PAINT HOUSING	RENOVATION COSTS : # _____ MAXIMUM TIME/LABOR FROM GREENHOUSE COSTS : # _____ ASSOCIATION MEMBERS TAX-CREDITS FOR WEATHERIZATION CITY GRANTS & LOANS FOR REPAIRS
land use: recreation commercial public private	RECREATIONAL, OPEN SPACE USES MOVE OFF SITE MAJOR LOSS OF PRIVATE OUTDOOR SPACES	
utilities waste energy	TEMPORARY GREENHOUSES SAVE ON HEAT RECYCLING , COMPOSTING WOOD STOVES , CUT POPLARS & BURN , COLLECT FIREWOOD DUMP OWN GARBAGE , CANCEL CITY SERVICE	LIVING A MORE ENERGY & WASTE CONSCIOUS LIFESTYLE SAVE MONEY ON ENERGY BILLS : # _____ SAVE MONEY ON WASTE COLLECTION : # _____ EXPENSES FOR WOODSTOVES : # _____
transportation	CARPOOLING & MASS TRANSIT WALKING & CYCLING	MONEY SAVED ON AUTOMOBILE GAS & MAINTENANCE # _____ SALE OF AUTOMOBILES : MONEY TO INDIVIDUALS

phase two

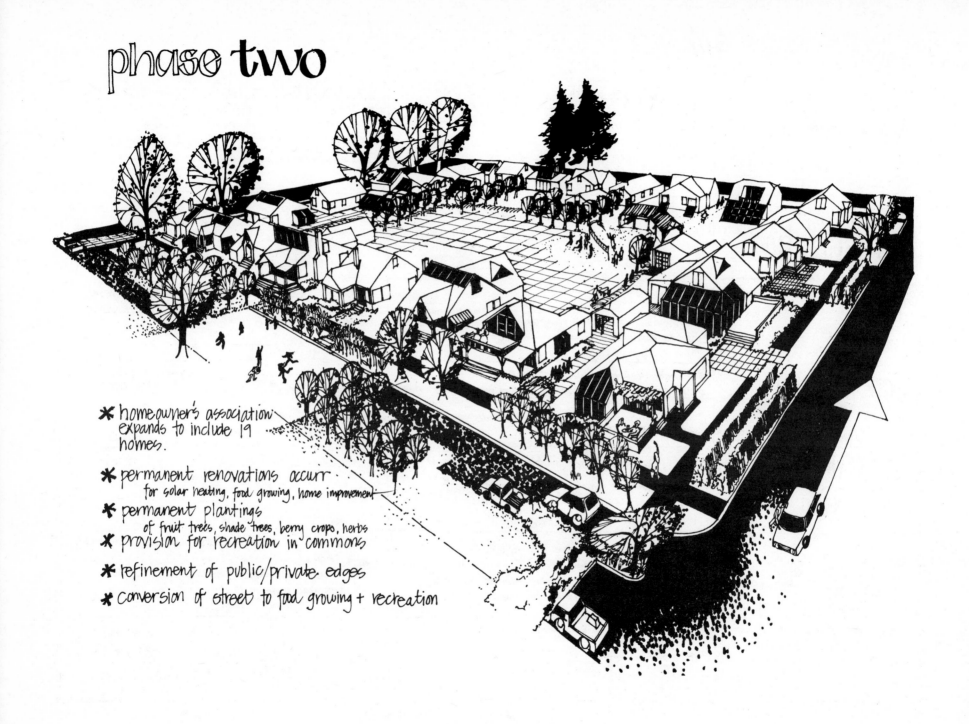

* homeowner's association expands to include 19 homes.

* permanent renovations occurr
    for solar heating, food growing, home improvement
* permanent plantings
    of fruit trees, shade trees, berry crops, herbs
* provision for recreation in commons

* refinement of public/private edges

* conversion of street to food growing + recreation

phase two

	physical changes	behind the scenes
land ownership	Houses remain individually owned w/ individual, private outdoor spaces Commodas is expanded and concentrated in center	19 houses have joined association, 5 houses have not • 14 share all land – deeded, bought & issued stock • 5 maintain individual control – bought into association Gradual aquisition,
financial organization		← Office built in a old garage, centrally located Financially stable, less pressure to cut down all costs Increased salaries for president & secretary On site employment
food	Fruit orchards, trees, berry, vegetables, agriculture moves to street side, too.	Long term production Save #____ on food bills
construction	Renovate housing for commercial/residential mix • Build community bldg. • Convert garages to other uses self-help & contracted • Permanent greenhouses – outdoor space • Agricultural/recreational structures – arbors, trellises, shelters • Solar retrofit – energy conservation – solar h.w. heating • Individual public/private spaces	Self-help and contracted Salvaged & recycled materials Shared help, labor
land use: recreation public commercial private	More recreation spaces & open spaces Public access, paths 3 commercial buildings w/residential above (cafe, offices, grocer) Mixing food production w/ recreation (ie. treehouses, picnicing) Gathering space for block, community, neighborhood ↔	Make money from lease of commercial buildings: #____ & retain use of land for association Emphasis on retaining co-operative spirit Development of individual private lands
utilities waste energy	• Solar hot water heating units • Central composting • Renovate for passive solar heating • Composting toilets	Cost of solar units: #____ can be shared by association Savings on energy: #____ per house, per month Tax credits & grants
transportation	• 2 streets narrowed & land converted, limited access • On-street parking • Bike/car shelters • Convert driveways to other uses •	New energy-efficient mass transit systems Fewer individually owned vehicles Association-owned vehicles Street vacation policy Some garages not used for autos

phase three

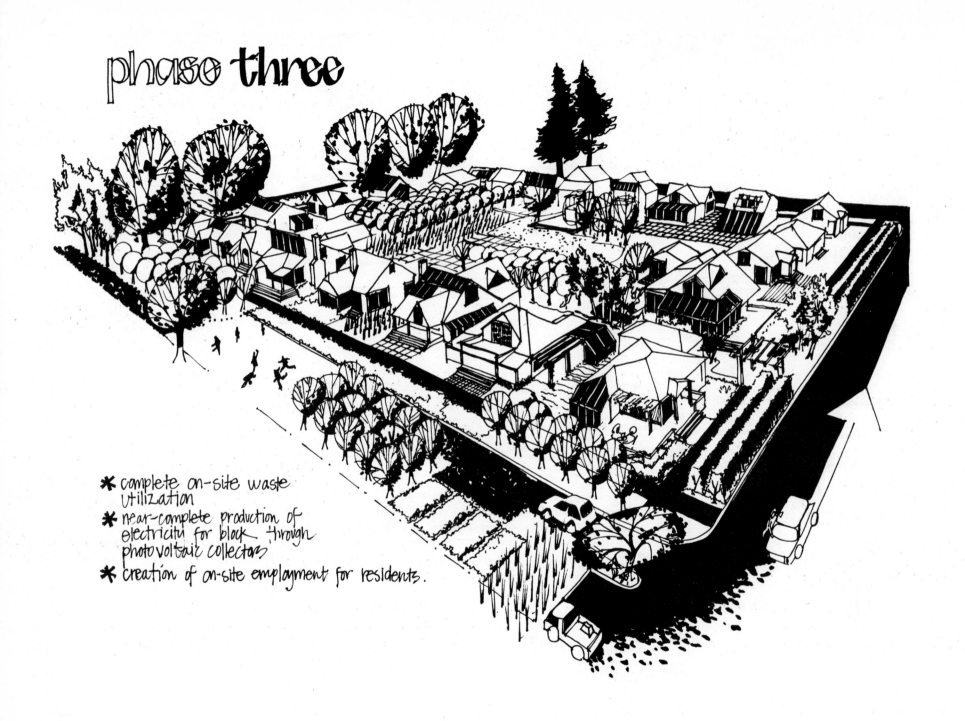

* complete on-site waste utilization
* near-complete production of electricity for block through photovoltaic collectors
* creation of on-site employment for residents.

phase three

	physical changes	behind the scenes
land ownership	Black homeowners association owns entire block, more houses leased for commercial, office purposes	Black homeowner's association owns entire block.
financial organization		Expenses cut significantly, so outside income needs reduced significantly. Less work, more time to spend as you wish.
food	Aquaculture facility in addition to cont'd production from fruit trees, berries, vegetables. Animals introduced: chickens & rabbits – replace cats & dogs.	Aquaculture facility commonly funded. Provides on site part time employment.
construction	Some homes converted to commercial & offices. Chicken coops & rabbit hutches constructed	Construction provides employment for short term. Maintenance & operation provide employment in long term.
land use: recreation commercial public private	Some homes converted to commercial & office use, with residences above perhaps. Chickens & rabbits	More diverse land use for greater stability and reduced energy use, & employment
utilities waste energy	Methane digester constructed to handle human and animal waste. End product feeds fish in aquaculture pond. Greywater system for entire block's waste.	
transportation	Few vehicles, shared in small groups. we own this truck together	Reduced need for automobile because of on-site employment, services, utilities, amenities

OR THE SAME BLOCK COULD DOUBLE ITS DENSITY LIKE THIS AND STILL MAINTAIN ITS FOOD PRODUCING HEART

BY CLUSTERING THE NEW HOUSING UNITS ON THE NORTH SIDE OF THE BLOCK.
THIS TECHNIQUE WILL ENABLE THE BLOCK UNIT AS A WHOLE TO CONSERVE
MAXIMUM ENERGY IN ITS SOUTH FACING 'U' SHAPE. HERE PATHWAYS INTO
THE BLOCK CENTER JOIN BLOCK TO BLOCK IN PEDESTRIAN "WAYS" AND
STREETS ARE CLOSED TO VEHICULAR TRAFFIC AND PLANTED IN FRUIT AND NUT
TREES.

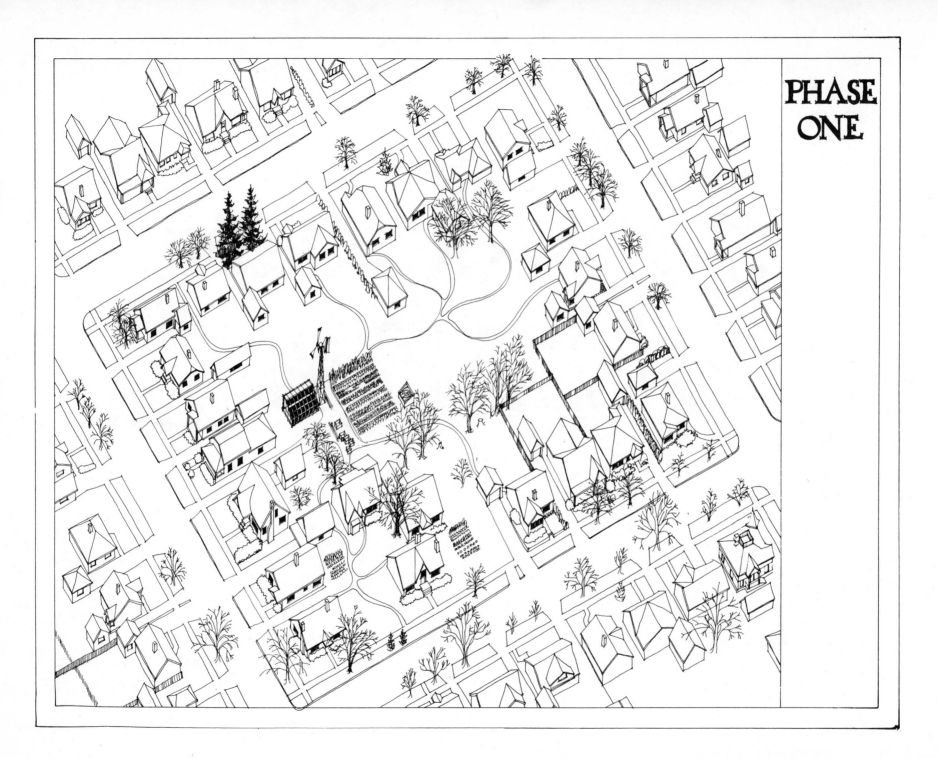

PHASE ONE

PHASE 1	YEAR ONE	YEAR TWO	YEAR THREE	YEAR FOUR
EXTERNAL PRESSURES ECONOMIC INFLATION POLLUTION SHORTAGES PARANOIA	·DEVELOPER THREAT	·LOAN PAYMENTS BEGIN ⟹	(PAYMENTS BEGIN HERE) (IF DEFERRED) ⟹	⟹ INCREASING ECONOMIC COSTS — BURDEN OF LOAN PAYMENTS

GROUP RESPONSE (DECISIONS)	· COOPERATIVE ORGANIZATION FORMS (10 INITIAL MEMBERS)	· DECISION TO GAIN CAPITAL TO PAY OFF LOAN
		
		· PLANS FOR USING CENTER OF BLOCK
			· DECISION TO MAXIMIZE FOOD PRODUCTION
			
			· DECISION TO USE NEW INCOME AFTER LOAN PAYMENTS ON: FOOD PRODUCTION / PLAY AREA
			
			· DECISION TO CREATE "SWEAT EQUITY" JOBS ON SITE TO WORK OFF INDIVIDUAL LOAN PAYMENTS
				· MAXIMUM GARDEN DEVELOPMENT (LOW CAPITAL INVESTMENT)
				· GROUP PLANS BLOCKS FUTURE
				  - DIVERSIFY VEGETATION
				  - PLANS FOR ENERGY CONSERVATION
				  - RE-ESTABLISH NATURAL ECO-SYSTEM
				· MINIMUM MONEY INTO BLOCK DEVELOPMENT THIS YEAR. MOST GOES TO PAY LOAN OFF QUICKER

ACTIONS	APPLY FOR LOAN
	
	LOAN APPROVED
	
	CENTER LOT PURCHASED
		· FARM HOUSE REHABILITATION TO ACCOMODATE SOLAR PANEL/WOOD STOVE DISTRIBUTORSHIP (STOCK ON CONSIGNMENT ... LOW INITIAL INVESTMENT)
		· LOW KEY SITE CLEAN UP
		  - PRUNING/WEEDING
		  - 3 POPLARS CUT DOWN (SUN ACCESS)
		· BEGIN BUYING FOOD COLLECTIVELY
		· TOOLS SHARED (TOOL LIBRARY) ⟹
			· 25% OF LOT IN INTENSIVE GARDENS 75% IN RED CLOVER
			· SITE DEVELOPMENT
			  - BUILD WORM BOXES
			  - COMPOSTING
			  - FOOD TREE PLANTING/GARDENING
			  - ANIMAL SHELTERS BUILT
			  - COLD FRAMES BUILT
			  - SMALL GREENHOUSE BUILT
			· BEGIN CONSTRUCTING PLAY AREA
			· PRIVATE WINDMILL BUILT
				· PANEL & STOVE INSTALLATION & SERVICE & ENERGY CONSERVATION CONSULTING
				
				· MORE GARDEN PLANTING

OBJECTS		· STOCK SUPPLY FOR DISTRIBUTORSHIP FROM EXTERNAL SUPPLIER
		  - SOLAR WATER HEATER SYSTEMS
		  - WOOD STOVES
		
		· FENCES/GATES
		· PATHS
			· INCOME SURPLUS FROM BUSINESS AFTER LOAN PAYMENTS
			· WORM BOXES
			· BEE HIVE
			· COMPOSTING BINS
			· RAISED BEDS
			· RABBITS/CHICKENS
			· FRUIT TREES/BERRIES
			· COLD FRAMES
			· SMALL GREENHOUSE
			· PLAY AREA
			· EXPERIMENTAL WINDMILL
				(INCREASING) ⟹

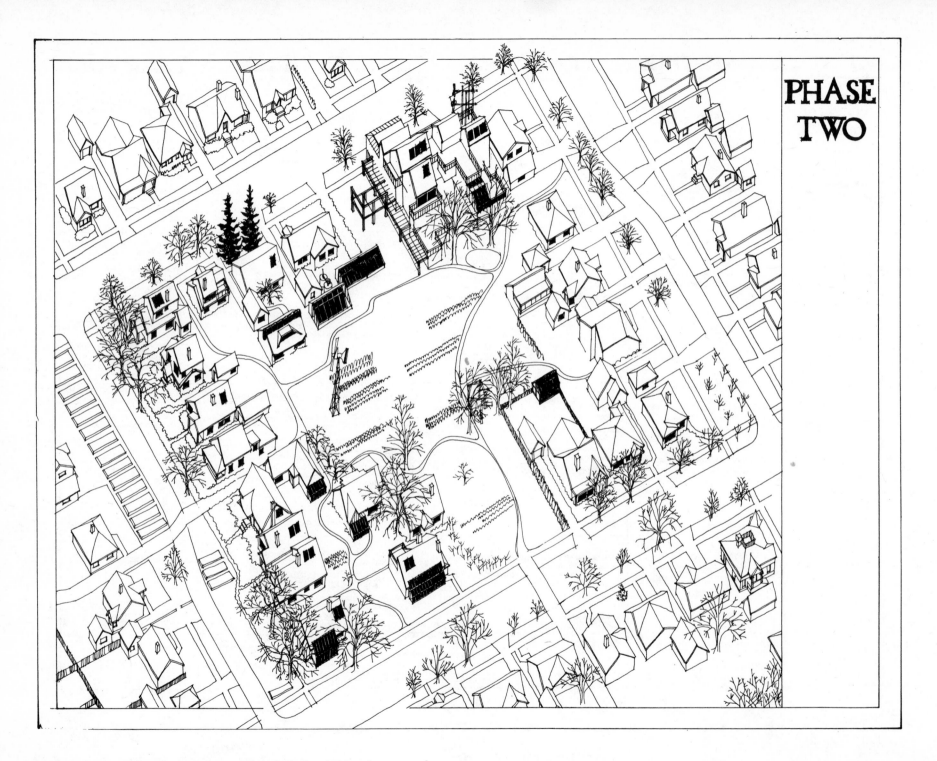

PHASE
TWO

PHASE 2	YEAR FIVE	YEAR SIX	YEAR SEVEN	YEAR EIGHT	YEAR NINE
EXTERNAL PRESSURES ECONOMIC INFLATION POLLUTION SHORTAGES PARANOIA ⬡	(LOAN PAYMENTS) ENERGY PRICES GO UP SHARPLY	• 1990 PLAN - DENSITY PRESSURE - HIGHER TAXES ON UNDEVELOPED LAND	• SOLAR COLLECTOR MARKET SATURATED LOCALLY (SALES DECLINE)	• EXTENDED MASS TRANSIT REZONING - SPLITS EXISTING LOTS IN HALF • INCREASING NUMBERS OF PEOPLE INTERESTED IN BLOCK	• ECONOMIC DEPRESSION - FOOD PRICES SKYROCKET • HOUSING SHORTAGE • 1990 PLAN DENSITY PRESSURE
GROUP RESPONSE (DECISIONS) ◇	(GROUP PLANS) • DECISION TO PUT MONEY INTO BLOCK DEVELOPMENT • CORNER GROUP DECIDES TO BECOME COLLECTIVE AND MOVE TWO HOUSES TO OPEN UP GARDEN SPACE	ONGOING PLANNING • DECISION TO GO COLLECTIVE AND JOIN LOTS	• SEARCH FOR ALTERNATIVE ECONOMIC BASE • DECISION TO CONSOLIDATE ELECTRIC BILLS AND SHARE A FEW WASHERS & DRYERS • DECISION TO EXPAND COLLECTIVE	MATRIX SEARCH • STUDY OPTIONS & PLANS FOR A BLOCK INDUSTRY - RECYCLING CENTER - HOUSING PANELS - AQUACULTURE • SEEKING UTILITY FREEDOM - COLLECT RAIN RUN OFF • DECISION WITH OTHER BLOCKS TO CLOSE ALMADEN AND TAYLOR TO THRU-TRAFFIC AND USE FOR PLANTING • PLAN TO SELL SOLAR COLLECTOR FRANCHISE	• FOCUS ON ENERGY AND FOOD AUTONOMY • PLAN CHOSEN - SET UP A HOUSING PANEL PRODUCTION CENTER WITH RECYCLING CENTER (LOAN APPLICATION) • DECISION TO CONCENTRATE HOUSING ON NORTH SIDE • DECISION TO USE GREENHOUSES PRIMARILY FOR FOOD FOR COLLECTIVE (LOSS OF INCOME)
ACTIONS ○	(DISTRIBUTORSHIP) • WORKSHOPS - CONSTRUCTION - GARDENING - RETROFITTING • MOVE TWO HOUSES • SOME MONEY ALLOTED TO SOCIAL EVENTS (SLUSH FUND)	• BUILD ANOTHER SMALL GREENHOUSE • COLLECT NATIVE TREES (FROM CLEARCUT AREAS) FOR NURSERY. ALSO GATHER FIREWOOD • BUILD AND INSTALL MORE SOLAR WATER HEATER UNITS • BUY ANOTHER BEEHIVE	• MERGE LOTS FOR COLLECTIVE OWNERSHIP • TWO NEW MEMBERS JOIN FROM W/IN BLOCK • CONSOLIDATE ELECTRIC BILLS • SELL REDUNDANT MAJOR APPLIANCES • SET UP YEARLY BUDGET FOR GARDENS • PRIVATE INFILL BUILT • ATTACHED GREENHOUSES ADDED • LARGE GREENHOUSE FOUNDATION BUILT • BUDGET FOR A RECYCLING CENTER	• ADDITION ON CORNER HOUSE AND CONVERSION TO COMMUNITY STORE - HOME CRAFTS SUPPLY STORE • EXPERIMENT WITH METHANE PRODUCTION • CARS START TO LOSE IMPORTANCE. PLUG IN TO CITY-WIDE MASS TRANSIT • FINISH LARGE GREENHOUSE • USE SHED @ FARM HOUSE FOR SORTING COLLECTING AND CLEANING UP MATERIALS	• SELL SOLAR COLLECTOR/ WOOD STOVE FRANCHISE • MAJOR LOAN OBTAINED WITH NEW INCREASED EQUITY FROM SITE IMPROVEMENTS AND LAND VALUE, TO BUY TWO NORTH LOTS AND BUILD TERRACED HOUSING • DIG POND AND BEGIN AQUACULTURE. TANKS IN GREENHOUSE • BUILD RECYCLING CENTER AND PANEL INDUSTRY SPACE
OBJECTS □	(INCREASING) SALES $ • MORE PLAY AREA ADDED	(LEVELING OUT) • SMALL GREENHOUSE • POTS & TREES FOR NURSERY • SOLAR WATER HEATER SYSTEMS • BEEHIVE • MORE CHICKENS	(DECREASING SALES) • FOUNDATION FOR LARGE GREENHOUSE • INFILL DWELLING UNITS • SOLAR WATER HEATER UNITS • ATTACHED GREENHOUSES ON DWELLINGS • ORCHARD PRODUCE	• LARGE GREENHOUSE • TANKS FOR METHANE PRODUCTION • COMMUNITY STORE FOR LOCALLY PRODUCED ITEMS • BUS STOP NEAR COMMUNITY STORE	• LARGE OPEN PLAN SPACE FOR RECYCLING AND PANEL MAKING • AQUA-CULTURE "SUN TUBB" • POND • FISH •

PHASE 3	YEAR TEN	YEAR ELEVEN	YEAR TWELVE	
EXTERNAL PRESSURES ⬡	(LOAN PAYMENTS) → (HOUSING SHORTAGE) → (1990 DENSITY PLAN) --→	ENERGY SHORTAGE ———→ (EXTERNAL)	ENERGY SHORTAGE ———————→ (DECREASING) ———→	
GROUP RESPONSE ◇	(FOCUS ON FOOD & ENERGY AUTONOMY) --→ • SKILLS AND CONCEPTUAL UNDERSTANDING READY TO PUSH AUTONOMY. • IMPROVED GARDEN EFFICIENCY IS A PRIORITY • PLANS TO BECOME AS SELF-SUFFICIENT AS POSSIBLE IN TERMS OF ENERGY.	• MATRIX SEARCH • BLOCK PLAN UPDATE FOR FUTURE	• CAPITAL SUFFICIENT FROM INDUSTRY SALES TO DEVELOP CAPITAL INTENSIVE FOOD & ENERGY DEVICES (APPROPRIATE TECHNOLOGY) • PUSH SELF SUFFICIENCY → • STUDY MAXIMUM DENSITY FEASIBILITY	
ACTIONS ○	(LOTS MERGING) → "BARN RAISING" SELF-HELP ON NEW INFILL DWELLINGS AND ADDITIONS • EXPERIMENTAL HYDROPONICS • MORE ON SITE JOBS • INCREASED METHANE PRODUCTION • ORCHARDS BEGIN FALL PRODUCTION → BEGIN OPERATION. BUILDING PANELS TO USE ON SITE AT FIRST	• MAXIMUM INTENSIVE USE OF OPEN LAND • BEGIN TERRACED HOUSING CLUSTER ON NORTH ABOVE PANEL WORKS 8 UNITS BUILT (BARN RAISING) • GARDENS ADDED AT WILL • VEGETATION BECOMING VERY DIVERSE • NURSERY PHASED OUT ———→	• STRUCTURAL EXPANSION OF ADDITIONAL TERRACED HOUSING • TAX PRESSURES TOO GREAT... SENIOR CITIZEN ON HOUSE ON SOUTH TRADES LOT FOR A UNIT IN TERRACED HOUSING – HOUSE TORN DOWN AND RECYCLED – MORE GARDEN SPACE • HYDROPONIC GARDENING SYSTEM IS EXPANDED ———→	
OBJECTS ▢	• WATER TOWERS FOR FIRE PROTECTION AND PUMPED STORAGE MECHANICAL CONVERSION • METHANE STORAGE TANKS • INCREASED POND SIZE • ORCHARD PRODUCE	• BRICK PATHWAYS • GARDEN PRODUCE AND ANIMAL PRODUCTS • TERRACED HOUSING UNITS WITH ROOF TOP GARDENS AND GREENHOUSES. SOLAR HEATED	• PHOTOVOLTAIC CELLS • VEGETABLES • EGGS • HERBS • FISH • RABBITS / CHICKENS • ENLARGED GARDENS • HYDROPONIC SYSTEMS	

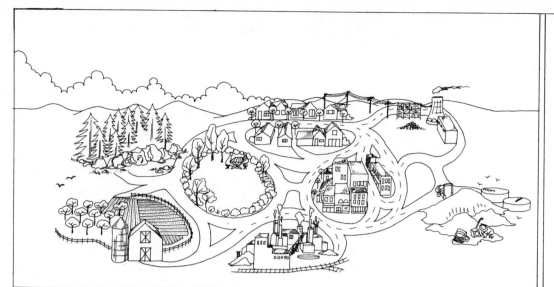

Separation of Functions

Specialization of land use

Long energy loops —

- reliance on long-distance transport.
- centralized energy supply.
- much waste of land, goods, energy.

Homogeneity of environment.

Lack of identity with one area.

Dependence on many suppliers.

Competition – ownership

Individualism – isolation.

Integration of Functions

Multiple land use – overlapping functions

Shortened energy loops —

- locally produced goods.
- local services available – decentralization.
- harnessing of local low-grade energy – sun, wind, wood.
- urban agriculture – local food supply.
- more efficient use of resources, less waste.
- wastes handled locally ⟳ recycled goods.
- composting of organic wastes.
- less reliance on motorized vehicles.
- human-powered transportation.

Diversity of environment, life-support systems.

Sense of belonging to and responsibility for area.

Local self-reliance, community inter-dependence.

Cooperation – sharing of ideas, energy, materials.

Community – neighborliness.

This sketch illustrates how neighbors might begin to link up their scattered block farms to a larger and more complex network. By growing a portion of their own food on the interiors of blocks and vacating certain streets for food growing (reclaiming through vacation), excess food can be sold or traded at the Farmer's Market. Bicycle transportation begins to supercede automobile transit.

The following section describes a prototype urban block farm in detail.

vegetables to market

FROM BLOCK FARM TO MARKET

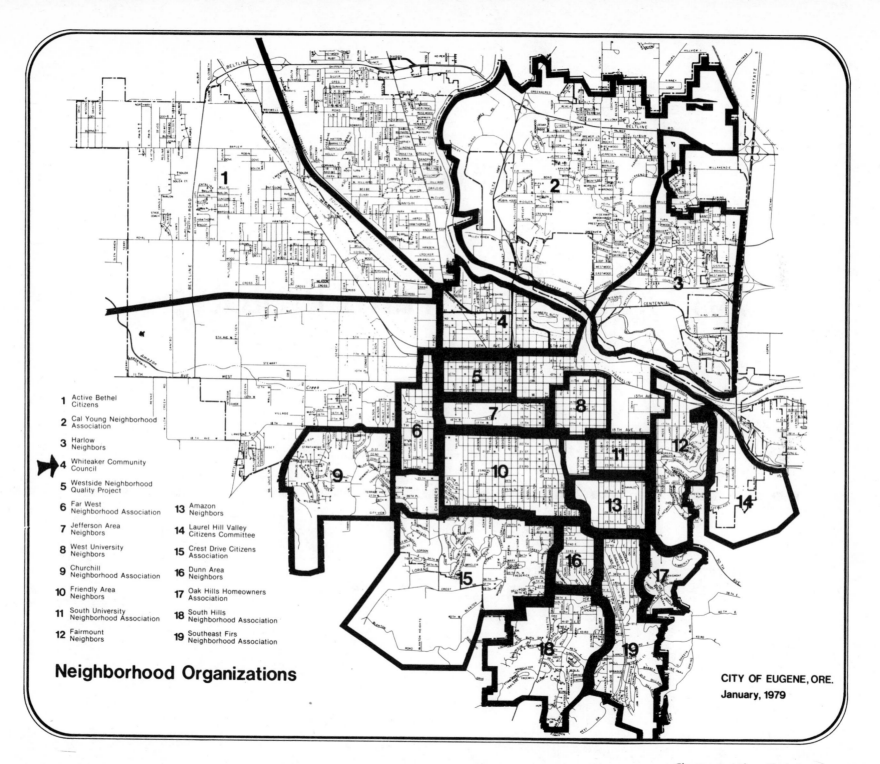

Neighborhood Organizations

1 Active Bethel
  Citizens

2 Cal Young Neighborhood
  Association

3 Harlow
  Neighbors

▶ 4 Whiteaker Community
  Council

5 Westside Neighborhood
  Quality Project

6 Far West
  Neighborhood Association

7 Jefferson Area
  Neighbors

8 West University
  Neighbors

9 Churchill
  Neighborhood Association

10 Friendly Area
   Neighbors

11 South University
   Neighborhood Association

12 Fairmount
   Neighbors

13 Amazon
   Neighbors

14 Laurel Hill Valley
   Citizens Committee

15 Crest Drive Citizens
   Association

16 Dunn Area
   Neighbors

17 Oak Hills Homeowners
   Association

18 South Hills
   Neighborhood Association

19 Southeast Firs
   Neighborhood Association

CITY OF EUGENE, ORE.

January, 1979

Chapter 5

An Urban Block Farm Prototype

AN URBAN BLOCK FARM PROTOTYPE AT THE UNIVERSITY OF OREGON URBAN FARM

OUTLINED BELOW ARE THE PRINCIPLES WE HAVE USED OVER ABOUT A FIVE-YEAR PERIOD OF TIME. THESE HAVE BEEN SUCCESSFUL IN ACHIEVING OUR GOALS OF COMMUNITY BUILDING, USE OF LOCALLY APPROPRIATE PLANTS AND MATERIALS, AND THE DEVELOPMENT OF A DEMONSTRATION OF MAXIMUM FOOD PRODUCTION OF UNDERUTILIZED URBAN LAND. WITH THE LAND RELEASED TO THE DEVELOPING URBAN FARM PROGRAM BY THE SCHOOL OF ARCHITECTURE AND ALLIED ARTS, AND WITH THE SPECIAL SUPPORT OF THE DEPARTMENT OF LANDSCAPE ARCHITECTURE, WE HAVE GRADUALLY TRANSFORMED THIS 1-1/2 ACRES FROM DERELICT TO A UNIVERSITY AND COMMUNITY RESOURCE. IT SHOULD BE ESPECIALLY NOTED, THAT THROUGH FIVE YEARS OF DEVELOPMENT, OUR TOTAL CASH OUTLAY HAS BEEN WELL UNDER $2,000, EXCLUDING LABOR...

SUCCESSFUL PRINCIPLES HAVE BEEN:

1. STRUCTURING THE LANDSCAPE...

1. MAINTAINING THE GARDEN AS THE HEART.

2. KEEPING THE CARS OFF THE LAND.

3. WELCOMING PEOPLE BY PLANTING AND BUILDING GATES, SUCH AS OUR "FIVE APPLE GATE" AND OUR "WIND-WATER PUMP".

4. ORGANIZING THE MATERIAL AND ENERGY FLOW SYSTEMS.

2. ENCOURAGING MULTIPLE AND COMPLEX PRODUCTIVITY AND CONTINUOUS MAINTENANCE BY:

1. WORKING TOWARD A BROAD AGE MIXTURE OF KIDS, TEENS, ADULTS AND SENIORS WORKING TOGETHER.

2. DEVELOPING A BROAD RANGE OF ACTIVITIES AND WORK TASKS.

3. VARYING THE PLANT TYPES TO INCLUDE TREES FOR WOOD, TREES FOR FOOD, SHRUBS FOR FLOWERS, SHRUBS FOR FOOD, AND PLANTS OF GREAT DIVERSITY INCLUDING HERBS AND DYES, AS WELL AS VEGETABLE CROPS MOST SUITED TO THE NORTHWEST.

4. SLOWING DOWN AND TAKING CARE...

5. UTILIZING CHINESE RAISED BED GARDENING.

6. GROWING YEAR 'ROUND.

7. PROVIDING ORGANIZED LEADERSHIP AND WORKING TOGETHER.

3. MANAGING ORGANIC "WASTE" AND REGENERATING THE SOIL... BY:

1. RECIEVING "GARBAGIO'S" PAYLOAD.

2. DEVELOPING COMPOSTING METHODS BY TRIAL AND ERROR.

3. USING EARTHWORMS FOR COMPOSTING.

4. UTILIZING LOCAL NATURAL RESOURCES.. BY:

1. DEVELOPING THE POTTERY KILN

2. WORKING WITH LOCAL WOOD.

5. PLANTING SYMBOLICALLY.. BY:

1. REMEMBERING THE BEAUTY OF FLOWERS.

2. REMEMBER-ING THE BEAUTY OF OUR FRIENDS.

3. PREPARING FOR TOMORROW'S HARVEST.

6. MAXIMIZING SOLAR GAIN AND STORAGE..... BY:

1. BUILDING A CHEAP ATTACHED SOLAR GREENHOUSE.

2. URBAN FARMING IN SOLAR GREENHOUSES.

3. ENCOURAGING SOLAR GREENHOUSES FOR WHITEAKER NEIGHBORHOOD.

7. NURTURING OUR RELATIONSHIPS WITH ANIMALS..... BY:

1. RAISING CHICKENS.

2. RAISING RABBITS.

8. MAINTAINING NATIVE PLANTS AND LOCAL NATURAL RESOURCES. BY:

1. REPLANTING WHAT WE BURN OR CUT: THE NATIVE PLANT NURSERY.

2. PRESERVING OUR BLACKBERRIES.

3. MONITORING OUR WATER QUALITY.

9. SHARING TOOLS AND BOOKS.. BY:

1. HOUSING THE UNIVERSITY OF OREGON ENVIRONMENTAL STUDIES CENTER.

2. MAINTAINING A COMMUNITY TOOL LIBRARY.

ROOT ISSUES: THE EDIBLE CITY AND THE URBAN FARM

The Urban Farm is a developing program devoted to the understanding and demonstration of environmentally-sound values and the operation of integrated food systems within the city. The site for coordinating this study involves an acre$^\pm$ of University land which contains an existing farmhouse, ceramics workshop, composting shelters and a greenhouse.

The preservation and enrichment of this land for ex-emplifying and experimenting with working alternatives of food production, energy conservation and generation, waste recycling, pest and soil management, and social ordering schemes can be seen as an investment in the future. This land provides a working example of im-proving the quality of our urban environment, es-pecially those areas under serious developmental pressures, where an appropriate balance between con-sumptive and productive land use may be addressed. The commitment to the preservation and renovation of historic farm structures is fundamental to this program's functions, since these historic structures were once the center of the Willamette Valley's economy. The farmhouse, when brought up to standard, can house a caretaker, classroom space, a small labora-tory, a resource library, a canning kitchen, and essentially provide a center for information exchange, thus bringing direct and tangible benefits to students, faculty, and community members at large. The farm can continue to demonstrate and experiment with or research alternatives which have minimal environ-mental impact and which contribute directly to the repair and enhancement of neglected land in a con-tinually urbanizing environment.

Urban agriculture, or bringing food production into the city, is becoming more important to city sur-vival. Due to rising food costs, the energy situa-tion, and renewed interest in soil ecology, human health, and nutrition, people desire more control over certain life support processes and may be en-couraged to escalate their concerns into action. Urban agriculture as an integrated food production system also allows an opportunity to recycle organic wastes at the household and neighborhood level, and increases perceptions regarding the balance of consumption with production and the cyclic nature of these processes. Additionally, the development of food production systems close to the area where the food can be consumed reduces energy costs of transportation, processing and storage. Also, the inherent smallness and controlability of local production facilities facilitates the implementation of appropriate technologies. Healthful outdoor recreation, exercise, and the ecological and aes-thetic value of preserving green space are other direct benefits. Community gardens as an example of urban agriculture are of growing interest because they give city dwellers access to land, reduce food and fuel costs, and utilize productive spaces that may otherwise go unused -- thus allowing a tangible small-scale experience to potentially generalize to larger scale land-use issues. Finally, urban agriculture as a process integrates biological, energetic, ecological, economic, and social concerns, and involves the exploration, research and develop-ment of alternative land use planning.

PROGRAM GOALS:

1. To promote agriculture as a "system and a process integrating biological, energetic, ecological, economic, and social concerns". NSF--Research Applied to National Needs, 1977.

   · To engage a maximum number of people in meaningful work through labor intensive methods with workers sharing responsible jobs (learning through direct experience).

   · To develop and research techniques and tools for ecological agriculture.

   · To research and document biological pest control and organic soil-building methods for urban areas.

   · To make healthy use of local resources and document that use.

   · To conserve energy and experiment with alternative energy systems in regard to understanding existing interactions of rural/city lifestyles.

2. To promote urban agriculture and self and community-reliant food production in the Eugene area.

   · To produce food and distribute it locally, preserving quality and freshness and equitable distribution.

   · To demonstrate alternative land use in the city.

   · To demonstrate maximum productivity on minimum land and to show that food production is a year-round activity in the southern Willamette Valley.

   · To display integrated garden systems, appropriate technology models, and alternative energy systems applicable to more self-reliant food production in and around the home in urban areas.

   · To provide for wide university and community access to information and technology via a resource library.

3. To develop a diverse ecosystem (the complex of a community and its environment functioning as one natural unit) as an educational model for studying more complex natural systems.

· To experiment with action-oriented, interdisciplinary education involving university and community people.

· To form a learning community through participation at the Urban Farm.

· To develop a closed system when possible and be appropriately interdependent with outside systems.

· To develop an integrated "mini-farm" made up of many cultures and study the interactions that take place.

In summation, we wish to encourage direct involvement with plants and the land through seasonal cycles with food webs that support plants, animals and people, and to stimulate variations of group processes in dealing with issues of "the commons" including individual and group work as they explore processes of summative gains (multi-disciplinary), and synergetic gain (interdisciplinary). We also wish to provide for "grounding" of environmental educational theory and steady state economic notions in physical form which are immediately accessible to students, faculty, and other community friends. And lastly, to relate class and studio work to direct demonstration which can supplement other professional public practice throughout the neighborhoods of Eugene.

Vegetable output declines in Oregon

PORTLAND (UPI) — Commercial vegetable production in Oregon in 1977 was 3 percent lower than in 1976 and value of the crops dropped by $400,000 to $75.1 million, the Oregon Crop and Livestock Reporting Service says.

The agency said a reduction in the acreage harvested and lower yields resulting from the drought caused production to drop to 793,150 tons.

Fresh market vegetable crops dropped 3 percent to 5.5 million hundredweight and vegetables harvested for processing dropped an equal percentage to 519,050 tons. Oregon ranks fifth in the nation in vegetable production for processing and seventh in fresh market vegetable production.

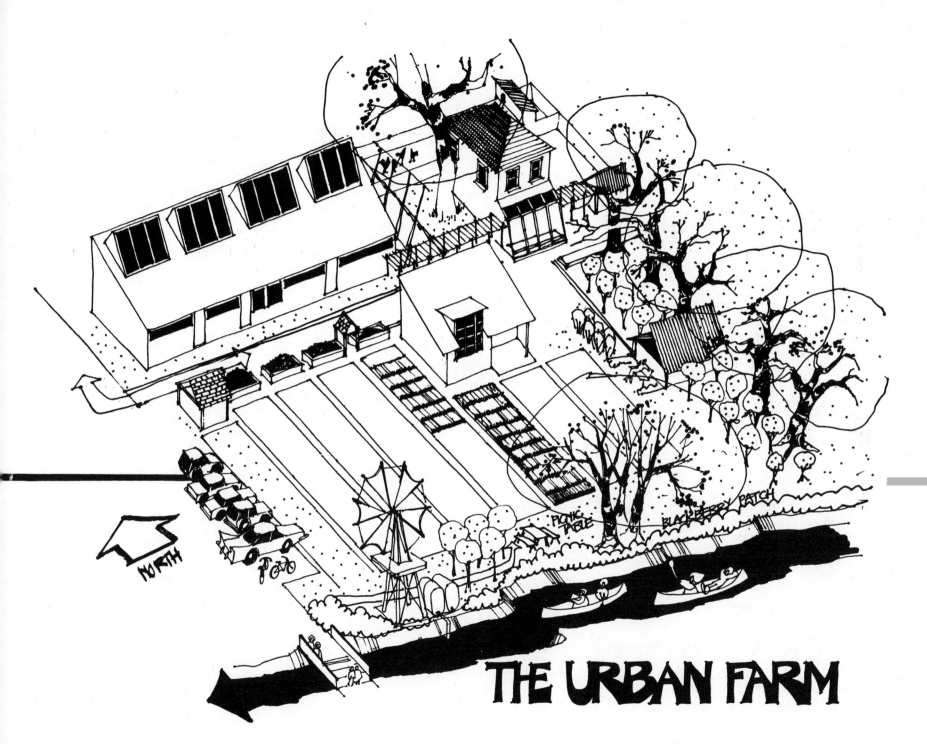

NORTH

PICNIC TABLE

BLACKBERRY PATCH

THE URBAN FARM

STRUCTURING THE LANDSCAPE

IS A FUNDAMENTALLY IMPORTANT PART OF THE URBAN FARM AND SOMETHING THAT HAS CONTRIBUTED CONSIDERABLY TO OUR SUCCESS. IN SHORT, WE HAVE FOUND IT ADVANTAGEOUS TO HAVE A VISUALLY CLEAR ORDER TO THE WAY IMPORTANT PARTS OF OUR MODEL URBAN BLOCK INTERIOR IS LAID OUT, AND TO MAINTAIN THAT ORDER CAREFULLY AND CONTINUOUSLY. IMPORTANT CONSIDERATIONS IN OUR LANDSCAPE STRUCTURING HAVE BEEN THE FOLLOWING:

THE GARDEN IS THE HEART...

1. <u>WE ORIENTED THE GARDEN TOWARD THE SOUTH.</u> WE HAVE CHOSEN TO ORGANIZE OUR 150 CHINESE RAISED BEDS WITH THE LONG AXIS OF EACH BED FACING DUE SOUTH, IN LONG ROWS, SEPARATED BY CONTINUOUSLY TROD PATHWAYS APPROXIMATELY 18" WIDE. AS SHOWN IN THE DRAWINGS AND PHOTOGRAPHS, THIS INITIAL STRUCTURING IMPLIES AN ORDER AND DESIGN TO THE GARDEN WHICH ACCOMPLISHES SEVERAL GOALS. ONE, <u>IT ENABLES FREE ACCESS</u> FOR FARM WORKERS AND FARM VISITORS (INCLUDING DOGS WHO TEND TO STAY <u>ON</u> THE PATHWAYS) AND PROVIDES SUFFICIENT CLUES AS TO LOCATION OF FOOD PRODUCING AREAS. TWO, <u>IT ENCOURAGES THE DOCUMENTATION AND RECORD-KEEPING</u> NECESSARY IN HIGH-YIELD, INTENSIVE, RAISED-BED AGRICULTURE IN ORDER TO PROVIDE SOIL REPLENISHMENT AND CROP ROTATION. THREE, <u>IT ENABLES OVERALL PLANTING COORDINATION AND PERSONAL ORGANIZATION</u> (WHERE <u>DID</u> THE BROCCOLI GO?) WHICH WE HAVE DISCOVERED LEADS TO A GREATER SENSE OF GROUP COMRADERIE AND COOPERATION.

2. <u>THE GARDEN IS SURROUNDED BY A CAREFULLY MAINTAINED STRIP OF LAND</u> WHICH SEPARATES PERIMETER ACTIVITIES FROM THOSE IN THE GARDEN ITSELF. THIS "SLUG DEMILITARIZED ZONE" BUFFERS UNWANTED PESTS FROM THE GARDEN ITSELF, INCLUDING SNAILS AND SLUGS WHICH ARE EATEN BY THE DUCK PATROL. WE MOW THIS GRASS AND COMPOST THE CLIPPINGS. IN ADDITION TO PROVIDING A GARDEN BUFFER, THIS ZONE APPROXIMATES THE "PRIVATE" BACK YARD SPACE OF AN URBAN BLOCK INTERIOR, ASSUMING THAT NEIGHBORS CAN COOPERATE BY "POOLING" A PORTION OF THEIR LOTS INTO A "COMMON" GARDEN. FOR EXAMPLE, WITH TYPICAL LOT DEPTHS UP TO 100 FEET, AND MOST BACKYARDS SIMPLE ECONOMIC LIABILITIES, YOU COULD CONSTRICT YOUR "PRIVATE" SPACE REQUIREMENTS TO, SAY, 20 FEET BEHIND YOUR HOUSE AND SHARE THE REST, DEVELOPING COOPERATION AND ECONOMIC POWER SIMULTANEOUSLY. REMEMBER, DIVIDED WE FALL....

2A. <u>GARDEN-SUPPORTING AND HOUSEHOLD-RELATED FUNCTIONS SHARE THIS STRIP OF LAND AROUND THE GARDEN.</u>

  •COMPOST BINS SHOULD BE DIRECTLY ADJACENT TO THE GARDEN (MORE ON THIS LATER) AND INCLUDED IN A GOOD PLACE TO FEED THE WORMS THE FOOD SCRAPS FROM THE GARDEN, AND FEED THE GARDEN THE COMPOST FROM THE WORMS.

  •SMALL ORCHARDS CAN FILL THIS STRIP AS AN ADJUNCT TO PRIVATE SPACES AS WELL AS INCREASE BLOCK-SCALE PRODUCTIVITY.

  •PICNIC TABLES, BARBEQUE PITS, AND SWINGS BELONG HERE TOO.

  •SMALL COTTAGE INDUSTRIES (WHICH HAVE TAKEN OVER UNUSED GARAGES) SUCH AS WOODWORKING AND FURNITURE-BUILDING SHOPS, POTTERIES, PROFESSIONAL OFFICES, DAY CARE FACILITIES, ETC. CAN BE INSERTED IN THESE ZONES QUITE EASILY. <u>MULTIPLE LAND USE EMPHASIZING DIVERSITY</u> PROVIDES RICHNESS TO AN AGRICULTURAL PLACE, AND BY-PRODUCTS (SUCH AS SAWDUST AND WOOD ASH) CAN BE ECOLOGICALLY BENEFICIAL TO THE FULL FUNCTIONING OF THE URBAN BLOCK FARM.

3. <u>BEYOND THIS STRIP OF LAND IS ANOTHER ZONE WHICH HOLDS HOUSES FOR PEOPLE AND OTHER ANIMALS</u> WHICH BUFFERS THE BIG GARDEN FROM THE STREET BY MASS AND DISTANCE. THIS ZONE CAN CONTINUOUSLY EVOLVE AND INCREASE HOUSING DENSITY BY MOVING OTHER HOUSES IN BETWEEN EXISTING ONES, ADDING ON ADDITIONS OR REPLACING SINGLE HOUSING WITH MULTIPLE HOUSING UNITS AS INDICATED BELOW.

IT IS IMPERATIVE IN THIS NOTION TO CONSIDER SOLAR ACCESS
RIGHTS, AND THE MAINTENANCE OF OPEN SPACE IN DIRECT PRO-
PORTION TO POPULATION DENSITY. FOOD PRODUCTION MAY BE THE
ONLY ECONOMICALLY AND SOCIALLY JUSTIFIABLE MEANS TO HOLD
THIS LAND OPEN FOR POSTERITY. MORE ON THIS LATER....

4. <u>BEYOND THIS STRIP IS THE STREET--HOME OF AUTOMOBILES.</u>

KEEPING THE CARS OFF OF THE LAND..

A. <u>KEEP THEM IN THE EXISTING STREETS.</u> --NEVER PAVE OVER GOOD
LAND FOR THEM. MOST PARKING LOTS NOW ARE
BUILT WITH SOIL STERILIZERS UNDER THEM WHICH
LAST A <u>LONG</u> TIME. STERILIZING GOOD AGRICUL-
TURAL SOIL IS A CRIME AGAINST LIFE.

B. <u>PREVENT THEM FROM CREEPING; ASSUME AN AGGRESSIVE COUNTER-
OFFENSIVE.</u> --BLOCK THEM OUT WITH POLES CUT FROM SLASH
PILES. LINE THE AREAS WHICH YOU DON'T WANT
INVADED WITH WELL-STAKED DOWN BORDERS.
DON'T MOVE THEM. ESTABLISH NEW PATTERNS OF
BEHAVIOR GRADUALLY BUT FIRMLY.

C. <u>DON'T COMPROMISE.</u> --EITHER DO IT OR DON'T DO IT. PAVE IT
ALL OR NONE. DON'T LET THEM NEAR PRODUCTIVE
SOIL. THEY DRIP OIL AND CREEP. GIVE THEM AN
INCH AND THEY WILL TAKE ALL THE LAND AROUND
THEM.

D. <u>IF YOU CAN, INVADE THEIR "TURF".</u> --CLOSE SOME STREETS.
DECIDE WHICH ONES YOU CAN LIVE WITHOUT.
PETITION YOUR NEIGHBORS ACROSS THE STREET.
ASK THE CITY TO ERECT TEMPORARY BARRIERS.
RECLAIM THE STREET FOR FOOD PRODUCTION AND
RECONDITION THE SOIL UNDERNEATH THE ASPHALT
YOU HAVE REMOVED.

5. <u>ENTRY TO THE BLOCK INTERIOR CAN BE WELCOMING BY PLANTING AND
BUILDING GATES.</u> AT THE URBAN FARM WE HAVE CHOSEN TO PLANT A
FIVE APPLE TREE GATE TO PASS THROUGH AND UNDER AS YOU HEAD
TO LUNCH AT THE PICNIC TABLE BY THE MILLRACE. OTHERS HAVE
CHOSEN TO BUILD A WELCOMING GATE AND EXCESS FOOD DISTRIBU-
TION TABLE NEAR THE OTHER LOGICAL ENTRY. BOTH OF THESE
GATES ARE INTENDED TO MARK ENTRY INTO ANOTHER KIND OF PLACE
FROM WHICH THE PERSON COMES, ONE THAT DEALS DIRECTLY WITH
TERRITORIALITY, PASSAGE, AND THE MAINTENANCE OF PUBLIC AND
SEMI-PUBLIC LAND USE AND MAINTENANCE.

A. <u>SYMBOLIC MARKERS IDENTIFY SPECIAL PLACES,</u> SUCH AS THE
WIND-WATER PUMP, WHICH ULTIMATELY WILL PUMP OUR IRRIGA-
TION WATER FROM THE MILLRACE. A GOOD INTERMEDIATE STEP
FOR US IS TO MONITOR THE WATER QUALITY (OIL POLLUTION
FROM SURFACE WATER RUN-OFF ON STREETS) <u>BEFORE</u> WE IRRI-
GATE OUR FOOD WITH IT.

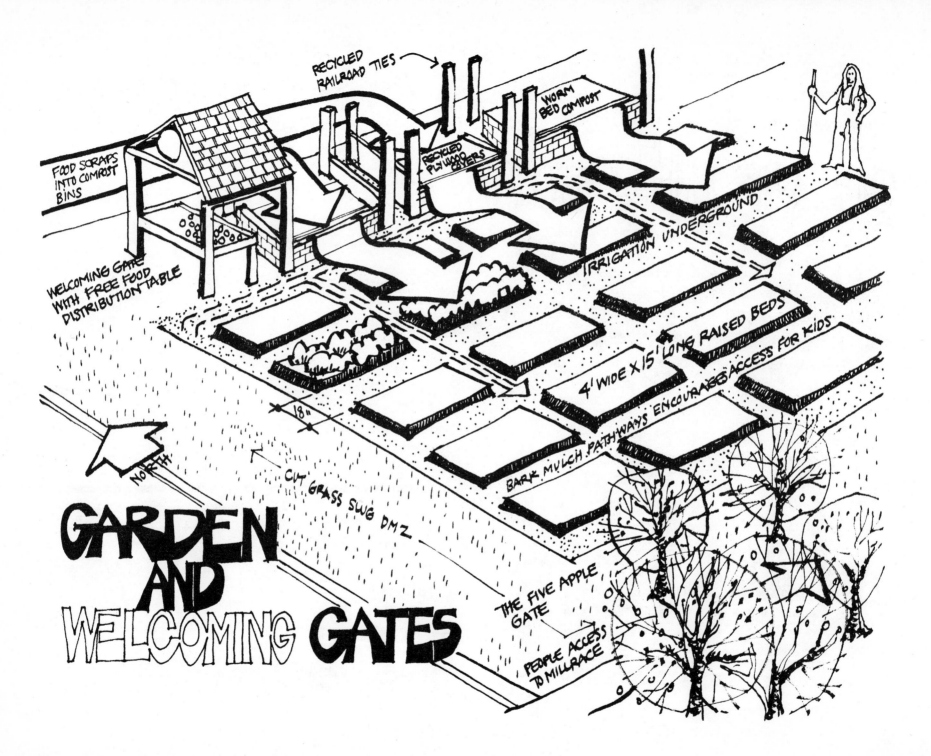

RECYCLED RAILROAD TIES

WORM BED COMPOST

FOOD SCRAPS INTO COMPOST BINS

RECYCLED PLYWOOD COVERS

IRRIGATION UNDERGROUND

WELCOMING GATE WITH FREE FOOD DISTRIBUTION TABLE

4' WIDE X 15' LONG RAISED BEDS

BARK MULCH PATHWAYS ENCOURAGE ACCESS FOR KIDS

18"

NORTH

CUT GRASS SLUG DMZ

GARDEN AND WELCOMING GATES

THE FIVE APPLE GATE

PEOPLE ACCESS TO MILLRACE

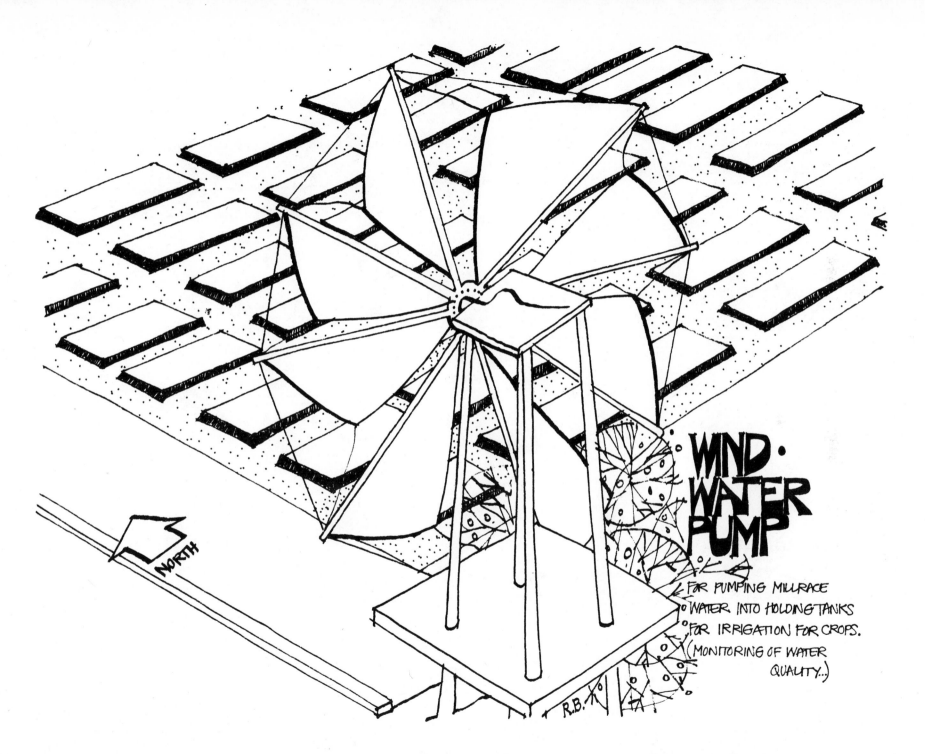

NORTH

WIND·
WATER·
PUMP

FOR PUMPING MILLRACE
WATER INTO HOLDING TANKS
FOR IRRIGATION FOR CROPS.
(MONITORING OF WATER
QUALITY...)

R.B.

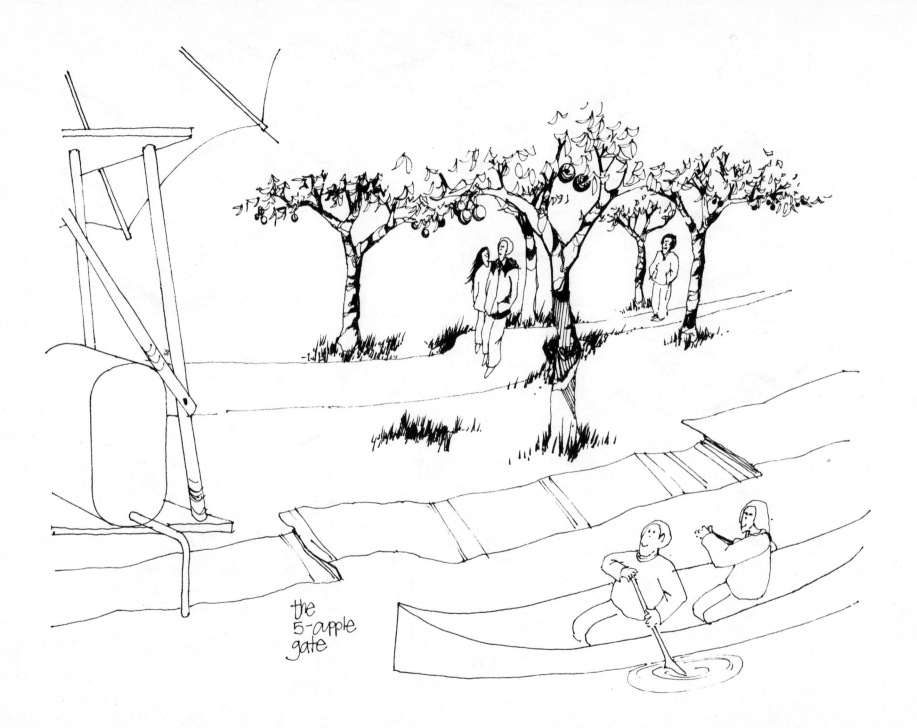

the
5-apple
gate

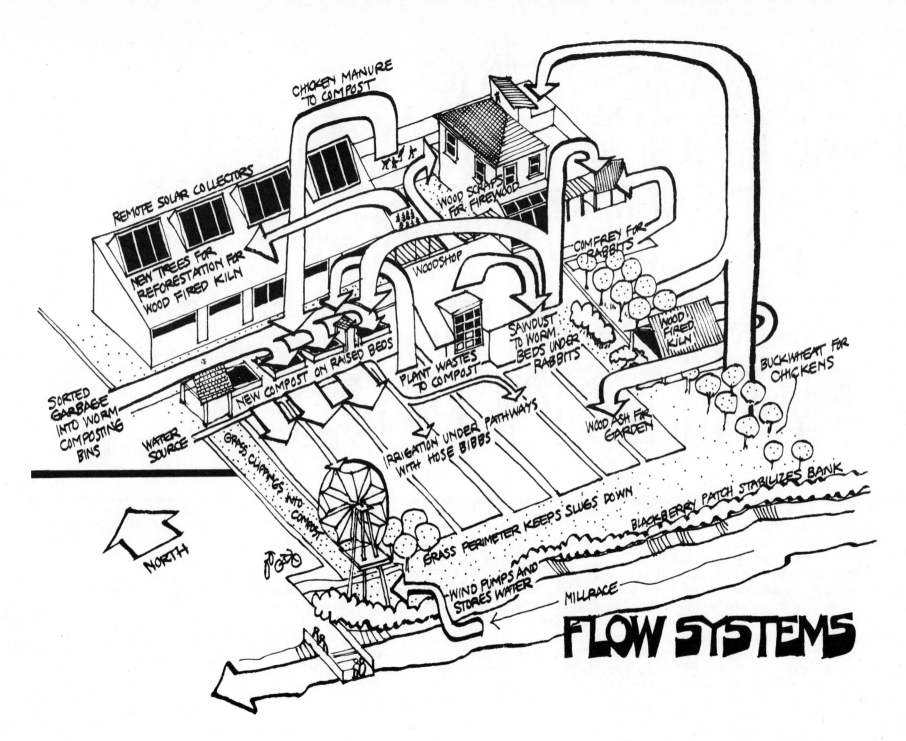

REMOTE SOLAR COLLECTORS

CHICKEN MANURE TO COMPOST

WOOD SCRAPS FOR FIREWOOD

COMFREY FOR RABBITS

NEW TREES FOR REFORESTATION FOR WOOD FIRED KILN

WOODSHOP

SAWDUST TO WORM BEDS UNDER RABBITS

WOOD FIRED KILN

BUCKWHEAT FOR CHICKENS

SORTED GARBAGE INTO WORM COMPOSTING BINS

WATER SOURCE

NEW COMPOST ON RAISED BEDS

PLANT WASTES TO COMPOST

WOOD ASH FOR GARDEN

GRASS CLIPPINGS INTO COMPOST

IRRIGATION UNDER PATHWAYS WITH HOSE BIBBS

NORTH

GRASS PERIMETER KEEPS SLUGS DOWN

BLACKBERRY PATCH STABILIZES BANK

WIND PUMPS AND STORES WATER

MILLRACE

FLOW SYSTEMS

ENCOURAGING MULTIPLE AND COMPLEX PRODUCTIVITY AND CONTINUOUS MAINTENANCE [DEVELOPING AN APPRECIATION FOR THE MUNDANE...]

The previous page indicates the material and energy flow systems that illustrate the Urban Farm principles of organization. Each part of the overall system is interconnected and mutually reinforces the function of the whole place. For example, local sorted organic garbage is delivered by small truck to the worm composting bins, which process the food scraps into rich humus-filled topsoil, which in turn is used to fertilize the raised beds. Vegetables are then grown, harvested (with the waste material going back into the compost bins), and eaten, generating human energy which manages the local ecosystem. Future dry composting toilets in living or educational places around the perimeter of the central garden provide rich fertilizer for the garden as do the animals grown on the block-scale farm site. Careful management and increased complexity of many diverse food crops adds to the smooth running and energy output of the Urban Farm ecosystem. Optimum variation of plant and animal types are continually being adjusted to the supportive capacity of the people involved. Maintenance of the Urban Farm system which models an urban block farm requires about four hours per week per person involved during the peak season (spring) and produces enough food to justify the labor output.

Throughout the 5-year evolution of the Urban Farm, several management strategies have been utilized. Simultaneous with the variation of these strategies have been varying levels of beauty and utility to the Urban Farm landscape. It is clear from this experience that the quality of managed cooperation is directly related to the quantity of food produced.

Initially, individuals utilized this place for individual private gardens randomly scattered and chosen, thus yielding solar rights conflicts, lack of group cohesiveness, redundancy in food produced (and therefore wasted), and failure to replace nutrients to the soil once the food had been harvested. I have called this "the mining stage" which was characterized by short-term involvement, maximization of self-gratification, and absence of consciousness of the cyclic nature of replenishment. Landscape structure consisted of randomly organized patches varying from year to year.

The second stage began with the organization of the land into a system of raised beds (described later in this document). This organization of the land enabled the labeling and numbering of particular "plots" which produced a greater sense of equity of concern for solar rights (crops shielding another person's garden's sunlight). Additionally, this "Jeffersonian grid stage" produced more food per acre and enabled the urban farmers to carry over their responsibilites to the summer season. Less food was wasted, individual-to-individual cooperation grew, small teams started to mutually support one another and connections to interconnected flow systems and cycles grew. Common work times for groups generated greater productivity of food and social cohesiveness.

The following stage, called "the family group stage", developed from the previous organizational model. Teams of five urban farmers developed ten raised beds as a "family" plot and coordinated the entire planning, planting, tending, harvesting and replenishing of each area independently, yet with some overall coordination to assure solar rights, minimize redundancy in crop types, integrate responsibilities for the animals with the gardens, and guarantee full system flow functioning. Food production was highest during this stage and the farm achieved a balanced homeostasis, with personal involvement generating four-season productivity.

The stage we are experimenting with now is "the full cooperative stage" which organizes all production by teams responsible for particular parts of the Urban Farm. This stage so far has yielded a certain degree of specialization of interest incompatible with the integrated learning goals of the Urban Farm.

CHINESE RAISED BEDS

AT THE URBAN FARM WE PRACTICE A COMBINATION OF THE CHINESE RAISED
BEDS METHOD AND THE BIODYNAMIC METHODS.

CHINESE RAISED BEDS:

THIS RAISED BED METHOD IS PRACTICED BY CHINESE FARMERS WHO HAVE WORKED
WITH PROBLEMS SIMILAR TO NORTHWEST GARDENERS: WET POORLY DRAINED SOIL,
LIMITED LAND AREA AVAILABLE FOR CULTIVATION, AND A RELATIVELY SHORT
GROWING SEASON. IN CHINA, FARMERS FOR OVER FOUR THOUSAND YEARS
FOUND THE INTENSIVE PRODUCTION OF RAISED BED METHODS MEETING THE
NEEDS OF THEIR GREAT POPULATION AND REPLENISHING THE FERTILITY OF
THE SOIL AT THE SAME TIME. RAISED BEDS HELP THE FARMER TO RAISE
LOTS OF FOOD ON A LITTLE LAND.

TO PREPARE THE SOIL FOR RAISED BED GARDENING, BEGIN BY REMOVING ROCKS
AND DEBRIS FROM THE SOIL. NEXT, OUTLINE A FOUR FOOT WIDE PLOT
(USING 4 CORNER STAKES AND STRING) MAKE THE BED 7 TO 25 FEET LONG,
OR WHATEVER LENGTH IS DESIRABLE TO YOU. MANURE AND COMPOST ARE DUG
INTO THE BED IN ENOUGH QUANTITY TO HELP RAISE THE BEDS EIGHT INCHES
ABOVE THE LEVEL OF THE ONE FOOT WIDE PATHWAYS. THEN RAISE THE BEDS
BY TURNING THE SOIL ONE SHOVEL'S LENGTH DEEP TO LOOSEN AND AERATE
IT. THE BEDS ARE RAKED SMOOTH ON TOP, MAKING THE TOP 3' WIDE. THE
BASE OF THE RAISED BEDS REMAINS 4' WHICH LEAVES A SLOPE ON THE SIDES
OF THE BED. A TRENCH IS PROVIDED AROUND THE PERIMETER OF EACH BED
TO CATCH WATER, SO THAT WATER DOES NOT RUN INTO THE PATHWAYS.

ONCE THE PERMANENT BEDS ARE ESTABLISHED, THEY ARE USED YEAR AFTER
YEAR, WITH ANNUAL ADDITIONS OF COMPOST AND OR MANURE. THE DRAINAGE
OF THE BEDS IS SUPERIOR, THE SOIL WARMS UP SOONER IN THE SPRING,
AND ALLOWS THE GARDENER OR FARMER TO PLANT EARLIER.

LIKE THE FRENCH INTENSIVE METHOD, THE CHINESE RAISED BED SAVES SPACE
OTHERWISE USED FOR PATHS AND SPACES BETWEEN ROWS, WHICH ALLOWS FOR
MORE PLANTS TO BE GROWN IN A SMALLER AREA.

THE BOOK, <u>BETTER VEGETABLES THE CHINESE WAY</u> BY PETER CHAN WITH
SPENCER GILL (GRAPHIC ARTS CENTER PUBLISHING CO., PORTLAND ,ORE. 1977)
GIVES THE DETAILS OF THIS METHOD, AND SHOWS HOW BEAUTIFUL SUCH PLANT-
INGS CAN BE. A COPY IS AVAILABLE FOR NEIGHBORS TO BORROW AT
PROJECT SELF RELIANCE, 315 MADISON.

THE BIODYNAMIC METHOD FRENCH INTENSIVE METHOD

THIS METHOD FIRST BECAME ACTIVE IN CALIFORNIA AND THE U.S. BY THE
INFLUENCE OF ALLAN CHADWICK. AS A YOUNG MAN, ALLAN STUDIED WITH
RUDOLPH STEINER, (FOUNDER OF THE BIODYNAMIC GARDENING METHOD IN THE
1920's) AND BECAME WELL ACQUAINTED WITH STEINER'S PHILOSOPHY OF THE
RELATEDNESS OF ALL LIVING THINGS. LATER ALLAN ADDED TO THIS METHOD
HIS OWN PRACTICAL TRAINING IN CLASSIC FRENCH HORTICULTURE.

BASICALLY, THE IDEA BEHIND INTENSIVE GARDENING (AS WITH THE CHINESE
METHOD) IS TO GET MAXIMUM PRODUCTION OUT OF A GIVEN GARDEN AREA. THE
METHOD IS AGAIN DONE IN RAISED OR ROUNDED PLANTING BEDS. IT DIFFERS
FROM OTHER RAISED BED METHODS OF GARDENING BY THE SPECIAL SOIL
PREPARATION IT REQUIRES.   CALLED DOUBLE DIGGING, THE PREPARATION
IS AS FOLLOWS:
        STARTING ON AN END OF THE GARDEN PLOT, DIG A TRENCH TWO SHOVELS
        FULL DEEP. LEAD THE DIRT FROM THIS INTO A WHEEL BARROW, TAKE IT
        TO THE OPPOSITE END OF THE PLOT. THEN, USING A SHARP SPACE OR

PICK-AXE BREAK UP THE SOIL IN THE BOTTOM OF THE TRENCH AND FILL
IT HALF WITH MANURE AND COMPOST. NEXT TO THE FIRST TRENCH, DIG
A SECOND ONE, USING SOIL FROM THE SECOND TRENCH TO FINISH FIL-
LING THE FIRST. AGAIN BREAK UP THE SOIL AT THE BOTTOM OF THE
TRENCH AND FILL IT HALF FULL WITH MANURE AND COMPOST. REPEAT
THIS PROCESS UNTIL YOU COME TO THE END OF THE GARDEN PLOT, USING
SOIL FROM THE FIRST TRENCH TO FILL THE LAST.

THE CHINESE AND THE BIODYNAMIC FRENCH INTENSIVE METHOD OPERATE IN THE
SAME WAY, IN THAT BOTH METHODS ELIMINATE A LOT OF GARDEN SPACE TAKEN
UP BY PATHS BETWEEN INDIVIDUAL ROWS, AND BECAUSE OF THE RICH DEEP
SOIL, ALLOWS FOR MUCH CLOSER SPACING BETWEEN PLANTS.

ACCORDING TO BOTH METHODS, ONCE YOU HAVE PLANTED YOU DO NOT WALK OR
CARRY EQUIPMENT ACROSS THE BEDS, AND YOU DO ALL OF YOUR WATERING,
WEEDING AND HARVESTING FROM THE SIDES. THIS HELPS WITH THE PROBLEM
OF COMPACTION.

THE PLANTS ARE GROWN SO THAT THEIR LEAVES JUST TOUCH ONE ANOTHER AT
MATURITY, THUS SHADING THE GROUND AND CONSERVING SOIL MOISTURE. SEEDS
ARE PLANTED IN DIAGONALLY OFFSET PATTERNS, FORMING A HEXAGON WITH ALL
SEEDS SPACED EVENLY TO GET MAXIMUM PRODUCTION. A NEWLY PLANTED BED
OF ALL THE SAME PLANTS WOULD LOOK LIKE THIS:

THE BOOK, <u>HOW TO GROW MORE VEGETABLES THAN YOU EVER THOUGHT POSSIBLE
ON LESS LAND THAN YOU CAN IMAGINE</u> BY JOHN JEAVONS (ECOLOGY ACTION OF
THE MID PENNINSULA, PALO ALTO, CAL. 1979) GIVES MORE DETAILS OF THIS
METHOD. A COPY IS AVAILABLE FOR NEIGHBORS IN THE PROJECT SELF RELAIN-
CE LIBRARY AT 315 MADISON, EUGENE, OREGON.

THE RAISED BED

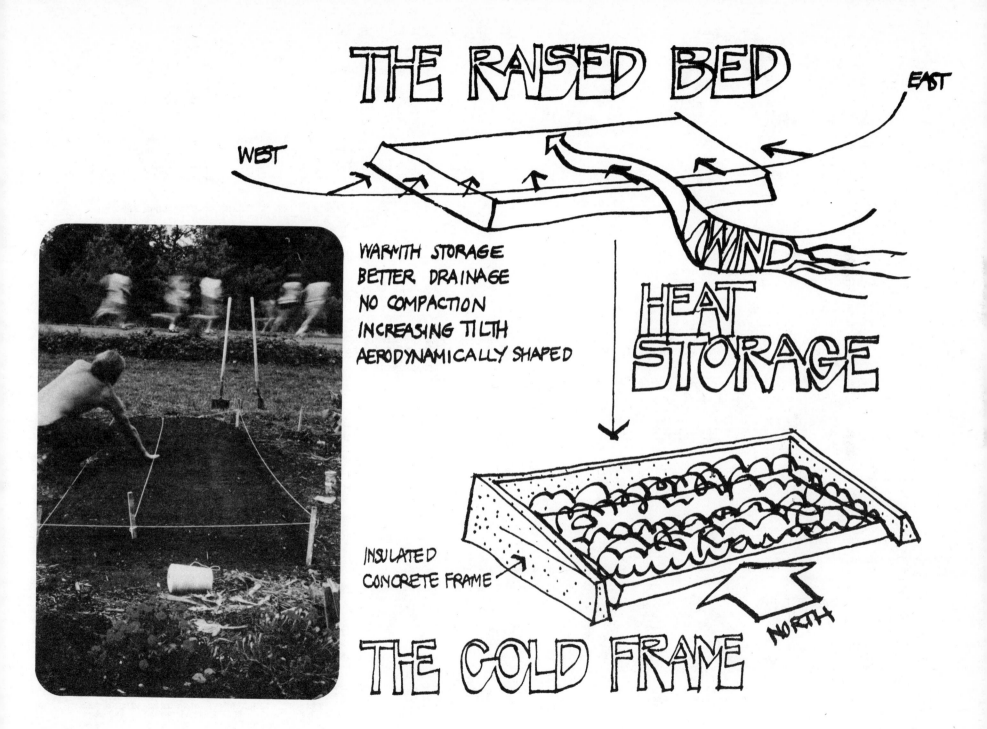

EAST

WEST

WIND

WARMTH STORAGE
BETTER DRAINAGE
NO COMPACTION
INCREASING TILTH
AERODYNAMICALLY SHAPED

HEAT STORAGE

INSULATED
CONCRETE FRAME

NORTH

THE COLD FRAME

SELECTING A SITE FOR YOUR GARDEN

AN IDEAL LOCATION FOR A GARDEN WOULD BE A SITE IN FULL SUN, WELL-PROTECTED FROM THE WIND, WITH A SOUTH-FACING SLOPE WITH GOOD DRAINAGE, NO STANDING WATER DURING THE WINTER, AND WITH LOOSE, DARK-COLORED FERTILE SOIL. NOT ALL GARDENERS WILL HAVE THE CONVENIENCE OF SUCH CONDITIONS. THE FOLLOWING MAJOR CONSIDERATIONS SHOULD BE HELPFUL IN SELECTING YOUR SITE:

1. LOCATING SUNNY LOCATIONS: MOST VEGETABLES GROW BEST IN FULL SUN, AND IDEALLY SHOULD HAVE SUN FROM 11:00 A.M. TO 4:00 P.M. THE SOUTH SIDE OF A BUILDING IS THE BEST LOCATION FOR FULL SUN. GARDENS CLOSE TO THE WEST AND SOUTH SIDES OF LOW STRUCTURES GROW WELL, ESPECIALLY WARMTH-LOVING SPECIES, BECAUSE THE STRUCTURE RADIATES HEAT LATE IN THE DAY. (AS A RULE, IT IS A WASTE OF TIME TO TRY TO GARDEN WITHIN 6-8 FEET OF A NORTHERN SIDE OF A ONE-STORY STRUCTURE.) THE EAST SIDE OF A STRUCTURE RECEIVES FULL MORNING SUN, BUT PLANT GROWTH MAY BE SLOWER THAN ON THE WEST SIDE WHERE AFTERNOON SUN ELEVATES THE SOIL AND AIR TEMPERATURES. ALL FRUIT AND NUT TREES PERFER FULL SUN, AND IF SUBJECTED TO SHADE OF EVEN MEDIUM DENSITY, WILL GROW SLOWLY AND BEAR POORLY.

2. PROTECTING FROM THE WIND AND FROST: STRONG WINDS CAN BATTER VEGETABLES, SLOWING DOWN GROWTH. WINDS ALSO DRY OUT SOIL IN SUMMER AND MAKE A GARDEN COOLER IN SPRING AND WINTER. THE GARDEN SHOULD BE AWAY FROM FENCES OR OTHER AIR TRAPS. FROST OCCURS IN LOW AREAS WHILE SLOPES ALLOW COLD AIR TO DRAIN AND WARM AIR TO RISE PAST THE GARDEN SITE.

3. ASSURING GOOD SOIL DRAINAGE: SELECT A PLOT THAT DOESN'T FLOOD, EVEN AFTER HEAVY RAINS. VEGETABLES REQUIRE NEAR-PERFECT DRAINAGE. THE SOIL MUST RETAIN ENOUGH WATER TO KEEP THA PLANTS SUPPLIED WITH MOISTURE (BUT MUST NOT BE SOGGY), AND BE POROUS ENOUGH TO ALLOW WATER AND AIR TO MOVE FREELY THROUGH IT. POORLY DRAINED SOIL CAN BE IMPROVED BY ADDING ORGANIC MATTER AND BY GROWING IN RAISED BEDS. HAVING A GARDEN ON A SLIGHT SLOPE WILL ENSURE GOOD DRAINAGE.

4. STAYING AWAY FROM TREES AND SHRUBS: LOCATE THE GARDEN AWAY FROM LARGE TREES AND SHRUBS AS MUCH AS POSSIBLE. ROOTS FROM NEARBY TREES AND SHRUBS COMPETE FOR WATER AND SOIL NUTRIENTS, AND THE ANNUAL VEGETABLE PLANTS GET ROBBED OF THE NUTRIENTS THEY NEED.

5. MAKING SURE WATER IS EASILY AVAILABLE: WATER AVAILABILITY IS AN IMPORTANT CONSIDERATION. TRY TO LOCATE THE GARDEN CLOSE TO A WATER FAUCET WHERE A HOSE CAN REACH ALL PARTS. A GARDEN THAT IS DIFFICULT TO WATER MAY NOT GET WATERED WHEN THE PLANTS NEED IT.

SOIL TESTING

TO GIVE YOU DIRECTION IN YOUR SOIL BUILDING EFFORT, AND TO PLAN YOUR GARDEN AND KNOW WHAT FERTILIZERS YOU ARE GOING TO NEED TO ASSURE YOUR SOIL LIFE AND HEALTHY PLANTS, IT IS BEST TO FIRST TEST YOUR SOIL FOR ITS LEVEL OF NITROGEN, PHOSPHOROUS, POTASH AND pH (THE ACIDITY OR ALKALINITY OF YOUR SOIL). A SOIL TEST CAN SAVE YOU MONEY SINCE IT OFTEN INDICATES THAT THE SOIL ALREADY CONTAINS SOME OF THE NUTRIENTS NEEDED FOR GOOD PLANT GROWTH.

HOW TO TAKE A SOIL SAMPLE:

1. DETERMINE HOW MANY DIFFERENT SOIL TESTS YOU ARE GOING TO NEED. WHAT YOU WANT IS TO TAKE A REPRESENTATIVE SAMPLE THAT APPLIES TO YOUR WHOLE GARDEN LOCATION.

   A. IF YOUR WHOLE GARDEN IS THE SAME KIND OF SOIL, ON THE SAME KIND OF SLOPE, HILLTOP, OR VALLEY, THEN TAKE 6 SAMPLINGS FROM DIFFERENT PLACES, MORE OR LESS EQUALLY SPACED APART, AND MIX THEM TOGETHER TO MAKE ONE SAMPLE.

   B. IF YOUR GARDEN LOCATION CONTAINS DIFFERENT TYPES OF SOIL, TAKE TWO OR THREE SAMPLINGS FROM ONE PARTICULAR SITE (ONE KIND OF SOIL) AND MIX THOSE SAMPLES TOGETHER. DO THE SAME FOR THE ADDITIONAL SITES (EACH DIFFERENT KIND OF SOIL) IN YOUR GARDEN LOCATION. KEEP THE DIFFERENT SITE SAMPLES SEPARATE.

2. WHEN TO TAKE A SAMPLE: THE BEST TIME TO TAKE YOUR SAMPLINGS IS IN THE FALL, ON A DAY THE SOIL IS FAIRLY DRY. SPRINGTIME IS THE MOST POPULAR TIME TO TEST, BECAUSE MANY GARDENERS ARE JUST BEGINNING TO PLAN FOR THEIR GARDENS, WHICH RESULTS IN A BUSY TIME FOR LABS, AND YOU MAY GET RECOMMENDATIONS BACK TOO LATE TO ACT UPON THEM.

3. TOOLS FOR TAKING A SOIL SAMPLE: A COMMON TOOL USED FOR TAKING A SAMPLE OF SOIL IS CALLED A SAMPLING PROBE. A SPECIALIZED TOOL IS NOT NECESSARY. JUST USE A GARDEN HAND TROWEL AND EXCAVATE A SMALL HOLE, THEN SHAVE A 1/2-INCH SLICE OF SOIL FROM THE SURFACE TO A DEPTH OF SIX TO EIGHT INCHES. THIS TECHNIQUE WILL GIVE YOU FAR MORE SOIL THAN NEEDED, BUT WILL BEST GIVE A SAMPLE REPRESENTATIVE OF THE SOIL IN THE GARDEN. MIX THE SAMPLING FROM ONE SITE TOGETHER WELL IN A BUCKET, AND KEEP ENOUGH TO FILL A SMALL PAPER BAG 1/2 FULL. IF THE SOIL IS WET, DRY IT OUT BEFORE MIXING. IF POSSIBLE, USE A PLASTIC BUCKET TO ELIMINATE FOREIGN ELEMENTS (LIKE RUST OR METAL FROM A TIN BUCKET) GETTING INTO THE SOIL AND INFLUENCING THE ANALYSIS. DO NOT HANDLE THE SOIL WITH YOUR HANDS. MAKE SURE ORGANIC MATTER SUCH AS ROOTS ARE NOT INCLUDED IN THE SAMPLES. ALSO AVOID TAKING A SAMPLE FOR TWO WEEKS AFTER ANY FERTILIZERS, MANURE, OR COMPOST HAS BEEN ADDED TO THE AREA.

TESTING YOUR SOIL SAMPLE

THE NEIGHBORHOOD GARDENER HAS TWO CHOICES OF HOW THE SOIL SAMPLE CAN BE TESTED: DO IT YOURSELF WITH A SOIL KIT, OR HAVE IT LABORATORY TESTED.

1. DOING IT YOURSELF: THERE ARE SEVERAL DIFFERENT KINDS OF DO-IT-YOURSELF SOIL TESTING KITS ON THE MARKET. CHECK AT LOCAL HARDWARE AND GARDEN SUPPLY STORES FOR THE BEST RECOMMENDED KITS AVAILABLE. JOHN JEAVONS, IN HIS BOOK HOW TO GROW MORE VEGETABLES, RECOMMENDS THE LAMOTTE SOIL KIT. THIS KIT MAY BE MORE COMPLETE THAN ONES WE HAVE USED IN THE PAST, AND MIGHT BE WORTH ASKING ABOUT.

2. LABORATORY TESTING: AGRICULTURAL COLLEGES IN ALMOST EVERY STATE HAVE SOIL TESTING SERVICES. AT OREGON STATE UNIVERSITY THEY WILL TEST YOUR SOIL FOR $5.00. RESULTS WILL SHOW EXACT LEVELS OF PHOSPHORUS, POTASSIUM, CALCIUM, AND MAGNESIUM, AND WILL MEASURE ACIDITY. THE TEST WILL IDENTIFY SOIL-BORNE PESTS, DISEASES, OR CHEMICAL RESIDUES. PRIVATE LABORATORIES ALSO TEST SOIL. THEY

OFTEN CHARGE MORE BUT USUALLY DO MORE (SOME SEND THEIR OWN PER-
SONNEL TO YOUR GARDEN TO TAKE THE SAMPLE). THE LABS WILL RUN
ANY TEST YOU ARE WILLING TO PAY FOR, INCLUDING ANALYSIS FOR
ORGANIC MATTER CONTENT, PESTICIDE CONTENT, OR FOR ANY OF THE
TRACE ELEMENTS YOU WISH TO SPECIFY. IT TAKES ABOUT SIX WEEKS
FOR YOU TO RECEIVE THE RESULTS OF THE SOIL TEST. IN HAVING ANY
SOIL TEST DONE, SPECIFY THAT YOU WOULD LIKE RECOMMENDATIONS
FOR ORGANIC MATERIALS TO ADD, SUCH AS FERTILIZERS AND SEE IF
THEY WILL COMPLY WITH THIS REQUEST.

YOUR RESULTS WILL OFTEN READ LIKE THE FOLLOWING: "USE 200 LBS.
OF LIMESTONE PER 1,000 SQUARE FEET. USE TEN LBS. OF 10-10-10
PER 1,000 SQUARE FEET AT PLANTING. SIDE DRESS WITH THREE LBS.
OF AMMONIUM NITRATE PER 1,000 SQUARE FEET DURING THE GROWING
SEASON." BUT WHAT DOES THIS MEAN IN ORGANIC TERMS? HERE IS
HOW THE EDITORS OF "ORGANIC GARDENING AND FARMING" SUGGEST IN-
TERPRETING SUCH TEST RECOMMENDATIONS: FOR A GARDEN PLOT OF
60x40 FEET (2,400 SQUARE FEET):

A. FOLLOW THE LIMESTONE APPLICATION RECOMMENDED OF 200 LBS.
   PER 1,000 SQ. FT. BY PUTTING DOWN ABOUT 500 LBS. OF
   CRUSHED LIMESTONE (FOR YOUR 2,400 SQ. FT.)

B. FORGET ABOUT TRYING TO FIND AN ORGANIC FERTILIZER WITH AN
   ANALYSIS OF 10-10-10. FOR NITROGEN, USE MANURE. THE
   OPTIMUM RATE FOR MANURING CROPLAND IS ABOUT 15 TONS PER
   ACRE. YOU WILL NEED ABOUT A TON FOR YOUR 2,400 SQUARE
   FEET (ABOUT 1/16 OF AN ACRE).

C. THE PHOSPHORUS AND POTASSIUM THE RECOMMENDATIONS CALL FOR
   --ALONG WITH SOME TRACE MINERALS--WILL UNDOUBTEDLY BE PRO-
   VIDED BY THE MANURE. IF YOU WANT MORE, APPLY 100 LBS.
   EACH OF ROCK PHOSPHATE AND GREEN SAND OR GRANITE POWDER
   ANNUALLY FOR THREE YEARS. THIS WILL PROVIDE FERTILIZING
   ACTION FOR UP TO SEVEN OR EIGHT YEARS.

START SMALL

FOR THE FIRST-YEAR NEIGHBORHOOD GARDEN, START SMALL. THERE IS A
TENDENCY TO GET SO EXCITED AND CARRIED AWAY AT FIRST, THEN FIND
LATER THAT THE GARDEN IS BECOMING A BURDEN OR A FINE WEED PATCH BE-
CAUSE IT TURNED OUT TO TAKE A LOT MORE TIME THAN YOU HAD ANTICIPATED.
GARDENING COOPERATIVELY WITH NEIGHBORS CAN HELP TO CUT DOWN ON ANY
ONE PERSON'S TIME IN A LARGER GARDEN. GARDENING BY MUSCLE AND BODY
POWER DOES TAKE TIME, BUT GREATLY BENEFITS YOUR BODY, THE SOIL, AND
THE PLANTS YOU ARE GROWING.

JOHN JEAVONS, MENTIONED ABOVE, GIVES A SERIES OF HELPFUL GARDEN
PLANS FOR THE FIRST YEAR THROUGH THE FOURTH OR FIFTH YEAR GARDEN.
HE SUGGESTS THAT THE FIRST-YEAR GARDENER BEGIN SMALL BY INCLUDING
IN THE GARDEN PLAN THE EASIEST CROPS TO GROW IN 100 SQ. FT. THE
SECOND YEAR, THE GARDEN PLAN EXPANDS TO DOUBLE THE SQUARE FOOTAGE
AND MORE DIFFICULT CROPS ARE ADDED, SUCH AS CAULIFLOWER, EGGPLANT,
HERBS, CANTALOUPE, MELONS, ETC. THE THIRD AND FOURTH YEARS, FRUIT
TREES, MORE HERBS, STRAWBERRIES AND ASPARAGUS ARE INCLUDED (THESE
PERMANENT PLANTINGS ARE NOW BEING PLACED IN SOILS THAT HAVE BEEN
WORKED FOR TWO YEARS). AFTER THREE OR FOUR YEARS OF EXPERIENCE,
YOUR GARDENING SKILLS MAY HAVE GROWN ENOUGH TO ENABLE YOU TO CON-
DENSE YOUR VEGETABLE GROWING FROM 200 SQ. FT. TO 100 SQ. FT.,
LEAVING 100 SQ. FT. FOR FRUIT TREES AND STRAWBERRIES, AND
FOR GRAINS (WHEAT, RYE, LENTILS, SOYBEANS, FIBERS SUCH AS COTTON
OR FLAX, OR SPECIAL INTEREST CROPS SUCH AS CHICKEN, GOAT OR BEE
FORAGE, GRAPES, BLUEBERRIES, BAMBOO, HERBS, NUT TREES AND SO ON).

THE FIRST YEAR GARDENER MAY SAY, "I WANT TO GROW ALL THE VEGGIES I
WANT TO EAT DURING THE SUMMER." BUT, WITHOUT GETTING CARRIED AWAY,
URBAN FARMERS CAN STILL GET GARDEN-FRESH, LOCAL PRODUCE ANY YEAR
WITHOUT HAVING TO GROW IT ALL YOURSELF. U-PICK, GLEANING, AND
EXCESS PRODUCE EXCHANGE PROGRAMS (SEE SECTION ON USING LOCAL RE-
SOURCES) TAKE CARE OF THIS NEED.

OUR CLIMATE AND GROWING SEASON

ANYONE WHO HAS LIVED IN THE WILLAMETTE VALLEY AND IN THE PACIFIC
NORTHWEST WILL GLADLY AGREE ON ONE COMMON FACT ABOUT OUR CLIMATE:
IT RAINS A LOT! OUR CLIMATE IS SIMILAR TO THAT OF SOUTHERN ENGLAND
AND PART OF JAPAN. THIS IS DUE TO THE JAPANESE CURRENT THAT SWEEPS
UP THE PACIFIC OCEAN AND LEAVES US WITH OUR MILD WINTERS, OUR
BOUNTIFUL RAINFALL, AND OUR PLEASANT GENTLE SUMMERS. THE CLIMATE
OF THE PACIFIC NORTHWEST (EXTENDING FROM BRITISH COLUMBIA SOUTH TO
THE CALIFORNIA BORDER, AND FROM THE PACIFIC COAST EAST TO THE
CASCADE MOUNTAINS) PROVIDES APPROXIMATELY 200 FROST-FREE DAYS PER
YEAR, ALONG WITH AN ABUNDANT (BUT VARIABLE) RAINFALL, AND A
GENERALLY HIGH LEVEL OF HUMIDITY. VARIATIONS IN THIS CLIMATE DO
OCCUR. FOR EXAMPLE, AS THE ELEVATION CLIMBS IN THE COASTAL RANGE,
THE NUMBER OF FROST-FREE DAYS IS FEWER; AND JUST EAST OF THE CAS-
CADE MOUNTAINS IS A RAIN SHADOW IN WHICH THE RAINFALL PER YEAR IS
LOWER.

HERE IN THE WILLAMETTE VALLEY, A REGION BETWEEN THE COASTAL AND THE
CASCADE RANGES, OUR WEATHER IS STRONGLY INFLUENCED BY THE OCEAN,
WHICH HAS GIVEN OUR CLIMATE A SPECIAL NAME, CALLED MARITIME CLIMATE.
THIS CLIMATE DETERMINES OUR PLANTING TIMES. OUR LAST FROSTS IN THE
SPRING ARE USUALLY BETWEEN THE FIRST OF APRIL AND MIDDLE OF MAY.
MANY OF THE VEGETABLES WE GROW CAN WITHSTAND A LITTLE FROST, AND
CAN THEREFORE BE PLANTED EARLY (AS LONG AS IT IS WARM ENOUGH FOR
THE SEEDS TO GERMINATE). FOR INSTANCE, WE CAN PLANT PEAS AS EARLY
AS THE FIRST OF FEBRUARY AND CONTINUE THE SPRING PLANTING SEASON
THROUGH THE MIDDLE OF MAY. WHEN THESE EARLY CROPS ARE HARVESTED,
WE CAN CONTINUE TO PLANT FROST-HARDY VEGETABLES THAT WILL GROW
DURING THE SUMMER AND MATURE IN THE FALL. OUR SUMMERS ARE GENERALLY
COOL AND WE GET ABOUT 60% OF THE POSSIBLE SUNSHINE DURING THE
SUMMER, WHICH IS GENERALLY ENOUGH TO RIPEN THE TOMATOES AND THE
CORN, YET LITTLE ENOUGH TO SATISFY THE REQUIREMENTS OF MANY OF THE
COOL WEATHER VEGETABLES, SUCH AS CABBAGE, LETTUCE AND VARIOUS OTHER
KINDS OF GREENS. WE CAN'T USUALLY GROW BIG LUSCIOUS WATERMELONS
OR PEANUTS WITHOUT A LITTLE HELP FROM COLDFRAMES AND OTHER "SEASON
EXTENDERS", BUT WE CAN HAVE BOUNTIFUL FEASTS OF PEAS, BEANS, SQUASH,
BROCCOLI, CABBAGE, PEPPERS, EGGPLANT, CORN AND POTATOES, PLUS MANY
OTHERS.

OUR FALL SEASON IS USUALLY LONG, AND WINTER TEMPERATURES ARE RARELY
VERY LOW. WE CAN PLANT FALL CROPS THAT DON'T RIPEN OR MATURE
QUICKLY, AND LEAVE THEM IN THE GROUND OVER WINTER TO GROW SLOWLY.
THIS WAY, THE URBAN FARMER IS ABLE TO HARVEST FRESH GREENS FOR A
WINTER SALAD OR SOUP WHEN NEEDED. THIS HELPS TO STRETCH THE
GROWING SEASON, THUS MAKING HAPPY YEAR-ROUND GARDENERS.

TO DEVELOP GOOD GARDENING HABITS, IT IS IMPORTANT TO UNDERSTAND OUR
CLIMATE AND KEEP A CONSTANT EYE AND EVEN RECORDS (USE A MAXIMUM/
MINIMUM THERMOMETER AND RAIN GAGE) ON THE WEATHER. IT IS THE
NATURE OF THE CLIMATE THAT DETERMINES WHERE WE PLANT OUR GARDENS,
WHEN WE PLANT, AND WHAT SORTS OF VEGETABLES WE CAN GROW MOST SUC-
CESSFULLY.

CROP ROTATIONS

AN ESSENTIAL PRACTICE FOR YEAR AROUND GARDENING IS CROP ROTATION BY FAMILY. GARDENING ROTATIONS ARE BASED ON SOIL REQUIREMENTS AND DISEASE PATTERNS OF COMMON VEGETABLES, ESPECIALLY THE BRASSICAS AND SOLANACEOUS PLANTS. NO FAMILY MEMBER SHOULD BE GROWN IN THE SAME PLACE ANOTHER FAMILY MEMBER HAS BEEN THE PRECEDING YEAR. THESE ARE THE FAMILIES:

SOLNACAE: TOMATOES, POTATOES, PEPPERS, EGGPLANTS. GENERALLY ALL REQUIRE HEAVY FERTILIZATION, AND THEY ARE ESPECIALLY HEAVY USERS OF NITROGEN AND POTASSIUM. THEY ALSO TEND TO PREFER ACID SOILS, ESPECIALLY THE EGGPLANTS AND POTATOES. POTATOES SHOULD NOT BE GROWN WHERE LIME HAS BEEN USED WITHIN A YEAR.

BRASSICAS: CABBAGE, BROCCOLI, BRUSSEL SPROUTS, CAULIFLOWER, KOHLRABI, MUSTARD, TURNIPS, RUTABAGA, KALE, RADISHES AND COLLARDS. GENERALLY ALL REQUIRE RICH SOILS WHICH ARE VERY WELL LIMED. THEY ARE HEAVY USERS OF NITROGEN AND POTASSIUM.

CORN: USES NITROGEN AND POTASSIUM BUT NOT VERY HEAVILY.

LEGUMES: BEANS, PEAS, CLOVER, VETCH, ALFALFA. THESE ARE ALL MODERATE USERS OF PHOSPHOROUS, BUT CONTRIBUTE A CONSIDERABLE AMOUNT OF NITROGEN TO THE SOIL.

GREENS: SPINACH, LETTUCE, CHARD, CELERY. USE NITROGEN VERY HEAVILY, BUT LITTLE ELSE.

ROOTS: CARROTS, BEETS, PARSNIPS. THESE USE PHOSPHOROUS, POTASSIUM MODERATELY.

CURCUBITS: SQUASHES, CUCUMBERS MELONS. USE NITROGEN AND POTASSIUM MODERATELY.

ALLIUMS: ONIONS, LEEKS, SCALLIONS, SHALLOTS, GARLIC. USE NITROGEN, POTASSIUM AND PHOSPHOROUS MODERATELY.

EACH FAMILY MEMBER HAS THE SAME KIND OF MINERAL REQUIREMENTS. IF, FOR EXAMPLE, CARROTS ARE FOLLOWED BY BEETS WHICH ARE FOLLOWED BY PARSNIPS IN ONE SPOT FOR A THREE YEAR PERIOD, EACH CROP WILL MAKE A HEAVY DEMAND FOR PHOSPHOROUS. BY THE THIRD ROOT CROP, ALL THE AVAILABLE PHOSPHOROUS IN THAT SPOT WILL BE EXHAUSTED. IF A ROTATION HAD BEEN PRACTICED, THERE WOULD HAVE BEEN TIME FOR ROCK PARTICLES TO BREAK DOWN AND RE-SUPPLY THE SOIL WITH PHOSPHOROUS.

ANOTHER GOOD REASON TO ROTATE IS THAT EACH FAMILY FALLS PREY TO THE SAME KIND OF SOIL BORNE DISEASES. FOR THE DISEASE-CAUSING BACTERIA TO REMAIN ACTIVE, A HOST MUST BE GROWING THERE. IF A HOST STAYS IN THE SAME SPOT FOR MORE THAN ONE SEASON, THE BACTERIA BUILD UP AND SERIOUS DISEASE PROBLEMS CAN RESULT. ROTATIONS AVOID THIS PROBLEM.

THE SAME IS TRUE FOR CERTAIN KINDS OF INSECTS. ROTATIONS INTERRUPT THEIR LIFECYCLE BY REMOVING THE MATERIAL ON WHICH THEY FEED. THIS PARTICULARLY APPLIES TO SOIL LIVING INSECTS LIKE NEMATODES, AND TO THOSE WHICH WINTER OVER IN DECAYING PLANT MATTER. PLAN YOUR ROTATIONS IN THREE OR FOUR YEAR PATTERNS. EACH ROTATION PATTERN SHOULD HAVE LEGUMES IN IT TO FURNISH NITROGEN FOR THE NEXT CROP PLANTED. ROTATION BECOMES MORE IMPORTANT AS YOU ADD LESS NUTRIENTS TO YOUR SOIL. HERE ARE SOME ROTATION PATTERNS FOR YOU TO TRY OUT.

SOLNACAE	LEGUMES	GREENS	ROOTS
BRASSICAS	ROOTS	LEGUMES	
ALLIUMS	BRASSICAS	LEGUMES	CORN

SEED COMPANIES

THE FOLLOWING LIST SUPPORTS SMALL LOCAL AND REGIONAL SEED COMPANIES (AND OTHERS) THAT PROVIDE UNTREATED OR ORGANICALLY-GROWN SEED AND/ OR PLANTS ADAPTED AND SELECTED ESPECIALLY FOR OUR CLIMATE AND CONDITIONS.

WHEN PURCHASING SEED, ORDER FROM MORE THAN ONE COMPANY IN CASE OF DELAY IN TRANSPORTATION. A SEED BUYING COOPERATIVE ALLOWS GARDENERS TO BUY SEED IN BULK AND DISTRIBUTE TO NEIGHBORS THE AMOUNT THEY NEED FOR JUST A FEW DOLLARS. CONSIDER STARTING ONE IN YOUR NEIGH- BORHOOD.

INCLUDED IS THE GRAHAM CENTER SEED DIRECTORY WHICH SUPPORTS THE USE OF OLD TRADITIONAL VARIETIES OF FRUIT, NUT AND VEGETABLES. OLD VARIETIES OF SEED, UNLIKE MODERN HYBRIDS, CAN BE SAVED AND REPLANTED. WITH THE BREEDING AND MARKETING OF "NEW, IMPROVED" VARIETIES, TRADITIONAL VARIETIES ARE BEING REPLACED. OLD FAMILY- OWNED SEED BUSINESSES HAVE BEEN AND ARE BEING BOUGHT UP BY LARGE MULTINATIONAL CHEMICAL AND DRUG FIRMS -- THE SAME COMPANIES THAT MANUFACTURE PESTICIDES, FERTILIZERS AND OTHER PRODUCTS BASED ON OIL CONSUMPTION. PUREX NOW OWNS FERRY-MORSE SEED COMPANY; SANDOZ (A SWISS CHEMICAL AND DRUG CONGLOMERATE) OWNS NORTHRUP- KING; AND ITT HAS JUST PURCHASED BURPEE SEED COMPANY. IN ADDI- TION, CELANESE, CIBA-GEIGY, MONSANTO, SHELL, PFIZER, UNION CARBIDE, AND UP-JOHN HAVE ALL RECENTLY BOUGHT SEED COMPANIES.

NOT INCLUDED IN THE FOLLOWING LIST ARE LARGE SEED COMPANIES. SEED FROM THESE COMPANIES CAN EASILY BE PURCHASED AT LOCAL GARDEN AND HARDWARE STORES AND NURSERIES. NOTE: BURPEE AND STOKES COMPANY CARRY UNTREATED SEED.

ABUNDANT LIFE SEED FOUNDATION
P.O. BOX 374
GARDINER, WASHINGTON 98334

> CATALOGUE $1.00. UNTREATED SEED, NO HYBRIDS; VEGETABLES, HERBS AND FLOWERS. OFTEN LOCALLY GROWN. SOLD IN EUGENE- AREA NATURAL FOOD AND CO-OP STORES, OR ORDER DIRECTLY. WILL DO TRADES, PASS ON INFORMATION, ETC.

CASA YERBA
STAR ROUTE 2
BOX 21
DAYS CREEK, OREGON 97429

> UNUSUAL HERB SEEDS. CATALOGUE 50¢.

FRIENDS OF THE TREES SOCIETY
P.O. BOX 567
MOYIE SPRINGS, IDAHO 83845

> ACTIVE PROMOTERS OF TREE-PLANTING AND CONSERVATION, THE SOCIETY ALSO DISTRIBUTES VERY INEXPENSIVELY-PRICED TREES. ACTIVITIES ARE ORIENTED TOWARDS THE NORTHWEST.

GRAHAM CENTER SEED DIRECTORY
ROUTE 3, BOX 95
WADESBORO, NORTH CAROLINA 28170

> BOOKLET, COLLECTION OF SMALL SEED COMPANIES THAT OFFER OLDER, TRADITIONAL SEED VARIETIES.

HERBS AND HONEY
ROUTE 2, BOX 205
MONMOUTH, OREGON 97361

> UNUSUAL HARDY HERBS, GOOD QUALITY. NO DELIVERIES (MUST PICK UP YOUR ORDER). ABUNDANT LIFE CARRIES SOME OF THEIR SEEDS.

HILLIER AND SONS
WINCHESTER, ENGLAND

> AN IMPORTANT SOURCE OF OLD VARIETIES, THE ~~BIGGING~~ BIGGEST SELECTIONS AVAILABLE ANYWHERE. SEND A DOLLAR OR TWO TO HELP DEFRAY THE COSTS OF A CATALOG AND POSTAGE. HILLIER AND SONS WILL SHIP TO NORTH AMERICA.

JOHNNY'S SELECTED SEEDS
ALBION, MAINE 04910

> CATALOGUE 50¢. UNTREATED, NO HYBRIDS, SEED GROWN ON OWN FARMS. EUROPEAN VARIETIES. VERY RELIABLE. BOOKLET, "GROW- ING GARDEN SEEDS" VERY HELPFUL.

MSK RARE AND NATIVE PLANT NURSERY
20066 15TH N.W.
SEATTLE, WASHINGTON 98117

NICHOLS GARDEN NURSERY
1190 N. PACIFIC HIGHWAY
ALBANY, OREGON 97321

> UNTREATED SEEDS, LARGE NUMBER OF HERBS AND GOURMET VEGETABLE SEED FROM AROUND THE WORLD. QUICK DELIVERY IN OREGON, (2-4 DAYS FROM RECEIVING ORDER).

ORGANICALLY GROWN SEEDS
VIDA, OREGON

> UNTREATED SEEDS, NO HYBRIDS. VARIETIES SUITED FOR THE PACIFIC NORTHWEST. (CARRIED BY DOWN-TO-EARTH IN EUGENE.)

RAINTREE NURSERY
265 BUTTS ROAD
MORTON, WASHINGTON 98356

> ORGANIC APPLE AND FRUIT TREES.

SUN & MOON HERB FARM
30973 KENADY LANE
COTTAGE GROVE, OREGON 97424

TERRITORIAL SEED COMPANY
P.O. BOX 27
LORANE, OREGON 97451

> STARTED IN 1979. CARRIES VEGETABLE SEED FOR OUR REGION (WILLAMETTE VALLEY).

TRUE SEED EXCHANGE
C/O KENT WHEALY
R.R. 1
PRINCETON, MISSOURI 64673

> OFFERS LIST, ORGANIZED BY STATE, OF PEOPLE WISHING TO SHARE SEEDS IN EXCHANGE FOR OTHER VARIETIES. (MARITIME NORTHWEST MEMBERS INVOLVED). $2 FOR LIST PLUS COMPANION PLANTING GUIDE AND MEMBERSHIP APPLICATION.

SEEDS AND VEGETABLE STARTS

VEGETABLE STARTS AND SEEDS CAN BE OBTAINED FROM ANY OF THE GARDEN SUPPLY STORES OR NURSERIES IN THE AREA. CHECK YOUR PHONE BOOK FOR LISTINGS.

PEOPLE LOOKING SPECIFICALLY FOR ORGANICALLY-GROWN STARTS, AND UNTREATED SEEDS, CAN CONTACT THE FOLLOWING:

THE COMMUNITY STORE   (345-1856)
444 LINCOLN
EUGENE, OREGON

DOWN-TO-EARTH   (345-7954)
HOME AND GARDEN STORE
740 E. 24
EUGENE, OREGON

FIFTH STREET GREENERY   (485-6394)
FIFTH STREET MARKET
5TH AND HIGH STREETS
EUGENE, OREGON

FUNKY'S FRUIT AND FLOWER COMPANY   (747-1881)
34072 SEAVEY LOOP ROAD
SPRINGFIELD, OREGON

SATURDAY MARKET
7TH AND OAK (IN
PARKING LOT)
EUGENE, OREGON   (CLOSED IN WINTER)

FOR OTHER POSSIBILITIES, CHECK WITH THE FOLLOWING MARKETS:

- THE KIVA
- GROWER'S MARKET
- WILLAMETTE PEOPLES FOOD CO-OP
- NEW FRONTIER MARKET
- SUNDANCE

MORE RESOURCES:

NEWSLETTERS, MAGAZINES AND JOURNALS

CASCADIAN REGIONAL LIBRARY (CAREL)
LANE BUILDING, BOX 1492
EUGENE, OREGON 97440

> A REGIONAL NETWORKING GROUP, PUBLISHES PERIODIC NEWSLETTERS, SPONSORS ANNUAL EQUINOX CONFERENCE.

CITY FARMER
318 HOMER STREET
VANCOUVER, B.C.
V6B 2V3

> AN INFORMATIVE AND USEFUL NEWSLETTER WRITTEN FOR THE HUNDREDS OF CITY FARMERS IN VANCOUVER AND ELSEWHERE. SUBSCRIPTION RATES: 5 ISSUES $2.50; 10 ISSUES $5.00; SUSTAINING $25.00.

CO-EVOLUTION QUARTERLY
P.O. BOX 428
SAUSALITO, CALIFORNIA 94965

> PUBLISHED QUARTERLY; $12.00/YEAR SUBSCRIPTION.

DUMP HEAP
371 IRWIN STREET
SAN RAFAEL, CALIFORNIA 94901

> A PUBLICATION ON URBAN LAND TRUSTS AND LAND REFORMS. SUBSCRIPTION RATES: 4 ISSUES/YEAR $5.00.

HARROWSMITH
CAMDEN EAST
ONTARIO, CANADA KOKIJO

> A MAGAZINE ON CITY FARMING.

LIFELINES
NUTRITION INFORMATION CENTER
239 S.E. 13TH AVENUE
PORTLAND, OREGON 97214

> A BI-MONTHLY NEWSLETTER ON NUTRITION, EVENTS AND FOOD ISSUES.

ORGANIC GARDENING AND FARMING
ORGANIC PARK
EMMAUS, PA. 18099

> SUBSCRIPTION RATES: 1 YEAR $9.00; 2 YEARS $16.50.

PROVENDER
NORTHWEST ALTERNATIVE FOOD NETWORK
1505 TENTH AVENUE
SEATTLE, WASHINGTON 98122

> A NEWSLETTER REGARDING PEOPLE AND ACTIVITIES INVOLVED WITH THE NORTHWEST ALTERNATIVE FOOD SYSTEM.

ORGANICS

CITY OF EUGENE   (687-5334)
PARKS MAINTENANCE DEPT.
201 CHESHIRE
EUGENE, OREGON

   LEAVES DELIVERED DURING FALL LEAF CLEANUP. POSSIBLE HORSE
   MANURE FROM 4-H PROJECT AT MORSE-RANCH. FREE.

MIKE DEGLER   (998-2115)
WEST OF CHESHIRE

   COW AND CHICKEN MANURES, DELIVERED. 15 CUBIC YARD TRUCK,
   CALL FOR PRICE.

EUGENE LIVESTOCK   (998-3353)
92380 HIGHWAY 99 N.
(SOUTH OF JUNCTION CITY)

   COW MANURE WITH SAND AND SAWDUST. FREE IF YOU LOAD; $5.00
   IF THEY LOAD.

EUGENE SEWER TREATMENT PLANT   (687-5293)
410 RIVER AVENUE
EUGENE, OREGON

   SEWER SLUDGE. (RIVER AVENUE RUNS EAST-WEST JUST SOUTH OF
   THE BELTLINE-RIVER ROAD AREA.) OPEN 8:00-4:00 WEEKDAYS;
   CAN PICK UP ON WEEKENDS TOO.

HEINRICH KAH CHICKEN RANCH   (726-8472)
84659 BRISTOW ROAD
PLEASANT HILL, OREGON

   CHICKEN MANURE WITH FIR SAWDUST. $1.00/SCOOP; $6/PICKUP
   LOAD.   FIFTEEN MINUTES FROM EUGENE.

KILMER-PAGE   (937-3094 OR 344-6997)
85381 MCBETH ROAD
EUGENE, OREGON

   CHICKEN MANURE. $5.00/TRUCK LOAD; THEY LOAD IT.

PAT PATTERSON   (687-4342)
LANE COUNTY EXTENSION SERVICE
950 W. 13
EUGENE, OREGON

   WORM CASTINGS. CALL FOR PRICE AND LOCATION.

SILVER DOLLAR HORSE RANCH   (726-6522 OR 746-9285)
3525 GARDEN
SPRINGFIELD, OREGON

   HORSE MANURE, HIGH IN CARBON, WITH CHIPS. FREE

SPRINGFIELD SEWAGE TREATMENT PLANT   (726-3760)
590 ASPEN STREET
SPRINGFIELD, OREGON

   SEWAGE SLUDGE. 8:30-11:30 MORNINGS; 12:30-4:00 AFTERNOONS,
   SEVEN DAYS A WEEK. CALL AHEAD FOR WEEKEND ARRANGEMENTS.
   BRING YOUR OWN SHOVEL.

THERE ARE MANY OTHER MANURE SOURCES THAT ARE NOT LISTED HERE, SUCH
AS INDIVIDUAL FARMERS. LET US KNOW IF YOU COME UP WITH OTHER
SOURCES, OR CALL THE URBAN FARM FOR NEW LISTINGS OF MANURE SOURCES.

WHEN TO PLANT: A PLANTING CALENDAR

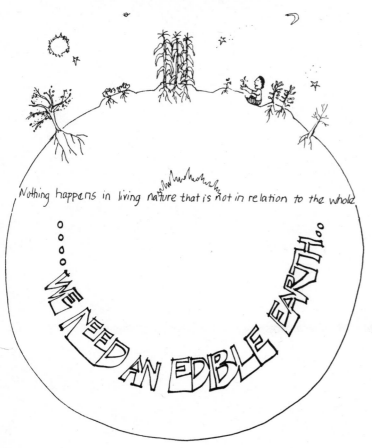

Nothing happens in living nature that is not in relation to the whole

WE NEED AN EDIBLE EARTH...

SPRING

March

WEATHER: "SPRING SETS FORTH THE GARDEN YEAR, SEEDS STIR IN THE WARMING SOIL, DAYS BEGIN TO LENGTHEN." (WHEATLY). THE VERNAL EQUINOX IS ABOUT MARCH 21, WHICH MEANS DAYS AND NIGHTS ARE OF EQUAL LENGTH, AND SOON NOTICEABLE CHANGES WILL APPEAR IN PLANT GROWTH. AN EXCESSIVELY WET WINTER USUALLY BRINGS A LATE SPRING WITH LUSH GROWTH IN BOTH THE ANIMAL AND VEGETABLE KINGDOMS. THE ZODIAC SIGN OF ARIES IS A TIME OF TENDER SHOOT VIGOR IN VEGETABLES AND WEED PLANTS, AND EXTRA WEED AND PEST CONTROL ARE NECESSARY. MARCH IN THE WILLAMETTE VALLEY USUALLY BRINGS WIND AS WELL AS A DECREASE IN MOISTURE.

INDOOR PLANTING: (STARTS IN HOUSE OR GREENHOUSE)

FIRST OF MONTH: EARLY CAULIFLOWER, LEEKS.

MID-MONTH: TOMATOES, CELERY/CELERIAC, MID-SEASON CABBAGE

GARDEN PLANTING: (OUTDOOR)

FIRST OF MONTH: RADISHES, RHUBARB, EARLY POTATOES, BRUSSEL SPROUTS, PEAS

MID-MONTH: RED CABBAGE, ROCKET; TRANSPLANT BROCCOLI AND EARLY CABBAGE; HARDEN OFF CABBAGE AND BROCCOLI PLANTED IN FEB.

GARDEN TIPS:

PREPARE GARDEN PLAN AND DESIGN FOR JULY THROUGH NOVEMBER.

ADD COMPOST AND ORGANIC FERTILIZERS TO GARDEN BEFORE CULTIVATING THE SOIL.

PREPARE RAISED BEDS (DO NOT WORK SOIL YET IF TOO WET AND STICKY)

COLLECT TOOLS NEEDED--SHARPEN AND REPAIR IF NECESSARY.

DON'T MISS THE PLANTING DATE!

MAKE COMPOST (BIOLOGICAL ACTIVITY IS AT ITS HIGHEST DURING SPRING AND AUTUMN).

PUT OUT SLUG TRAPS.

WEEDS BECOME ABUNDANT AFTER SPRING RAINS. GET READY TO WEED WITH ADEQUATE TOOLS, GLOVES, WEEDING BOXES OR BASKETS (THIS WILL ALSO HELP ELIMINATE SLUGS.)

WEATHER: LENGTHENING DAYS, PLANTS GROW RAPIDLY. APRIL OFTEN BRINGS RAIN AND STORMS TO THE WILLAMETTE VALLEY. THE ZODIAC SIGN OF TAURUS MEANS A GOOD TIME FOR PLANTING ROOT CROPS AND FLOWERS.

INDOOR PLANTING:

FIRST OF MONTH: EGGPLANT, PEPPERS

MID-MONTH: BRUSSEL SPROUTS, TOMATOES, MID-SEASON CAULIFLOWER

GARDEN PLANTING:

WHOLE MONTH: SPINACH, KHOLRABI, CHARD, LATE PEAS; TRANSPLANT BROCCOLI, MID-SEASON CABBAGES.

MID-MONTH: TRANSPLANT FIRST EARLY CAULIFLOWERS NOW THROUGH END OF APRIL; HARDEN OFF EARLY CAULIFLOWERS, ONIONS.

END OF MONTH: TRANSPLANT SWEET SPANISH ONIONS, HARDEN OFF CABBAGES PLANTED MID-MARCH.

GARDEN TIPS:

DON'T MISS THE PLANTING DATES.

WATER AFTER WEEDING.

PLANT BULBS, TUBERS, DAHLIAS, PINK AMARYLLIS, AGAPANTHIS, PERUVIAN LILY FOR LATE SUMMER BLOOMS.

SPRING PRUNING OF TREES; BEGIN TRAINING NEW SHRUB AND TREE BRANCHES BEFORE THEY BECOME TOO STIFF TO BEND (SEE TREE CARE GUIDE).

IF WINTER RAINS HAVE BEEN LIGHT, WATER THE GARDEN. (LOOK FOR MOISTURE SHORTAGE SIGNS: LIMPNESS OF NEW GROWTH, FAILURE OF PLANTS TO MATURE, DROPPING OF FLOWER STALKS).

PREPARE ANY SOIL NOT YET PREPARED (IF DRY ENOUGH TO WORK IT).

MAKE COMPOST, IF NOT YET DONE.

Cucumbers thrive in the shade provided by corn

WEATHER: SPRING DAYS CONTINUE TO LENGTHEN. GEMINI IS THE ZODIAC SIGN FOR MAY--A GOOD TIME FOR GROWTH OF HERBS, GRAIN, VINES AND CORN. THERE ARE INCREASING PERIODS OF HOT SUNNY AND WINDY WEATHER AT THIS TIME IN THE WILLAMETTE VALLEY.

GARDEN PLANTING:

WHOLE MONTH: SCALLIONS, CHARD, LETTUCE, TURNIPS, CARROTS, RADISHES, BEETS, STORAGE-TYPE ONIONS; TRANSPLANT CAULI-FLOWERS, MID-SEASON CABBAGES, AND BROCCOLI.

FIRST OF MONTH: EARLY BUSH BEANS (UNDER CLEAR PLASTIC), CAULIFLOWER, BRUSSEL SPROUTS; HARDEN OFF LEEKS STARTED IN MARCH (UNTIL THEY HAVE STOCKY STALKS AND A GOOD ROOT SYSTEM...MAY TAKE UNTIL THE END OF MAY UNTIL READY FOR TRANSPLANTING. KEEP PRUNED TO 3" TALL).

MID-MONTH: ALL SQUASHES (THROUGH JUNE 1), EARLY SWEET CORN (THROUGH MID-JUNE), CUCUMBERS (IF GROUND WARM ENOUGH); HARDEN OFF CAULIFLOWER STARTED MID-APRIL, BRUSSEL SPROUTS STARTED MID-APRIL, AND PEPPERS STARTED FIRST OF APRIL.

END OF MONTH: EARLY CHINESE CABBAGE, CAULIFLOWER, BUSH AND POLE BEANS (THROUGH MID-JUNE), PARSNIPS (THROUGH MID-JUNE), LATE POTATOES (TIL MID-JUNE), OVERWINTERING CABBAGES; TRANSPLANT TOMATOES.

GARDEN TIPS:

APHIDS WILL BECOME FEWER AS SUMMER SETS IN, BUT CHEWING CREATURES WILL BE PLENTIFUL.

MAKE NEST BOXES FOR SUMMER RESIDENT BIRDS. WRENS AND TITMICE WILL STAY IN THE GARDEN IF HOUSING IS PROVIDED, ENSURING ON-THE-SPOT CONTROL OF APHIDS AND ALL KINDS OF WORMS.

MAKE COMPOST, IF NOT ALREADY DONE.

SUMMER

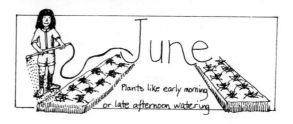

Plants like early morning or late afternoon watering

WEATHER: THIS SUMMER MONTH BRINGS THE LONGEST DAYS OF THE YEAR. GEMINI IS THE ZODIAC SIGN FOR THE FIRST PART OF MAY, CANCER FOR THE LAST PART. THIS IS A TIME OF GROWTH FOR HERBS, GRAIN, VINES AND CORN.

GARDEN PLANTING:

FIRST OF MONTH: LAST DATE FOR MOST WINTER CROP HARVESTS. CABBAGE (FALL AND WINTER VARIETIES), BRUSSEL SPROUTS BROCCOLI (PURPLE SPROUTING AND ST. VALENTINE), CAULI-FLOWER, LETTUCE, CARROTS, CUCUMBERS; SOYBEANS, MELONS (CROSS YOUR FINGERS); TRANSPLANT CAULIFLOWER, BRUSSEL

SPROUTS, PEPPERS, CELERY/CELERIAC, EGGPLANT; HARDEN OFF
CELERY/CELERIAC STARTED MID-MARCH.

MID-MONTH: LAST SWEET CORN, OVERWINTERING CABBAGES, LAST
BUSH/POLE BEANS, PARSNIPS; TRANSPLANT LATE CABBAGE NOW
THROUGH MID-JUNE.

GARDEN TIPS:

BEGIN TO HARVEST FRUITS AND VEGETABLES.

GIVE PLANTS SUPPLEMENTAL WATER THROUGHOUT THE SEASON.

ESTABLISH REGULAR ROUTINE OF MAINTENANCE/CARE--COORDINATE
GARDEN WORK WITH OTHERS.

"IF THE PEOPLE WORK HARD THEN THE EARTH WON'T BE LAZY."
(CHAN)

July

Dill planted by cabbage
repels some harmful insects.

WEATHER: THIS SECOND MONTH OF SUMMER TENDS TO BE CLOUDY, COOL
AND OCCASIONALLY RAINY. MID-SUMMER'S ZODIAC SIGN IS LEO,
A TIME OF GOOD GROWTH FOR AROMATIC HERBS AND FLOWERS, WEEDS,
AND A TIME OF PEST CONTROL.

GARDEN PLANTING:

FIRST OF MONTH: RUTABAGAS, FLORENCE FENNEL (TIL MID-JULY),
BROCCOLI (ITALIAN GREEN SPROUTING), CAULIFLOWER, LAST
HEAD LETTUCE; TRANSPLANT CAULIFLOWER.

MID-MONTH: BEANS, PEAS (60-DAY VARIETIES), KOHLRABI, BEETS
(SMALL VARIETIES), TURNIPS, COLLARD GREENS, CARROTS
(FALL AND OVER-WINTERING VARIETIES), FALL PEAS, CAULI-
FLOWER, SPINACH; TRANSPLANT THE LAST LATE FALL CABBAGES.

END OF MONTH: KOHLRABI (THROUGH MID-AUGUST), KALE (THROUGH
JULY), LATE CHINESE CABBAGE, OVERWINTERING BROCCOLI,
ONIONS (EARLY BULB TYPE), ENDIVE/ESCAROLE, RADISHES
(ORIENTAL), SPINACH, SWISS CHARD; TRANSPLANT LATE BROCCOLI.

GARDEN TIPS:

THIS IS A CRITICAL MONTH FOR GIVING ADEQUATE WATER TO YOUR
GARDEN AND ADEQUATE PEST CONTROL.

KEEP UP WITH WEEDING AND CULTIVATING. MULCH IF YOU WANT TO
REDUCE WEED GROWTH AND CONSERVE WATER.

August
A replanting time
for a winter garden

WEATHER: AUGUST BRINGS THE FULL RIPENESS OF THE YEAR. MANY WARM
DAYS ARE YET TO COME, DAYS ARE GROWING SHORTER WHICH IS A
SIGN OF AUTUMN APPROACHING. AUGUST IS A REPLANTING TIME.
BY MONTH'S END MANY ANNUALS, BOTH FLOWER AND VEGETABLE, WILL
HAVE PASSED THEIR HIGH SEASONAL YIELD.

GARDEN PLANTINGS:

WHOLE MONTH: OVERWINTERING CABBAGE FOR SPRING HARVEST,
TURNIPS, RADISHES,CAULIFLOWER AND SPINACH FOR FALL AND
OVERWINTERING, LAST CHINESE CABBAGE, TURNIPS,
LATE KOHLRABI.

MID TO END OF MONTH: ONIONS (BUNCHING AND EARLY BULB TYPES),
ROCKET, LETTUCES, MUSTARD GREENS, CHINESE CABBAGES,
CORN SALAD, SPINACH, BROCCOLI (ITALIAN GREEN SPROUTING),
WINTER CRESS.

END OF MONTH: LAST TURNIPS AND RADISHES, CABBAGES (APRIL,
HISIPI, JERSEY WAKEFIELD), CARROTS (FIREBRAND).

GARDEN TIPS:

PLAN AND PLANT NOW FOR THE GARDEN YOU WILL ENJOY DURING LATE
WINTER AND EARLY SPRING, AND WORK TO KEEP THE PRESENT GARDEN
GLOWING.

CONTINUE JULY'S MAINTENANCE WORK.

SPEND A LITTLE TIME UNDER A SHADY TREE DRINKING LEMONADE.
ENJOY JUST LOOKING AT THE BEAUTIFUL GARDEN. GARDENING ISN'T
ALL WORK.

FALL

September

WEATHER: SEPTEMBER IS THE FIRST MONTH OF AUTUMN AND THE HARVEST
MOON. IT IS A TIME WHEN VEGETABLES, FRUIT AND NUT TREES MA-
TURE, RESULTING IN AN ABUNDANT HARVEST. THE AUTUMNAL EQUI-
NOX (SEPTEMBER 22) IS WHEN THE DAYS AND NIGHTS ARE OF EQUAL
LENGTH. THE DAYS WILL BEGIN TO BE SHORTER, THE MOISTURE IN-
CREASES, AND THE WEATHER IS LIKE AN INDIAN SUMMER.

GARDEN PLANTING:

WHOLE MONTH: CABBAGES (APRIL, HISIPI, JERSEY WAKEFIELD),
ONION SETS, ENDIVE, SPINACH, CORN SALAD (IN WARM MICRO-
CLIMATE ONLY), LETTUCE AND RADISH, GARLIC, TURNIPS.

FIRST OF MONTH: LAST TRANSPLANTING OF OVERWINTERING BROCCOLI,
CABBAGES, AND CAULIFLOWERS.

GARDEN TIPS:

THIS IS THE MONTH TO HARVEST TOMATOES, GREEN BEANS AND OTHER
RIPE AND READY VEGETABLES.

BUTTERNUT, ACORN AND GREEN HUBBARD SQUASH AND PUMPKINS ARE
A LATE AUTUMN DELIGHT AND SHOULD BE RIPE FOR HALLOWEEN.

HARVEST SQUASHES WHEN THE VINES BEGIN TO WITHER AND STORE THEM
IN A COOL, AIRY PLACE. APPLES AND PEARS ALSO STORE WELL.
COLD STORAGE SAVES ON REFRIGERATION, CANNING OR FREEZING.

SEPTEMBER IS THE FOOD PRESERVATION SEASON (ACTUALLY JUNE
THROUGH NOVEMBER, WITH SEPTEMBER A PEAK TIME). CAN, FREEZE
AND/OR DRY VEGETABLES AND FRUITS. CALL PROJECT SELF-RELIANCE
(343-2711) FOR EQUIPMENT AND INFORMATION.

ASK PROJECT SELF-RELIANCE ABOUT FOOD PRESERVATION CENTERS IN
WHITEAKER, AND NEIGHBORHOOD GLEANING AND U-PICK PROGRAMS FOR
EXCESS AND SOMETIMES-FREE FOOD.

PLAN AND PLANT A WINTER SALAD GARDEN OF HARDY VARIETIES THAT
WILL REMAIN SWEET AND TENDER THROUGH THE COLD MONTH AHEAD.
TRY COLORFUL RHUBARB, CHARD, AND DANDELION AS A TASTY ADDI-
TION TO YOUR WINTER SALAD OR SOUP.

WHEN SUMMER ANNUALS BEGIN TO MILDEW, REPLACE THEM NOW FROM
SEED, WITH SNAPS, STOCK, PRIMULA, AND OTHER FAVORITE SHORT-
DAY, BLOOMING ANNUALS FOR LATE FALL/EARLY SPRING FLOWER GARDEN.

MAKE AUTUMN COMPOST (BIOLOGICAL ACTIVITY IS AT ITS HIGHEST
FOR COMPOST IN AUTUMN AND SPRING).

MULCH GARDEN WITH SPOILED HAY, LEAVES, STRAW, WOOD CHIPS,
AS MUCH AS A FOOT THICK THROUGHOUT FALL.

WATCH FOR SLUGS, USE TRAPS (SEE INSECT SECTION).

WEATHER: AUTUMN PLANTINGS HAVE THE ADVANTAGE OF WARM SOILS,
COOLER NIGHTS, AND LOVELY INDIAN SUMMER DAYS FOR DELIGHTFUL
GROWING AND GARDENING WEATHER.

GARDEN PLANTINGS: GARLIC, ONION SETS, SHALLOTS, FAVA BEANS, LONG
POD PEAS, JERUSALEM ARTICHOKES, CHICKWEED.

GARDEN TIPS:

CONTINUE TO UTILIZE THE ABUNDANT HARVEST IN THE NEIGHBORHOOD
BY STORING AND PUTTING UP FOR THE WINTER.

REMOVE OLD VEGETABLES AND PREPARE THE SOIL FOR MORE FALL
PLANTING; ADD MANURES, COMPOST (LIME IF NEEDED); TILL UNDER
WITH SPADE OR GARDENING FORK.

PULL GREEN TOMATO VINES BEFORE FROST AND HANG IN GARAGE, OR
WRAP IN NEWSPAPER TO RIPEN.

TO LENGTHEN THE HARVEST FOR PICKING VEGETABLES, TIE 3-6
STAKES TOGETHER AND DRAPE NIGHTLY WITH CANVAS TO FORM A TEE-
PEE OVER BEANS AND TOMATO PLANTS, TO PROTECT THEM FROM FROST
DAMAGE.

MAKE AUTUMN COMPOST, IF NOT ALREADY DONE.

SLUGS CAN BE A PROBLEM IN THE FALL WHEN TRYING TO GET THE LATE
PLANTINGS UP AND GROWING. USE SLUG TRAPS.

PUMPKINS SHOULD BE READY FOR HALLOWEEN.

WEATHER: NOVEMBER IS LAVISH WITH BURNISHED COLORS, AND USUALLY
BRINGS BOUNTIFUL RAIN. NOVEMBER'S ZODIAC SIGNS ARE SCORPIO
AND SAGITTARIUS. PLANTS WILL BECOME TALLER AND SPREAD QUICKLY.
DAYS BECOME INCREASINGLY SHORTER.

GARDEN PLANTINGS: PLANT THE SAME AS OCTOBER.

GARDEN TIPS:

IF THERE IS NO BOUNTIFUL RAINFALL, GIVE PLANTS ADEQUATE BUT
NOT EXCESSIVE WATER. SERIOUS DAMAGE RESULTS TO EVEN DORMANT
PLANTS IF SOIL BECOMES TOO DRY TO SUPPLY MOISTURE TO PLANT
TISSUE.

PREPARE GARDEN FOR WINTER; CHECK THE STAKES OF TREES, RENEW,
LOOSEN, OR OTHERWISE MAKE SUPPORTS SECURE AND COMFORTABLE.

MAKE COLD FRAMES, IF YOU HAVE NOT ALREADY DONE SO.

MAKE COMPOST, IF YOU HAVE NOT ALREADY DONE SO.

WINTER

December

A time for pruning

WEATHER: WINTER IS APPROACHING, ALONG WITH THE SHORTEST DAY OF THE YEAR. IT IS WITH GOOD REASON THE AMERICAN INDIANS CALL DECEMBER "LONG NIGHT MOON", BECAUSE WINTER COMES OFFICIALLY THIS MONTH. THE GARDEN IS REACHING ITS DORMANT STAGE, NOT ACTIVELY GROWING.

GARDEN TIPS:

PRUNE FRUIT TREES, SHRUBS.

PUT COLD FRAME OVER LESS-HARDY GARDEN PLANTS.

AS SOON AS THE GROUND HAS FROZEN, MULCH STRAWBERRIES AND OTHER PLANTS THAT NEED WINTER PROTECTION.

ALL HAND TOOLS SHOULD BE CLEANED AND STORED (SEE TOOL SECTION).

January

Planning for a spring garden

WEATHER: DAYS ARE GETTING SHORTER. BY THE END OF JANUARY THERE ARE SIGNS OF DAYS BEGINNING TO LENGTHEN AGAIN.

GARDEN PLANTINGS: CRESS, FAVA BEANS (IF WEATHER PERMITS), PLANT BEETS, CARROTS, LETTUCE, ONION SETS, RADISHES IN HOT FRAMES.

GARDEN TIPS:

SEND AWAY FOR SEED CATALOGUES.

PRUNE TREES THAT FLOWER IN THE SUMMER.

PUT IN SOME EARLY SEED ORDERS.

PROTECT PLANTS FROM FROST DAMAGE IF NOT ALREADY DONE, BY MULCHING THICKLY WITH LEAVES AND HOLD IN PLACE WITH EVER-GREEN BOUGHS OR WIRE NETTING.

February

A time to start such seeds as cabbage and brussel sprouts in flats.

WEATHER: THIS IS THE LAST MONTH OF WINTER. THE CROCUSES POP UP SHOWING SIGNS OF SPRING. ALTHOUGH WE MAY FIND SIGNS OF DE-PARTING WINTER AND RETURNING SPRING, FEBRUARY IS OFTEN A STORMY MONTH.

GARDEN PLANTINGS:

WHOLE MONTH: MUSTARD/ORIENTAL GREENS IN HOT FRAME; PREPARE ROOT BED FOR ASPARAGUS; TRANSPLANT ASPARAGUS ROOT.

FIRST TO MID-MONTH: TURNIPS, FAVA BEANS, PEAS.

MID TO END OF MONTH: PLANT PARSLEY, LEEKS, BROCCOLI (ITALIAN GREEN SPROUTING, AND NINE-STAR PERENNIAL), BRUSSEL SPROUTS, CELERY AND CELERIAC IN COLD FRAME OR GREENHOUSE. PLANT TURNIPS, ROCKET, EARLY PEAS (UNTIL MID-MARCH), AND FAVA BEANS OUTDOORS IN THE GARDEN.

GARDEN TIPS:

SEND AWAY FOR SEED VARIETIES AND BULBS.

BEGIN PLANNING FOR COLORFUL AND ABUNDANT SPRING GARDEN.

SHARPEN AND PREPARE TOOLS, IF YOU HAVE NOT DONE THIS ALREADY. (ASK ABOUT TOOL LIBRARIES FOR LENDING TOOLS--PROJECT SELF-RELIANCE OR THE URBAN FARM, AND OTHERS).

MOLES AND GOPHERS ARE BEGINNING TO BE ACTIVE IN THE GARDEN.

PRUNE ROSES.

PLANT SPRING BULBS.

WINTER GARDENING

START SOME BEETS, SWISS CHARD, SPINACH, COLLARD GREENS, CARROTS, KALE, LETTUCE, CABBAGE AND OTHER "COOL WEATHER" CROPS. (PUT IN ABOUT 40 SQUARE FEET OF SWISS CHARD AND BEETS, AND YOU CAN PROVIDE DAILY GREENS YEAR-ROUND FOR A FAMILY OF FOUR.)

IF YOU CONSIDER ADDING ABOUT SIX COLLARD GREEN PLANTS, A DOZEN OR SO SPINACH PLANTS, SOME KALE AND CABBAGE TO THE ABOVE "GREENS PATCH", YOU WILL HAVE TO GIVE BAGS OF GREENS AWAY WEEKLY!

ONCE YOU GET THE PLOT IN AND PRODUCING, THE LEAFY VEGETABLES WILL CONTINUE TO PRODUCE FOR ABOUT TWO YEARS -- IF -- YOU CONSTANTLY PRUNE THE EDIBLE LEAVES BEFORE THEY GET ABOUT 8 INCHES LONG. THIS WILL STIMULATE THEM TO KEEP PRODUCING, INSTEAD OF GOING TO SEED. THIS CONSTANT PRUNING IS THE KEY -- DON'T LET THE PLANTS GO TO SEED. REMEMBER, PLANTS WORK SEVEN DAYS A WEEK PROVIDING ENERGY FOR YOU TO USE.

SEE THE LIST AT THE END OF THIS SECTION FOR WINTER-HARDY VEGETABLE VARIETIES FOR THE PACIFIC NORTHWEST.

THE NEIGHBORHOOD GARDENER EXPERIENCES GREAT DELIGHT IN HARVESTING FOOD FROM THE GOOD EARTH IN EXCHANGE FOR HIS AND HER GENTLE CARE. NO NEED TO RETREAT TO THE SUPERMARKET FOR PRODUCE COME WINTER TIME! SUMMER'S HARVESTS PROVIDE AN ABUNDANCE OF FOOD, SO A GARDENER CAN STOCK UP FRUIT AND VEGETABLES TO BE PUT ON THE PANTRY SHELF FOR A WINTER'S DAY. HERE IN THE WILLAMETTE VALLEY WE HAVE A LONG GROWING SEASON WITH MODERATE WINTERS, WHICH ENABLES US TO HAVE SOMETHING GREEN FOR SALAD OR THE SOUP POT TWELVE MONTHS OF THE YEAR. OUR REGION RESEMBLES WEATHER CONDITIONS FOUND IN NORTH-WESTERN EUROPE. WESTERN EUROPEAN GARDENERS HAVE BEEN WINTER GARDENING FOR GENERATIONS. THEY HAVE EVOLVED HUNDREDS OF DIFFERENT VARIETIES AND TECHNIQUES TO ENSURE THEMSELVES SUCCESSFUL CROPS, MANY OF WHICH ARE STILL HERE TODAY FOR YOUNGER GARDENERS TO LEARN FROM.

FOR WILLAMETTE VALLEY URBAN FARMERS, THERE ARE TWO VERY GOOD BOOKS AVAILABLE TO HELP US LEARN HOW TO EXTEND THE GROWING SEASONS AND PROVIDE VARIETIES GROWN BEST IN OUR CLIMATE. THEY ARE, WINTER GARDENING IN THE MARITIME NORTHWEST, BY BINDA COLEBROOK, AND GROWING ORGANIC VEGETABLES WEST OF THE CASCADES, BY STEVE SOLOMON. THESE SOURCES OFFER IMPORTANT PLANNING CONSIDERATIONS FOR GROWING A WINTER GARDEN, SOME OF WHICH ARE AS FOLLOWS:

> A PROPER SITE: THE BEST SITE IS A GENTLE SLOPE ON THE SOUTH SIDE OF THE BUILDING. A BUILDING REFLECTS AND HOLDS HEAT; THE SLOPE AIDS DRAINAGE; TERRACES MAKE A WARM MICROCLIMATE, AND THE WIND PROTECTION FROM THE BUILDING HELPS TO KEEP WARM POCKETS OF AIR IN THE GARDEN. EXTRA WARMTH MEANS EXTRA GROWTH ON YOUR WINTER VEGETABLES; HENCE, MORE FOOD FOR YOU.

> A WELL-DRAINED SOIL: IF YOUR SUMMER GARDEN IS IN A LOW SPOT WHICH COULD ACCUMULATE THE WINTER RAINS IN PUDDLES, START LOOKING FOR ANOTHER WINTER GARDEN SPOT. YOUR PLANTS WILL SUFFER, ROOTS WILL ROT AND THE PLANTS MAY DIE IF WATER IS ALLOWED TO STAND ON THE SOIL DURING HEAVY RAINS. A PERFECT SOIL FOR WINTER GARDENING IS DARK, FULL OF HUMUS, SANDY AND WELL DRAINED. (PLANTING VEGETABLES IN RAISED BEDS HELPS KEEP THE SOIL WELL DRAINED.) THIS KIND OF SOIL HEATS UP FAST AND HOLDS THE HEAT LONGER WHICH HELPS PREVENT THE SOIL FROM FREEZING EASILY. WORKING COMPOST AND FRESH MANURES INTO THE SOIL BEFORE SUMMER HARVEST WILL ALSO ADD WARMTH AS IT DECOMPOSES.

THE BOOK WINTER GARDENING IN THE MARITIME NORTHWEST BY BINDA COLEBROOK (TILTH ASSOC. ARLINGTON WA., 1977) IS VERY HELPFUL TO URBAN FARMERS IN PLANNING AND PLANTING A WINTER GARDEN. THE BOOK ALSO PROVIDES INFORMATION ON WINTER VARIETIES, WHERE TO SEED AND A PLANTING CALENDAR. A COPY IS AVAILABLE FOR NEIGHBORHOOD USE AT PROJECT SELF RELIANCE, 315 MADISON STREET.

GREEN MANURE AND COVER CROPS

FOR THE PARTS OF THE GARDEN THAT ARE NOT GOING TO BE PLANTED IN LATE SUMMER WITH WINTER GREENS, A GREEN MANURE CROP CAN BE PLANTED IN THE OTHERWISE IDLE GROUND. GREEN MANURING CAN BE BETTER THAN MULCHING IN THE WINTER BECAUSE THE ACTIVITY OF THE ROOTS ADDS BOTH ORGANIC MATTER AND KEEPS NUTRIENTS IN THE ROOT AREA WHERE THEY AREN'T LEACHED BY THE ENDLESS RAIN.

THERE ARE MANY GOOD GREEN MANURE CROPS, INCLUDING LEGUMES, GRASSES AND HERBS. THE BEST CROPS FOR WINTER ARE FAST GROWING MEMBERS OF THE GRASS FAMILY; BARLEY AND OATS, OR THE LEGUME FAMILY; CLOVER, VETCH, AUSTRALIAN WINTER PEAS OR BROAD BEANS, AND RYE WILL GERMINATE DOWN TO 40 DEGREES F. AT LEAST, AND WILL CONTINUE TO GROW AT 32 F. WINTER PEAS CAN USUALLY BE PLANTED AS LATE AS OCTOBER, THOUGH EARLIER SOWINGS WILL ADD MORE NITROGEN TO THE SOIL (ESPECIALLY IF YOU HAVE REMEMBERED TO ADD INOCULANT FOR NITROGEN FIXING BACTERIA.) VETCH PLANTED BY AUGUST HAS BEEN SHOWN TO CONTRIBUTE 100 LBS. OF NITROGEN PER ACRE TO THE SOIL.

A GREEN MANURE CROP IS USUALLY PLANTED IN EARLY FALL SO THAT IT WILL BE HALF GROWN BY SPRING. THE ENTIRE CROP IS THEN TILLED IN THE GROUND A MONTH BEFORE PLANTING TIME WHEN THEY ARE ONE HALF TO THREE FOURTHS GROWN AND STILL SUCCULENT. IN REGIONS WITH SUB ZERO WINTER TEMPERATURES, IT IS BEST TO SEED THE CROP BETWEEN STANDING VEGETABLES IN LATE SUMMER SO PLANTS CAN ROOT BEFORE A HEAVY FROST.

IF YOU MUST DELAY TURNING THE CROP UNDER BECAUSE THE SOIL IS TOO WET, KEEP THE CROP DOWN TO AN EASILY HANDLED SIZE BY CUTTING IT DOWN WITH A SCYTHE, SHEARS OR A ROTARY MOTOR. THE CROP DOES NOT HAVE TO BE MATURED TO BE TURNED UNDER. ALTHOUGH THE TOP GROWTH MAY BE SPARSE, THE WELL-DEVELOPED ROOT SYSTEM WILL ADD A SUBSTANTIAL AMOUNT OF ORGANIC MATTER AS IT DECAYS.

THIS METHOD OF FERTILIZING AND PREPARING THE SOIL IS A VERY BENEFICIAL WAY OF ADDING NITROGEN AND OTHER NUTRIENTS TO THE SOIL. IT ALSO LEAVES THE GARDEN ALL PREPARED FOR THE NEXT GARDEN INSTEAD OF HAVING TO ROTOTIL AND/OR DIG UP A GARDEN WHICH HAS BEEN LEFT TO GROW OUT OF CONTROL WITH THE WEEDS OVER THE WINTER.

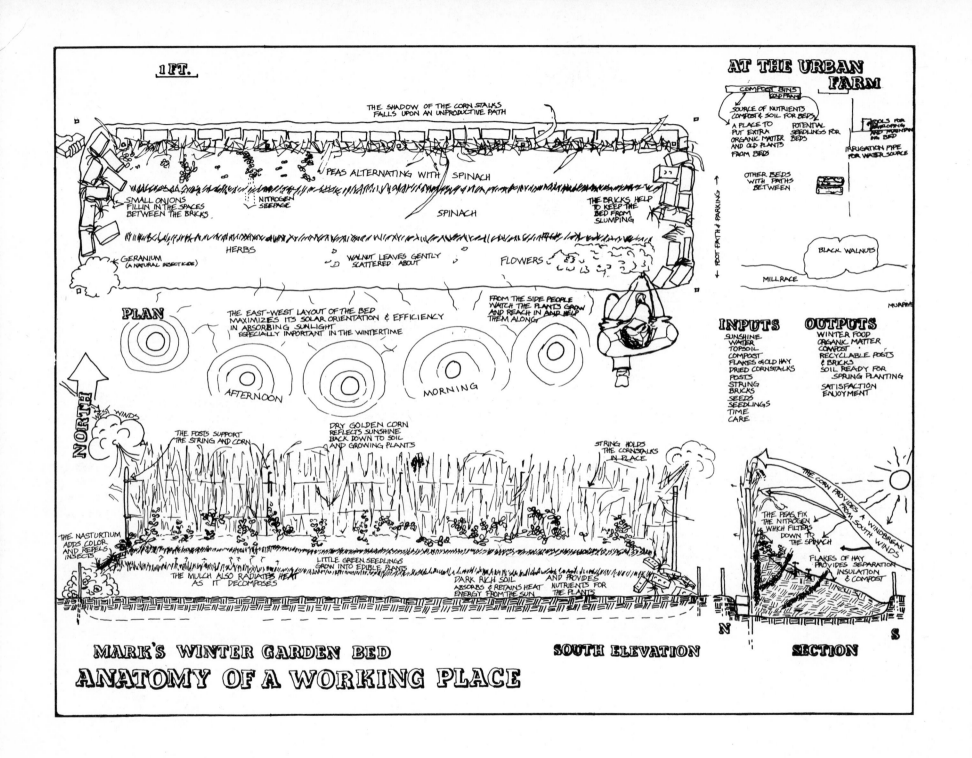

1 FT.

THE SHADOW OF THE CORN STALKS
FALLS UPON AN UNPRODUCTIVE PATH

PEAS ALTERNATING WITH SPINACH

SMALL ONIONS
FILLIN IN THE SPACES
BETWEEN THE BRICKS

NITROGEN
SEEPAGE

SPINACH

THE BRICKS HELP
TO KEEP THE
BED FROM
SLUMPING

HERBS

WALNUT LEAVES GENTLY
SCATTERED ABOUT

FLOWERS

GERANIUM
(A NATURAL INSECTICIDE)

PLAN

THE EAST-WEST LAYOUT OF THE BED
MAXIMIZES ITS SOLAR ORIENTATION & EFFICIENCY
IN ABSORBING SUNLIGHT
ESPECIALLY IMPORTANT IN THE WINTERTIME

FROM THE SIDE PEOPLE
WATCH THE PLANTS GROW
AND REACH IN AND HELP
THEM ALONG

AFTERNOON

MORNING

NORTH

WEST WINDS

THE POSTS SUPPORT
THE STRING AND CORN

DRY GOLDEN CORN
REFLECTS SUNSHINE
BACK DOWN TO SOIL
AND GROWING PLANTS

STRING HOLDS
THE CORNSTALKS
IN PLACE

THE NASTURTIUM
ADDS COLOR
AND REPELS
INSECTS

THE MULCH ALSO RADIATES HEAT
AS IT DECOMPOSES

LITTLE GREEN SEEDLINGS
GROW INTO EDIBLE PLANTS

DARK RICH SOIL
ABSORBS & RETAINS HEAT
ENERGY FROM THE SUN

AND PROVIDES
NUTRIENTS FOR
THE PLANTS

AT THE URBAN FARM

COMPOST BINS
COLD FRAME

SOURCE OF NUTRIENTS
COMPOST & SOIL FOR BEDS

A PLACE TO
PUT EXTRA
ORGANIC MATTER
AND OLD PLANTS
FROM BEDS

POTENTIAL
SEEDLINGS FOR
BEDS

TOOLS FOR
DEVELOPING
AND MAINTAIN-
ING BED

IRRIGATION PIPE
FOR WATER SOURCE

OTHER BEDS
WITH PATHS
BETWEEN

FOOT PATH & PARKING

BLACK WALNUTS

MILLRACE

MURRAY

INPUTS
SUNSHINE
WATER
TOPSOIL
COMPOST
FLAKES OF OLD HAY
DRIED CORNSTALKS
POSTS
STRING
BRICKS
SEEDS
SEEDLINGS
TIME
CARE

OUTPUTS
WINTER FOOD
ORGANIC MATTER
COMPOST
RECYCLABLE POSTS
& BRICKS
SOIL READY FOR
SPRING PLANTING

SATISFACTION
ENJOYMENT

THE CORN PROVIDES A WINDBREAK
FROM SOUTH WINDS

THE PEAS FIX
THE NITROGEN
WHICH FILTERS
DOWN TO
THE SPINACH

FLAKES OF HAY
PROVIDES SEPARATION
INSULATION
& COMPOST

N

S

MARK'S WINTER GARDEN BED

SOUTH ELEVATION

SECTION

ANATOMY OF A WORKING PLACE

MANAGING ORGANIC WASTE AND REGENERATING THE SOIL THROUGH COMPOSTING

The essence of healthy soil is humus, small microorganisms, and the accompanying organic tilth that guarantees future soil productivity and water permeability in urban areas. Maintaining and improving urban soil health which supports urban plant health guarantees a future supply of clean water, clean soils, clean plants and clear air in direct proportion to the quantities involved. Utilizing urban "waste" such as "problematic" fall leaf drop (expensive and ecologically degrading to haul "away") and organic garbage is essential in maintaining and improving closer-looped nutrient cycling. The attendant energy benefits to urban dwellers are important in several ways. Decreasing larger scale economies or urban land fills and their attendant rate increases point the way toward "taking care of your own". Block scale soil improvement programs can be accomplished by garbage recycling in efficient and clean systems (based on the earthworm). Block scale waste recycling stimulates the opportunity for greater social integration while decreasing garbage bills to urban residents, while systemically alleviating (or minimizing) continued need for urban landfills in unsuited areas. Urban animal manures, green plant manures, wood ash residue and leaf drop excess, properly composted can lead to greater health and stability of urban ecosystems. Several methods suggested for use are the "indore technique" and composting with worms, which follow....

SOIL

THE HEART OF URBAN AGRICULTURE

CARE OF THE SOIL IS WHAT MAKES A TRUE "STEWARD" OUT OF THE AGRICUL-
TURALIST. THE URBAN FARMER IS NOT ONLY INTERESTED IN GROWING FOOD
FOR HUMAN CONSUMPTION. HE OR SHE IS ALSO COMMITTED TO INCREASING
THE WORLD'S LIMITED STOCK OF FERTILE, ORGANIC SOIL. THE URBAN FAR-
MER HAS A SPECIAL CALLING BECAUSE MOST OF THE SITES AVAILABLE FOR
AGRICULTURE USES IN CITIES HAVE SOILS DEPLETED OF ORGANIC MATTER
AND THE NUTRIENTS REQUIRED BY PLANTS. HOWEVER, MANY URBAN AREAS,
INCLUDING EUGENE, WERE LOCATED ON THE VERY BEST LAND AVAILABLE LO-
CALLY FOR AGRICULTURE BECAUSE THAT IS WHERE EARLY SETTLERS BUILT
THEIR FARMS, WHICH LATER MERGED INTO TOWNS. MUCH OF EUGENE'S SOIL,
EXPECIALLY IN THE OLDER RESIDENTIAL AREAS IS RIVER BOTTOM CLAYEY
LOAM, WHICH IN A VERY SHORT TIME CAN BE BUILT UP INTO WONDERFUL
TILTH FOR GARDEN SOILS. TO BUILD WASTED SOIL INTO RICH GARDEN LOAM
WITH HIGH TILTH (THE QUALITY OF CULTIVATED SOIL) DOES NOT REQUIRE
SPECIALIZED KNOWLEDGE OR EVEN VERY MUCH TIME. BUT THERE ARE A FEW
CONCEPTUAL AND PRACTICAL STEPS THAT MAKE THE PROCESS EASIER AND
QUITE ENJOYABLE.

FIRST OF ALL, CONSIDER A MODEL OF A BALANCED ECOSYSTEM WHERE THE
SOIL REGENERATES ITSELF AND MAINTAINS A RELATIVE BALANCE OF NUTRI-
TIONAL CONTENT SUCH AS THE CLIMAX FORESTS OF THE PACIFIC N.W.
SEVERAL FACTORS ARE NECESSARY TO MAINTAIN THIS BALANCE, AND APPLY
EQUALLY WELL TO THE URBAN FARMER. FIRST, THERE IS GREAT DIVERSITY
OF BIOLOGICAL ACTIVITY IN THE SOIL ITSELF, DECOMPOSING (COMPOSTING)
THE LITTER AND DEADFALL OF THE FOREST FLOOR. SECOND, THERE IS A
DIVERSITY OF PLANTS INTERMIXED AND FILLING DIFFERENT TASKS IN MAIN-
TAINING SOIL HEALTH. THIRD AND MOST IMPORTANT, IT IS THE COMPLEX
INTERACTION BETWEEN PLANTS, MICROORGANISMS AND LARGER ANIMALS THAT
KEEPS THE COMPONENTS OF THE FOREST ECOSYSTEM FLOWING, ONE RESULT OF
WHICH IS HEALTHY SOIL. WHETHER IN THE CITY GARDEN OR RURAL FARM,
THESE THREE FACTORS; THE DIVERSITY OF PLANTS, ANIMALS AND THEIR
INTERACTION, HAS THE SAME RESULT OF BUILDING UP SOIL. WE LIVE IN A
TIME WHEN MANY BELIEVE THAT YOU CAN "FEED" THE SOIL WITH ELEMENTS,
WHETHER CHEMICALLY OR ORGANICALLY DERIVED, THAT ARE VOID OF ALL BIO-
LOGICAL ACTIVITY. FOR THIS REASON FEW PEOPLE KNOW VERY MUCH ABOUT
THE COMPLEXITY OF BIOLOGICALLY ACTIVE SOIL.

ONE ALL IMPORTANT DIFFERENCE BETWEEN THE FOREST ECOSYSTEM AND THE
AGRICULTURAL ECOSYSTEMS IS THAT HUMAN BEINGS ARE AN ADDED FOURTH
FACTOR. WE CAN TRANSFORM ECOSYSTEMS FROM ONE KIND TO ANOTHER BY
ADDING OR SUBTRACTING PLANTS AND ANIMALS OR BY ENHANCING RO INHIB-
ITING THE NATURAL CYCLES OF A PARTICULAR ECOSYSTEM. WE WILL TALK
ABOUT HOW IT IS POSSIBLE TO ENHANCE THE BIOLOGICAL ACTIVITY OF THE
SOIL THROUGH A VARIETY OF METHODS. IT IS A VAST SUBJECT AREA THAT
CAN ATTAIN GREAT REFINEMENT, FOR EXAMPLE BY KNOWING EXACTLY WHAT
PLANTS AND ANIMALS AFFECT THE SOIL IN PARTICULAR WAYS. THIS IS THE
GOAL TO STRIVE FOR, BUT WE WILL STICK TO AN ELEMENTARY STEP-BY-STEP
PROCEDURE FOR SOIL BUILDING.

FIRST OF ALL, IT IS NECESSARY TO KNOW WHAT YOUR SOIL IS LACKING THAT
PLANTS ABSOLUTELY MUST HAVE. IT HAS BEEN KNOWN FOR A TIME NOW THAT
PLANTS CANNOT SURVIVE LONG OR REGENERATE WITHOUT THE AVAILABILITY OF
NITROGEN, PHOSPORUS AND POTASSIUM (N P K) AND A HOST OF TRACE ELE-
MENTS. IT WOULD BE WISE TO HAVE THE SOIL TESTED BY THE EXTENSION
SERVICE TO FIND OUT AT LEAST TO WHAT EXTENT THESE NUTRIENTS ARE A-
VAILABLE. IT IS POSSIBLE TO INTERPRET IF YOUR GARDEN SOIL HAS ENOUGH
MINERAL CONTENT TO KEEP THE PLANTS FAIRLY WELL NOURISHED BY THE
"DYNAMIC ACCUMULATORS" THAT MIGHT BE GROWING IN THE SOIL. A DYNAMIC
ACCUMULATOR IS A PLANT THAT TENDS TO ACCUMULATE OUTSTANDING QUAN-

TITIES OF A PARTICULAR NUTRIENT FROM THE SOIL. DYNAMIC ACCUMULATORS
ADDED TO YOUR COMPOST PILE INCREASE ITS NUTRIENT AND HUMUS LEVEL.
HERE ARE A FEW EXAMPLES.

BRACKEN FERN--ACCUMULATES LARGE AMOUNTS OF POTASSIUM IN ITS FRONDS.

BORAGE--ACCUMULATES POTASSIUM

STINGING NETTLES--GREAT ACCUMULATORS OF SILICA AND TENDS TO CLEANSE
        THE SOIL OF DISEASE PATHOGENS. ALSO HIGH IN IRON.

FOX GLOVE--ANOTHER SILICA ACCUMULATOR (A RARE VIRTUE IN PLANTS)
        COLLECTS LARGE AMOUNTS OF MANGANESE. IT'S ALSO DEER PROOF,
        USE IT ALONG SIDE A DEER FENCE, THEN USE THE LEAVES FOR COM-
        POST.

LAMBS QUARTERS--ACCUMULATES IRON, PHOSPHOROUS AND CALCIUM. IF YOU
        DON'T HAVE ANY LAMB'S QUARTERS GROWING AS WEEDS IN YOUR
        GARDEN, YOU MAY HAVE SOIL PROBLEMS.

PURSLANE--ACCUMULATES CALCIUM, IRON AND PHOSPHOROUS.

MULLEINS--ACCUMULATES IRON, MAGNESIUM AND SULPHER.

COMFREY--ACCUMULATES NITROGEN, AND IS INVALUABLE IN PREVENTING
        DISEASES IN SOLANUM SPECIES (POTATOES, EGGPLANT FAMILY)
        MAKE COMFREY COMPOST FOR YOUR POTATO PATCH.

BEFORE YOU BEGIN TO PREPARE THE SOIL, TAKE
A GOOD LOOK AT IT. SOIL IS COMPOSED OF MIN-
ERAL PARTICLES AND HUMUS (DECAYED ORGANIC
MATTER). IN THE SPACES BETWEEN THE PARTICLES
ARE WATER, AIR AND "DISSOLVED" NUTRIENTS, ALL
OF WHICH ARE ESSENTIAL TO PLANT GROWTH.
SOILS ARE CLASSIFIED BY THESE TEXTURES:

CLAY MADE OF MINUTE PARTICLES THAT BIND
    TIGHTLY TOGETHER, EXCLUDING AIR AND
    WATER. CLAY SOILS ARE USUALLY RED-
    DISH BROWN OR GRAY IN COLOR. THEY
    WILL STICK TOGETHER IN CLUMPS WHEN
    YOU SQUEEZE THEM.

SAND MADE OF LARGER, IRREGULAR SHAPED
    PARTICLES WITH LARGE SPACES BETWEEN
    THEM THAT CAUSE RAPID WATER DRAIN-
    AGE AND LEACHING OF NUTRIENTS.
    SAND IS PALE BROWN, YELLOW, OR RED
    WITH LITTLE OR NO ORGANIC MATTER.
    IT FEELS GRITTY, AND DOES NOT HOLD
    A SHAPE.

LOAM IS NEARLY A PERFECT COMBINATION OF
    DIFFERENT SIZE PARTICLES ABLE TO
    RETAIN HUMUS (SOURCE OF NUTRIENTS)
    WATER AND AIR. LOAM IS DARK BROWN
    TO BLACK IN COLOR, FULL OF TINY
    BITS OF VEGETABLE MATTER AND GRAINY
    PARTICLES. IT CRUMBLES THROUGH YOUR
    FINGERS AND WON'T FORM A TIGHT BALL.

COMPOST

THE FINISHED PRODUCT OF A COMPOSTING SYSTEM IS A DARK BROWN, CRUMBLY SUBSTANCE KNWON AS HUMUS. THIS SUBSTANCE, WHEN APPLIED TO SURFACE SOILS, CREATES A NUMBER OF LIFE ENRICHING BENEFITS.

THE NUTRIENT COMPOSITION OF HUMUS CAN VARY DEPENDING UPON THE MATERIALS AND/OR MINERAL ADDITIVES PUT INTO THE BATCH OF COMPOST, AND THE PARTICULAR METHOD USED. COMPOSTING HAS BEEN LIKENED TO THE PROCESS OF MAKING A FINE BREAD--IT IS A BLEND OF ART AND SCIENCE.

FOLLOWING IS A LIST OF THE IMPORTANT BENEFICIAL FUNCTIONS THE USE OF HUMUS CAN PROVIDE TO SURFACE SOILS AND THE PLANTS GROWING THERE.

1. PROVIDES READILY AVAILABLE NUTRIENTS WHICH SUPPORT HEALTHY PLANT GROWTH.
2. ALLOWS PLANT TO DEVELOP MORE EXTENSIVE ROOT SYSTEMS.
3. INCREASES SOIL POROSITY--ALLOWS WATER TO DRAIN INTO THE SOIL.
4. INCREASES THE SOIL'S WATER RETENTION--MINIMIZES NEED FOR WATERING.
5. DEVELOPS SOIL AGGREGATES AND PORE SPACE FOR OXYGEN AND OTHER VITAL GAS EXCHANGES NECESSARY FOR PLANT ROOTS.
6. HELPS TO PREVENT WIND AND WATER EROSION.
7. SUPPORTS SOIL MICROORGANISMS AND OTHER SOIL FAUNA (EARTHWORMS FOR EXAMPLE).
8. DECREASES INCIDENCE OF PLANT DISEASE.
9. HELPS MAINTAIN WARMER SOIL TEMPERATURES ON COLD DAYS AND COOLER SOIL TEMPERATURES ON HOT DAYS.
10. REDUCES OR ELIMINATES THE NEED FOR USING NONRENEWABLE, SYNTHETIC CHEMICAL FERTILIZERS.

COMPOSTING MICROBES, LIKE ALL LIVING ORGANISMS, NEED AN ENVIRONMENT WHICH MEETS CERTAIN FUNDAMENTAL LIFE SUPPORT CONDITIONS. THESE CONDITIONS ARE ESSENTIAL IN ORDER TO SUPPORT THEIR BIOLOGICAL PROCESSES. COMPOSTING IS A NATURAL, _LIVING_ PROCESS; YOUR SUCCESS WILL DEPEND UPON KEEPING THESE ORGANISMS ALIVE, GROWING AND THRIVING. BELOW ARE SOME POINTERS ON HOW TO MEET THEIR BASIC ENVIRONMENTAL NEEDS.

THERE ARE FOUR CATEGORIES OF COMPOST INGREDIENTS WHICH ARE ABSOLUTELY ESSENTIAL. EACH TYPE INGREDIENT SHOULD BE MADE AVAILABLE IN ADEQUATE AMOUNTS TO ALL PARTS OF THE COMPOST BATCH. THESE ARE:

1. GOOD SOIL OR FRESHLY PULLED WEEDS: THIS IS WHERE THE COMPOSTING ORGANISMS COME FROM IN THE FIRST PLACE. THE SOIL AROUND THE ROOTS OF FRESHLY PULLED WEEDS SHOULD BE RICH IN THESE ORGANISMS.

2. ORGANIC WASTE MATERIALS: THESE MATERIALS ARE THE SOURCE OF MOST MINERALS AND NUTRIENTS WHICH FEED THE COMPOSTING ORGANISMS AND PROVIDE THE SOIL ENRICHING QUALITIES OF THE HUMUS. DO NOT INCLUDE RAW OR COOKED MEATS, OILS--EITHER VEGETABLE OR MINERAL, GREASE OR FATS, PLASTIC MATERIALS, SYMTHETIC FIBERS OR PAPER WITH INK ON IT. THE PRESENSE OF THESE ITEMS WILL RETARD OR INHIBIT THE DECOMPOSITION PROCESS. TABLE 1 SHOWS A LIST OF COMMON ORGANIC WASTES WHICH ARE GOOD COMPOSTABLES.

3. ORGANIC MATTER WITH A HIGH NITROGEN CONTENT: ADDING A CERTAIN AMOUNT OF ORGANIC MATERIALS WHICH HAVE CHARACTERISTICALLY HIGH NITROGEN CONTENT (2-15%) IS VITALLY IMPORTANT TO OBTAINING GOOD COMPOSTING RESULTS. NITROGEN IS AN ESSENTIAL NUTRIENT FOR THE DECOMPOSING ORGANISMS. ADDITIONALLY, NITROGEN IS FUNDAMENTAL IN PROMOTING THE RAPID GROWTH OF NITROGEN FIXING BACTERIA WHICH ARE RESPONSIBLE FOR PRODUCING THE HEAT NECESSARY TO BEGIN THE WASTE DECOMPOSITION PROCESS. KITCHEN WASTES, HAIR, ANIMAL URINE AND MANURE, ALFALFA, GRASS CLIPPINGS AND BLOOD MEAL ARE EXAMPLES OF HIGH NITROGEN ORGANICS.

4. WATER: MOST COMPOSTING ORGANISMS NEED MOIST CONDITIONS TO OPTIMALLY FUNCTION. MANY OF THE ORGANIC MATERIALS ADDED TO COMPOST WILL BE DRY OR MODERATELY DAMP AT BEST. THIS WILL MEAN THAT PERIODIC WATERING OF THE COMPOST WILL BE NECESSARY. MAINTAINING A BALANCE IS BASIC TO SUCCESSFUL COMPOSTING. DO NOT CREATE SOGGY CONDITIONS--IDEALLY YOUR COMPOST SHOULD FEEL LIKE A MOIST SPONGE.

BEFORE DECIDING WHICH COMPOSTING TECHNIQUE IS RIGHT FOR YOU IT IS HELPFUL TO KEEP IN MIND A FEW BASIC CONSIDERATIONS.

IN YOUR LIVING SITUATION....

HOW MUCH SPACE DO YOU HAVE TO SET UP A COMPOSTING SYSTEM?

THE AMOUNT OF SPACE NEEDED CAN RANGE FROM AS LITTLE AS ONE ADDITIONAL GARBAGE CAN TO A 15 x 15 FOOT PLOT OR MORE. THE CHOICE IS BASICALLY UP TO YOU. IT ALL DEPENDS ON THE AMOUNT OF ORGANIC WASTE YOU WISH TO COMPOST AND UPON THE SPECIFIC METHOD USED.

A FAMILY OF THREE WILL PRODUCE APPROXIMATELY 30 LBS. OF HOUSEHOLD ORGANICS PER MONTH--NOT A LARGE AMOUNT TO COMPOST. YARD TRIMMINGS WILL ADD TO THIS AMOUNT CONSIDERABLY AND IF YOU ARE A GARDENER YOU WILL HAVE A GREATER AMOUNT OF ORGANIC MATTER TO PUT INTO A COMPOSTING SYSTEM AND A GREATER INTEREST IN THE OUTPUT OF YOUR SYSTEM.

HOW MUCH TIME AND ENERGY DO YOU WISH TO SPEND MAINTAINING THE COMPOST SYSTEM?

THIS CAN RANGE FROM AN HOUR OR LESS A WEEK TO 3-4 HOURS A WEEK. AGAIN, THE CHOICE IS UP TO YOU. THE SPECIFIC COMPOSTING METHOD USED AND THE VOLUME OF WASTE COMPOSTED ARE KEY FACTORS IN THIS CONSIDERATION.

WHAT TYPE LOCATION BEST SUITS A COMPOSTING SITE?

GENERALLY SPEAKING, SHADY LOCATIONS ARE BEST SUITED FOR COMPOSTING. THIS PREVENTS YOUR COMPOST FROM OVER-DRYING AND PRESERVES SUNNY AREAS FOR GORWING PLANTS. OFTEN THE NORTH SIDE OF A BUILDING, FENCE OR HEDGE IS A PERFECT SPOT. TWO ADDITIONAL THINGS TO CONSIDER IN SITING YOUR COMPOSTING SYSTEM ARE:

1. IS YOUR LOCATION CLOSE TO YOUR GARDENING AREA OR FLOWER BEDS?

2. IS YOUR LOCATION ACCESSIBLE FOR HAULING IN COMPOSTABLES THAT YOU MAY WISH TO GATHER FROM YOUR NEIGHBORHOOD OR THE COUNTRY?

THE LONG INDORE METHOD

NO DISCUSSION ON COMPOSTING IS COMPLETE WITHOUT MENTION OF THE "INDORE TECHNIQUE". THIS METHOD WAS DEVELOPED BY SIR ALBERT HOWARD AND TAKES ITS NAME FROM A REGION IN INDIA WHERE THE PROCESS WAS DEVELOPED. BEHIND ALL OTHER COMPOSTING METHODS, LIES THE SUCCESS OF THE ORIGINAL METHOD. ORGANIC GARDENERS HAVE GAKEN HOWARD'S CORE OF COMPOST RESEARCH AND PRODUCED BEAUTIFUL COMPOST AND BEAUTIFUL GARDENS.

ONE OF THE MORE POWERFUL DISCOVERIES IN THE DEVELOPMENT OF THIS METHOD LIES IN THE LAYERING TECHNIQUE. THIS LAYERING TECHNIQUE, DESCRIBED BELOW, SHOULD BE APPLIED TO ALL COMPOSTING METHODS REVIEWED IN THIS GUIDE EXCEPT THE "DRUM" METHOD. THE LAYERING METHOD ASSURES THAT ALL OF THE ESSENTIAL COMPOST INGREDIENTS DESCRIBED IN THE PREVIOUS SECTION WILL BE AVAILABLE IN ALL PARTS OF THE PILE.

1. START YOUR INDORE PILE AT GROUND LEVEL BY LAYERING DOWN SOME ABSORBENT MATERIAL TO ACT AS A BASE, SUCH AS HAY, STRAW, SAWDUST SUNFLOWER OR CORN STALKS.

2. SPREAD DOWN A FIVE OR SIX INCH LAYER OF GREEN ORGANIC MATTER SUCH AS LEAVES, GARDEN WASTE, ETC.

3. LAYER THE PILE TWO OR THREE INCHES HIGH WITH HIGH NITROGEN CONTENT MATERIALS (REFER TO TABLE 1) SUCH AS MANURE, COMFREY, ETC.

4. SPRINKLE WITH A CAP OF SOIL OR FRESHLY PULLED WEEDS.

5. SPRINKLE WITH WATER (DO NOT SOAK).

6. REPEAT THIS PROCESS UNTIL THE PILE IS FIVE TO SEVEN LAYERS HIGH.

WHEN YOU HAVE FINISHED, THE PILE SHOULD BE FIVE TO TEN FEET WIDE AND FIVE FEET HIGH, THE LENGTH IS OPTIONAL, IN ANY CASE, THE PILE SHOULD BE NO SMALLER THAN ONE CUBIC YARD. A SMALLER PILE WILL NOT HOLD THE HEAT WELL.

MAINTENANCE

TURN THE PILE AFTER SIX WEEKS AND NINE WEEKS, PLACING THE OUTSIDE AND TOP MATERIALS INTO THE BOTTOM CENTER OF THE PILE. COMPOST WILL BE FINISHED AFTER THREE TO SIX MONTHS. INNER TEMPERATURE OF 150 TO 160 DEGREES SHOULD BE ATTAINED DURING THE PROCESS.

SOIL
GREEN VEGETABLES
TWIGS & BRANCHES
DRY VEGETATION
LOOSENED SOIL

ADVANTAGES OF THE LONG INDORE METHOD

1. NO BUILDING OR CONSTRUCTION MATERIALS NEEDED

2. ONLY WORK INVOLVED IS THE INITIAL CONSTRUCTION OF THE PILE AND INFREQUENT TURNING.

3. NO EQUIPMENT NEEDED EXCEPT PITCHFORK OR SHOVEL

4. AN EFFICIENT SET-UP FOR THE LARGE VOLUME URBAN COMPOSTER

DISADVANTAGES OF THE LONG INDORE METHOD

1. TAKES THREE MONTHS OR LONGER TO PRODUCE FINISHED HUMUS

2. HUMUS PRODUCED IS NOT UNIFORM IN CONSISTENCY

3. CAN ATTRACT RODENTS AND NEIGHBORHOOD PETS

4. MAY BE CONSIDERED UNSIGHTLY

5. MANY OF THE NUTRIENTS MAY EVAPORATE OR LEACH INTO SOIL UNDER PILE.

THE "BIN" SHORT INDORE METHOD

double compost bin

THIS METHOD IS A VARIATION OF THE LONG INDORE METHOD, COMMONLY REFERED TO AS THE FOURTEEN DAY METHOD. THE DIFFERENCES ARE 1. USE WOOD SLATE OR CONCRETE BLOCK BOX TO CONTAIN THE PILES, 2. MATERIALS ARE SHREDDED (OR FINELY CHOPPED UP) AND 3. PROCESS TAKES LESS TIME.

CONTAINERIZED COMPOSTING IS DESIRABLE IN MANY URBAN SETTINGS BECAUSE THEY REDUCE THE EYESORE CONCERNS AS WELL AS INHIBIT NEIGHBORHOOD OR RODENTS FROM BECOMING AN OFFENSE TO THE COMMUNITY. THE FRONT OF THE BINS MAY BE MADE OUT OF REMOVABLE BOARDS, ALLOWING EASY ENTRY AS THE CONTENTS ARE EMPTIED. SIDES AND FLOORS OF THE BINS SHOULD BE AS TIGHT AS POSSIBLE SO THAT BITS OF ORGANIC MATTER CANNOT FALL THROUGH AND PROVIDE OVERLOOKED FLY BREEDING MATERIAL UNDERNEATH OR OUTSIDE THE BINS. A TIGHTLY FITTED PLYWOOD HINGED TOP CAN BE USED TO KEEP OUT UNWANTED RAIN DURING THE WINTER.

THE ORGANIC MATERIALS BEING USED (KITCHEN FOOD SCRAPS, GRASS AND YARD CLIPPINGS, LEAVES, MANURES, URINE AND SOIL) CAN BE FINELY SHREDDED OR CHOPPED BEFORE LAYERING (SEE LONG INDORE METHOD). THE SHREDDING PROCESS INCREASES THE SURFACE AREA OF THE ORGANIC MATTER THAT IS EXPOSED TO THE COMPOSTING ORGANISMS. THIS ACCELERATES THE DECOMPOSITION CYCLE.

THIS PROCESS REQUIRES MORE FREQUENT TURNING--AT LEAST TWICE WEEKLY. TURN THE COMPOST DAILY IF YOU WANT FINISHED COMPOST IN TEN TO FOURTEEN DAYS. THIS METHOD CAN ALSO BE DONE WITHOUT THE BIN, USE A FREE GROUND AREA THREE TO FOUR FEET WIDE AND NINE TO TWELVE FEET LONG.

ADVANTAGES OF THE "BIN" SHORT INDORE METHOD

1. QUICK PROCESS; CAN PRODUCE READY TO USE HUMUS IN AS LITTLE AS TEN DAYS.

2. UNIFORM HEATING AT HIGHER TEMPERATURES

3. REDUCES ODOR, FLIES AND RODENT PROBLEMS

4. PRODUCES HUMUS WHICH IS UNIFORM IN TEXTURE.

DISADVANTAGES OF THE "BIN" SHORT INDORE METHOD (FOR SOME PEOPLE)

1. PILE MUST BE TURNED OFTEN, REQUIRING PHYSICAL EXERTION AND TIME.

2. A SHREDDING DEVICE IS NEEDED (COMMERCIAL UNITS ARE COSTLY AND USE FOSSIL FUEL BASED ENERGY SOURCES). A CONVENTIONAL LAWN MOWER OR SHARP SPACE CAN BE USED.

3. TIME IS NEEDED FOR SHREDDING OR CHOPPING.

4. CONSTRUCTION OF BINS REQUIRES TIME, LABOR AND MATERIALS (CAN USE RECYCLED <u>FREE</u> MATERIALS)

THE ANEROBIC METHOD

THE ONLY MATERIALS NEEDED TO BEGIN THIS METHOD ~~IS~~ ARE A LARGE SHEET OF PLASTIC AND SOME ORGANIC WASTE. BLACK PLASTIC MAY BE BEST BECUASE BLACK ENCOURAGES THE HEATING NECESSARY. A PLASTIC SHEET WILL WORK, HOWEVER. ANOTHER OPTION IS USING WASHABLE <u>REUSABLE</u> PLASTIC BAGS.

THIS IS HOW YOU DO IT. LAY THE PLASTIC ON THE GROUND. LAYER ALL COMPOSTABLES ON OR IN THE PLASTIC. INGREDIENTS SHOULD BE DAMPENED IF NOT THOROUGHLY MOIST. THE TOP SHOULD BE PULLED OVER THE PILE AND DIRT BANKED UP AROUND THE EDGES TO SEAL OFF AIR. IF A BAG IS USED, IT SHOULD BE SECURELY TIED AT THE TOP. AS SOON AS AIR IS KEPT OUT, YOUR COMPOST IS BEGINNING TO WORK.

THIS PROCESS CAN ALSO BE USED INSIDE A BOX, DRUM OR GARBAGE CAN. THE IMPORTANT THING IS TO SEAL COMPOST OFF FROM AIR.

ADVANTAGES OF THE ANAEROBIC METHOD

1. THE METHOD TAKES THE SMALLEST OUTPUT FOR MATERIALS.

2. REQUIRES NO TURNING.

3. NO SMELL (SEALED FROM THE AIR).

4. TAKES ONLY FOUR TO EIGHT WEEKS.

DISADVANTAGES OF THE ANAEROBIC METHOD

1. HIGH HEAT (120 TO 150 DEGREES) IS NOT ACHIEVED BY THIS PROCESS, THUS REDUCED SANITIZING OCCURS IN RELATION TO AEROBIC SYSTEM.

2. CAN PRODUCE ODORS IF NOT PROPERLY COVERED OR SEALED.

THIS METHOD WILL PROBABLY BE A GOOD CHOICE FOR THE WILLAMETTE VALLEY AREA BECAUSE IN THE ANAEROBIC PROCESS, THE HUMUS BULK IS ABOUT THE SAME AS THE ORIGINAL COMPOST BULK. AEROBIC METHODS LOSE FO'RTY TO EIGHTY PERCENT OF THE COMPOST BULK THROUGH BACTERIAL BREAKDOWN, EVAP- ORATION AND LEACHING INTO SOIL UNDER THE PILE. BULK IS WHAT WE NEED HERE IN THIS AREA TO AERATE OUR HEAVY, CLAY SOIL**S**. AEROBIC COMPOST- ING SYSTEMS ALLOW FO':RTY TO NINETY PERCENT OF AVAILABLE NITROGEN TO BE LOST IN EVAPORATION (THAT SLIGHTLY AMMONIA-TYPE SMELL) AND IN MOISTURE LEACHING. THE PLASTIC UNDER THE COMPOST IN THIS ANAEROBIC PROCESS PREVENTS SUCH NUTRIENT LOSSES. THE HUMUS OF ANAEROBIC COM- POSTING ALSO WILL UNDERGO SOME AEROBIC BREAKDOWN AND SO SHOULD BE TILLED IMMEDIATELY INTO SOIL SO THAT NITROGEN LOSS TO AIR IS REDUCED.

"THROUGH INTENSIVE PROPAGATION AND USE OF DOMESTICATED
EARTHWORMS, USING THE MATERIALS WHICH ARE IMMEDIATELY
AT HAND AND IN CHEAP ABUNDANCE, THE EARTHWORM CULTUR-
IST CAN BUILD FERTILE TOPSOIL TO MEET ALL HES REQUIRE-
MENTS. WHETHER FOR FOOD OR PLEASURE, FOR PROFIT OR
FOR SUBSISTENCE, HE CAN INCREASE PRODUCTION BEYOND
ANYTHING HE HAS THOUGHT POSSIBLE BEFORE."

--THOMAS J. BARRETT
"HARNESSING THE EARTHWORM"

USING WORMS AND COMPOSTING
(VERMICULTURE AND ANNELIDIC CONSUMPTION)

NOEL PRCHAL

nature's finest

methane gas

landfills

casting

fertilizer

natural gas

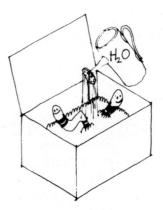

H₂O

AT THE URBAN FARM, THE MANURE EARTH-
WORM (THOSE LOWLY CREATURES OF THE
EARTH) ARE USED IN A LOW-TECHNOLOG-
ICAL APPROACH FOR ECOLOGICALLY SAFE
CONVERSION OF BIODEGRADABLE SOLID
WASTES INTO A NUTRIENT-RICH SOIL
SUPPLEMENT WHICH CAN BE READILY
ASSIMILATED INTO PLANT FEEDER ROOT
SYSTEMS.

THE SCIENCE OF RAISING EARTHWORMS
IS KNOWN AS "VERMICULTURE". THE
CONVERSION OF ORGANIC WASTES INTO
CASTINGS, OR EXCRETIONS FROM THE
EARTHWORM, IS KNOWN AS "ANNELIDIC
CONSUMPTION". SOME VALUABLE BENE-
FITS OF THIS NEIGHBORHOOD ALTERNA-
TIVE SOLID WASTE MANAGEMENT PROGRAM
ARE:

- NO AIR POLLUTION

- NO GROUNDWATER AND SURFACE WATER
  CONTAMINATION

- NO SOIL STERILITY

(ALL OF THE ABOVE ARE ASSOCIATED
WITH PRESENT SANITARY LANDFILLS.)

- RECYCLING OF ORGANIC WASTES AND
  CONVERSION INTO A NUTRIENT-RICH
  SOIL SUPPLEMENT FOR ORGANIC
  GARDENING AND FOOD PRODUCTION
  (CONTRIBUTING TO SELF-SUFFICIENCY
  WITHIN LOCAL NEIGHBORHOODS)

- REPLACES CHEMICAL FERTILIZERS,
  NOW EXTENSIVELY USED AND DERIVED
  FROM SCARCE AND EXPENSIVE FOSSIL
  FUELS

- LOW SET-UP COSTS AND LOW MAINTEN-
  ANCE

- VIRTUALLY NO ODOR, PESTS, OR
  RODENT PROBLEMS THAT ARE RELATED
  TO EXPOSED COMPOST PILES (MAINLY
  BECAUSE THE BEDS ARE COVERED AND
  ENCLOSED)

THERE ARE SOME SHORT AND SIMPLE
FACTS WE SHOULD UNDERSTAND ABOUT
MANURE WORMS AND THE WASTE STREAM
TO ASSURE OURSELVES AN EFFICIENT,
EASILY-MANAGED NEIGHBORHOOD OR-
GANIC WASTE CONVERSION PRACTICE.

- MANURE WORMS' NATURAL HABITAT IS
  MANURE/REFUSE PILES (HIGH ORGANIC
  CONTENT). THE "RED WIGGLER"
  COMPRISES 80-90% OF COMMERCIALLY
  PRODUCED MANURE WORMS. THEY
  ADAPT WELL TO DOMESTICATION AND
  TEMPERATURE VARIANCES. THEY EAT
  VIGOROUSLY AND RETAIN HIGH HARDI-
  NESS AND BREEDING CHARACTERISTICS
  IF FED, WATERED AND MAINTAINED
  WELL.

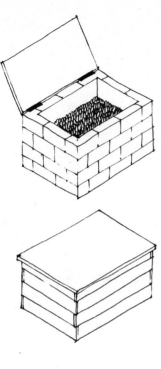

Rabbit manure

- NEED ADEQUATE <u>MOISTURE</u> (NOT SOGGY) FOR RESPIRATION, DIGESTION, EXCRETION AND MOVEMENT.

- WORMS ARE <u>LIGHT-SENSITIVE</u> (TO ULTRAVIOLET RAYS). WHEN EXPOSED FOR ONE HOUR, WILL BECOME PARALYZED; IN TWO HOURS WILL DIE. (KEEP BED COVERED WHEN NOT MAINTAINING BEDS.)

BED CONSTRUCTION

- MAY USE CONCRETE BLOCK, SCRAP LUMBER, BRICKS, OR THE NATURAL GROUND. (INITIALLY CONSTRUCTION OF 3' WIDTH X 4' LENGTH X 2' HEIGHT WILL HOUSE 1000 WORMS).

- AT THE URBAN FARM WE USE CONCRETE BLOCK BECAUSE OF SUMMER/WINTER INSULATING VALUES, RODENT-PROOF CONSTRUCTION, QUICK SET-UP AND RELOCATION, STRUCTURAL STABILITY.

- THE BOTTOM BOARD CAN BE CONCRETE BLOCK, BRICK RUBBLE, 3/4" LUMBER, WIRE MESH, OR NATURAL GROUND (WELL DRAINED).

- SCRAP LUMBER (1" X 6" OR 2" X 6") PREFERABLY RAISED OFF THE GROUND IS AN INEXPENSIVE METHOD.

- PLYWOOD (1/2") FOR A TOP COVER, HINGED AT BACK.

BEDDING

THE WORMS INITIALLY LIVE IN THIS ENVIRONMENT, LAY EGG CAPSULES IN THIS MATERIAL, AND IS THE INITIAL FOOD SOURCE FOR THE WORMS. AS THEY CONSUME THIS MATERIAL, THEY EXPELL THEIR CASTINGS HERE AND GRADUALLY MOVE UPWARD TO THE NEW FOOD THAT HAS BEEN ADDED. (THE MANURE WORMS ARE TOP FEEDERS). SHREDDED WET CARDBOARD AND RABBIT MANURE ARE IDEAL FOR INITIAL BEDDING MATERIAL.

FEEDING

IDEALLY THE FOOD SHOULD BE HIGH IN <u>CELLULOSE</u> AND <u>CARBOHYDRATE</u> MATTER (PLANT MATTER WASTES). PREFERENCE OF FOOD: A) DAIRY MANURE, B) RABBIT MANURE, C) SHREDDED WET CARDBOARD, D) WOOD/PAPER PULP (WET NEWSPRINT), E) ORGANIC WASTES SUCH AS CHOPPED KITCHEN SCRAPS, SHREDDED GARDEN FOLIAGE, DECIDUOUS LEAVES, ORCHARD WASTES, F) SHEEP AND GOAT MANURE, G) WOOD STOVE ASHES, H) SEWER SLUDGE, IF HEAVY METALS OR FROM HOME SEPTIC TANK SYSTEM, I) PEAT MOSS--CHECK ACIDITY, J) CRUSHED

DECIDUOUS LEAVES, K) SHREDDED DECIDUOUS BARK, L) CRUSHED COTTONSEED/SUNFLOWER HULLS, M) CHOPPED HAY/STRAW, N) GRASS CLIPPINGS, O) WET NON-RESINOUS WOOD SHAVINGS.

- AERATE THESE FOOD AND BEDDING WASTES ONCE A MONTH, AFTER PLACEMENT IN WORM BED, WITH 5-6 SPLINE PITCHFORK.

- WORMS HAVE NO TEETH. THEY ABSORB FOOD DIRECTLY INTO GIZZARD, CROP AND STOMACHS. <u>THEY MUST HAVE MOIST, PARTIALLY DECOMPOSED FOOD</u>, CHOPPED AND SHREDDED AS MUCH AS POSSIBLE. A MEDIUM-SIZED PORTABLE HAND GRINDER, MACHETE KNIFE, OR SHOVEL WILL WORK WELL. PLACE YOUR CHOPPED ORGANICS IN ONE CORNER OF THE BED OR IN A SEPARATE COMPOST BIN FOR INITIAL DECOMPOSITION, THEN ADD TO WORM BED <u>AFTER IT PASSES THE HEATING STAGE</u>. (MINUTE AMOUNTS OF ALCOHOL BY-PRODUCT AND METHANE GAS ARE HARMFUL TO WORM'S DIGESTIVE TRACT.)

hand grinder

- MAINTAIN A pH BETWEEN 6.8 AND 7.2 (NEAR NEUTRAL) WITHIN BED. LITMUS PAPER OBTAINED FROM DRUG STORES IS YOUR TEST STRIP. PLACE SAMPLE INTO BEDDING/FEED -- IF BLUE/PURPLE, IT'S ALKALINE, IF RED/PINK, IT'S ACIDIC. IF TOO ACIDIC, ADD CALCIUM CARBONATE (LIMESTONE FLOUR--SAME TYPE USED TO CHALK FOOTBALL FIELDS). CONTACT SPORTING GOODS DEALER OR SCHOOL MAINTENANCE FOR LOCAL DEALERS--APPROXIMATELY $3.00 PER 50 POUND SACK. DO NOT USE AGRICULTURAL SLAKED (OR UNSLAKED) OR DEHYDRATED LIME (PHOSPHATE CONTENT IS TOO HIGH). YOU CAN USE CRUSHED OYSTER SHELLS IF AVAILABLE.

- DO NOT USE CITRUS, PINE, OAK, BLACK WALNUT, SEQUOIA, CYPRESS, OR CEDAR LEAVES OR NEEDLES; THEY GIVE OFF TANNIC ACID OR RESINOUS SAP.

- PREFERABLY, MOST MANURE (WITH THE EXCEPTION OF RABBIT MANURE), IF LESS THAN TWO YEARS OLD, SHOULD BE <u>LEACHED</u> (SOAKED IN WATER) AHEAD OF TIME TO REDUCE ACIDITY AND SALT CONTENT.

- ANIMAL FATS/OILS COAT THE DIGESTIVE TRACT; DON'T USE.

- POULTRY MANURE (FRESH) IS TOO HOT (NITROGEN CONTENT) AND BURNS THE INTESTINAL TRACT OF THE WORMS. IT ALSO PACKS AND SEALS THE BED

SURFACE (LESS OXYGEN AND WATER
INTAKE THROUGH BED).

- BOTH CARDBOARD AND NEWSPRINT ARE
EXCELLENT FOOD SOURCES, BUT ARE
VALUABLE RECYCLABLE ITEMS. RE-
FRAIN FROM USING UNLESS MATERIAL
IS FOUND VERY SOGGY.

- MOST OF YOUR LEFTOVER KITCHEN
SCRAPS CAN BE ADDED DIRECTLY TO
BED, BUT CHOP SOLID PARTICLES.
DO NOT ADD BANANA PEELS BECAUSE
OF CHEMICAL SPRAYS ON SKINS.

ENVIRONMENTAL FACTORS

THE BED TEMPERATURE SHOULD BE BE-
TWEEN 55° AND 100°F (IDEALLY 60°-
75°) FOR THE WORMS TO BE ACTIVE
FEEDERS. THEIR WASTE CONSUMPTION
DROPS CONSIDERABLY BETWEEN DECEM-
BER AND FEBRUARY DUE TO LOW BED
TEMPERATURES. INSULATE THE BED
WELL (PREFERABLY PLACE INSIDE
BUILDING) AND FEED 1 FOOT OF MANURE
IN WINTER TO KEEP CONSUMPTION PRO-
GRESSING. EXCESSIVE MOISTURE
DURING WINTER ALSO KEEPS BED COLD.
WORMS WILL BECOME DORMANT IN LOW
TEMPERATURES AND WILL NOT TOLERATE
A SOLID FREEZE OF THE BED.

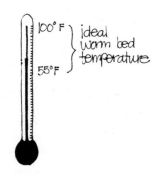

ideal
worm bed
temperature

PESTS

- CENTIPEDES EAT EARTHWORMS. CON-
SULT COUNTY EXTENSION AGENCY FOR
ORGANIC PESTICIDE.

- MICE/RATS EAT EARTHWORMS' FOOD
AND ARE A HEALTH PROBLEM. PLACE
CRUSHED GLASS UNDER BED'S BOTTOM
BOARD.

- JAPANESE BEETLES DAMAGE STRUCTURE
OF BEDDING, CARRY SAND INTO IT.
USE A MICRO-ORGANISM KNOWN AS
MILKY SPORE DISEASE (ORGANIC).

- MOLES AND GOPHERS EAT EARTHWORMS.
PLACE A BOTTOM BOARD OR BRICK
UNDER BED. CONSULT COUNTY EXTEN-
SION AGENCY FOR ORGANIC PESTICIDE.

- MITES, SPIDERS, GRASSHOPPERS AND
INSECTS MAY OCCUR. WATER BED,
AND WHEN THEY COME TO THE SURFACE
DRESSING, TORCH THEM.

REPRODUCTION

EARTHWORMS ARE BISEXUAL--BOTH MALE
AND FEMALE ORGANS. TWO WORMS COPU-
LATE AND EXCHANGE SPERM. EACH WORM
FORMS EGG CAPSULES IN THE CLITELUM
(BAND NEAR THE HEAD). CAPSULES ARE
SHED IN CASTINGS, WITH NUTRIENTS
FOR WORM DEVELOPMENT TIGHTLY SEALED

ideal
reproduction
temperature

INSIDE. THE TEMPERATURE OF THE BED
MUST BE BETWEEN 50° AND 80°F FOR THE
CAPSULES TO HATCH, PLUS ADEQUATE
MOISTURE TO SOFTEN THE CAPSULE SHELL.
AN AVERAGE OF FOUR WORMS ARE IN EACH
CAPSULE. THE GESTATION PERIOD IS

: 7-10 DAYS FOR MATURE MANURE
WORM TO PRODUCE EGG CAPSULE

: 14-21 DAYS FOR CAPSULES TO
HATCH (MOISTURE/TEMPERATURE OK)

: 60-90 DAYS FOR YOUNG WORMS TO
DEVELOP INTO MATURE WORMS.

REGENERATION

WHEN A WORM IS SEVERED INTO TWO POR-
TIONS (THROUGH CARELESS HANDLING), THE
HEAD END WILL GROW A NEW TAIL PRO-
VIDING THE VITAL ORGANS IN SEGMENTS
11-36 ARE INTACT. (A MATURE WORM HAS
APPROXIMATELY 30-90 SEGMENTS.)

HARVESTING

IF YOUR BED BECOMES TOO CROWDED WITH
WORMS YOU MUST DIVIDE YOUR BED THE
SAME WAY YOU HARVEST THE BED. LET
THE WORMS CONSUME ALL THE ORGANIC
WASTE PLACED IN THE LAST FEEDING.
THEN STARVE THE WORMS A FEW DAYS.
PLACE SOME LIGHT MANURE (PREFERABLY
RABBIT MANURE) ON TOP OR JUST PLACE
SOME RAW GARBAGE ON TOP. COVER NEW
LAYER OF FEED AND ENTIRE TOP OF BED
WITH BLACK PLASTIC. A FEW DAYS
LATER REMOVE THE BLACK PLASTIC, A
PORTION AT A TIME AND REMOVE THE TOP
5"-6" FROM THE BED (95% OF THE WORMS
WILL BE IN THIS AREA). PLACE THE
WORMS INTO ANOTHER PREPARED BEDDING
MATERIAL OR TEMPORARILY STORE IN A
LARGE, COVERED BOX UNTIL YOU REMOVE
ALL THE CASTINGS. REPLACE WORMS IN
BED.

CASTINGS

THIS SOIL SUPPLEMENT GENERALLY CON-
TAINS LESS THAN 5% NPK (.5-2% NITRO-
GEN; .06-.68% PHOSPHATE PHOSPHORUS;
.10-.68% POTASSIUM; .58-3.5% CALCIUM)
AND OTHER ESSENTIAL MICRO-NUTRIENTS.
FOR POTTING MIX FOR INDOOR PLANTS YOU
MAY WANT TO COMBINE WITH VERMICULITE,
PEAT MOSS, SAND, AND DRIED POULTRY
MANURE. BY ITSELF, THE CASTING'S RAW
NITROGEN IS LOWER THEN MANURE SO IT
WON'T BURN PLANT ROOTS.

APPLY CASTINGS DIRECTLY AS A TOP
DRESSING TO YOUR ORGANIC GARDEN PRIOR
TO PLANTING SEASONS.

tail end
becomes
nitrogen
fertilizer

EARTHWORM BENEFITS TO GARDEN SOILS

BECAUSE YOU HAVE NOW PLACED CASTINGS
IN YOUR GARDEN YOU UNDOUBTEDLY
BROUGHT A FEW EGG CAPSULES ALONG
WITH YOU. WHEN THEY HATCH, THE
MANURE WORMS, NIGHTCRAWLERS, AND
COMMON GARDEN WORMS WILL NOW FORAGE
ON THE SOIL AND ORGANIC MATTER
THROUGHOUT THE SOIL. DO NOT TILL
THE GARDEN, BUT SPADE YOUR PLOT OR
BEDS, AND ONLY USE ORGANIC PEST CON-
TROL PRACTICES. OTHERWISE YOU DE-
STROY THE VALUABLE CONTRIBUTIONS
WORMS HAVE PERFORMED IN YOUR GARDEN:

- THEY ASSIST IN SOIL PARTICLE
  AGGREGATION (BONDING) AND IN
  CREATING WATER-STABLE GRANULES
  (WITH EXCRETIONS AND BODY SECRE-
  TIONS.)

- THEY AID IN SOIL TURNOVER (700#
  OF CASTINGS/ACRE PER DAY IN OPEN
  FIELDS.)

- SOIL MOISTURE, POROSITY, DRAINAGE
  ABILITY, AND INFILTRATION ARE IN-
  CREASED.

- SOIL AERATION IS INCREASED.

- THEY AID ORGANIC MATTER DECOMPO-
  SITION AND SOLUBILIZE PLANT NU-
  TRIENT ELEMENTS.

- THEY PROVIDE SOIL NUTRIENTS SUCH
  AS NITROGEN FROM EXCRETIONS AND
  DEAD EARTHWORMS.

- THEY INCREASE AVAILABLE PHOSPHORUS,
  MOLYBDENUM, EXCHANGEABLE CALCIUM,
  MAGNESIUM, AND POTASSIUM.

WHERE TO ACQUIRE EARTHWORMS

MAIL ORDER RETAIL PRICES AND PERHAPS LOCAL SOURCES FOR "RED
WIGGLER" MANURE WORMS (BEDRUN) ARE $6/1000; $13/5000; $24/10,000
$100/50,000 (AIR FREIGHT OR UPS INCLUDED). THE WORMS ARE PACKAGED
IN DAMP PEAT MOSS. SOME SOURCES ARE:

LOCAL

- TERRACE HILL FARM
  GARY OLSON/JOHN LONG
  90704 HILL ROAD
  SPRINGFIELD, ORE 97477
  (747-0415/726-0722)

  BOB SMITH
  EUGENE, OREGON
  (726-0855)

- NANCY SCHAFFER
  135 E. 33RD
  EUGENE, OREGON
  (344-2365)

MAIL ORDER

- BUD KINNEY-OG   (CALIFORNIA'S
  BOX 133   OLDEST AND
  VINA   LARGEST FARM)
  CALIFORNIA 96092
  (916-839-2328)

- CENTEX EARTHWORM FARM
  ROUTE 1, BOX 227
  BURNET, TEXAS 78611

SOME RESOURCES...

EARTHWORMS FOR ECOLOGY AND PROFIT, VOL. I, "SCIENTIFIC EARTHWORM
  FARMING" AND VOL. II, "EARTHWORMS FOR ECOLOGY", BY RONALD E.
  GADDIE, SR. AND DONALD DOUGLAS, BOOKWORM PUBLISHING COMPANY,
  BOX 3037, ONTARIO, CANADA 91761

LET AN EARTHWORM BE YOUR GARBAGE MAN, BY HOME, FARM AND GARDEN
  RESEARCH ASSOCIATES, AND HENRY HOPP, GARDEN WAY PUBLISHING
  COMPANY, CHARLOTTE, VERMONT, 1973

WHAT EVERY GARDENER SHOULD KNOW ABOUT EARTHWORMS, BY HENRY HOPP,
  GARDEN WAY PUBLISHING COMPANY, CHARLOTTE, VERMONT,1973

HARNESSING THE EARTHWORM, BY THOMAS J. BARRET, WEDGEWOOD PRESS,
  1959

EARTHWORM FEEDS AND FEEDING, BY CHARLIE MORGAN, SHIELDS PUBLICA-
  TIONS, 1961

THE EARTHWORM BOOK, BY JENX MINNICH, RODALE PRESS, EMMAUS, PA.

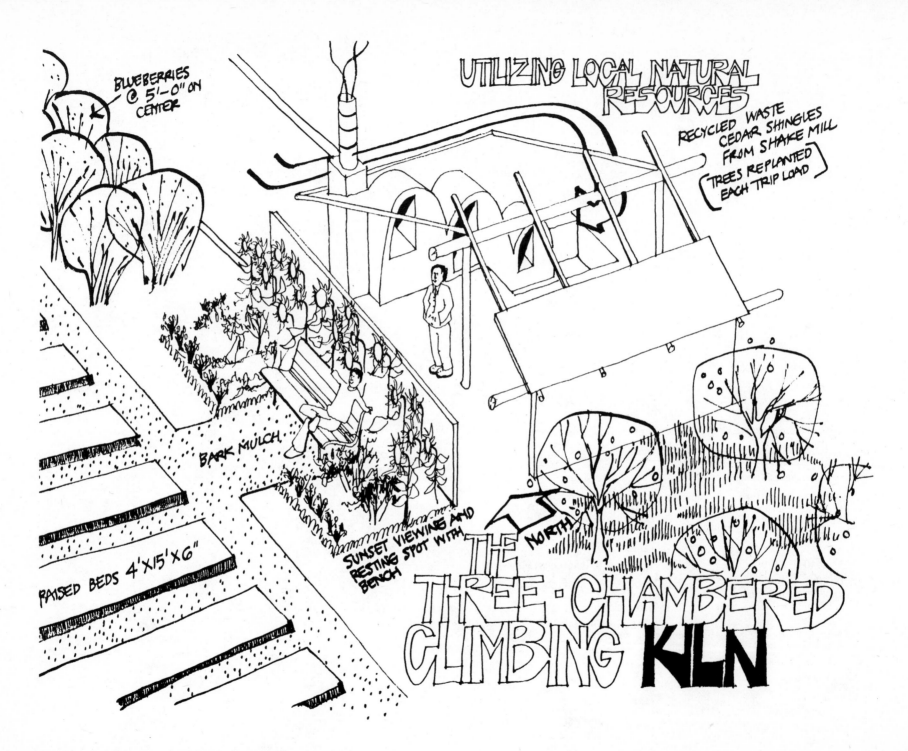

BLUEBERRIES @ 5'-0" ON CENTER

UTILIZING LOCAL NATURAL RESOURCES

RECYCLED WASTE CEDAR SHINGLES FROM SHAKE MILL [TREES REPLANTED EACH TRIP LOAD]

BARK MULCH

RAISED BEDS 4'X15'X6"

SUNSET VIEWING AND RESTING SPOT WITH BENCH

NORTH

THE THREE-CHAMBERED CLIMBING KILN

DEVELOPING THE POTTERY KILN

- CHRIS GUM

THIS WOOD FIRED POTTERY KILN WAS BUILT AT THE URBAN FARM OVER THE SUMMER OF 1978 BY A SMALL GROUP OF STUDENTS. THE HISTORIC CLIMBING CHAMBER KILN DESIGN WAS CHOSEN TO STUDY FUEL ECONOMY AND HOMEMADE REFRACTORIES. THIS CHAMBER KILN IS DESIGNED TO BE FUEL EFFICIENT. WASTE GASSES WHICH ARE NORMALLY ALLOWED TO GO OUT THE CHIMNEY ARE UTILIZED SEVERAL TIMES OVER BY THE CHIMNEY-LIKE EFFECT CREATED BY THE CLIMBING CHAMBERS. THE FIRE IS STARTED IN THE MAIN FIREBOX. THIS FIREBOX IS FED UNTIL TEMPERATURE IS REACHED IN THE FIRST CHAMBER. THEN THE STOKING IS MOVED TO THE SECOND CHAMBER FIREBOX. THIS SECOND CHAMBER WHICH HAS BEEN PREHEATED BY THE WASTE GASSES (2/3RDS OR MORE DONE) IS FINISHED RELATIVELY QUICKLY AND THE STOKING MOVES ON TO THE THIRD CHAMBER UNTIL IT IS FINISHED ALSO. THE AMOUNT OF PREHEATING IN THE SECOND AND THIRD CHAMBERS AND THE EFFECTIVE PREHEATED AIR SYSTEM SUBSTANTIATED OUR HYPOTHESIS THAT THIS KILN DESIGN WAS VERY FUEL-EFFICIENT.

THE RESULTS OF OUR STUDY OF HOMEMADE REFRACTORIES WERE LESS CONCLUSIVE. WE FOUND THAT OUR MATERIAL MIX (FIRECLAY, GROG AND SAWDUST) WORKED BUT THAT THE MIXTURE WAS NOT AS EFFICIENT AN INSULATOR AS WE HAD EXPECTED, AND THAT SOME PHYSICAL PROBLEMS SUCH AS CRACKING AND CRUMBLING NEEDED TO BE RESOLVED.

☐ BRICK

▦ HOMEMADE REFRACTORIES

THE WOOD

THE ENERGY REQUIRED TO FIRE THE POTTERY KILN IS THE POTTER'S MOST OBVIOUS ENERGY NEED. THE MAJOR FUELS CURRENTLY AVAILABLE ARE ELECTRICITY, OIL, GAS, AND WOOD. OF THE FOUR, WOOD APPEARS TO BE THE BEST WAY TO GO FOR SEVERAL RELATED REASONS.

ECONOMICALLY, WOOD IS BECOMING MORE COMPETITIVE, AS FUEL COSTS RISE. IN THE LONG RUN, THIS TREND WILL CONTINUE AS AVAILABILITY OF FOSSIL FUELS DECREASES AND THE COST OF RECOVERY INCREASES. WOOD, HOWEVER, IS A RENEWAL RESOURCE. THE AVAILABILITY OF WOOD DEPENDS UPON HOW WE MANAGE OUR FORESTS. A PLENTIFUL FUTURE

REQUIRES THAT WE PLANT AND DEVELOP GREAT REFORESTATION PROGRAMS NOW. THE HUMAN ELEMENT OR REASON INVOLVED IN WOOD FIRING IS THAT BY FIRING WITH WOOD WE ASSUME MORE DIRECT RESPONSIBILITY FOR OUR ACTIONS. WE HAVE BECOME RESPONSIBLE FOR PLANTING, NURTURING, TRANSPLANTING, RECOVERING AND USING OUR FUEL.

SECOND, THE SCALE IS TANGIBLE; IT IS QUITE POSSIBLE FOR EVERY INDIVIUAL TO BE INVOLVED WITH BOTH THE GROWING AND THE USING OF OUR FUEL. THE SATISFACTION TO BE DERIVED FROM THIS INVOLVEMENT AND RESPONSIBILITY SEEMS DESIRABLE AND HEALTHY. BY CONTRAST, OUR PRESENT RELATIONSHIP WITH FOSSIL FUEL PRODUCTION AND USE INVOLVES PRIMARILY USING. OUR PRESENT SITUATION FEELS UNBALANCED BECAUSE THERE IS SO LITTLE SENSE OF GIVE AND TAKE.

A THIRD RELATED REASON FOR FIRING WITH WOOD CONCERNS OUR ECOLOGY AND ENVIRONMENT. THERE IS AN IMBALANCE OF OUR USE AND MISUSE OF OUR ENVIRONMENT. WE PAY CERTAIN PRICES FOR THE WAYS WE USE OUR RESOURCES, SUCH AS AIR OR WATER QUALITY, AND WE NEED TO CONSIDER HOW OUR USAGES AFFECT OUR ENVIRONMENT. WHAT IS OUR IMPACT? THEN, WE NEED TO CHOOSE THE ALTERNATIVES WHICH HAVE THE LEAST AND THE GENTLEST IMPACT. THE IMPACT IS DIRECTLY RELATED TO THE DIFFICULTY AND COMPLEXITY OF RECOVERY. HOW MUCH TIME, EFFORT, AND ENERGY DID WE EXPEND RECOVERING "NEW" ENERGY AND WHAT EFFECTS DO THE RECOVERY AND USE OF THAT ENERGY HAVE ON THE ENVIRONMENT.

THE SOLUTION TO THESE INVOLVED QUESTIONS BEGINS WITH OUR ABILITY TO MAXIMIZE THE RESPONSIBLE USE OF OUR LOCAL RESOURCES. OUR LOCAL RESOURCES OFTEN HOLD THE POTENTIAL TO BOTH SATISFY OUR NEEDS AND TO HELP BALANCE OUR OTHER NEED FOR A HEALTHY ENVIRONMENT. WE NEED TO EXPLORE AND DEVELOP THIS POTENTIAL MORE THAN WE HAVE. WOOD FIRING, PARTICULARLY HERE IN THE NORTHWEST, APPEARS TO BE A DIRECT AND HEALTHY WAY IN WHICH A POTTER CAN CONTRIBUTE TO THE QUALITY OF LIFE THAT SURROUNDS US.

PLANTING SYMBOLICALLY

THIS BOOK DONATED IN MEMORY
OF LYNN MATHEWS

The new orchard at the Urban Farm was planted in memory of Lynn Mathews by 100 people working together for two hours on a Saturday morning. The conversion of a part of the landscape, long neglected, into an understory of new trees beneath the old and nestled close to them enabled us to participate in the cycles of natural processes. Once the trees were planted for the utilitarian purpose of replenishing the old and fading orchard, bee hives were placed among them to assure their constant pollination.

Simultaneously, a fund was established to update the Eugene Public Library with books relating to ecological agriculture, bearing the above logo. It was and remains our hope that planting seeds and starts and saplings and books will continue a tradition of concern for each other, the plants, and the soils which comprise our environment. The successional concept enables us to see change yet stabilize processes which have continuing utility and beauty.

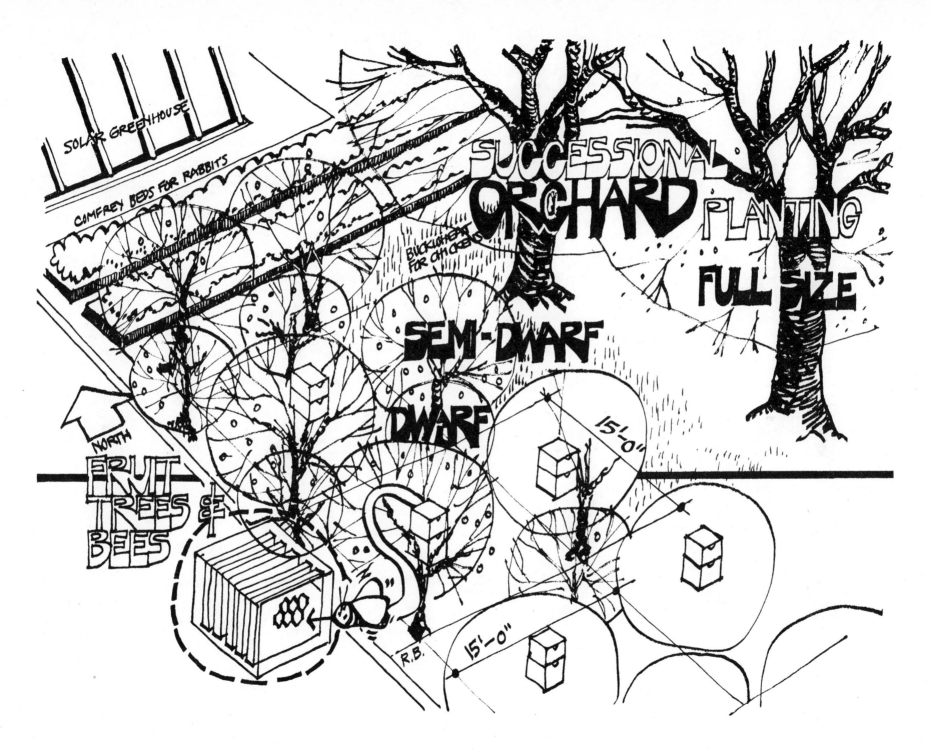

SOLAR GREENHOUSE

COMFREY BEDS FOR RABBITS

SUCCESSIONAL ORCHARD PLANTING

FULL SIZE

BUCKWHEAT FOR CHICKENS

SEMI-DWARF

DWARF

15'-0"

NORTH

FRUIT TREES & BEES

R.B.

15'-0"

MAXIMIZING SOLAR GAIN AND STORAGE...

CAN ALSO BE EXTENDED FROM THE GARDEN PRINCIPLES DISCUSSED EARLIER TO ENCLOSING PLANTS AND OTHER ORGANISMS INSIDE WIND SHELTERS WHICH ACCEPT THE SUN AND HOLD ITS ENERGY. AS YOU REMEMBER IN THE EARLIER SECTION ON RAISED BEDS, THE ORIENTATION SHOULD BE DUE SOUTH WHETHER FREE-STANDING OR ATTACHED TO YOUR HOUSE.

AT THE URBAN FARM WE HAVE DESIGNED A LOW-COST SOLAR GREENHOUSE FOR THE FRONT OF THE FARMHOUSE. IT IS BEING BUILT IN STAGES AS MATERIALS AND LABOR BECOME AVAILABLE (IT TAKES A WHILE IF YOU ARE BUILDING CHEAPLY), AND WE INTEND TO COMPLETE WHAT WE CONSIDER TO BE

A CHEAP... ATTACHED SOLAR GREENHOUSE

FOR THE URBAN FARMHOUSE.

- 4" DIAM. UNTREATED FENCE POSTS (@ $1.97) (8) $ 15.76
- 1 GAL. CREOSOTE TO TREAT (PAINT ON) FENCEPOSTS   10.00
- TWO 97# BAGS CEMENT (MIX W/ ROAD GRAVEL & EARTH)
- 2 x 4's
- OLD USED DOUBLE HUNG WINDOWS
- 50-GAL. DRUMS PAINTED BLACK
- USED BROKEN BRICK

WHILE THE ABOVE BREAKDOWN IS NOT COMPLETE YET (WE ARE STILL IN THE "FORAGING AND GATHERING" STAGE), IT DOES REFLECT OUR BASIC LOW-COST PHILOSOPHY. ACQUIRE YOUR MATERIALS FIRST! DESIGN SECOND! DEVELOP GENERAL SPECIFICATIONS AND SEARCH FOR MANY WAYS TO ACCOMPLISH THESE GOALS, THEN HONE IN ON THE SPECIFICS AS TIME, TRANSPORTATION, AND MATERIALS PERMIT. REMEMBER, ON FIXED OR DECLINING INCOMES, IT SEEMS LIKE A GOOD IDEA TO:

- RE-CYCLE MATERIALS AND HUMAN ENERGY
- CONSERVE RESOURCES
- USE INGENUITY

THE WAY WE BUILT WHAT WE HAVE DRAWN OUT HERE HAS BEEN GUIDED BY THESE THREE PRINCIPLES. INITIALLY, WE ACQUIRED SOME 4" DIAMETER UNTREATED FENCE POSTS 8'-9' LONG AND USED THESE FOR THE MAIN FRAMING. THESE POSTS WERE SET DIRECTLY IN CONCRETE FOOTINGS ABOUT 18" DEEP AFTER PAINTING CREOSOTE ON THE PARTS TO KEEP THE WOOD FROM ROTTING---EXPEDIENT, CHEAP, AND SEMI-PERMANENT. THE FRAMING WAS THEN USED (OF RECYCLED 2"x4"S) TO TIE THE VERTICAL POSTS TO THE HOUSE AND TO EACH OTHER, AND RAFTERS (ALSO OF 2"x4"S) WERE ANGLED UP TO FRAME INTO THE FRONT OF THE HOUSE. 50-GALLON DRUMS (USED TO STORE HEAT IN THE WATER INSIDE THEM) WERE ACQUIRED FROM CHEF FRANCISCO FOR $2.00 EACH. DURING THE FIRST WINTER, OR AS FINANCES DICTATE, SHEET POLYETHELENE COULD BE USED AS A COVERING FOR SOME DEGREE OF PLANT PROTECTION; HOWEVER, WE HAVE CHOSEN TO AVOID PLASTIC AS MUCH AS POSSIBLE. INSTEAD, WE ARE FRAMING INTO OUR "CRUDE" POLE AND 2"x4" FRAMEWORK SETS OF OLD, RECYCLED DOUBLE-HUNG WINDOWS (SINGLE PANE), AVAILABLE FROM LANE SALVAGE (ON AIRPORT ROAD JUST OFF HWY. 99) WHICH, IN TURN, WILL BE COVERED BY MOVABLE HINGED INSULATING PANELS. THUS, CHEAPLY ENCLOSED SPACE CAN ACT AS A HEAT TRAP DURING THE DAY AND CLOSE UP LIKE A FLOWER AT NIGHT AND HOLD IN THE HEAT. HEAT THEN WILL RADIATE OFF THE WARM 50-GALLON DRUMS AT NIGHT AND WARM THE TRAYS OF PLANT STARTS SITTING ON TOP OF THE DRUMS. WE PLAN ON TEMPERED GLASS PANELS OVERHEAD ("SECONDS" FROM SLIDING GLASS DOOR OUTLETS) WHICH WILL NOT SHATTER UPON IMPACT AS UNTEMPERED GLASS ALWAYS DOES, AND WE WILL PROBABLY LATER CONSIDER REPLACING ALL OF THE GLASS WITH THESE SAME TEMPERED SHEETS. AT OUR LAST ESTIMATE, THESE PATIO DOOR SHEETS WERE ABOUT $12-$15 EACH. A FUTURE CONSIDERATION IS THE USE OF SHATTER-PROOF AUTOMOBILE WINDSHIELDS AS GLAZING MATERIAL, BUT THIS IS DISCUSSED FURTHER IN THE LATTER SECTION ON PROTOTYPE BLOCK FARMS. RECYCLED BRICK FLOOR IS USED WHERE WALKING SPACE IS LOCATED; OTHERWISE THE SOIL IS TURNED OVER AND AERATED TO MAKE ONE LONG AND WIDE RAISED BED WITHIN THE GREENHOUSE. IN LIEU OF CONCRETE PERIMETER FOUNDATIONS, OLD RECYCLED RAILROAD TIES ARE USED FOR RESTING THE GREENHOUSE FRAMEWORK UPON, AROUND WHICH THE SOIL IS BERMED AND PLANTED WITH FLOWERS.

THE OBJECT OF ALL THIS IS TO:

1. REDUCE HEATING COSTS AND LOWER CASH OUTFLOW (ENERGY CONSERVATION)

2. UTILIZE LOCAL MATERIALS WHERE POSSIBLE AND RECYCLE INDUSTRIAL WASTE MATERIALS (SUCH AS GLASS AND 50-GALLON DRUMS) AS WELL AS MORE COMMON "WASTE MATERIALS".

3. PROVIDE AN EXTENSION OF THE SEASONS (FALL AND SPRING) IN ORDER TO HAVE A WARM MICROCLIMATE BY UTILIZING WASTE HEAT (FROM ENTERING AND LEAVING THE HOUSE) AS WELL AS THE SOLAR GAIN FROM THE WINTER SUN ITSELF.

4. ENABLE DIRECT INVOLVEMENT WITH FOOD PRODUCTION YEAR 'ROUND!

IN THE FOLLOWING TWO SECTIONS, "URBAN FARMING IN SOLAR GREENHOUSES" BY AMITY FOUNDATION OF EUGENE, AND "SOLAR GREENHOUSES FOR WHITEAKER NEIGHBORHOOD" BY LARRY PARKER OF WHITEAKER NEIGHBORHOOD PROJECT SELF RELIANCE, WE HOPE TO REVIEW BASIC PRINCIPLES OF SOLAR GREENHOUSES, EXPAND THOSE TO INCLUDE FISH LIVING IN THE WATER HEAT STORAGE TANKS AND GENERALLY PROVIDE A GUIDE TOWARD BECOMING INVOLVED BY STARTING SMALL AND GROWING....

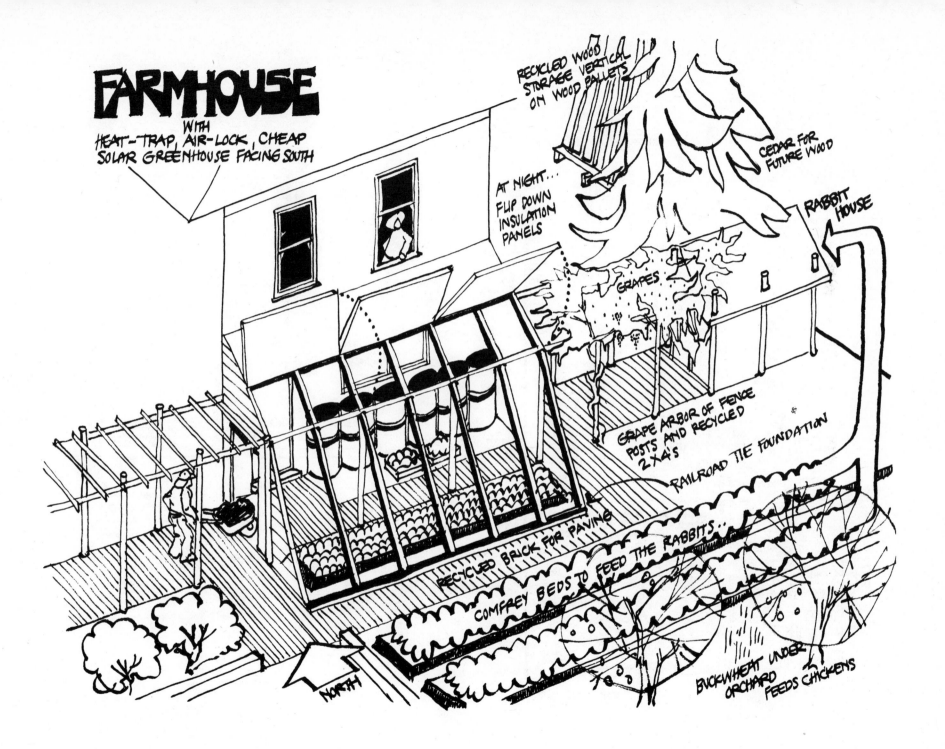

FARMHOUSE

WITH
HEAT-TRAP, AIR-LOCK, CHEAP
SOLAR GREENHOUSE FACING SOUTH

RECYCLED WOOD STORAGE VERTICAL ON WOOD PALLETS

CEDAR FOR FUTURE WOOD

AT NIGHT...
FLIP DOWN INSULATION PANELS

RABBIT HOUSE

GRAPES

GRAPE ARBOR OF FENCE POSTS AND RECYCLED 2×4'S

RAILROAD TIE FOUNDATION

RECYCLED BRICK FOR PAVING

COMFREY BEDS TO FEED THE RABBITS...

NORTH

BUCKWHEAT UNDER ORCHARD FEEDS CHICKENS

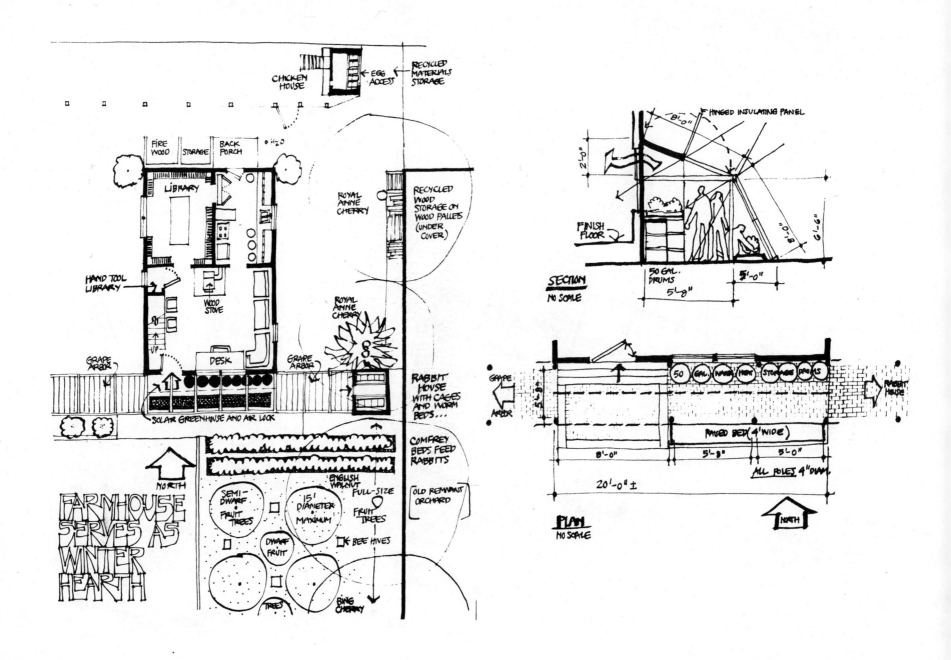

CHICKEN HOUSE

EGG ACCESS

← RECYCLED MATERIALS STORAGE

FIRE WOOD | STORAGE | BACK PORCH

H₂O

LIBRARY

ROYAL ANNE CHERRY

RECYCLED WOOD STORAGE ON WOOD PALLETS (UNDER COVER)

HAND TOOL LIBRARY

WOOD STOVE

ROYAL ANNE CHERRY

UP

GRAPE ARBOR

DESK

GRAPE ARBOR

RABBIT HOUSE WITH CAGES AND WORM BEDS...

SOLAR GREENHOUSE AND AIR LOCK

COMFREY BEDS FEED RABBITS

NORTH

FARMHOUSE SERVES AS WINTER HEARTH

ENGLISH WALNUT

OLD REMNANT ORCHARD

SEMI-DWARF FRUIT TREES

15' DIAMETER MAXIMUM

FULL-SIZE FRUIT TREES

DWARF FRUIT

BEE HIVES

TREES

BING CHERRY

HINGED INSULATING PANEL

8'-0"

2'-0"

6'-6"

8'-0"

FINISH FLOOR

SECTION
NO SCALE

50 GAL. DRUMS

5'-8"

5'-0"

GRAPE ARBOR

50 GAL. WATER STORAGE DRUMS

RABBIT HOUSE

5'-8"

RAISED BED (4' WIDE)

8'-0"

5'-8"

5'-0"

ALL POLES 4" DIAM.

20'-0" ±

PLAN
NO SCALE

NORTH

URBAN FARMING IN SOLAR GREENHOUSES

AMITY FOUNDATION

URBAN FARMERS AND GREENHOUSE HOBBYISTS ARE INCREASINGLY RECOGNIZING THAT SOLAR GREENHOUSES AND OTHER APPROPRIATE TECHNOLOGIES CAN MAKE THEIR GARDENING MORE ECONOMICAL, EFFICIENT AND FUN. AMONG THE ADVANTAGES OF ADOPTING SOLAR TECHNOLOGIES ON THE INDIVIDUAL, NEIGHBORHOOD OR COMMUNITY LEVEL ARE LOCALIZATION OF FOOD PRODUCTION, SEASON EXTENSION FOR VEGETABLE AND PLANT PRODUCTION, SPACE HEAT SUPPLEMENTATION IF THE GREENHOUSE IS RETROFITTED OR ATTACHED TO AN EXISTING BUILDING, AND POSSIBLE EXPERIMENTATION WITH OTHER INTEGRATED SYSTEMS, SUCH AS THE RAISING OF EDIBLE FISH IN THE GREENHOUSE HEAT STORAGE WATER. BUT BEFORE WE LOOK AT THESE AND OTHER ADVANTAGES IN GREATER DEPTH, LET'S ANSWER A FEW BASIC QUESTIONS: WHAT IS A SOLAR GREENHOUSE? ARE THERE DIFFERENT TYPES AND DESIGNS? HOW DO I DECIDE WHAT IS THE BEST FOR ME?

THE SOLAR GREENHOUSE DEFINED

THREE DESIGN FEATURES DISTINGUISH A SOLAR GREENHOUSE FROM A CONVENTIONAL ALL-GLASS OR PLASTIC GREENHOUSE:

·THE GLAZED SURFACE FACES SOUTH TO CAPTURE THE LOW ANGLE WINTER SUN, WHEN HEAT NEEDS ARE GREATEST.

·AREAS NOT RECEIVING DIRECT WINTER SUN, SUCH AS THE NORTH WALL AND ROOF, ARE SOLID AND WELL-INSULATED TO MINIMIZE HEAT LOSS.

·EXCESS SOLAR ENERGY IS STORED IN MASSIVE OBJECTS LIKE ROCK, EARTH OR WATER. THIS STORED HEAT IS LATER RELEASED INTO THE GREENHOUSE AT NIGHT OR DURING CLOUDY PERIODS.

THE PURPOSE OF THESE FEATURES IS TO INCREASE THE INPUT OF SOLAR ENERGY AND TO STORE IT AS HEAT. A CONVENTIONAL GREENHOUSE DOES NOT HAVE A WAY OF STORING SOLAR ENERGY AND USES ELECTRICITY, GAS OR OIL FOR SUPPLEMENTAL HEATING. IDEALLY, A SOLAR GREENHOUSE NEEDS NOTHING OTHER THAN THE SUN FOR LIGHT AND HEAT.

GREENHOUSE TYPES AND USES

SOLAR GREENHOUSES COME IN ALL SIZES, SHAPES AND CONFIGURATIONS; ONE OF THE EXCITEMENTS OF WORKING WITH SOLAR LIES IN THE FACT THAT CONVENTIONAL DESIGN CONCEPTS ARE BEING CONSTANTLY CHALLENGED. OUR DISCUSSION HERE WILL BE LIMITED TO THE PASSIVE SOLAR DESIGNS, I.E. THOSE WHICH DO NOT REQUIRE ADDITIONAL ENERGY INPUTS TO COLLECT OR DISTRIBUTE HEAT GLEANED FROM THE SUN.

1. WINDOWBOX DESIGNS ARE SUITED FOR THOSE WHO MAY NOT HAVE SPACE, ORIENTATION OR RESOURCES TO ATTACH A LEAN-TO GREENHOUSE TO THEIR HOME. THESE WINDOWBOX STRUCTURES CAN BE PLACED OVER ANY SOUTH-FACING WINDOW WHICH RECEIVES SUNLIGHT MOST OF THE YEAR, AND WILL INCREASE BOTH USABLE SUNLIGHT AND HEAT CONSERVATION.

2. GREENHOUSE/COLDFRAME HYBRIDS COMBINE THE BEST OF TWO TECHNOLOGIES FOR LOW-COST SEASON EXTENSION AND PROTECTION FROM UNSEASONAL COLD SPELLS, RADIATION FROSTS AND THE LIKE. INSULATED COLDFRAMES SAVE ENERGY AND PROVIDE ADDITIONAL GARDENING SPACE FOR AT LEAST THREE SEASONS IN MOST PARTS OF THE UNITED STATES.

3. ATTACHED (LEAN-TO, RETROFITTED) DESIGNS MAXIMIZE ENERGY EFFICIENCES BY VENTING THEIR EXCESS HEAT INTO THE STRUCTURE, USUALLY RESIDENTIAL, TO WHICH THEY ARE ATTACHED. OVERALL EFFICIENCY DEPENDS UPON PRE-EXISTING HOME INSULATION AND ENERGY CONSERVATION, AS WELL AS THE PROPER CONNECTION AND VENTING BETWEEN THE HOME AND THE GREENHOUSE. ATTACHED GREENHOUSES, IN THIS SENSE, ACT AS SOLAR COLLECTORS FOR THE HOME OR STRUCTURE. THE SPACE-SAVING NATURE OF ATTACHED GREENHOUSES MAKE THEM ATTRACTIVE FOR URBAN AREAS.

4. FREESTANDING GREENHOUSE DESIGNS PROVIDE A MAXIMUM OF FLEXIBILITY REGARDING SIZE, SITING AND ORIENTATION. HOWEVER, THEY DO NOT INCORPORATE THE SPACE HEATING POTENTIAL THAT ATTACHED DESIGNS PROVIDE. SINCE PRODUCTION CAPACITY IS RELATED TO SIZE, FREE-STANDING GREENHOUSES ARE PARTICULARLY APPLICABLE FOR NEIGHBORHOOD BLOCK FARM AND/OR COMMUNITY USE.

CHOICE OF GREENHOUSES SHOULD BE GUIDED BY AN INDIVIDUAL'S OR GROUP'S PLANNED USE FOR THE GREENHOUSE, BY SPACE REQUIREMENTS AND AVAILABILITY, BY SITE CHARACTERISTICS, AND BY THE PARTICULAR GEOGRAPHIC AND CLIMATIC FEATURES OF THE AREA IN QUESTION. MANY INTERESTING AND COMPREHENSIVE BOOKS NOW AVAILABLE AT ALMOST ANY BOOKSTORE OR LIBRARY COVER THIS TOPIC IN DETAIL; A BRIEF BIBLIOGRAPHY IS INCLUDED AT THE END OF THIS SECTION. ORIENTATION TO THE SUN, GLAZING ANGLE, INSULATION (WALLS, ROOF, FOUNDATION, ETC.) VAPOR BARRIER REQUIREMENTS, VENTING AND HEAT STORAGE MEDIUM, VOLUME AND PLACEMENT WITHIN THE GREENHOUSE ARE ONLY A FEW OF THE DESIGN PARAMETERS WHICH MUST BE CONSIDERED IN PLANNING A SUCCESSFUL SOLAR GREENHOUSE.

ADVANTAGES OF SOLAR GREENHOUSES

AS URBAN FARMERS, THE DESIRE TO MOVE FOOD PRODUCTION CLOSER TO HOME IS A BASIC PREREQUISITE. SOLAR GREENHOUSES AID IN DOING THIS, BY LETTING THE URBAN FARMER START PLANT STARTS, EXTEND GROWING SEASONS, EXPERIMENT WITH CROPS REQUIRING SPECIAL CONDITIONS AND CARE, TRY NEW VARIETIES, AND EXPERIMENT WITH INNOVATIVE TECHNIQUES SUCH AS COMBINING VEGETABLE PRODUCTION WITH SMALL-SCALE AQUACULTURE FOR DIETARY PROTEIN.

ENERGY CONSERVATION

URBAN FARMING, AND ESPECIALLY THE USE OF SOLAR GREENHOUSES AND COLD-FRAMES, IS A DEFINITE ENERGY CONSERVER. AS STUDIES CONDUCTED BY THE FARALLONES INSTITUTE INDICATE, VEGETABLES IDEAL FOR AND MOST LIKELY TO BE GROWN AT HOME ARE ALSO THE MOST ENERGY-INTENSIVE FOR LARGE-SCALE AGRIBUSINESS PRODUCTION (FARALLONES INSTITUTE, 1979). GRAINS AND OTHER CROPS WHICH REQUIRE COMPARATIVELY FEWER FOSSIL FUEL ENERGY INPUTS PER CALORIE OF FOOD PRODUCED ARE BEST GROWN ON FARMS, WHILE VEGETABLES FOR FRESH OR PRESERVED HOME CONSUMPTION ARE IDEALLY GROWN RIGHT IN THE BACKYARD OR IN THE SOLAR GREENHOUSE.

"...FOR THIS REASON, GROWING YOUR OWN VEGETABLES CAN REPRESENT REAL SAVINGS IN FUEL ENERGY FOR THE NATION AS A WHOLE. THIS ENERGY CONSERVATION FACTOR IS GREATLY INCREASED WHEN RESIDENTS OF COLD-WINTER AREAS PRODUCE SOME OF THEIR VEGETABLES DURING THE WINTER

MONTHS IN SOLAR GREENHOUSES THAT USE LITTLE
OR NO AUXILIARY HEAT. IN SOME AREAS, VEGE-
TABLES EATEN DURING THE WINTER MAY REPRESENT
THE GREATEST ENERGY INVESTMENT, GIVEN THE
ENERGY USE INVOLVED IN THE LONG-DISTANCE
SHIPPING FROM THE FARMS IN THE WEST TO THE
COUNTRY'S NORTHERN AND EASTERN CITIES."
(FARALLONES INSTITUTE, 1971, P. 48)

ADD TO THIS THE ENERGY SAVINGS REALIZED THROUGH THE VENTING OF
HEAT FROM AN ATTACHED SOLAR GREENHOUSE, AND THE URBAN FARMER CANNOT
HELP BUT REALIZE POTENTIAL ENERGY SAVINGS BY INCORPORATING SOLAR
TECHNOLOGIES INTO THE HOME AND LIFESTYLE.

SEASON EXTENSION AND CROP PRODUCTION/VARIETY

SOLAR GREENHOUSES ALSO EXTEND GROWING SEASONS ON BOTH ENDS, ESPEC-
IALLY FOR COLD-TOLERANT CROPS WHICH DEMAND ONLY LOW LIGHT INTENSITY.
THE ECOTOPE GROUP OF SEATTLE, WASHINGTON REPORTS THAT SUCH USE OF
SOLAR GREENHOUSES CAN EXTEND SEASONAL VEGETABLE PRODUCTION IN THE
MARITIME NORTHWEST FROM FOUR TO NINE MONTHS WITHOUT ANY ADDITIONAL
INPUT OF ARTIFICIAL LIGHT (ECOTOPE GROUP, 1979). CROP PRODUCTION
CAN BY MAXIMIZED BY MATCHING THE NEEDS OF SPECIFIC PLANTS WITH
CYCLES IN GREENHOUSE PERFORMANCE. VARIOUS CROPPING METHODS WHICH
EMPHASIZE INTENSIVE PLANTINGS AND INTERCROPPING WORK WELL IN SOLAR
GREENHOUSES. AND THE URBAN FARMER GAINS ADDITIONAL TIME AND
SEASONAL LATITUDE FOR TRYING OUT NEW TECHNIQUES. GREENHOUSE GARDEN-
ING GENERATES SPECIFIC PLANS FOR PLANT ROTATION, HARVEST AND PHASING
WITH THIS EXPERIENCE PROVIDING KNOWLEDGE AND CONFIDENCE WHICH IS
APPLICABLE TO ALL PHASES OF GARDENING.
ACCELERATED PRODUCTION AND SUSTAINED YEILDS ARE TWO MORE OF THE
BY-PRODUCTS. AND WHAT TRULY INVOLVED AND SINCERE URBAN FARMER
ENJOYS PUTTING THAT GARDENING TIME INTO "COLD STORAGE" FOR THE
WINTER MONTHS? USE OF A SOLAR GREENHOUSE ASSURES THAT THE OPPOR-
TUNITY FOR "GARDENING THERAPY" WILL BE AVAILABLE ALL YEAR ROUND.

AQUACULTURE AND OTHER INNOVATIONS

WHILE TRYING OUT NEW GARDENING TECHNIQUES, THE URBAN FARMER MAY
WISH TO EXPERIMENT WITH SOME OF THE OTHER INNOVATIVE USES OF SOLAR
GREENHOUSES ALSO. RAISING FISH IN THE HEAT STORAGE WATER OF THE
GREENHOUSES IS ONE OF THESE OPTIONS AND AN ATTRACTIVE ONE FOR
ANYONE WHO IS INTERESTED IN DIETARY SUPPLEMENTATION, PRODUCING
FOOD CLOSER TO HOME AND INDIVIDUAL SELF-RELIANCE. RAISING FISH
IS A HIGHLY ENERGY-EFFICIENT WAY TO PRODUCE PROTEIN; IN A WELL-
MANAGED AQUACULTURE SYSTEM, FISH WILL PRODUCE MORE PROTEIN PER
POUND OF FEED THAN ANY ANIMAL ON LAND. THE VALUE OF ADDITIONAL
FOOD, AND MORE VARIETY, AT LOW COST IS A DEFINITE BENEFIT. IF
A SOLAR GREENHOUSE IS DESIGNED PRIMARILY FOR PLANTS, THE PRODUCTION
OF FISH CAN BE A SIDE-PRODUCT, BECAUSE WATER IS MORE EFFICIENT
THAN ANY OTHER MATERIAL FOR SOLAR HEAT STORAGE. WHY NOT USE THAT
WATER TO GROW FISH? FISH ARE CONVENIENT FOR URBAN FARMING SINCE
THEY ARE EASY TO CARE FOR, DO NOT MAKE NOISE, DO NOT SMELL, REQUIRE
COMPARITIVELY LITTLE TIME FOR FEEDING AND NO WATERING AT ALL.
IN SHORT, FISH MAKE IDEAL URBAN FARM ANIMALS FOR PRODUCTION AND
CONVENEINCE REASONS. AMITY HOUSE'S FISH HOUSE, ONE
EXAMPLE OF AN INTEGRATED SOLAR AQUACULTURE SYSTEM, IS DESCRIBED IN
A PUBLICATION ENTITLED FISH FARMING IN YOUR SOLAR GREENHOUSE (SEE
BIBLIOGRAPHY, PORTIONS EXERPTED BELOW). AFTER SUCCESSFULLY CONSTRUCT-
ING A GREENHOUSE WHICH EFFECTIVELY COLLECTS AND STORES SOLAR HEAT
AND INCORPORATES DESIGN FEATURES FOR FISH FARMING (TWO 1700-GALLON
FISH TANKS, WATER FILTRATION SYSTEMS AND A SAVONIUS ROTOR WINDMILL
FOR SUPPLEMENTARY WATER AERATION), SPECIAL ATTENTION WAS GIVEN TO
PROVIDING THE RIGHT ENVIRONMENTAL CONDITIONS FOR FISH:

"RAISING FISH IS LIKE RAISING ANY CROP. FOR
INSTANCE, IN ORDER TO GROW TOMATOES OR LETTUCE
YOU NEED TO KNOW WHAT THESE PLANTS NEED AND
PREPARE YOUR GROWING AREA AND BE READY WITH
ANY TOOLS AND MATERIALS YOU WILL NEED AS TIME
GOES BY. THE SAME KIND OF PLANNING AND MAN-
AGEMENT IS NEEDED FOR RAISING FISH. YOU NEED
A BASIC KNOWLEDGE OF A FISH'S LIFESTYLE---
THE KIND OF ENVIRONMENTAL NEEDS THAT HAVE TO BE
SATISFIED IN ORDER FOR THEM TO BE HEALTHY AND
GROW WELL."
--HEAD AND SPLANE, 1979

FISH NEEDS, ALTHOUGH PERHAPS CONFUSING AT FIRST TO THE NOVICE,
ARE REALLY NOT THAT DIFFICULT TO UNDERSTAND. PROVIDING PURE
WATER MUST BE FOREMOST ON THE SOLAR AQUACULTURIST'S MIND AND
HE/SHE MUST MONITOR AND CONTROL WATER TEMPERATURE, OXYGEN LEVELS
AND TOXIC SUBSTANCES LIKE AMMONIA. DECAYING SUBSTANCES (UNEATEN
FOOD, FISH FECES, PLANT MATTER, ETC.) CONTRIBUTE TOXINS TO THE
WATER ALONG WITH THOSE GENERATED BY FISH METABOLISM. FINALLY,
WATER QUALITY NEEDS MUST BE MESHED WITH PLANS FOR WATER VOLUME AND
STORAGE FEATURES IN ORDER TO ACCOMPLISH OPTIMAL, SIMULTANEOUS SOLAR
HEATING AND AQUACULTURE USES.
NATURAL WATER PURIFICATION PROCESSES (DIFFUSION, PHOTOSYNTHESIS AND
NITRIFICATION) SHOULD BE SUPPLEMENTED BY AERATION, PARTIAL WATER
CHANGES AND FILTRATION IF YOUR SOLAR FISH FARM IS TO PRODUCE MORE
THAN TWO OR THREE FISH DINNERS EACH MONTH. AERATION CAN BE ACCOM-
PLISHED IN A NUMBER OF WAYS, DEPENDING ON SCALE: WITH AQUARIUM
PUMPS AND AIR STONES, WITH DIAPHRAGM AIR COMPRESSORS, WITH MECHANICAL
AGITATORS OR WITH SUPPLEMENTARY AERATION BY WIND POWER. WATER
CHANGES MUST BE LIMITED SINCE THE WATER SERVES AS A HEAT STORAGE
MEDIUM AND RAPID TEMPERATURE CHANGES MAY STRESS THE FISH. MOST
EXPERIMENTERS ADVISE PARTIAL WATER CHANGES OVER SEVERAL DAYS, OF
BETWEEN 10 AND 25% OF THE TOTAL WATER VOLUME EACH WEEK.

PARTICULATE REMOVAL IS BEST ACCOMPLISHED BY SIPHONING, SETTLING IN
A SEDIMENTATION TANK OR MECHANICAL FILTRATION. PURIFICATION BY
PLANTS (PHYTOPLANKTON, DUCKWEED AND WATER HYACINTH IN PARTICULAR)
WORKS WELL IN SYSTEMS WITH LOW STOCKING DENSITIES AND EMPHASIS ON
HIGH SEASONAL PRODUCTIVITY IN THE SUMMER. BIOLOGICAL FILTRATION
SYSTEMS CONSIST OF A MEDIUM ON WHICH BACTERIA COLONIZE AND "EAT"
ORGANIC COMPOUNDS DISSOLVED IN THE FISH TANK WATER. TRICKLE FIL-
TERS OF THIS TYPE, BUILT OF PALLETS AND FILLED WITH OYSTER SHELL
OR CRUSHED LIMESTONE, SIMULTANEOUSLY AERATE, PURIFY AND BUFFER THE
WATER. WHEN USED WITH A SEDIMENTATION TANK AND A SMALL ELECTRIC
PUMP, A TRICKLING FILTER SYSTEM CAN SUPPORT A 1,000-GALLON TANK
PRODUCING 15 TO 20 POUNDS OF FISH PER MONTH (SEE APPENDIX C.)

ONCE THE GREENHOUSE IS CONSTRUCTED AND PROVISIONS ARE MADE TO
ASSURE ADEQUATE WATER QUALITY, THE URBAN FISH FARMER FACES A
NUMBER OF ADDITIONAL CHOICES: WHAT TYPES OF FISH TO PUT IN THE
SOLAR GREENHOUSE WATER, HOW TO GET THE FISH, HOW TO TRANSPORT AND
STOCK THEM ONCE THEY ARE ACQUIRED, AND HOW TO BEST MANAGE THE FISH
FOR MAXIMUM PRODUCTION, MINIMAL DISEASE AND MORTALITY, AND CONVEN-
IENCE. ACCORDING TO HEAD AND SPLANE:

"FISH THAT ARE PUT IN THE GREENHOUSE WATER MUST
NOT ONLY BE SUITED TO THE CONDITIONS OF TEMPER-
ATURE AND WATER QUALITY, THEY SHOULD ALSO NOT COST
AN ARM AND A LEG. WHEN SELECTING FISH CONSIDER
WHETHER YOU WILL BE ABLE TO BREED THEM. IF YOU
CAN'T MAINTAIN YOUR OWN STOCK IT MIGHT BECOME TOO
EXPENSIVE TO BUY NEW STOCK EVERY YEAR. SOME SPECIES
OF FISH CAN BE EASILY CAUGHT FROM LOCAL PONDS AND
STREAMS AND BREEDING THESE MAY NOT BE NECESSARY
IF YOU ARE WILLING TO SPEND A LITTLE TIME EACH
SEASON CATCHING YOUNG FISH FOR THE GREENHOUSE.

WHATEVER FISH YOU CHOOSE, WE WOULD RECOMMEND RAISING MORE THAN ONE KIND OF FISH TOGETHER. THIS IS CALLED POLYCULTURE AND IF SEVERAL SPECIES, WITH DIFFERENT FEEDING HABITS, ARE MIXED TOGETHER, IT WILL USUALLY RESULT IN MORE POUNDS OF FISH THAN IF JUST ONE SPECIES IS STOCKED. THIS IS BECAUSE THE DIFFERENT FEEDING HABITS OF THE SEVERAL SPECIES STOCKED RESULT IN BETTER UTILIZATION OF FEED.

FOR EXAMPLE, IF A FISH THAT FEEDS PRIMARILY NEAR THE WATER SURFACE IS STOCKED BY ITSELF, THAT PORTION OF THE FOOD THAT IS NOT EATEN AS IT SINKS TO THE BOTTOM IS WASTED AND CAUSES FURTHER WATER QUALITY PROBLEMS AS IT ROTS. INCLUDING A BOTTOM FEEDER WITH THE SURFACE FEEDER RESULTS IN MORE OF THE FEED BEING BEING CONSUMED AND MORE WEIGHT OF FISH BEING PRODUCED. AN ADDITIONAL GAIN IS REALIZED BECAUSE OF THE IMPROVED WATER QUALITY. FISH IN GOOD QUALITY WATER PRODUCE MORE WEIGHT GAIN PER UNIT OF FEED THAN THOSE HIGHLY STRESSED BY BEING KEPT IN POOR WATER.

A CARP OR CATFISH SPECIES MIXED WITH TILAPIA HAS PROVEN TO BE A PRODUCTIVE POLYCULTURE SYSTEM. THE TILAPIA WILL FILTER SMALL PARTICLES OF DISINTEGRATED FOOD AND ALGAE FROM THE WATER, WHILE THE CARP OR CATFISH WILL FEED ON MATERIAL THAT ACCUMULATES ON THE BOTTOM.

ANOTHER SUCCESSFUL POLYCULTURE MIX HAS BEEN THE CARP, TILAPIA AND GOURAMI."

A SAMPLE DESCRIPTION OF THE TILAPIA SPECIES TAKEN FROM THE AMITY HANDBOOK PROVIDES SOME BASIC INFORMATION ON THE TYPE OF FISH WHICH ARE WELL-SUITED FOR SOLAR GREENHOUSE GROWING. ADDITIONAL INFORMATION IS AVAILABLE IN THE AQUACULTURE REFERENCES ATTACHED TO THE ARTICLE.

TILAPIA

JAVA TILAPIA (TILAPIA MOSSAMBICA
TILAPIA ZILLII)
BLUE TILAPIA (TILAPIA AUREA)
NILE TILAPIA (TILAPIA NILOTICA)
TEMPERATURE: BEST WATER TEMP. FOR FEEDING AND GROWTH IS BETWEEN 82-86 DEGREES F., BUT MOST SPECIES WILL GROW IN TEMPERATURES FROM 64-90 DEGREES F. WILL NOT USUALLY SURVIVE IN WATER BELOW 50 DEGREES F.
OXYGEN: CAN SURVIVE LOW OXYGEN LEVELS, BUT DO BEST WHEN OXYGEN IS AT LEAST THREE PPM.
AMMONIA: KEEP NH_3 BELOW 0.15 PPM FOR GOOD GROWTH.
PH: GROW WELL BETWEEN 6.5-8.5 pH.

ORIGINALLY FROM AFRICA, TILAPIA HAVE BECOME AN IMPORTANT SOURCE OF FOOD FOR MANY CULTURES THROUGHOUT THE WORLD. OVER 14 SPECIES ARE RAISED IN BOTH FRESH AND SALT WATER. TILAPIA ARE ROBUST FISH. THEY ARE RESISTANT TO DISEASE, TOLERANT OF LOW OXYGEN AND HIGH AMMONIA LEVELS, GROW FAST, AND TASTE GOOD.

TILIPIA ARE HERBIVOROUS, OR PLANT EATING, FISH. SOME ARE ADAPTED TO FEED ON LARGE PLANTS AND OTHERS ON TINY ALGAE. BY FEEDING LOW ON THE FOOD CHAIN, TILAPIA CAN BE MUCH MORE PRODUCTIVE THAN MEAT-EATING FISH SUCH AS BASS OR TROUT. IT COSTS A LOT LESS TO FEED AN ANIMAL A PLANT PROTEIN DIET THAN A MEAT PROTEIN DIET.

JAVA TILAPIA (TILAPIA MOSSAMBICA)

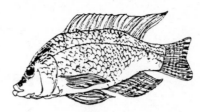

TILAPIA NOT ONLY GROW FAST, THEY REPRODUCE FAST. THEY BECOME SEXUALLY MATURE AT A SMALL SIZE AND REPRODUCE EVERY MONTH. TOO MANY FISH IN A LIMITED AMOUNT OF WATER SPACE CAUSES OVER-CROWDING. THIS RESULTS IN COMPETITION AMONG THE FISH, NOT ONLY FOR SPACE BUT ALSO FOR FOOD WHICH IN TURN RESULTS IN SMALLER, STUNTED FISH. UNLESS THE FRY ARE REMOVED, YOU COULD END UP WITH SEVERAL THOUSAND ONE-OUNCE FISH INSTEAD OF 300 ONE-POUND FISH. TO SOLVE THIS PROBLEM, A PREDATOR FISH CAN BE RAISED IN THE SAME WATER TO EAT THE TILAPIA FRY. OR, DIFFERENT SPECIES OF TILAPIA CAN BE CROSSED TO PRODUCE ALL MALE OFFSPRING. THIS IS CALLED HYBRIDIZATION. MALES ARE DESIRABLE BECAUSE THEY GROW FASTER AND LARGER THAN THE FEMALES.

FISH TO BE USED IN SOLAR AQUACULTURE CAN BE CAUGHT FROM LOCAL PONDS AND STREAMS OR BOUGHT FROM TROPICAL FISH STORES AND LICENSED DEALERS. BUYING AND TRANSPORTING FISH IS REGULATED IN MOST STATES BY THE APPROPRIATE FISH AND WILDLIFE MANAGEMENT AGENCY. THE URBAN FISH FARMER SHOULD BECOME KNOWLEDGEABLE OF THE REGULATIONS BY CONTACTING THIS AGENCY. SOMETIMES THE AGENCY CAN PROVIDE LISTS OF LICENSED FISH DEALERS AS WELL AS INFORMATION ON OUT-OF-STATE FISH PURCHASE AND TRANSPORTATION.

TRANSPORTING, STOCKING AND MANAGING FISH IN A SOLAR AQUACULTURE SYSTEM REQUIRES A LITTLE BIT OF HOMEWORK. FOR EXAMPLE, FISH CAN BE EASILY STRESSED IF THE MODE OF TRANSPORTATION TAKES TOO LONG, INVOLVES CONTAINERS WHICH ARE TOO SMALL FOR THE TYPE OR NUMBER OF FISH, OR IS AFFECTED BY CHANGES IN WATER TEMPERATURE OR AIR SUPPLY. EXCESSIVE HANDLING OR EXPOSURE TO ANY UNUSUAL CHANGE CAN ALSO AFFECT THE FISH. WHEN STOCKING, IT IS IMPORTANT TO GRADUALLY ADJUST WATER TEMPERATURE AND TO ASSURE THAT THE TANKS OR CONTAINERS ARE PREPARED TO PREVENT AMMONIA LEVEL BUILD-UP. THE FISH SHOULD BE DISINFECTED AS THEY ARE STOCKED.

THESE SUBJECTS, ALONG WITH FISH FEEDING (AMOUNT AND FREQUENCY), USE OF HOMEGROWN DIETS, PREVENTION OF DISEASE AND HARVEST ARE DISCUSSED IN MORE DETAIL IN THE AMITY FOUNDATION FISH FARMING HANDBOOK. THIS PUBLICATION SHOULD BE REQUIRED READING FOR ANYONE WHO IS SERIOUSLY CONSIDERING A SOLAR GREENHOUSE/AQUACULTURE PROJECT. COPIES OF THIS PUBLICATION AND INFORMATION ON THE FISH HOUSE, INCLUDING TOURS AND WORKSHOPS, ARE AVAILABLE FROM AMITY FOUNDATION, P.O. BOX 7066, EUGENE, OREGON 97401, (503) 484-7171.

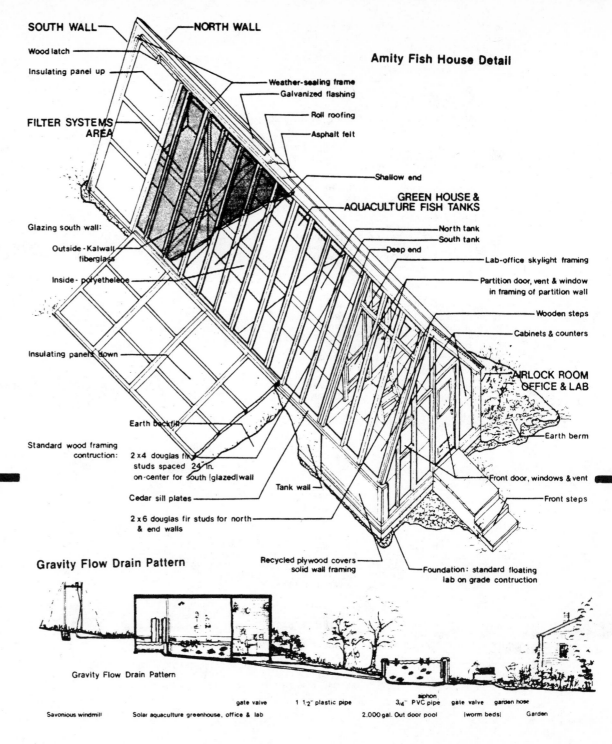

Amity Fish House Detail

SOUTH WALL — NORTH WALL

Wood latch

Insulating panel up

Weather-sealing frame
Galvanized flashing
Roll roofing
Asphalt felt

FILTER SYSTEMS AREA

Shallow end

GREEN HOUSE & AQUACULTURE FISH TANKS

North tank
South tank
Deep end

Glazing south wall:

Outside - Kalwall fiberglass

Inside - polyethelene

Lab-office skylight framing

Partition door, vent & window in framing of partition wall

Wooden steps

Cabinets & counters

Insulating panels down

AIRLOCK ROOM
OFFICE & LAB

Earth berm

Earth backfill

Standard wood framing contruction:

2 x 4 douglas fir studs spaced 24 in. on-center for south (glazed) wall

Front door, windows & vent

Cedar sill plates

Tank wall

Front steps

2 x 6 douglas fir studs for north & end walls

Recycled plywood covers solid wall framing

Foundation: standard floating lab on grade contruction

Gravity Flow Drain Pattern

Gravity Flow Drain Pattern

Savonious windmill   Solar aquaculture greenhouse, office & lab

gate valve   1 1/2" plastic pipe

siphon

3/4" PVC pipe   gate valve   garden hose

2,000 gal. Out door pool   (worm beds)   Garden

SOME RESOURCES

ECOTOPE GROUP. 1979. A SOLAR GREENHOUSE GUIDE FOR THE PACIFIC NORTHWEST. LYNNWOOD, WASHINGTON. AD PRO PRINTING.

FARALLONES INSTITUTE. 1979. THE INTEGRAL URBAN HOUSE: SELF RELIANT LIVING IN THE CITY. SAN FRANCISCO. SIERRA CLUB BOOKS.

HEAD, WILLIAM AND JON SPLANE. 1979. FISH FARMING IN YOUR SOLAR GREENHOUSE. EUGENE OREGON. NORTHWEST WORKING PRESS.

LANE COUNTY. OFFICE OF APPROPRIATE TECHNOLOGY. UNDATED. SOLAR ENERGY FOR LANE COUNTY. EUGENE, OREGON.

ENCOURAGING SOLAR GREENHOUSES FOR WHITEAKER NEIGHBORHOOD

— LARRY PARKER, PROJECT SELF-RELIANCE

ALMOST ANY WAY YOU LOOK AT IT, A SOLAR GREENHOUSE MAKES A LOT OF SENSE. IT CAN PROVIDE BASIC NECESSITIES AND LUXURY AT THE SAME TIME. IT CAN SAVE MONEY ON FOOD AND FUEL, AND IT CAN BE A SOURCE OF INCOME FOR A FAMILY OR A COMMUNITY. A PROPERLY-DESIGNED AND WELL-BUILT SOLAR GREENHOUSE IS A MULTI-PURPOSE STRUCTURE IN WHICH YOU CAN 1) EXTEND YOUR GROWING SEASON TO VIRTUALLY YEAR-ROUND, 2) PRODUCE HEAT FOR YOUR HOME, 3) PRE-HEAT WATER BEFORE IT GOES TO YOUR GAS OR ELECTRIC WATER HEATER, AND 4) DRY FRUITS AND VEGETABLES TO PRESERVE THEM.

A SOLAR GREENHOUSE CAN ALSO BE A DELIGHTFUL ADDITION TO YOUR PER-SONAL LIVING SPACE, WITH AN INVIGORATING ATMOSPHERE OF SUNLIGHT AND GROWING PLANTS. IT CAN PROVIDE YOU WITH HEALTHY LEISURE ACTI-VITY, AND IN THE PROCESS OF MANAGING YOUR GREENHOUSE YOU WILL GROW IN APPRECIATION OF THE SEASONAL CYCLES, THE FREE ENERGY OF THE SUN, AND THE INTERDEPENDENCE OF NATURE AND HUMAN BEINGS. THE SUPERIOR NUTRITIONAL VALUE OF FOOD GROWN AT HOME WILL ALSO ADD TO YOUR QUALITY OF LIFE; AND THE POTENTIAL SAVING OF FOSSIL FUELS WILL AFFECT YOUR CHILDREN'S QUALITY OF LIFE IN THE FUTURE.

JUST ABOUT ANYONE CAN BUILD A SOLAR GREENHOUSE USING ORDINARY BUILDING MATERIALS AND TOOLS, AT A COST VARYING FROM UNDER $100 TO OVER $2000 (LABOR NOT INCLUDED). AT THE COMMUNITY SCALE, SEVERAL SOLAR GREENHOUSES COVERING UP TO 6000 SQUARE FEET HAVE BEEN BUILT IN VARIOUS PARTS OF THE COUNTRY, PROVIDING FOOD AND JOBS FOR THE PEOPLE WHO COOPERATE IN THE CONSTRUCTION AND MAINTEN-ANCE. IN THE WHITEAKER NEIGHBORHOOD, DEMONSTRATION SOLAR GREEN-HOUSES HAVE BEEN BUILT FREE-STANDING AND ATTACHED TO DWELLINGS, USING VARIOUS COMBINATIONS OF NEW AND RECYCLED MATERIALS.

DESIGN

WHATEVER THEIR SIZE AND CONSTRUCTION, ALL SOLAR GREENHOUSES HAVE THREE MAIN DESIGN CRITERIA: A LARGE SOUTH-FACING GLAZED AREA, A "THERMAL MASS" SUCH AS WATER CONTAINERS, MASONRY, OR ROCKS, TO ABSORB SOLAR HEAT, AND INSULATION IN THE UNGLAZED WALLS AND ROOF TO HOLD THE HEAT. THIS MEANS THAT UNDER NORMAL CONDITIONS THE ONLY HEAT SOURCE FOR THE GREENHOUSE IS SUNLIGHT. TO DECREASE HEAT LOSSES AT NIGHT, THE SOUTH GLAZING CAN BE COVERED WITH MOVABLE INSULATION, AS IN THE AMITY AND URBAN FARM EXAMPLES ABOVE.

THERE IS STILL MUCH TO LEARN ABOUT THE DESIGN AND OPERATION OF SOLAR GREENHOUSES IN THE WILLAMETTE VALLEY, AND THERE IS VERY LITTLE LOCAL EXPERIENCE WITH YEAR-ROUND GROWING IN A SOLAR GREEN-HOUSE. BUT THE BEST WAY TO LEARN IS TO BUILD A GREENHOUSE AND BEGIN USING IT.

FIRST, CHECK THE SOUTH SIDE OF YOUR HOUSE OR THE SUNNY PART OF YOUR YARD. ARE THERE ANY MAJOR OBSTRUCTIONS TO DIRECT SUNLIGHT, SUCH AS LARGE TREES OR BUILDINGS? IF YOUR "SOLAR ACCESS" IS GOOD DURING THE FALL, WINTER AND SPRING, YOU CAN BEGIN PLANNING YOUR GREENHOUSE AS AN ATTACHED OR FREE-STANDING MODEL, DEPENDING ON YOUR SITUATION.

A FREE-STANDING GREENHOUSE CAN HAVE AN ATTACHED COMPOST BIN WHICH VENTS HEAT AND BENEFICIAL CARBON DIOXIDE TO THE GREENHOUSE, AND USUALLY WILL NOT REQUIRE A BUILDING PERMIT. AN ATTACHED MODEL IS EASIER TO BUILD, AND EXCESS HEAT CAN FLOW DIRECTLY INTO YOUR HOUSE THROUGH WINDOWS, DOORS AND VENTS, AS IN THE URBAN FARM GREENHOUSE ABOVE. BE CAREFUL NOT TO COVER THE ONLY WINDOW TO A BEDROOM, AS THIS IS A VIOLATION OF FIRE SAFETY CODES. IF YOU ARE BUILDING A SUBSTANTIAL GREENHOUSE ADDITION, YOU WILL PROBABLY HAVE TO GET A BUILDING PERMIT. A SIMPLE, CLEAR DRAWING OF YOUR DESIGN WILL BE SUFFICIENT FOR THE CITY BUILDING DEPARTMENT, AND WILL HELP YOU TO CALCULATE YOUR MATERIAL NEEDS.

ZONING AND BUILDING CODES

GREENHOUSES ARE SUBJECT TO VARIOUS ZONING AND BUILDING CODE RE-QUIREMENTS, DEPENDING ON THEIR SIZE, TYPE, LOCATION AND CONSTRUC-TION. FOUR BASIC TYPES OF GREENHOUSES ARE ILLUSTRATED HERE FOR PURPOSES OF CODE DISCUSSION.

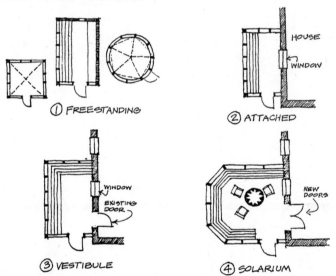

① FREESTANDING ② ATTACHED ③ VESTIBULE ④ SOLARIUM

GREENHOUSES MUST MEET THE SETBACK REQUIREMENTS OF THE EUGENE CODE WHICH REQUIRES A MINIMUM FRONT YARD OF 15 FEET (SEC. 9.538(6)(A)) AND A MINIMUM SIDE YARD OF 5 FEET (SEC. 9.542(1)). IN THE WINTER A GREENHOUSE MAY FALL INTO DISREPAIR AND CREATE A NUISANCE. THE EUGENE CODE (SEC. 8.365(6)) REQUIRES THE OWNER OR OCCUPANT TO MAINTAIN THE BUILDING AND YARD IN A NEAT CONDITION.

BUILDING PERMITS FOR TEMPORARY GREENHOUSES ARE GRANTED IF SETBACK REQUIREMENTS AND RESISTANCE TO WIND FORCES ARE SATISFIED. IF YOU ARE UNSURE AS TO WHETHER YOU NEED A BUILDING PERMIT, CALL OR VISIT THE BUILDING DIVISION AND DESCRIBE YOUR PROJECT.

IF YOU ARE IN FIRE ZONE 2 AND YOU WISH TO BUILD A "SOLARIUM" TYPE OF GREENHOUSE, CHECK WITH THE BUILDING DIVISION FOR SPECIAL RE-QUIREMENTS.

AN ATTACHED GREENHOUSE MAY BE VENTED TO THE INTERIOR OF THE HOUSE. THE GREENHOUSE MAY COVER AN EXISTING WINDOW IF LIGHT AND VENTILATION ARE PROVIDED TO THE ROOM BY ALTERNATE MEANS, AS REQUIRED BY THE UNIFORM BUILDING CODE (UBC), SEC. 1405. A WINDOW REQUIRED FOR EXIT FROM A SLEEPING ROOM (SEC. 1404) MAY NOT BE COVERED UNLESS ANOTHER APPROVED EXIT IS PROVIDED.

CONSTRUCTION

A LIGHTWEIGHT GREENHOUSE MAY COLLAPSE OR OVERTURN IF IT IS NOT STRUCTURALLY ADEQUATE TO RESIST THE LATERAL AND UPLIFT FORCES OF WIND. THE STRUCTURAL FRAME OF THE GREENHOUSE MUST BE RIGIDLY BRACED AND SECURED IN THE GROUND. UBC SEC. 2311(H) SPECIFIES DESIGN LOADS FOR WIND FORCES ON THE GREENHOUSE.

THE ILLUSTRATIONS SHOW TWO POSSIBLE GREENHOUSE FOUNDATIONS:

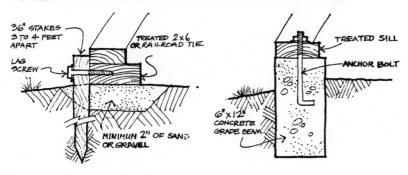

NOTE: SILLS SHOULD NOT BE TREATED WITH "PENTA", WHICH WILL HARM PLANTS. USE COPPER NAPTHENATE OR ACCEPTABLE SUBSTITUTE.

THE GREENHOUSE STRUCTURE MUST BE ABLE TO SUPPORT A LIVE (SNOW) LOAD OF 20 POUNDS PER SQUARE FOOT. IF THE GREENHOUSE IS ATTACHED TO AN EXTERIOR WALL, THE CONNECTION TO THE WALL MUST BE ADEQUATE FOR THE LOAD.

COVERING MATERIALS

FOR A PERMANENT GREENHOUSE, WINDOW GLASS OR AN APPROVED PLASTIC IS SUFFICIENT. GLASS USED IN A DOOR AND WITHIN 12" OF A DOOR MUST BE SAFETY GLAZING, AS REQUIRED BY UBC CHAPTER 54. POLYETHYLENE MAY BE USED AS AN INNER LAYER FOR DOUBLE GLAZING.

PLASTICS APPROVED BY I.C.B.O.:
- ALSYNITE/STRUCTOGLAS FIBERGLAS
- FIBERGLAS TYPE 8012 PANELS
- FILON
- HIGH RIB, BUTLERIB, CORRUGATED LITE
- LASCOLITE
- ORNYTE FIBERGLAS
- PLEXIGLAS ACRYLIC

PLASTICS MUST BE FASTENED AND SUPPORTED IN A MANNER ADEQUATE FOR DESIGN LOADS. FOR EXAMPLE, CORRUGATED FIBERGLASS SHOULD BE SUPPORTED BY STRUTS PERPENDICULAR TO CORRUGATIONS, AS SHOWN.

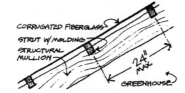

RESOURCES...

THE NUMBER OF SOLAR GREENHOUSE REFERENCE BOOKS IS CONSTANTLY INCREASING, AND MANY ARE AVAILABLE ON LOAN FROM THE WHITEAKER RESOURCE CENTER. YOUR EXACT DESIGN WILL BE DEPENDENT ON THE PLACEMENT OF THE GREENHOUSE AND THE MATERIALS YOU CHOOSE. IF YOU PLAN TO USE RECYCLED MATERIALS, BEGIN COLLECTING 2'x4' AND 2'x6' FRAMING LUMBER, SIDING BOARDS OR PLYWOOD, AND WATER CONTAINERS SUCH AS 5-GALLON CANS OR 50-GALLON OIL DRUMS. OLD WINDOWS MAY BE USED FOR GLAZING, AND RAILROAD TIES MAKE A GOOD FOUNDATION FOR A SMALL GREENHOUSE. LOOK FOR SALES ON INSULATION, AND CONSIDER ALSO ADDING INSULATION IN YOUR ATTIC AND UNDER YOUR FLOOR IF IT IS INSUFFICIENT NOW, SINCE ANY SOLAR HEAT GAIN IS EFFECTIVELY WASTED IF YOUR HOUSE LETS IT LEAK AWAY TOO FAST.

IF YOU RENT YOUR HOME, YOU CAN DESIGN YOUR GREENHOUSE TO BE PORTABLE OR DETACHABLE. AN ALTERNATIVE IS TO JOIN WITH NEIGHBORS IN A COMMUNITY GREENHOUSE PROJECT, WITH LABOR, COSTS, AND BENEFITS SHARED BY ALL PARTICIPANTS. IF YOU OWN YOUR HOME, YOU CAN CLAIM A STATE INCOME TAX CREDIT AMOUNTING TO 25% OF THE CONSTRUCTION COST, UP TO $1000, IF YOUR GREENHOUSE PROVIDES AT LEAST 10% OF YOUR HOME'S TOTAL ENERGY REQUIREMENTS. TO QUALIFY, YOU MUST APPLY FOR THE TAX CREDIT BEFORE YOUR GREENHOUSE IS INSTALLED.

FOR STATE TAX CREDIT INFORMATION CALL: 1-800-452-7813 (DEPT. OF ENERGY), OR 1-800-452-2838 (DEPT. OF REVENUE).

LOCAL DEMONSTRATION GREENHOUSES (SEE NEXT PAGES)

1. 1420 BRIARCLIFF LANE--AN ATTACHED SOLAR GREENHOUSE, 10'x18', ON THE HOME OF RON McMULLIN AND ELAINE PERKO, BUILT BY VOLUNTEERS WITH FUNDS FROM WHITEAKER PROJECT SELF-RELIANCE. CALL 343-2711 OR 688-1596 FOR MORE INFORMATION.

2. 1309 W. 4TH--A FREE-STANDING SOLAR GREENHOUSE, 9'x12', BUILT WITH A $200 GRANT TO NANCY COSPER. CALL 343-2711 OR 687-0295.

3. ROYAL AVENUE AT BERNTZEN--A FREE-STANDING SOLAR GREENHOUSE, 8'x12', AT PETERSEN PARK. BUILT BY VOLUNTEERS USING ALL RECYCLED OR BARTERED MATERIALS. CALL 343-2711 OR 689-1446 FOR MORE INFORMATION.

4. 301 N. ADAMS, RIVER HOUSE--A FREE-STANDING SOLAR GREENHOUSE CONVERTED FROM A 9'x12' SHED AT THE EUGENE PARKS AND RECREATION COMMUNITY GARDENS.

attached solar greenhouse

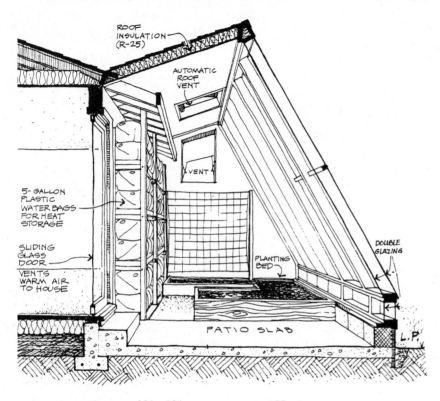

ROOF INSULATION (R-25)

AUTOMATIC ROOF VENT

VENT

5-GALLON PLASTIC WATER BAGS FOR HEAT STORAGE

SLIDING GLASS DOOR

VENTS WARM AIR TO HOUSE

PLANTING BED

DOUBLE GLAZING

PATIO SLAB

L.P.

OUTSIDE DIMENSIONS: 10' x 18'; FLOOR AREA: 175 SQ. FEET

GLAZING AREA: SOUTH 175 SQ. FEET; WEST 15 SQ. FEET

GLAZING MATERIALS: 3/16" GLASS W/ 8 MIL VINYL; FILON W/ 4 MIL POLY.

GROWING SPACE: 50 SQ. FT. BED, SHELVES AND HANGERS

HEAT STORAGE: 360 GAL. H_2O IN 5-GAL. PLASTIC CONTAINERS;
40 GAL. IN 5-GAL. CANS

COST: APPROX. $900 (LABOR DONATED)

THE FIRST DEMONSTRATION SOLAR GREENHOUSE IN WHITEAKER WAS BEGUN ON
"SUN DAY", MAY 19, 1979 IN A PUBLIC WORKSHOP SPONSORED BY PROJECT
SELF-RELIANCE. ALL FUNDS CAME FROM COMMUNITY DEVELOPMENT BLOCK
GRANTS (U.S. DPT. HUD) THROUGH THE CITY OF EUGENE. DESIGN WAS BY
TOM LAVELLE, RON McMULLIN (THE RESIDENT) AND LARRY PARKER. THE
LARGE SOUTH GLAZING IS DESIGNED TO MAXIMIZE SOLAR GAIN IN THE HEAT-
ING SEASON, AND THE PATIO DOORS LET WARMED AIR PASS INTO THE HOUSE.
THE GREENHOUSE WAS BUILT OVER AN EXISTING PATIO, AND CLEAR GLAZING
WAS USED OVER THAT PORTION TO RETAIN THE VIEW OF SPENCER'S BUTTE.
BOTH NEW AND RECYCLED MATERIALS WERE USED; ROCK WOOL INSULATION FOR
THE ROOF CAME FROM AN OLD DAIRY BUILDING IN DOWNTOWN EUGENE, AND
THE SCREW-CAP PLASTIC "BAGS" FOR WATER STORAGE CAME FROM SACRED
HEART HOSPITAL, WHERE THEY ARE NORMALLY DISCARDED BY THE THOUSANDS.
THIS GREENHOUSE WILL BE USED BY NEIGHBORS WITH PLOTS IN THE NORTH
POLK COMMUNITY GARDEN.

portable solar greenhouse

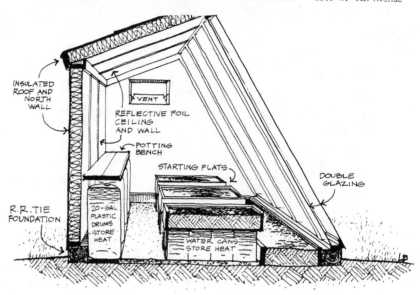

INSULATED ROOF AND NORTH WALL

VENT

REFLECTIVE FOIL CEILING AND WALL

POTTING BENCH

STARTING FLATS

DOUBLE GLAZING

R.R. TIE FOUNDATION

20-GAL. PLASTIC DRUMS STORE HEAT

WATER CANS STORE HEAT

L.P.

OUTSIDE DIMENSIONS: 9'4" x 12'; FLOOR AREA: 100 SQ. FEET

GLAZING AREA: SOUTH 100 SQ. FEET; WEST 4 SQ. FEET

GLAZING MATERIAL: FILON AND 4 MIL POLYETHYLENE

GROWING SPACE: RAISED BED, LARGE STARTING FLATS, SHELF AND HANGERS

HEAT STORAGE: 300 GAL. H_2O IN 20-GAL. PLASTIC CONTAINERS

COST: APPROX. $250 (LABOR DONATED)

THIS SMALL SOLAR GREENHOUSE WAS BUILT AS A DEMONSTRATION FOR THE
1979 HARVEST WHIT-A-FAIRE, AND NOW RESTS AT THE RENTED HOME OF
NANCY COSPER AND FAMILY, CLEARLY VISIBLE FROM POLK STREET. A $200
GRANT FROM PROJECT SELF-RELIANCE HELPED TO BUY MATERIALS. RECYCLED
MATERIALS USED INCLUDED RAILROAD TIES FOR THE FOUNDATION, 2'x4'
LUMBER, PLYWOOD, INSULATION, A DOOR, AND WATER STORAGE CONTAINERS.
THE GREENHOUSE WAS DESIGNED TO BE PORTABLE; THE FIVE MAJOR PANELS
WERE PRE-FABRICATED AND BOLTED TOGETHER. THE LARGE GLAZED (SOUTH)
PANEL PROVED TO BE TOO LARGE TO LEGALLY TRANSPORT, AND SHOULD BE
MADE IN TWO SECTIONS. PLANS FOR THIS GREENHOUSE ARE AVAILABLE FROM
PROJECT SELF-RELIANCE.

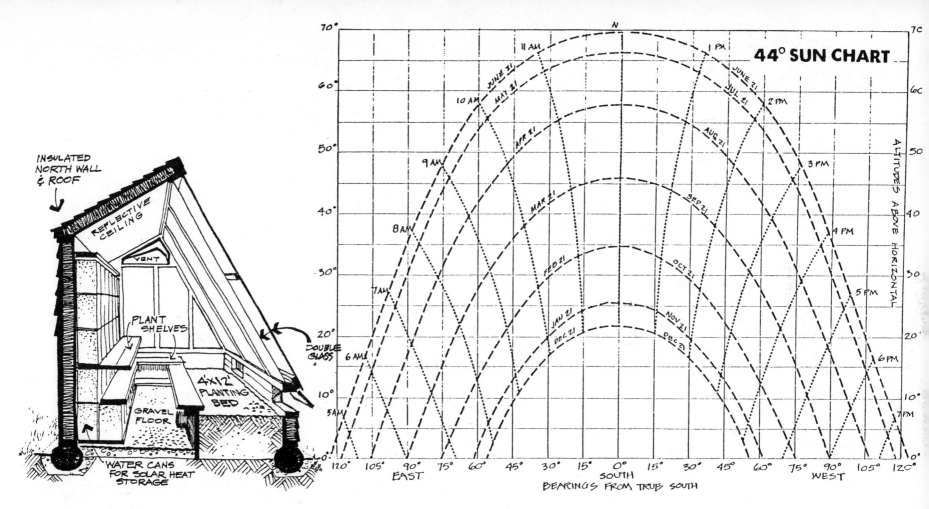

44° SUN CHART

ALTITUDES ABOVE HORIZONTAL

BEARINGS FROM TRUE SOUTH

EAST — SOUTH — WEST

INSULATED NORTH WALL & ROOF

REFLECTIVE CEILING

VENT

PLANT SHELVES

4'x12' PLANTING BED

GRAVEL FLOOR

DOUBLE GLASS

WATER CANS FOR SOLAR HEAT STORAGE

OUTSIDE DIMENSIONS: 8'x 12'; FLOOR AREA: 90 SQ. FEET
GLAZING AREA: SOUTH WALL 90 SQ. FT; EAST WALL 25 SQ. FT.
GROWING SPACE: 43 SQ. FT. BED, SHELVES AND HANGERS
HEAT STORAGE: 350 GALLONS H2O, IN 5-GALLON CANS
COST: APPROX. $25

THE PETERSEN PARK SOLAR GREENHOUSE WAS BEGUN IN THE SUMMER OF 1978 AS PART OF THE BETHEL ENERGY EFFICIENCY PROGRAM (BEEP). THE LOW COST OF THE STRUCTURE RESULTED FROM THE USE OF RECYCLED MATERIALS AND DONATED LABOR. ABOUT 20 PEOPLE PARTICIPATED IN THE CONSTRUCTION, WORKING ON SUNDAY AFTERNOONS. ALL LUMBER, INSULATION (R-11 FIBERGLASS) AND ROOFING WERE SALVAGED FROM BUILDINGS IN EUGENE AND SPRINGFIELD. THE FIBERGLASS GLAZING (TEDLAR-COATED FILON) WAS OBTAINED VIA A TRADE WITH PROJECT SELF-RELIANCE FOR WINDOW GLASS WHICH WAS A LABOR EXCHANGE WITH WESTSIDE AUTO GLASS. THE 5-GALLON CANS WERE PICKED UP AT VARIOUS NATURAL FOOD STORES IN EUGENE. TELEPHONE POLES USED IN THE FOUNDATION WERE FOUND ON THE SITE.

DESIGN AND CONSTRUCTION WERE COORDINATED BY REBA WEST AND LARRY PARKER. THE GREENHOUSE IS A GIFT TO THE BETHEL NEIGHBORHOOD, AND WILL BE ACCESSIBLE TO RESIDENTS, SCHOOL GROUPS, AND OTHER INTERESTED FOLKS. USE OF THE GREENHOUSE WILL BE COORDINATED BY THE PETERSEN BARN COMMUNITY CENTER.

CHICKENS

CHICKENS ARE INTEGRAL TO THE URBAN FARM ECOSYSTEM. THEY PROVIDE
EGGS AND MEAT, WHILE PRODUCING QUALITY MANURE AND CLEANING UP GARDEN
AND KITCHEN SCRAPS. THEY ARE RELATIVELY EASY TO RAISE AND CARE FOR
AND REQUIRE ONLY A LITTLE SPACE.

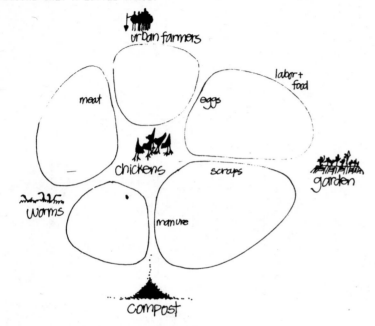

chicken ecosystem

BEFORE ONE BEGINS BUYING BABY CHICKS AND BUILDING HUTCHES, SOME OF
THE LEGAL TECHNICALITIES SHOULD BE EXAMINED. THE FOLLOWING IS A
SUMMARY OF LOCAL ORDINANCES THAT AFFECT THE RAISING OF CHICKENS.

LANE COUNTY (OUTSIDE OF INCORPORATED CITIES). CHICKENS MAY BE
RAISED IN ANY AREAS ZONED RURAL OR AGRICULTURAL, OR RA (SUBURBAN).
THEY ARE NOT ALLOWED IN R-1 OR ANY COMMERCIAL OR INDUSTRIAL DIS-
TRICTS. YOU MAY HAVE A MAXIMUM OF ONE CHICKEN PER 500 SQUARE FEET
OF YOUR LOT, PLUS UP TO 3 YOUNG (UNDER SIX MONTHS) PER FULLY-GROWN
CHICKEN. THEY MUST BE IN A FENCED YARD, WITH THEIR HOUSE ON THE
REAR HALF OF THE LOT, NO LESS THAN 70 FEET FROM THE FRONT PROPERTY
LINE, AND NO LESS THAN 50 FEET FROM A RESIDENCE.

SPRINGFIELD: BASICALLY THE SAME RULES AS IN THE COUNTY, WITH THE
ADDITIONAL RULE THAT NO CHICKENS MAY BE KEPT ON A LOT LESS THAN
20,000 SQUARE FEET IN AREA. THEY ARE ALLOWED ONLY IN AREAS ZONED
RA OR AG.

EUGENE: POULTRY MAY BE RAISED IN ANY AREAS ZONED RA OR AG (A VERY
SMALL PORTION OF THE CITY). NO RESTRICTIONS ARE PUT ON LOT SIZE,
NUMBER OF BIRDS, OR LOCATION OF COOPS.

THE CHIEF OBJECTION TO CHICKENS IN THE CITY ARE: NOISE, SANITATION,
AND ODOR. ALL OF THESE PROBLEMS CAN BE CORRECTED IF SOME SIMPLE
RULES ARE FOLLOWED:

1. ROOSTERS ARE OFTEN SPECTACULAR TO WATCH AND THEIR CROW IS REMI-
   NISCENT OF EARLY MORNINGS ON THE HOMESTEAD. HOWEVER, MOST
   ROOSTERS ARE SOMEWHAT CARELESS ABOUT THEIR TIMING AND CAN WAKE
   EVERY ADJACENT NEIGHBOR AT LATE HOURS SIMPLY PRACTICING FOR
   DAYBREAK. ALTHOUGH CLAIMS HAVE BEEN MADE ABOUT THE INCREASED
   NUTRITION OF FERTILIZED EGGS, UNFERTILIZED HOME-GROWN EGGS ARE
   PROBABLY JUST AS HEALTHY AND CERTAINLY EQUALLY TASTY. YOU
   SHOULD NOT KEEP A ROOSTER IN YOUR FLOCK, TO ELIMINATE IRATE
   NEIGHBORS.

2. SANITATION IS ANOTHER COMPLAINT OFTEN VOICED CONCERNING CHICK-
   ENS. RATS WILL EAT CHICKEN FOOD AND FLIES BREED IN THEIR
   BEDDING. THE RATS CAN BE ELIMINATED BY USING TRAPS AND KEEP-
   ING ALL EXCESS FOOD CAREFULLY SEALED IN GARBAGE CANS. THE
   FLY POPULATION CAN BE REDUCED BY PRACTICING CONSISTENT MAINTEN-
   ANCE OF THE HUTCH AND PEN. EACH WEEK, THE HUTCH AND NESTING
   PENS SHOULD BE CLEANED, AND THE STRAW REPLACED. THE OUTSIDE
   PEN AREA SHOULD BE CONTINUOUSLY MULCHED WITH GRASS, STRAW, AND/OR
   SAWDUST. MAGGOTS THAT DEVELOP WILL QUICKLY BE EATEN BY THE
   CHICKENS. AT LEAST 8-10 INCHES OF LITTER SHOULD BE MAINTAINED
   AT ALL TIMES. CHICKENS RAISED ON THE GROUND WITH THIS DEEP
   LITTER ALSO TEND TO BE HEALTHIER.

3. ODOR IS SELDOM A PROBLEM IF THE HUTCH IS CLEANED REGULARLY AND
   THE PEN MAINTAINED IN A DEEP LITTER SYSTEM. THE LITTER CAN BE
   REMOVED EVERY THREE MONTHS AND COMPOSTED FOR THE GARDEN.

HOW TO START

THERE ARE HUNDREDS OF CHICKEN VARIETIES AVAILABLE. SOME ARE BRED
EXCLUSIVELY FOR EGG PRODUCTION, OTHERS FOR MEAT. THE MOST UTILI-
TARIAN BREEDS ARE THOSE THAT PRODUCE A LARGE QUANTITY OF EGGS AND
ARE LARGE ENOUGH TO PROVIDE A MEAL. RHODE ISLAND REDS, NEW HAMP-
SHIRE REDS, SEX-LINKS AND BARRED ROCKS ARE EXCELLENT BIRDS FOR AN
URBAN FARM. THEY LAY 250+ EGGS PER YEAR, AND WHEN THEIR TIME
COMES, THEY MAKE A FINE STEW. THESE BIRDS ARE ALSO VERY GENTLE,
UNLIKE SOME BREEDS.

CHICKENS ARE RANKED ACCORDING TO AGE. YOUNG CHICKENS FROM DAY-OLD
TO ABOUT 3 MONTHS ARE CALLED CHICKS. PULLETS ARE CHICKENS FROM
ABOUT 2-6 MONTHS. HENS ARE OLDER, LAYING CHICKENS. YOU CAN BUY
DAY-OLD CHICKS OR PULLETS AT LOCAL FEED AND SEED STORES DURING THE
SPRING AND SUMMER. DAY-OLD CHICKS ARE THE LEAST EXPENSIVE, RANGING
IN PRICE FROM .60 TO $1.00 APIECE. SEX-LINK CHICKS ARE PURCHASED
AS HENS OR ROOSTERS. THE OTHERS ARE SOLD MIXED, USUALLY ABOUT
20-40% BEING ROOSTERS, WHICH ARE BUTCHERED AT 10-12 WEEKS FOR MEAT.

FEED

CHICKENS WILL EAT ALMOST ANYTHING. KITCHEN SCRAPS, GARDEN GREENS,
AND COMMERCIAL FEED WILL KEEP THEM HEALTHY AND PRODUCING EGGS.
WHEN THEY ARE YOUNG, A SPECIAL CHICK FEED IS BEST, BUT AFTER TEN
WEEKS OR SO, AN ALL-PURPOSE CHICKEN FEED IS BEST. AN OCCASIONAL
HELPING OF CRUSHED OYSTER SHELL OR CRUSHED EGG SHELL WILL HELP
KEEP CALCIUM LEVEL HIGH FOR GOOD EGG QUALITY. BE CERTAIN NONE OF
THE SCRAPS YOU FEED THEM IS ROTTED, AS CHICKENS CAN DIE FROM FOOD
POISONING. CLEAN WATER IS BEST SUPPLIED BY USING COMMERCIAL
WATERING DEVICES.

THE YOUNG CHICKS MUST BE KEPT WARM AND CONTINUOUSLY SUPPLIED WITH
FRESH UNCONTAMINATED WATER. COMMERCIAL BROODERS ARE AVAILABLE,
BUT A HOME-BUILT MODEL IS EASY TO CONSTRUCT. A SMALL CROCK OF
WATER AND FOOD, CHANGED AND CLEANED DAILY WILL BE SUFFICIENT FOR
7-15 CHICKS. THE CHICKS ARE KEPT IN THE BROODER FOR ABOUT
WEEKS AND THEN SET INTO THE PEN. THE AVERAGE HEN WILL BEGIN LAY-
ING EGGS AT ABOUT SIX MONTHS.

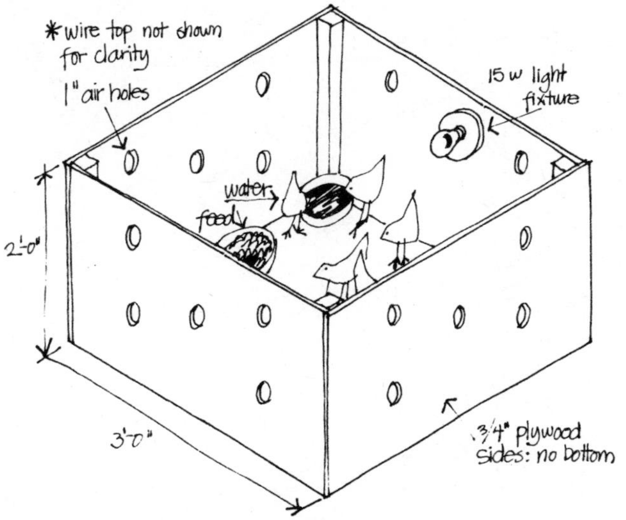

※ wire top not shown
for clarity
1" air holes

15 w light
fixture

water
food

2'-0"

3'-0"

3/4" plywood
sides: no bottom

chicken brooder
holds up to 20 chickens

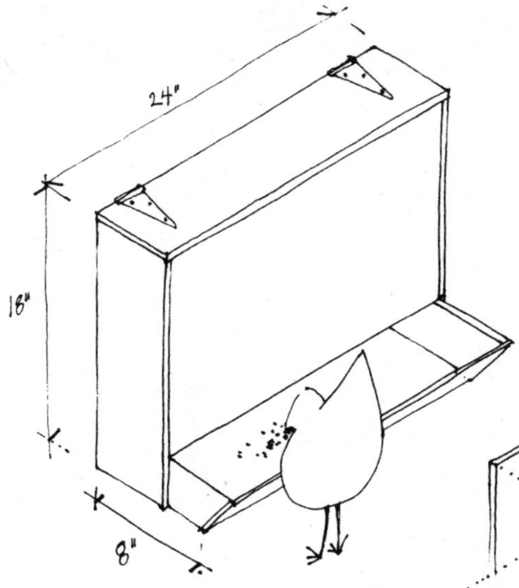

24"

18"

8"

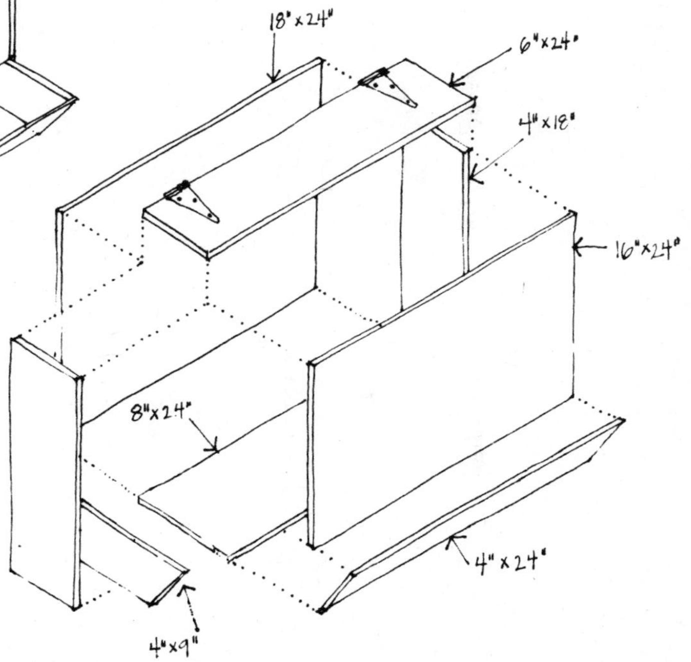

commercial waterer
set on bricks

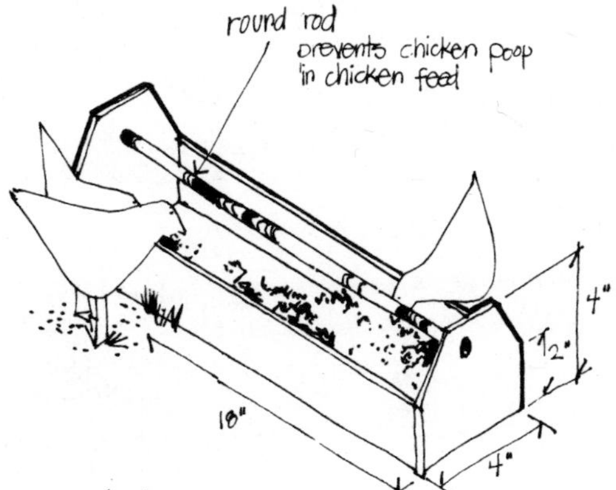

round rod
prevents chicken poop
in chicken feed

18"

4"

4"

2"

chicken feeder

chicken feeder
· constructed of 1×4 stock and 3/4" plywood
· use wood screws — 1¼" #8

18"×24"

6"×24"

4"×18"

16"×24"

8"×24"

4"×24"

4"×9"

THE CHICKEN PEN AND HUTCH CAN BE BUILT IN A VARIETY OF STYLES.
THE BASIC REQUIREMENTS ARE: 2.5 SQUARE FEET PER CHICKEN, A ROOSTING
BAR FOR SLEEPING, A COVERED HUTCH FOR SECURITY AND WEATHER PROTEC-
TION, AND NEST BOXES FOR EGG-LAYING. IN URBAN AREAS, DOGS ENJOY
EATING OR CHASING CHICKENS, SO ADEQUATE PROTECTION IS NECESSARY.

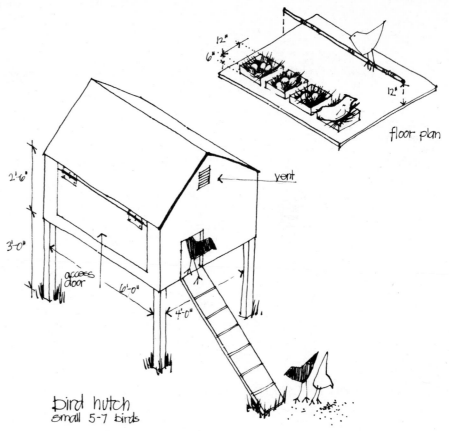

floor plan

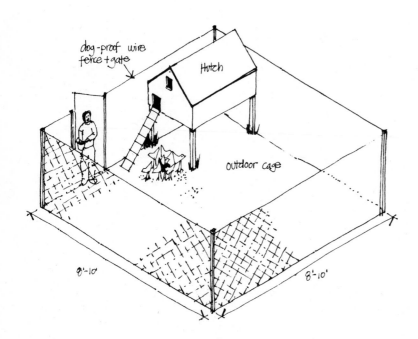

dog-proof wire
fence + gate

Hutch

outdoor cage

8'-10'   8'-10'

bird hutch
small 5-7 birds

resources

"BROODING CHICKS", EXTENSION CIRCULAR 854, OREGON STATE UNIVERSITY
    EXTENSION SERVICE, CORVALLIS, OREGON

"FEEDING LAYING AND BREEDING HENS", EXTENSION CIRCULAR 960, OREGON
    STATE UNIVERSITY EXTENSION SERVICE CORVALLIS, OREGON

HOW TO RAISE RABBITS AND CHICKENS IN AN URBAN AREA, ECOLOGY CENTER,
    2179 ALLSTON WAY, BERKELEY, CALIFORNIA 94704

POULTRY, CARNATION MILLING DIVISION, 6400 GLENWOOD, P.O. BOX 2917,
    SHAWNEE-MISSION, KANSAS 66201

"RAISING CHICKENS IN THE CITY", MARY BECK, UNPUBLISHED WORKING PAPER,
    EUGENE, OREGON, 1977

THE HOMESTEADERS HANDBOOK TO RAISING SMALL LIVESTOCK, JEROME D.
    BELANDER, RODALE PRESS, INC. BOOK DIVISION: EMMAUS, PA.

RABBITS

RABBITS AND THEIR BY-PRODUCTS ARE INTEGRAL COMPONENTS OF THE URBAN FARM. THE GOAL OF RABBIT PRODUCTION IS TO PROVIDE NUTRITIOUS CHEMICALLY-FREE MEAT PROTEIN AT A MINIMUM COST ECONOMICALLY AND ENVIRONMENTALLY.

RABBIT MEAT IS HIGHER IN PROTEIN THAN ANY OTHER DOMESTICALLY RAISED LIVESTOCK. IT ALSO IS THE LOWEST IN SATURATED FAT, CALORIES AND WATER. THE PRODUCTION OF THIS TASTY AND NUTRITIOUS MEAT IS A SURPRISING 1000 TIMES MORE EFFICIENT THAN THE AVERAGE BEEF CATTLE OPERATION. THIS IS QUITE IMPORTANT IN OUR OVER-CROWDED, UNDERNOURISHED PLANET. RABBITS EAT ALFALFA, A GRASS THAT IS NOT EASILY USED FOR HUMAN CONSUMPTION.

RABBITS ARE ALSO EXTREMELY VALUABLE TO THE GARDEN ECOSYSTEM. AN AVERAGE DOE AND HER YOUNG WILL PRODUCE UP TO SIX CUBIC FEET OF QUALITY MANURE PER YEAR. THIS MANURE CONTAINS MANY IMPORTANT ELEMENTS. ITS COMPOSITION IS ABOUT 89% ORGANIC MATTER, 6.5% WATER, 2.3% NITROGEN, 1.3% PHOSPHORIC ACID, AND 0.7% POTASH. THIS MATERIAL CAN BE ADDED TO THE COMPOST HEAP OR APPLIED DIRECTLY TO THE GARDEN SOIL WITHOUT FEAR OF BURNING SMALL PLANTS.

MANURE CAN ALSO BE USED TO RAISE WORMS. THE WORMS FEED ON THE MANURE AND IN THE PROCESS PRODUCE WORM CASTINGS, A VALUABLE COMPOST/SOIL AMENDMENT. THE WORMS CAN THEN BE FED TO THE CHICKENS AND FISH AS A SOURCE OF PROTEIN.

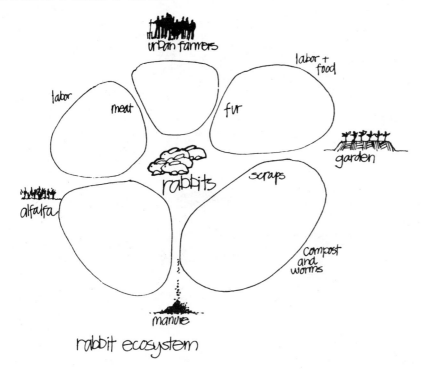

rabbit ecosystem

THE EXPERIENCE OF THE URBAN FARM

WE HAVE BEEN RAISING RABBITS AT THE URBAN FARM AND IN OUR OWN BACK YARD FOR OVER THREE YEARS. DURING THIS TIME, WE HAVE EXPERIMENTED AND RESEARCHED A NUMBER OF SYSTEMS, WITH SOME SUCCESSES, AND SOME FAILURES. THE INFORMATION WE PROVIDE HERE IS BASED ON OUR EXPERIENCES WITH, AND KNOWLEDGE OF, RABBITS. IT IS BY NO MEANS COMPREHENSIVE IN SCOPE NOR INTENDED TO BE AN ABSOLUTE DOCUMENT. HOWEVER, WITH THIS INTRODUCTION, YOU SHOULD BE ABLE TO GET A FEEL FOR THE ENERGY REQUIRED AND BENEFITS RECEIVED FROM RAISING RABBITS FOR MEAT AND THEIR BY-PRODUCTS.

THE FIRST THING INVOLVED IN SETTING UP A SMALL RABBITRY IS DECIDING HOW MANY AND WHAT KIND OF RABBIT YOU WANT. THREE FEMALES (DOES) AND ONE MALE (BUCK) IS SUFFICIENT FOR A FAMILY OF FOUR TO HAVE TWO OR MORE RABBITS A WEEK FOR THE DINNER TABLE, OR ABOUT 240 POUNDS OF MEAT ANNUALLY. RABBIT VARIETIES ARE NUMEROUS, BUT FOR MEAT, HEALTH, AND EASE OF CARE, THE NEW ZEALAND RABBIT IS THE BEST. MOST, IF NOT ALL, COMMERCIAL RABBITRIES THAT ARE IN BUSINESS TO PRODUCE HIGH-QUALITY MEAT USE THE NEW ZEALANDS. THE WHITE COLOR IS THE MOST COMMON, AS THE PELT IS WORTH A FEW CENTS MORE, BUT COLORED RABBITS CAN ALSO BE FOUND AND ARE JUST AS GOOD.

THE LIFE CYCLE OF THE RABBIT IS FAIRLY SIMPLE. A DOE IS BRED BY A BUCK, OF COURSE, AND 31 DAYS LATER SHE WILL GIVE BIRTH TO 6-10 LITTLE BABIES, CALLED KITTENS. THE KITTENS LIVE ENTIRELY ON MOTHER'S MILK FOR THE FIRST THREE WEEKS, AND SLOWLY BEGIN TO EAT SOLID FOOD. BY THE TIME THEY ARE EIGHT WEEKS OLD, THE KITTENS WILL WEIGH 4+ POUNDS, AND BE READY FOR THE FREEZER. IF YOU KEEP SOME OF THE YOUNG DOES, THEY WILL BE READY TO BREED AT FIVE AND ONE-HALF MONTHS. AND THE CYCLE REPEATS. A DOE CAN BE REBRED AT A VARIETY OF TIMES, AS GENERALLY THEY ARE ALWAYS IN HEAT -- HENCE THE RABBIT'S INFAMOUS REPUTATION. SEE THE BREEDING SCHEDULE CHART FOR DETAILS.

VEGETARIANS AND RABBITS

SOME MAY ASK, BUT WHY EAT MEAT? OF COURSE, THIS IS A VIABLE QUESTION. OFTEN IT IS BASED ON THE ASSUMPTION THAT WE SHOULD NOT KILL TO LIVE. THIS ARGUMENT IS INTERESTING, AND IF BASED ON RELIGIOUS CONVICTIONS IS DIFFICULT TO QUIBBLE WITH. HOWEVER, IF ONE CONSIDERS RIPPING A CARROT FROM THE GROUND OR GENTLY SLICING A HEAD OF LETTUCE SOMETHING LESS THAN KILLING, THE PHILOSOPHICAL FOUNDATION APPEARS FAULTY. DRAWING THE LINE BETWEEN PLANT AND ANIMAL INVITES A DISCUSSION ON WHAT IS LIFE, AND WHO MIGHT BE BETTER OR HIGHER THAN WHAT. TO AVOID THIS ARGUMENT, BUT NOT DISMISS IT, ANOTHER ASPECT MIGHT BE EXAMINED.

SOME CONTEND THAT EATING MEAT SUCH AS BEEF, PIG, OR COMMERCIAL CHICKEN IS INEFFICIENT AND THEREFORE A DRAIN ON THE WORLD'S FOOD SUPPLY, AS THESE ANIMALS CONSUME FOODSTUFFS THAT HUMANS MIGHT EAT. THIS IS INDEED TRUE. RABBITS, HOWEVER, DO NOT EAT GRAINS OR GRASSES THAT HUMANS MIGHT CONSUME. THEY THRIVE ON ALFALFA -- A GRASS HUMANS CANNOT DIGEST, AND WHICH CAN BE GROWN ON MARGINAL LANDS NOT SUITED TO THE PRODUCTION OF INTENSIVE ROW CROPS. THE RESULT IS THE PRODUCTION OF A HIGH-GRADE PROTEIN FROM A LOW-GRADE SOURCE. RABBITS ARE ALSO EXTREMELY EFFICIENT IN TERMS OF POUNDS OF PROTEIN PRODUCED PER POUND OF FEED CONSUMED. EUROPEANS HAVE KNOWN THIS FOR SOME TIME, AND RABBIT MEAT IS ONE OF THE TOP MEAT SOURCES FOR THAT SPACE-SHORT, CROWDED CONTINENT.

WE ARE NOT ADVOCATING THAT EVERYONE EAT RABBITS, SIMPLY TO UNDERSTAND THEIR IMPORTANCE IN THE URBAN FARMING FOOD CHAIN.

HOW TO BREED

BREEDING RABBITS IS FAIRLY SIMPLE. PLACE THE DOE IN THE BUCK'S CAGE AND HE WILL DO ALL THE WORK IN A MATTER OF SECONDS. THE ACT IS SUCCESSFUL WHEN THE BUCK FALL OFF BACKWARDS OR TO THE SIDE IN OBVIOUS BLISS. HE MAY EVEN SCREAM. SOMETIMES THE DOE IS NOT INTERESTED. SHE WILL SHOW THIS BY NOT LIFTING HER TAIL, THUS MAKING THE BREEDING PROCESS IMPOSSIBLE. IN THIS CASE, TRY AGAIN THE NEXT DAY. DO NOT LEAVE THE DOE WITH THE BUCK UNATTENDED. UNHAPPY DOES HAVE BEEN KNOWN TO REMOVE THE BUCK'S VITAL ORGANS IN A FIT OF ANGER. AFTER A SUCCESSFUL MATING, RETURN THE DOE TO HER CAGE. MARK DOWN 31 DAYS FROM THIS DATE ON HER CALENDAR. YOU CAN EXPECT LITTLE KITTENS THEN. A BOX WITH STRAW OR DRIED GRASSES SHOULD BE PLACED IN HER CAGE ON THE 28TH DAY. SHE WILL MAKE A NEST IN IT FOR THE KITTENS.

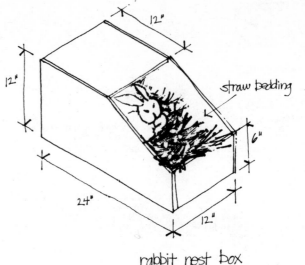

straw bedding

rabbit nest box
built of 3/4" plywood or 1×12 stock

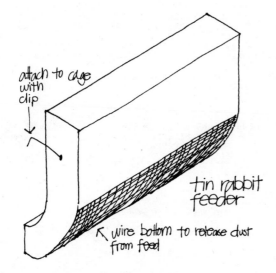

attach to cage with clip

tin rabbit feeder

wire bottom to release dust from feed

HOW TO FEED

WE RECOMMEND FEEDING RABBITS COMMERCIAL FEED AVAILABLE IN LOCAL FEED STORES, AND FEEDING A SMALL AMOUNT OF ALFALFA HAY EVERY DAY FOR A GOOD FIBER DIET. THERE ARE A NUMBER OF FOODS A RABBIT WILL ALSO EAT, SUCH AS GRASS, STRAW, COMFREY, LETTUCE, CARROTS, ETC., AND IF YOU WANT TO DO SO, IT IS ACCEPTABLE. HOWEVER, WE RECOMMEND YOU RESEARCH THE POSSIBILITIES BY READING THE SUGGESTED LITERATURE THOROUGHLY TO KNOW ALL OF THE PROBLEMS INVOLVED. WE FEEL 5 OUNCES A DAY OF RABBIT PELLETS AND SOME ALFALFA HAY IS A NUTRITIOUS DIET AND SIMPLE TO MANAGE. DO NOT OVERFEED A RABBIT, AS IT WILL EVENTUALLY MAKE THEM STERILE AND/OR KILL THEM.

DOES AND KITTENS

WHEN KITTENS ARE OLD ENOUGH TO EAT SOLID FOOD, CONTINUE TO INCREASE THEIR PELLET RATION TO GIVE THEM ALL THEY CAN EAT IN ONE DAY. WHILE THE MOTHER IS NURSING SHE CAN HAVE 6 OUNCES OF FEED PER DAY. SOME ALFALFA HAY IS ALSO DESIRABLE, BUT ONLY ENOUGH SO THEY CLEAN IT UP EACH DAY.

WATER

WATER CAN BE PROVIDED IN A VARIETY OF WAYS, BUT IT MUST ALWAYS BE AVAILABLE. THE EASIEST METHOD INVOLVES SIMPLY WIRING A LARGE METAL CAN TO THE CAGE SIDE AND KEEPING IT FULL OF WATER. AN EARTHEN CROCK IS ALSO USEFUL AND EASIER TO KEEP CLEAN. AN AUTOMATIC SYSTEM IS BEST, AS IT KEEPS THE RABBITS CONSTANTLY SUPPLIED, AND ELIMINATES UNSANITARY DRINKING WATER PROBLEMS. LITTLE RABBITS TEND TO SWIM IN LARGE OPEN CONTAINERS.

ceramic water crock

HOUSING SYSTEMS FOR THE HOME RABBITRY

HOUSING FOR RABBITS CAN RANGE FROM A SIMPLE WOODEN BOX FOR SHELTER
TO ELABORATE METAL CAGES WITH AUTOMATIC WATERING SYSTEMS. EACH SYS-
TEM HAS ITS ADVANTAGES AND LIABILITIES. EXPLAINED HERE ARE THREE
SYSTEMS WHICH RANGE IN COMPLEXITY AND EFFICIENCY.

1. THE ON-GROUND MOVABLE CAGE: THIS IS THE SIMPLEST OF MANAGEMENT
   TECHNIQUES. IT INVOLVES A SIMPLE BOX COVERED WITH CHICKEN WIRE
   THAT IS OPEN ON THE BOTTOM. THE CAGE IS SIMPLY MOVED EVERY DAY
   OR SO AS THE RABBIT DEVOURS THE GRASS FLOOR. THIS IS ADEQUATE
   FOR A DOG-FREE NEIGHBORHOOD, BUT IN MOST URBAN SITUATIONS, THE
   RABBITS WOULD SOON DISAPPEAR.

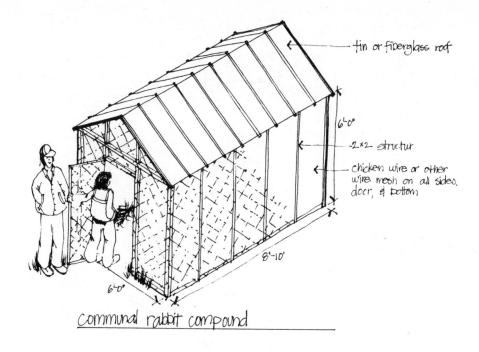

tin or fiberglass roof

2×2 structur

chicken wire or other
wire mesh on all sides,
door, & bottom

6'-0"

8'-10"

6'-0"

communal rabbit compound

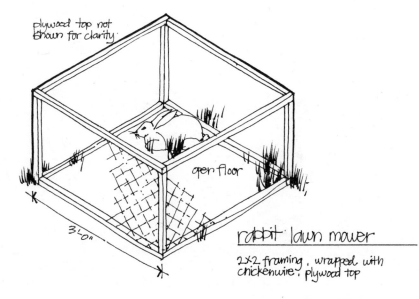

plywood top not
shown for clarity

open floor

3'-0"

rabbit lawn mower

2×2 framing, wrapped with
chickenwire; plywood top

2. THE ENCLOSED COMMUNAL CAGE: THIS SYSTEM EMPLOYS A LARGE CAGE
   THAT IS ENCLOSED ON ALL SIDES WITH AN ADEQUATE ROOF PROVIDED
   FOR RAIN PROTECTION. BALED STRAW IS ADDED TO A DEPTH OF ABOUT
   FOUR FEET, AND FOOD AND WATER ARE PROVIDED WITH A FEEDER AND
   DAILY WATER BOWL. TWO RABBITS, MALE AND FEMALE, ARE PLACED IN
   THE CAGE AND ALLOWED TO LIVE SEMI-NATURALLY. THEY BURROW INTO
   THE STRAW BALES FOR WARMTH, COMFORT, AND PRIVACY. AFTER A
   YEAR OR SO ALL THE RABBITS ARE REMOVED AND USED FOR MEAT. THE
   CAGE IS THOROUGHLY CLEANED AND THE STRAW IS REPLACED. THE CYCLE
   IS REPEATED YEAR AFTER YEAR.

   THE PRINCIPLE PROBLEMS WITH THIS SYSTEM ARE: LOSS OF YOUNG FROM
   DISEASE OR INFERIOR CARE, LOW CONTROL OVER INFECTIOUS DISEASES,
   LOW EFFICIENCY IN TERMS OF POUNDS OF FEED AND POUNDS OF MEAT,
   AND MOST SERIOUSLY, THE UNCONTROLLED BREEDING PATTERNS WHICH
   RESULT IN AN INFERIOR GENETIC STOCK.

   ALTHOUGH THE RABBITS DO INDEED LIVE IN A SEMI-NATURAL ENVIRON-
   MENT, IT IS STILL A CONTAINMENT. THE INTER-BREEDING THAT
   OCCURS INTENSIFIES ANY GENETIC PROBLEMS IN SUCCEEDING GENERA-
   TIONS. THE ADVANTAGE IS LOW MAINTENANCE, AND ONLY A ONCE-A-
   YEAR HARVEST CHORE.

3. THE HANGING WIRE CAGE: THIS SYSTEM IS PROBABLY THE BEST FOR THE HOME RABBITRY. THE CARETAKER HAS CONTROL OVER THE BREEDING PATTERNS, DISEASE PROBLEMS AND SANITATION. THE HIGHEST EFFICIENCIES ARE POSSIBLE WITH THIS SYSTEM.

THE CAGES, USUALLY BUILT IN GROUPS OF THREE, ARE HUNG FROM SUPPORTING RAFTERS WITH STURDY WIRE. THIS ARRANGEMENT KEEPS THE RABBITS AWAY FROM THE GROUND AND PARASITE INFESTATIONS. ALSO IT ALLOWS FOR EASY ACCESS TO THE RABBITS AND TO THEIR MANURE FOR USE IN THE GARDEN. THE CAGE CAN BE HUNG OUTSIDE UNDER A SEMI-ENCLOSED STRUCTURE OR INSIDE A SMALL BARN. (SEE ILLUSTRATIONS.)

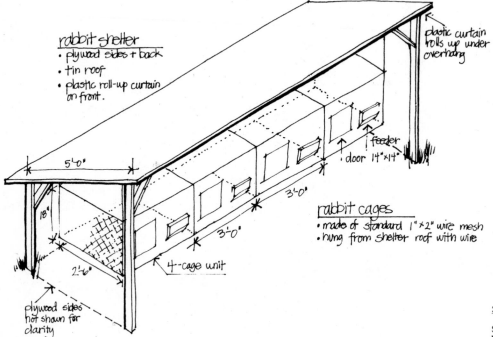

rabbit shelter
• plywood sides + back
• tin roof
• plastic roll-up curtain on front.

5'-0"
18"
2'-6"
3'-0"
3'-0"

plastic curtain rolls up under overhang

feeder
door 14"×14"

rabbit cages
• made of standard 1"×2" wire mesh
• hung from shelter roof with wire

4-cage unit

plywood sides not shown for clarity

HANDLING LIVE RABBITS

DESPITE A RABBIT'S GENTLE DISPOSITION, THOSE HIND LEGS ARE EXTREMELY POWERFUL. A FEW SWIFT KICKS WITH THOSE TOENAILS AND YOU WILL SOON REALIZE THE IMPORTANCE OF HANDLING THEM PROPERLY. WITH ONE HAND, GRASP THE LOOSE SKIN ON THE BACK OF THE NECK, AND WITH THE OTHER SUPPORT THE REAR END AT THE THIGHS, HOLDING THE RABBIT CLOSE TO YOUR BODY.

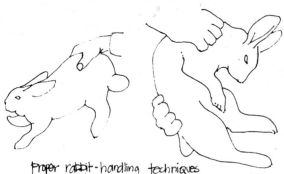

Proper rabbit-handling techniques

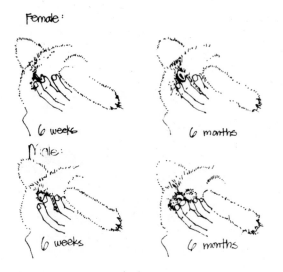

Female:

6 weeks
6 months

Male:

6 weeks
6 months

SEXING RABBITS

SEXING YOUNG RABBITS, UNDER SIX WEEKS OLD, IS SOMEWHAT DIFFICULT, BUT WITH A LITTLE PRACTICE IT CAN BE MASTERED. GENTLY PICK UP THE RABBIT AS DESCRIBED ABOVE. ARTFULLY PLACE THE RABBIT ON ITS BACK TO EXPOSE THE GENITAL AREA. BE CAREFUL THE RABBIT DOES NOT SCRATCH YOU.

THE GENITALS OF THE FEMALE RESEMBLE A SLIT IN YOUNGER AND OLDER RABBITS. THE MALE RABBIT'S GENITALS ARE EASY TO IDENTIFY ON OLDER RABBITS, AS THE PARTS ARE QUITE OBVIOUS. THE YOUNGER MALE'S EQUIPMENT RESEMBLES A SHORT SHAFT WITH A SMALL HOLE.

DISEASE AND RABBITS

RABBITS ARE NOT PRONE TO A GREAT MANY DISEASES. HOWEVER, SOME OF THE COMMON PROBLEMS CAN BE FATAL. RABBITS DISLIKE DRAFTS AND TEMPERATURES ABOVE 90 DEGREES FAHRENHEIT. COLDS CAN DEVELOP IN THESE SITUATIONS AND, UNLIKE HUMANS, THEY CAN DIE FROM THE INFECTION. A RABBIT WITH A COLD SHOULD BE ISOLATED FROM THE REST OF THE HERD AS IT IS CONTAGIOUS. ANTIBIOTICS ARE EASILY AVAILABLE AT MOST FEED AND SEED STORES. THEY ARE ADMINISTERED THROUGH THE WATER SUPPLY. COLDS ARE THE MOST COMMON DISEASE PROBLEMS YOU MAY ENCOUNTER. HOWEVER, IT IS WISE TO EXAMINE SOME OF THE BOOKS AND PAMPHLETS MENTIONED IN THE REFERENCE SECTION TO FAMILIARIZE YOURSELF TO THE OTHER PROBLEMS.

HARVESTING

THE RABBITS ARE HARVESTED AT 8 WEEKS OF AGE. AT THIS POINT THEY SHOULD WEIGH APPROXIMATELY 4 POUNDS LIVE WEIGHT. SLAUGHTERING RABBITS IS NOT ONE OF THE MOST PLEASANT JOBS THE URBAN FARMER HAS TO PERFORM, BUT WITH SOME FORETHOUGHT AND DETERMINATION IT SHOULD BE RELATIVELY EASY AND CONSIDERATE.

slaughtering, skinning & butchering a rabbit

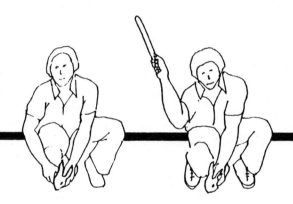

HANG THE ANIMAL FROM ITS HIND FEET WITH THE STOMACH TOWARDS YOU. CUT THE SKIN AROUND EACH OF THE LEGS CAREFULLY AND BEGIN PEELING THE SKIN DOWN TOWARDS THE HEAD. SIMPLY CONTINUE PEELING THE SKIN AND CUTTING CONNECTIVE TISSUE AS YOU GO. CUT THE SKIN AWAY FROM THE HEAD WHEN YOU GET TO THAT POINT. THE SKIN WILL BE IN ONE PIECE, AND CAN NOW BE HUNG ON THE STRETCHER TO DRY. CONSULT "RAISING RABBITS THE MODERN WAY" BY BOB BENETT, FOR ADVICE ON TANNING AND USING PELTS.

A SHARP KNIFE, A BOARD WITH TWO NAILS FOR HANGING THE RABBIT, A BUCKET FOR THE WASTE, AND A STRETCHER FOR THE SKIN ARE NEEDED. FIRST, REMOVE THE ANIMAL FROM ITS CAGE AND PLACE IT GENTLY ON THE GROUND. QUICKLY STRIKE THE HEAD FIRMLY JUST ABOVE THE EARS WITH A HEAVY STICK OR IRON BAR. THEN CAREFULLY SLICE THE RABBIT'S NECK TO CUT THE JUGLAR VEIN. HOLD THE RABBIT UPSIDE DOWN OVER THE BUCKET UNTIL IT HAS CEASED MOVEMENT.

BUNNY HOP NEWS, EMERALD EMPIRE RABBIT BREEDERS ASSOCIATION, P.O. BOX 2321, EUGENE, OREGON 97402.

HOW TO RAISE RABBITS AND CHICKENS IN AN URBAN AREA, ECOLOGY CENTER, 2179 ALLSTON WAY, BERKELEY, CALIFORNIA 94704.

RABBITS, CARNATION-ALBERS, 6400 GLENWOOD, BOX 2917, SHAWNEE-MISSION, KANSAS 66201.

"RABBITS", A MAGAZINE, COUNTRYSIDE PUBLICATIONS, LTD., 19 EAST WATERLOO, WISCONSIN 53594.

MAINTAINING NATIVE PLANTS AND LOCAL NATURAL RESOURCES

Replanting what we burn or cut, preserving our blackberries and riparian vegetation and monitoring our water quality, are all aspects of both maintaining and nurturing local natural resources.

The concept and execution of the native plant nursery at the Urban Farm was born out of discussions about the cycles of life. Initially a response to a fuel question for the ceramics kiln (see Section on--Utilizing Local Natural Resources), this notion of promulgating locally adapted native plants and taking an active role in their sponsorship, has resulted in several extensions of this idea...

First, proposals for URBAN FORESTRY came out of workshops with both neighborhood and business groups to promote the economic development impact of shifting from an urban landscape based primarily on ornamental and exotic plants to those more suited to our bio-region. These discussions have also evolved into efforts utilizing native tree stock for large scale visual and pollutant buffering of urban freeways.

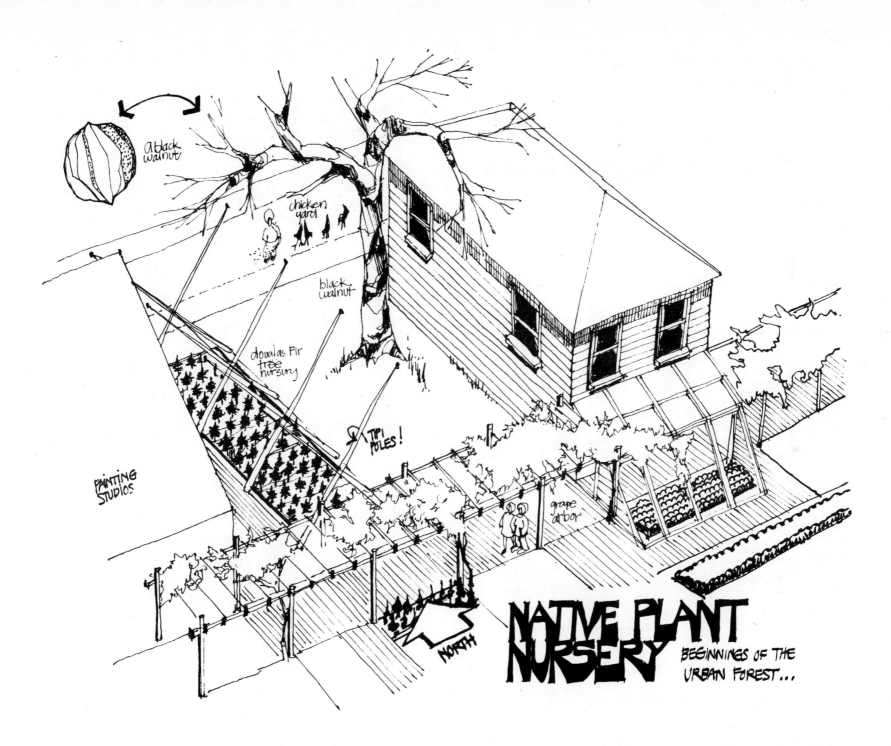

A black walnut

chicken yard

black walnut

douglas fir tree nursery

TIPI POLES!

PAINTING STUDIOS

grape arbor

NORTH

NATIVE PLANT NURSERY

BEGINNINGS OF THE URBAN FOREST...

SHARING TOOLS AND BOOKS

IT IS NECESSARY TO TAKE CARE OF YOUR TOOLS TO INSURE THAT THEY PER-
FORM WELL. WELL TENDED TOOLS WILL LAST FOR A LONG TIME. YOU MAY WANT
TO KEEP A BUCKET OF DRY SAND NEAR YOUR TOOL STORAGE AREA. AT THE
END OF A DAY'S USE, PUT THE SHOVEL OR SPADE INTO THE SAND, PUSH AND
PULL IT UP AND DOWN SEVERAL TIMES, AND IT WILL COME CLEAN. FEBRUARY
IS THE TIME TO SHARPEN AND RECORD THE CONDITION OF THE GARDEN TOOLS.
YOU CAN WORK THE EDGES OF THE SHOVEL AND HOE BLADES WITH A FILE
AND WHETSTONE TO KEEP THEM AS SHARP AS A GOOD KITCHEN KNIFE FOR
EASY DIGGING. AFTER CLEANING YOU CAN USE A SILICONE SPRAY ON METAL
TO PREVENT RUST AND BOILED LINSEED OR MINERAL OIL ON THE WOODEN PARTS
TO PREVENT SPLINTERING. WHEN ANY MOIST DIRT STICKS TO A TOOL, RE-
MOVE IT AFTER USE TO PREVENT RUST.

YOU MAY FOR SOME REASON DECIDE NOT TO INVEST IN TOOLS OF YOUR OWN.
YOU CAN SHARE OR TRADE TOOLS WITH NEIGHBORS OR MAKE USE OF THE VAR-
IOUS TOOL LIBRARIES HERE. THE FOLLOWING ARTICLE WILL GIVE YOU SOME
IDEAS OF THE TOOL LIBRARIES HERE AND WHAT THEY HAVE TO OFFER.

A FRIEND OF MINE WENT TO THE LIBRARY THE OTHER DAY AND CAME HOME
WITH A WHEELBARROW, A FOOD DRYER, AN ELECTRIC DRILL AND AN EXTEN-
SION LADDER.

WHAT? YOU DIDN'T KNOW THE LIBRARY OFFERED THOSE KINDS OF THINGS?
WELL, TO BE HONEST, YOU WON'T FIND THEM AT THE EUGENE PUBLIC LIB-
RARY OR YOUR NEIGHBORHOOD BOOKMOBILE. BUT THERE ARE SEVEN PLACES
IN EUGENE WHERE YOU CAN CHECK OUT TOOLS, TOOLS OF EVERY DISCRIPTION,
AND TAKE THEM HOME FOR THREE DAYS TO A WEEK JUST LIKE YOU WOULD
WAR AND PEACE OR DR. ATKIN'S DIET REVOLUTION.

THE HIGH COST OF BUYING OR RENTING TOOLS KEEPS A LOT OF US FROM
LEARNING NEW SKILLS AND DOING IMPORTANT THINGS FOR OURSELVES, SUCH
AS GROWING OUR OWN FOOD AND PRESERVING IT, DOING OUR OWN CARPENTRY
AND REPAIRS, CUTTING OUR OWN FIREWOOD, FIXING CARS AND BICYCLES, EVEN
DEVELOPING OUR OWN PHOTOGRAPHS. HOW MANY TIMES A YEAR WILL YOU
USE A $1,000 ROTOTILLER, OR A $150 CIRCULAR SAW? IN THE CASE OF
THESE MAJOR TOOLS INVESTMENTS, TOOL LIBRARIES PROVIDE AN ALTERNA-
TIVE TO BUYING AND STORING YOUR OWN.

MOST OF THE LOCAL TOOL LIBRARIES BEGAN WITH THE IDEA THAT ACCESS
TO TOOLS HELPS CREATE A MORE SELF-RELIANT COMMUNITY. "IT'S A
MOVEMENT THAT I WOULD LIKE TO SEE GROW", SAYS GINNY BOLGER, WHO CAME
TO EUGENE FROM WISCONSIN IN SEPTEMBER, AND IS NOW PROGRAM DIRECTOR
FOR THE AMITY FOUNDATION, SPONSOR OF THE FRIENDLY AREA TOOL LIBRARY.
"I THINK EVENTUALLY PEOPLE ARE GOING TO BE FORCED BY ECONOMIC CIR-
CUMSTANCES TO BECOME MORE SELF-RELIANT. THERE WON'T BE AS MUCH TRAV-
ELING AROUND, AND PEOPLE WILL BE MORE LOCALIZED--THEY'LL PUT MORE
ENERGY INTO THE PLACE THEY'RE LIVING. I'M OPTIMISTIC THAT PEOPLE
WILL HAVE TO START LEARNING TO DO MORE FOR THEMSELVES."

HOWEVER, GINNY ADMITS THE LIBRARIES NEVER WILL HAVE EVERY TOOL A
PERSON MIGHT NEED. BUT WHILE THEY MAY NEVER PUT REPAIRMEN OUT OF
BUSINESS, OR MAKE A BIG DENT IN THE TOOL RENTAL TRADE, SOME OF THE
LIBRARIES DO SHOW PROMISE OF BRINGING THE NEIGHBORHOOD PEOPLE TO-
GETHER AND OF SERVING AS A BASE FOR NEIGHBORHOOD ACTIVITIES INVOLVING
THOSE TOOLS.

THE FRIENDLY, PETERSEN PARK BARN AND WHITEAKER TOOL LIBRARIES EITHER
ARE NOW OR PLAN TO BE, NEIGHBORHOOD CONTROLLED. THE WHITEAKER TOOL

LIBRARY ALREADY OFFERS A SHOP TO WORK IN, AND COORDINATOR CLAUDIA
GRUBBS THINKS A NEIGHBORHOOD BICYCLE REPAIR SHOP WOULD BE A GOOD
IDEA. IN FACT, SAYS GRUBBS, "SOME PEOPLE THINK OF THE BICYCLE IT-
SELF AS A TOOL, AND HAVE SUGGESTED A BICYCLE DROP-OFF WHERE YOU
COULD PICK UP A BICYCLE AND THEN DROP IT OFF LATER FOR ANOTHER PER-
SON TO USE."

IF A BIKE CAN BE CONSIDERED A TOOL, WHAT LIMIT IS THERE TO WHAT A
TOOL LIBRARY CAN INCLUDE? "I THINK OF A TOOL AS ANY IMPLEMENT THAT
HELPS YOU DO SOMETHING BETTER," SAYS AMITY'S GINNY BOLTER. "THAT
COULD INCLUDE THINGS LIKE TYPEWRITERS AND SEWING MACHINES." LI-
BRARIES ALREADY OFFER SUCH UNTRADITIONAL TOOLS AS FOOD DRYERS,
PRESSURE COOKERS, FOOD MILLS, A SOIL TEST KIT, A MOVIE CAMERA, AND
DARKROOM EQUIPMENT. AND WHITEAKER'S LARRY PARKER SAYS HE'S CONSID-
ERED THE PURCHASE OF A VACUUM CLEANER AND A FLOUR GRINDER.

MOST TOOLS, HOWEVER, ARE OF THE GARDENING, CARPENTRY AND AUTOMOTIVE
VARIETY, AND ARE EITHER PURCHASED THROUGH GRANTS OR DONATED BY THE
CITY OR BY INDIVIDUALS. RICHARD WEINMAN OF THE COUNTY'S SENIOR
SERVICES DIVISION HAS BEEN INSTRUMENTAL IN GIVING SEVERAL LIBRARIES
AN INITIAL DONATION OF TOOLS. WEINMAN HAS AIDED THE WHITEAKER AND
PETERSEN PARK BARN LIBRARIES. OUTSIDE THE EUGENE AREA HE'S WORKED
WITH A TOOL LIBRARY AT THE HUMAN RESOURCE CENTER AT COTTAGE GROVE,
ONE AT MID-LANE MULTI-SERVICES IN VENETA, AND A SMALL ONE OPERATING
THROUGH THE COBURG ACTIVITIES CLUB IN COBURG.

FOR THE MOST PART, LIBRARIES AIDED BY SENIOR SERVICES HAVE GARDEN-
ING AND FOOD PROCESSING EQUIPMENT, BUT LACK CARPENTRY TOOLS. "I'D
BE HAPPY TO GIVE ADVICE TO NEIGHBORHOOD GROUPS WANTING TO START TOOL
LIBRARIES," SAYS WEINMAN, "AND WILL HELP WITH TOOLS IF POSSIBLE."
HE'S ESPECIALLY INTERESTED IN FINDING A SPONSOR AND A LOCATION FOR
A LIBRARY IN THE SPRINGFIELD AREA; IF YOU HAVE ANY SUGGESTIONS, CALL
687-4038.

THE ACTION NOW TOOL LIBRARY AT THE UNIVERSITY OF OREGON RECEIVED ITS
FIRST TOOLS AS A DONATION FROM A NOW-DEFUNCT UNIVERSITY PROGRAM. IT
BUYS NEW TOOLS EACH YEAR FROM STUDENT INCIDENTAL FEES, WHICH HAS
RESULTED IN A LARGER COLLECTION THAN OTHER LIBRARIES.

"WE BUY TOOLS THAT ARE REQUESTED OFTEN" SAYS KATIE MCCOUGHLIN, THE
LIBRARY'S WORK-STUDY STAFF PERSON. "FOR EXAMPLE, SEVERAL PEOPLE
ASKED FOR A SET OF BOX WRENCHES, AND THE LIBRARY RECENTLY BOUGHT IT.
OCCASIONALLY SOMEONE WILL ASK FOR A TOOL NO ONE HAS EVER HEARD OF,
AND WE HAVE ONE TOOL THAT HAS ONLY BEEN CHECKED OUT ONCE--IT'S A
MURDEROUS LOOKING THING CALLED A BUSH-HOOK USED FOR CUTTING DOWN
HEAVY UNDERBRUSH." KATIE SAYS PEOPLE CAN TAKE OUT AS MANY TOOLS
AS THEY WISH: "ONE FELLOW FILLED OUT TWO PAGES WORTH."

AS MIGHT BE EXPECTED, ALL THE LIBRARIES HAVE TROUBLE WITH TOOLS
STAYING OUT TOO LONG. A FEW CHARGE DEPOSITS AND FINES TO MAKE IT
MORE CERTAIN THAT THE TOOL WILL BE RETURNED ON TIME. "SOMETIMES
ITS HARD TO TRACK DOWN TOOLS BECAUSE OUR NEIGHBORHOOD IS SOMEWHAT
TRANSIENT", SAYS WHITEAKER'S CLAUDIA GRUBBS, "BUT WE'VE HAD NO
BIG PROBLEMS. THERE SEEMS TO BE A BELIEF IN THIS SOCIETY THAT
PEOPLE ARE DISHONEST, BUT WE'VE FOUND IT'S GENERALLY NOT TRUE, THEY
ARE JUST FORGETFUL."

FRIENDLY CHARGES SMALL RENTALS ON TOOLS SO THEY'LL HAVE MONEY TO
REPLACE THEM WHEN NECESSARY. UNLIKE OTHER LIBRARIES WHICH RELY ON
VOLUNTEERS, MAINTENANCE AT FRIENDLY IS IN THE CHARGE OF LUIS MAS,
A SENIOR COMMUNITY SERVICE PROGRAM EMPLOYEE. NOT ONLY DOES LUIS
CHOOSE NEW TOOLS AND REPAIR OLD ONES, HE TRAVELS TO PEOPLE'S HOMES
AND ADVISES THEM ON PROJECTS AND TOOL USE.

NATURALLY, SEASONS DICTATE THE TYPE OF TOOL THAT'S REQUESTED. FALL
IS THE TIME FOR CHAIN SAWS AND AXES. THE URBAN FARM, WHICH HAS
PRIMARILY GARDEN TOOLS, IS NOT BUSY DURING THE WINTER, WHEN MOST
WORK WILL BE DONE INDOORS, WHILE THE ACTION NOW LIBRARY IS QUITE

BUSY THIS TIME OF YEAR WITH FIFTY TO SEVENTY FIVE PEOPLE DROPPING IN EACH DAY.

THE LIBRARIES AT THE COMMUNITY ENERGY BANK AND EASTSIDE COMMUNITY SCHOOL ARE UNDER-UTILIZED YEAR AROUND, PROBABLY BECAUSE FEW PEOPLE KNOW THAT THEY HAVE TOOLS AVAILABLE.

FOR THAT MATTER, NONE OF THE TOOL LIBRARIES ARE USED TO CAPACITY. "PEOPLE STILL DON'T THINK ABOUT COMING TO THE WHITEAKER TOOL LIBRARY WHEN THEY NEED SOMETHING," SAYS LARRY PARKER. "IT SEEMS UNREAL TO PEOPLE THAT CHECKING OUT A TOOL CAN BE AS EASY AS IT IS, BUT ONCE PEOPLE GET USED TO THE TOOL LIBRARY LOTS OF COMMUNITY-ORIENTED THINGS CAN GROW OUT OF IT. ALREADY THERE'S TALK ABOUT COMMUNITY GARDENS, A FOOD CENTER AND A BICYCLE SHOP. WE CAN START OUT WORKING WITH PEOPLE'S BASIC NEEDS AND EXPAND FROM THERE--IT WOULD LEAD TO A WHOLE DIFFERENT APPROACH TO COMMUNITIES."

VICKIE STEA

WHITEAKER TOOL LIBRARY

315 MADISON (IN THE JEFFERSON ELEVATOR) 343-2711
DAYS/HOURS: TUES. - SAT., 1 TO 5 P.M.
USERS: ANYONE, BUT PREFERENCE GIVEN TO WHITEAKER RESIDENTS.
FEES: NONE
PROCEDURE: NUMBERED TOOLS ARE SIGNED OUT, USUALLY FOR ONE WEEK MAXIMUM. NEED CURRENT I.D.
TOOLS: MORE THAN ONE HUNDRED TOOLS, SEVERAL OF EACH KIND. THERE ARE SHOVELS, RAKES, HOES, HAYFORKS, TROWELS, WEED WHIPS, SOIL TEST KIT, HOSES, WHEELBARROW, SPRINKLER, BUCKETS, SPADES, LAWN MOWERS, STEP LADDER, SAWS, HAMMERS, WEDGES, SCREWDRIVERS, ELECTRIC DRILLS, CHISELS, NAIL REMOVER, STAPLE GUNS, CAULKING GUNS, TIN SNIPS, RADIAL ARM SAW, PIPE WRENCHES AND MORE. THERE ARE ALSO FOOD DRYERS PRESSURE COOKERS, ENAMEL CANNING KETTLES, JAR LIFTERS, BOOKS AND A BIKE TRAILER.

URBAN FARM TOOL LIBRARY

URBAN FARM (OFF FRANKLIN BLVD. BEHIND B.J. KELLY'S) 686-3647
DAYS/HOURS: TUES. 2:30 -4:30, THURS. 8:30 - 11:30
USERS: ANYONE
FEES: NONE
PROCEDURE: SIGN OUT THE TOOLS WITH CURRENT I.D. ONE WEEK LENDING.
TOOLS: MAINLY GARDENING EQUIPMENT, INCLUDING SHOVELS, HOES, SPRINKLERS, TROWELS, SPADING FORKS, WHEELBARROW, LADDERS AND MORE. ALSO TWO FOOD DRYERS, CANNERS AND FOOD MILLS.

PETERSEN PARK BARN TOOL LIBRARY

PETERSON PARK BARN COMMUNITY CENTER, 870 BERNTZEN ROAD, 689-1446.
DAYS/HOURS: CENTER HOURS ARE UNTIL 9 P.M. DAILY, STAFFED BY PARKS AND RECREATION.
USERS: ANYONE
FEE: $5 DEPOSIT GIVES YOU A CARD IN THE FILES.
PROCEDURE: INDICATE WHICH TOOLS YOU TAKE; CHECK OUT TIME VARIABLE.
TOOLS: THEY HAVE ABOUT A DOZEN TOOLS, INCLUDING SHOVELS, A HOE, POTATO FORK, RAKES AND A WHEELBARROW. ALSO FOOD PROCESSING EQUIPMENT LIKE A PRESSURE COOKER, FOOD STRAINER AND FOOD DRYER.

COMMUNITY ENERGY BANK (CEB)

454 WILLAMETTE ST., 485-8133
DAYS/HOURS: 12 - 5 P.M. MONDAY, WEDNESDAY AND FRIDAY.
USERS: MEMBERSHIP REQUIRED
FEE: $9 DEPOSIT PER YEAR OR $4.50 EVERY SIX MONTHS.
PROCEDURE: CEB IS A SKILLS EXCHANGE, BUT MEMBERS INDICATE WHEN THEY HAVE TOOLS THAT THEY CAN ALLOW TO BE USED. ALL INFORMATION IS IN A FILE OPEN TO MEMBERS. USE OF A TOOL HAS TO BE NEGOTIATED WITH ITS OWNER.
TOOLS: THERE ARE AS MANY TOOLS AS THERE ARE SKILLS, BUT TO NAME A FEW, DRAFTING TOOLS, HYDRAULIC JACK, BEEKEEPING EQUIPMENT, BICYCLE REPAIR TOOLS, CANNING EQUIPMENT, CARPENTRY AND GARDENING TOOLS, SEWING MACHINES AND TRUCKS.

EASTSIDE COMMUNITY SCHOOL

1328 E. 22ND AVE. 687-3284.
DAYS/HOURS: MON.-FRIDAY. 8 AM - 5 PM, ASK FOR VIC HANSEN OR ELLIE DRAPER
USERS: ANYONE, BUT RESIDENTS OF NEIGHBORHOODS SURROUNDING EDISON-EASTSIDE WILL BE GIVEN PREFERENCE.
FEE: NONE
PROCEDURE: CALL AT LEAST ONE DAY IN ADVANCE TO SET THINGS UP. WILL NEED I.D. CARD AND TELEPHONE NUMBER. THE DARKROOM COMES EQUIPPED WITH CHEMICALS AND PAPER (FOR A SMALL FEE) IF YOU DON'T HAVE YOUR OWN. LENGTH OF TIME TO USE TOOLS IN NEGOTIABLE.
TOOLS: LOTS OF HAND TOOLS, ELECTRIC SANDER, DRILL PRESS, ELECTRIC JIGSAW, TABLESAW AND DRILL. ALSO HAVE THREE 35 MM. CAMERAS, SUPER 8 MOVIE CAMERA, MOVIE PROJECTORS, MISCELLANEOUS GARDENING TOOLS, LEATHERWORKING TOOLS, AND ;USE OF THE DARKROOM AND DEVELOPING EQUIPMENT. YOU CAN USE THE WOODSHOP TOO.

FRIENDLY AREA TOOL LIBRARY:

1846 PEARL ST. 484-7171
DAYS/HOURS: 3 - 7 PM. MON.-WED., 9-3 PM.   SATURDAY
BOTH ADDRESS AND HOURS ARE SUBJECT TO CHANGE IN THE NEAR FUTURE, WHEN THE LIBRARY WILL HOPEFULLY BE TAKEN OVER BY THE FRIENDLY AREA NEIGHBORS. IT IS CURRENTLY OPERATED BY AMITY FOUNDATION, WHICH WILL GIVE REFERRALS TO THE NEW LOCATION AFTER THE LIBRARY CHANGES HANDS.
USERS: MEMBERSHIP REQUIRED AND PREFERENCE GIVEN TO RESIDENTS OF THE FRIENDLY AREA NEIGHBORHOOD.
FEE: REGISTRATION FEE $6. DEPOSIT $12 FOR THE DURATION OF MEMBERSHIP. REGISTRATION AND RENTAL FEES CAN BE PAID OFF BY VOLUNTEER LABOR AT $3 PER HOUR.
PROCEDURE: THERE ARE RENTAL FEES ON EACH TOOL, FROM 10¢ PER WEEK TO $5 PER WEEK, WITH AN AVERAGE FEE OF 25¢. CHECKOUTS FOR ONE WEEK, AND CAN BE RENEWED BY PHONE. FINES FOR OVERDUE TOOLS.
TOOLS: MORE THAN 100 TOOLS. FRIENDLY HAS AN AMAZING ASSORTMENT OF TOOLS, INCLUDING ALL KINDS OF GARDEN AND CARPENTRY TOOLS, PLUS A CEMENT MIXER, AN AIR COMPRESSOR, A SHREDDER, BIG ROTOTILLER, PLUMBING TOOLS, CHALKLINE, EXTENSION CORDS, FERTILIZER SPREADER, GLASS CUTTER, WEED EATER, TORCH KIT AND A POST-HOLE DIGGER. COMPLETE LIST AVAILABLE FROM THE LIBRARY.

U. OF O. ACTION NOW TOOL LIBRARY

1555 AGATE ST. 686-3702
DAYS/HOURS: 12:30 TO 6:30 MON.-WED., 12:30 - 5:30 THRUS.-FRI., 10:30-1:30 SAT.
USERS: STUDENTS AND THE GENERAL PUBLIC.
FEE: STUDENTS ARE FREE WITH I.D.; NON-STUDENTS PAY $2.50 PER QUARTER OR $10 PER YEAR PLUS 25¢ RENTAL ON EACH TOOL. THREE DAY CHECK OUT AND FINES FOR OVERDUE TOOLS. STUDENT I.D. IS KEPT UNTIL THE TOOL IS RETURNED.

REGULAR READINGS

RAIN
2270 N.W. IRVING
PORTLAND, OREGON 97210

A JOURNAL OF APPROPRIATE TECHNOLOGY, PUBLISHED 10 TIMES YEARLY BY THE RAIN UMBRELLA, INC.

SERIATIM, JOURNAL OF ECOTOPIA
EL CERRITO, CALIFORNIA 94530

A JOURNAL OF APPROPRIATE TECHNOLOGY AND SELF RELIANCE. SUBSCRIPTION RATES: 1 YEAR $9.00. PUBLISHED QUARTERLY.

SMALL FARM ENERGY PROJECT NEWSLETTER
CENTER FOR RURAL AFFAIRS
P.O. BOX 736
HARTINGTON, NEBRASKA 68739

SMALL FARMER'S JOURNAL
P.O. BOX 197
JUNCTION CITY, OREGON 97448

PUBLISHED QUARTERLY FOR THE FAMILY FARM, WITH INFORMATION ON FARM EQUIPMENT, LIVESTOCK, CROPS, ETC. SUBSCRIPTION RATES: ONE YEAR (4 ISSUES) $10.00.

TILTH
RT. 2, BOX 190-A
ARLINGTON, WASHINGTON 98223

A JOURNAL ON BIOLOGICAL AGRICULTURE AND ACTIVITIES IN THE NORTHWEST. SUBSCRIPTION RATES: 1 YEAR $8.00.

URBAN FARMER
RIVER HOUSE
301 NORTH ADAMS
EUGENE, OREGON 97402

A MONTHLY NEWSLETTER ON COMMUNITY GARDENS IN EUGENE, AND OTHER GARDENING INFORMATION. PUBLISHED BY THE EUGENE PARKS AND RECREATION COMMUNITY GARDENS PROGRAM.

WILLAMETTE VALLEY OBSERVER
99 W. 10TH, SUITE 216
EUGENE, OREGON 97401

A WEEKLY PUBLICATION ON LOCAL ISSUES, INCLUDING A WEEKLY GARDEN COLUMN, AND SPRING GARDENING SECTION. 25¢ PER COPY.

BOOKS

BELANGOR, JEROME, P., THE HOMESTEADER'S HANDBOOK ON RAISING SMALL LIVESTOCK; RODALE PRESS: EMMAUS, PA.

CHAN, PETER, AND GILL SPENCER, BETTER VEGETABLE GARDENS THE CHINESE WAY; GRAPHIC ARTS CENTER: PORTLAND, OREGON. 1977

COLEBROOK, BINDA, WINTER GARDENING IN THE MARITIME NORTHWEST; TILTH ASSOCIATION: ARLINGTON, WASHINGTON. 1977

CUTHERTSON, TOM, ALAN CHADWICK'S ENCHANTED GARDEN; E.P. DUTTON: NEW YORK. 1978

DADANT, C.P., FIRST LESSONS IN BEEKEEPING; JOURNAL PRINTING CO.: ILLINOIS. 1976

DARLINGTON, JEANNIE, GROW YOUR OWN; THE BOOK PEOPLE: CALIFORNIA. 1970

ECOLOGY CENTER, HOW TO RAISE RABBITS AND CHICKENS IN URBAN AREAS; BERKELEY, CALIFORNIA

EL MIRASON EDUCATIONAL FARM, AGRICULTURE IN THE CITY; COMMUNITY ENVIRONMENTAL COUNCIL, INC.: CALIFORNIA. 1976

FALGE, PAT AND ARNOLD LEGGET, THE COMPLETE GARDEN; OLIVER PRESS: WILLITS, CALIFORNIA. 1975

GROSSET, DUNLAP, THE WISE GARDEN ENCYCLOPEDIA; WM. H. WISE & CO., INC.: NEW YORK. 1977

HECKEL, ALICE (ED), THE PFEIFFER GARDEN BOOK, BIODYNAMICS IN THE HOME GARDEN; BIO-DYNAMIC LITERATURE PUB.: WYOMING, RHODE ISLAND.

HYLTON, H. WILLIAM (ED), BUILD IT BETTER YOURSELF, RODALE PRESS, INC.: EMMAUS, PA. 1977

JEAVON, JOHN, HOW TO GROW MORE VEGETABLES THAN YOU EVER THOUGHT POSSIBLE ON LESS LAND THAN YOU CAN IMAGINE; ECOLOGY NATIONAL MIDPENINSULA: PALO ALTO, CALIF. 1979

JOHNSTON, ROBERT J.R., GROWING GARDEN SEEDS, ALBION, MAINE. 1976

KEOPH, PETERSON, SCHAUMANN, BIO-DYNAMIC AGRICULTURE; BIO-DYNAMIC LITERATURE PUB: WYOMING, RHODE ISLAND.

KEOPH, H., WHAT IS BIO-DYNAMIC AGRICULTURE; BIO-DYNAMIC LITERATURE PUB: WYOMING, RHODE ISLAND.

LAPPÉ, M. FRANCES; JOSEPH COLLINS, FOOD FIRST; BALLANTINE BOOKS: INSTITUTE FOR FOOD AND DEVELOPMENT POLICY: NEW YORK. 1978

LERZA, CATHERINE, AND MICHAEL JACOBSON, FOOD FOR PEOPLE NOT FOR PROFIT; BALLANTINE BOOKS: NEW YORK. 1975

LOGSDON, GENE, SMALL-SCALE GRAIN RAISING, RODALE PRESS, INC.: EMMAUS, PA. 1977

MERRILL, RICHARD, RADICAL AGRICULTURE, HARPER COLOPHOR BOOKS: SAN FRANCISCO, CALIFORNIA. 1976

MILLER, DOUGLAS, VEGETABLE AND HERB SEED GROWING FOR THE GARDEN AND SMALL FARMER, BULLKILL CREEK: HERSEY, MICHIGAN.

MINNICH, JERRY AND MARJORIE HUNT (EDS), ORGANIC GARDENING AND FARMING, THE RODALE GUIDE TO COMPOSTING; RODALE PRESS: EMMAUS, PA. 1979

OLKOWSKI, HELGA AND WILLIAM, THE CITY PEOPLE'S BOOK OF RAISING FOOD; RODALE PRESS: EMMAUS, PA.

PHILBRICK, H. AND DEVIN ADAIR, COMPANION PLANTS AND HOW TO USE THEM; GARDEN WAY PUB. 1966

PHILBRICK, H. AND J., THE BUG BOOK; GARDEN WAY PUB. 1974

RODALE BOOKS, INC.: EMMAUS, PA.
- THE BASIC BOOK OF ORGANIC GARDENING
- ENCYCLOPEDIA OF ORGANIC GARDENING

·HOW TO GROW FRUITS AND VEGETABLES BY THE ORGANIC METHOD
·ORGANIC WAY TO PLANT PROTECTION
·BEST IDEAS FOR ORGANIC VEGETABLE GROWING
·THE RODALE GUIDE TO COMPOSTING
·THE GARDENER'S GUIDE TO BETTER SOIL
·STARTING RIGHT WITH SEEDS
·SMALL-SCALE GRAIN RAISING (GENE LOGSDON)

SEVERN, JILL, GROWING VEGETABLES IN THE PACIFIC NORTHWEST;
    MADRONE PUBLISHERS, INC.: SEATTLE, WASHINGTON. 1978

SOLOMON, STEVE, GROWING ORGANIC VEGETABLES WEST OF THE CASCADES;
    TERRITORIAL SEED COMPANY PUB: LORANE, OREGON. 1979

SOPER, JOHN, STUDYING THE AGRICULTURE COURSE; BIO-DYNAMIC LITERA-
    TURE: WYOMING, RHODE ISLAND.

TETRAULT, J., AND SHERRY THOMAS, COUNTRY WOMEN; ANCHOR/DOUBLEDAY
    PUB: GARDEN CITY, NEW YORK. 1976

WHEATLY, MARGARET, GARDENING ROUND THE YEAR; WOODBRIDGE PRESS:
    SANTA BARBARA, CALIFORNIA. 1977

SHARING RESOURCES

IT'S THE PEOPLE THAT MAKE A COMMUNITY. THE FOLLOWING LIST INCLUDES
SOME OF THE ORGANIZATIONS THAT CAN HELP COMMUNITIES BECOME MORE
SELF-RELIANT IN FOOD PRODUCTION.

IN EUGENE

AMITY FOUNDATION   (484-7171)
2760 RIVERVIEW
EUGENE, OREGON 97403

    RESEARCH AND INFORMATION ON AQUACULTURE (FISH AS A SOURCE OF
    PROTEIN), SOLAR GREENHOUSES, AND WORMS. MANAGES A TOOL LI-
    BRARY IN THE FRIENDLY NEIGHBORHOOD.

APROVECHO INSTITUTE   (345-5981 OR 1-929-6925)
359 POLK STREET
EUGENE, OREGON 97402

    PROVIDES WORKSHOPS AND CONSULTATION ON GREENHOUSES, COMMUNITY
    GARDENS (IN CORVALLIS), LORENA COOKSTOVES, HAYBOX COOKERS,
    SOLAR WATER HEATERS. INTERNATIONAL APPROPRIATE TECHNOLOGY
    NETWORKING.

COALITION FOR COMMUNITY SELF-RELIANCE   (484-7171) OR (343-2711)

    A COALITION OF LOCAL APPROPRIATE TECHNOLOGY GROUPS IN AND
    AROUND EUGENE. OFFERS WORKSHOPS AND INFORMATION ON URBAN
    AGRICULTURE, APPROPRIATE TECHNOLOGY, AND SELF RELIANCE.
    FOR INFORMATION CALL AMITY FOUNDATION, OR PROJECT SELF-
    RELIANCE.

COMMUNITY FOOD BANK   (687-4038)
LANE COUNTY SENIOR SERVICES
135 E. 6
EUGENE, OREGON 97401

    REDISTRIBUTE DONATED AND GLEANED FOODS TO LOW-INCOME PEOPLE
    AND SENIOR CITIZENS IN LANE COUNTY.

COMMUNITY GARDENS PROGRAM   (687-5329)
EUGENE PARKS AND RECREATION
RIVER HOUSE
301 N. ADAMS
EUGENE, OREGON 97402

    COORDINATES CITY-SPONSORED COMMUNITY GARDEN AREAS THROUGHOUT
    EUGENE.

EDIBLE CITY RESOURCE CENTER   (686-3647, MESSAGES)
EUGENE, OREGON

    PROVIDES CONSULTATION AND INFORMATION ON NEIGHBORHOOD COM-
    MUNITY SELF RELIANCE IN FOOD PRODUCTION AND DISTRIBUTION
    SYSTEMS.

LANE COUNTY EXTENSION SERVICE   (687-4243)
950 W. 13
EUGENE, OREGON 97402

> PROVIDES EDUCATION, TRAINING AND ADVICE ON GARDENING AND
> AGRICULTURE. ASK ABOUT MASTER GARDENING PROGRAM.

OREGON APPROPRIATE TECHNOLOGY   (683-1613)
BOX 5388
EUGENE, OREGON 97405

> PROVIDES CONSULTATION, EDUCATION AND RESEARCH IN AREAS OF
> SOLID WASTE MANAGEMENT, SOLAR GREENHOUSES, COMPOSTING,
> BUILDING DESIGN AND OTHER APPROPRIATE TECHNOLOGIES.

ORGANICALLY GROWN COOPERATIVE   (686-3176)
1640 E. BEACON
EUGENE, OREGON 97404

> A PRODUCER'S COOPERATIVE OF LOCAL FARMERS. INFORMATION EX-
> CHANGE AND RESEARCH CONCERNING LAND MANAGEMENT, CROP PRO-
> DUCTION, MARKETING, AND THE USE OF APPROPRIATE TECHNOLOGIES
> IN ORGANIC AGRICULTURE.

OREGON FOOD ACTION COALITION   (344-0009)
1414 KINCAID
EUGENE, OREGON 97403

> A COALITION TO WORK ON FOOD AND AGRICULTURE STATE POLICY
> GUIDELINES FOR OREGON. ISSUES INCLUDE FAMILY FARMS, FOOD
> QUALITY AND HEALTH, FEDERAL FOOD ASSISTANCE PROGRAMS, AND
> OREGON'S INTERNATIONAL TRADE OF FOOD STUFFS.

PEOPLE TO PRESERVE AGRICULTURAL LAND   (485-3366, EVES.)
547½ E. 13
P.O. BOX 1815
EUGENE, OREGON 97403

> PROVIDES ADVICE AND INFORMATION ON LAND USE ISSUES AND
> URBAN SERVICES, TO PREVENT URBAN DEVELOPMENT OF THE AREA'S
> PRIME AGRICULTURAL LAND.

QUEENRIGHT BEEKEEPER'S ASSOCIATION   (342-8015)
BOX 1255
EUGENE, OREGON 97440

> PROVIDES INFORMATION AND TRAINING, BULK BUYING OF BEE-
> KEEPING EQUIPMENT. NEW MEMBERS WELCOME.

URBAN FARM
DEPT. OF LANDSCAPE ARCHITECTURE
UNIVERSITY OF OREGON
EUGENE, OREGON 97403

> DEMONSTRATION MODEL FOR URBAN AGRICULTURE IN NEIGHBORHOOD
> BLOCKS. INCLUDES TOOL LIBRARY, GARDENING INFORMATION,
> COOPERATIVE BIODYNAMIC RAISED-BED GARDENING, WORM COMPOST-
> ING, FRUIT ORCHARDS AND ANIMALS. GARDEN WORK DAYS OPEN
> TO COMMUNITY. VOLUNTEERS WELCOME.

OUTSIDE OF EUGENE

ECO-ALLIANCE   (1-753-2101)
2555 N.E. HYW. 99 W.
CORVALLIS, OREGON 97330

> A GROUP PROVIDING INFORMATION AND RESEARCH ON ALTERNATIVE
> WASTE MANAGEMENT SYSTEMS.

FARALLONES INSTITUTE RURAL CENTER
15290 COLEMAN VALLEY ROAD
OCCIDENTAL, CALIFORNIA 95465

> OFFERS WEEKEND AND 5-WEEK CLASSES IN BIO-INTENSIVE HORTI-
> CULTURE, EDIBLE LANDSCAPING, NUTRITION, AND OTHER APPROPRIATE
> TECHNOLOGIES.

TILTH
RT. 2, BOX 198
ARLINGTON, WASHINGTON 98223

> PROVIDES INFORMATION ON ALL ASPECTS OF NORTHWEST AGRICULTURE,
> FROM CULTURAL PRACTICES TO ECONOMICS. EMPHASIS ON BIOLOGICAL
> AGRICULTURE.

WASTE TRANSFORMATION, INC.
P.O. BOX 1236
CORVALLIS, OREGON 97330

> PROVIDES INFORMATION ON ALTERNATIVE WASTE TRANSFORMATION
> AND COMPOSTING SYSTEMS, AS PUBLISHED IN THEIR JOURNAL,
> "WASTE TRANSFORMATION."

Section III

Neighborhood Transformation Principles Applied:

The Whiteaker Neighborhood

This portion of the work involves the direct application of theory to practice, a bridge important in spanning planning and design with the implementation of these demonstration projects. Many thanks to Cindy Girling and Jim Klein, Graduate Teaching Fellows in the Department of Landscape Architecture at the University of Oregon and the following student-neighborhood catalysts:

Bob Downing, Doug Graham, Don High, Gary Hoyt, Liz Lardner, Laurel Lyon, Alan Pardee, Fred Patch, Noel Prchal, Garth Ruffner, Marc Russell, Tam Shawgo, Charlie Sundberg, Paul Swenby, Frank Theis, Mary Truax, Florian Trummer, Jeff Wilson

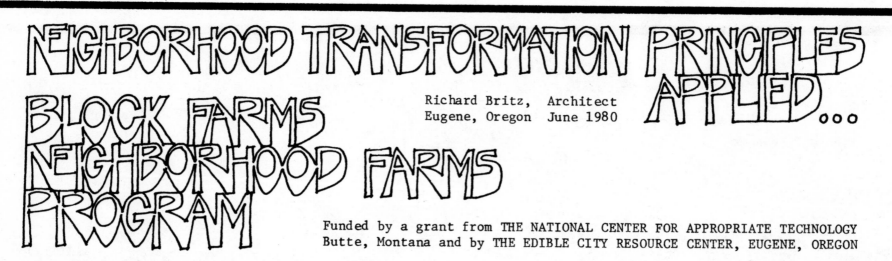

NEIGHBORHOOD TRANSFORMATION PRINCIPLES APPLIED...

BLOCK FARMS NEIGHBORHOOD FARMS PROGRAM

Richard Britz, Architect
Eugene, Oregon   June 1980

Funded by a grant from THE NATIONAL CENTER FOR APPROPRIATE TECHNOLOGY Butte, Montana and by THE EDIBLE CITY RESOURCE CENTER, EUGENE, OREGON

THE BLOCK

Watch this space!

Chapter 6

Whiteaker's Urban Experiment

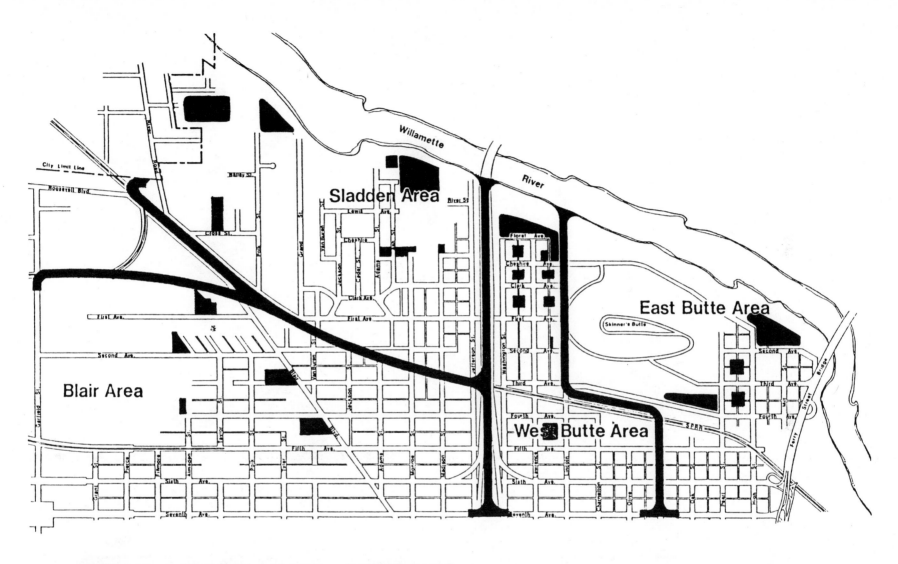

Willamette

River

City Limit Line

Roosevelt Blvd.

Sladden Area

East Butte Area

Skinner's Butte

Blair Area

West Butte Area

PROJECT LOCATIONS : CLUSTERS
BLOCK FARMS, NEIGHBORHOOD FARMS

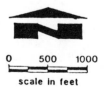

N

0   500   1000

scale in feet

WHITEAKER'S URBAN EXPERIMENT

Maureen Good, Project Coordinator

Don Corson and Sam Sadler, Energy Coordinators

Jim McCoy, Housing Coordinator

Sandy Boysen, Health Coordinator

Michael Baechler, Recycling Coordinator

Richard Britz, Food Coordinator

Report prepared by Brian Livingston, Project Evaluator

Dec. 1, 1979

The Whiteaker neighborhood in Eugene, Oregon includes 6000 people,
more than 65% of whom qualify as low-income by HUD definitions. About
29% of the area's residents are elderly, and almost three-fourths of
those fall beneath the low-income guidelines. The neighborhood
exhibits many of the classic symptoms of urban neglect, and presents
a setting for a significant experiment in new policies.

Whiteaker Community Council, 341 Van Buren, Eugene, OR 97402, 343-7713.

Overview

A serious attempt to provide a model for urban problem-solving, through an integrated, comprehensive approach.

The problems facing urban areas have been the subject of numerous studies, reports, speeches, and pilot projects. Billions of dollars have been expended on demonstration programs and pet theories --but urban problems in housing, health care, and food and energy supply still persist.

In July 1979, a neighborhood group received a grant for $146,000 from the National Center for Appropriate Technology, an agency based in Butte, Montana. The Whiteaker Community Council, a democratically-governed body representative of Eugene, Oregon's central city area, is using these funds to test the feasibility of managing urban re-development on a locally-controlled scale, rather than a massive one. The funds are being spent to plan a new urban strategy that uses available skills and technology to put people to work producing essentials for their own community. This is the midway report on that effort.

Five areas have been selected for concentration, all because of their essential nature, with these goals:

ENERGY -- Develop an energy plan that will have maximum benefit for low-income households, and create jobs locally.

HOUSING -- Reverse the spiraling costs of housing while making new and rehabilitated housing units available.

HEALTH -- Reduce health costs for needy families, while improving the quality of care and emphasizing overall wellness.

FOOD -- Make local food resources available to inner-city residents, lowering total food costs in the process.

RECYCLING -- Take advantage of valuable materials that would otherwise go to waste, and create jobs by re-processing and selling those materials.

Each of these areas has a specific plan, with goals and methods of accomplishment spelled out. The five components are designed to be model plans that can be widely used in cities across the country. A major element in this approach is that each component's work supports the efforts of the others, thereby assisting in their success. However, each component can stand alone as a separate part of a community economic development plan.

Overall, the neighborhood is assembling $2.5 million in private capital and public resources to make all of its components operable and capable of future self-sufficiency. The number of jobs this would create and the impact on low-income residents will be detailed in final reports to be published at the end of the initial grant period, March 1, 1980.

Energy

The rapidly increasing cost of energy strikes hardest at those with low incomes. The impact is particularly hard on renters, who are not able to take advantage of energy conservation subsidy programs designed for homeowners.

More than 35% of U.S. households consist of renters, and of that number over 55% are considered low-income. Yet there are few, if any, programs that are effective to help these households lower their energy bills, because of the conflicting interests between landlords and tenants. The tenant in most cases has little motivation to make capital improvements for home insulation on someone else's property, even if it would eventually result in savings from reduced energy bills. The landlord, on the other hand, has no incentive to spend dollars for weatherization if he or she is not paying the unit's energy costs or can pass them through to tenants. And few agencies have been able to reconcile these divergent viewpoints with the pressing need so save as much energy as possible within the community.

Faced with this dilemma, the Energy component decided that its top priority would be to make energy programs available to renters. Numerous studies have shown that weatherization and insulation are the most cost-effective ways to lower residential heating bills. But three problems had to be solved before this approach could be taken toward rental housing: 1) how to convince property owners to weatherize; 2) how to raise loan money for such improvements; and 3) how to pay back such loans.

1. It was found that there are three basic strategies to reach landlords: education of tenants; persuading them directly; or through a locally-enforced energy efficiency code. The first approach, using materials hand-delivered by outreach workers to renters who had returned a neighborhood energy questionnaire, was fairly simple and well-received. But it was not very effective since there is little tenants can do by themselves to conserve energy inexpensively, and they have little influence with property owners. The second approach, contacting landlords directly, produces some results but is time-consuming and ultimately discouraging.

The most adequate and equitable approach thus appears to be a mandatory weatherization program with readily available financing. A comprehensive energy-efficiency requirement for all housing would accomplish the tasks of providing adequate, warm shelter and saving energy, while being fair in spreading costs and benefits evenly. This approach was successfully inaugurated by the city of Portland, Oregon in 1979, which now requires the installation of any weatherization that has a payback period of ten years or less.

2. To make financing available to property owners (again using the Portland model), the city is establishing a quasi-independent, non-profit corporation to arrange lending through financial institutions. Owners must upgrade when their property is sold.

There are other ways to raise the necessary loan money to make a

---

The Energy component intends to lower energy costs for low-income households by:

1) developing a locally-mandated program to weatherize all rental housing, and

2) creating an Energy Services Co. especially to meet the needs of the affected community.

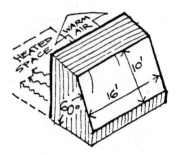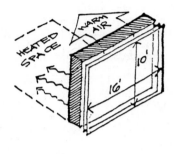

Fig. 1: EFFECTIVE RETROFITS. A simple addition such as this south-facing sun room (A) exhibits the "greenhouse effect," warming the adjacent living space; where there is limited room, a thermal wall with double glazing in front of a dark masonry surface (B) collects heat almost as well. Either can provide more than 10% of home heating needs with average daily temperatures of 39° F. in January.

community-wide energy code possible. One of the most promising approaches involves a municipal utility raising sufficient revenue through the sale of bonds. The revenue is used for low-interest loans to property owners, the loans are gradually paid back by various means, and the resulting energy savings keep the utility's rates from climbing due to the high cost of developing new energy resources. Investor-owned utilities, of course, could use private capital to do the same thing.

A potential barrier to this means of financing is the National Energy Conservation Policy Act of 1978, which has a provision prohibiting utilities from being involved in financing mechanisms. However, the U.S. Dept. of Energy is encouraging requests for waivers of the NECPA prohibition.

3. If funds were available for rental weatherization, how would the financing be paid back? Two payback mechanisms are being studied. One method would add a surcharge to the utility bill of each individual rental unit. This charge would cover the cost of weatherization, but would be calculated to be less than or equal to the unit's energy savings, so the occupant would reap the benefits of the improvement. The other method is for the principal and interest of the loan to be repaid in one sum at the time the building changes ownership, with property owners increasing rents to cover costs. There are pros and cons to either method, and a final conclusion has not been reached on which method would be more advantageous for low-income households.

The Energy component is preparing a separate feasibility study for a neighborhood Energy Services Co. This enterprise would take advantage of the new cost-effectiveness of insulation, storm windows, and solar water heating, and provide jobs for low-income residents. If demand warranted, a full-line energy store could be established.

PROJECT
INTEGRATION

The Energy component affects:

The Housing component by weatherizing low-income housing;

The Health component to help prevent respiratory and other illnesses related to cold and drafty shelter;

The Recycling component in its energy analyses of solid waste utilization;

The Food component in designing sun rooms that double as urban greenhouses.

Chapter 6: Whiteaker's Urban Experiment   203

Housing

The Housing component intends to lower housing costs by organizing a cooperative corporation to eliminate unnecessary housing expenses.

Housing has become a stark nightmare for many urban centers. In many great metropolitan areas, housing units are simply abandoned as uninhabitable. In cities where the situation is not yet that critical, housing costs threaten to rise to the point where few can afford even minimal standards of shelter.

The cost of a two-bedroom house in Eugene, Ore., a metropolitan area under 500,000, has skyrocketed from $20,000 in 1976 to $45,000 today—a 125% increase over the three-year period. Rental housing, of course, has reflected these increases as property values soar. And the rental market will get worse and worse for low-income families as middle-income wage-earners are priced out of the mortgage market. To qualify for financing on the small house above at nearly any financial institution, a family would be required to have an income of $23,400 per year (compared with the county's median family income of $17,700). Whereas rental housing used to be a temporary stopgap until new families had a chance to settle into their own home, it has become a bleak inevitability for many people today.

The transience and lack of self-management attendant with rental units creates other social problems, as well. Youth rootlessness, capital flight from unstable urban areas, and disruption of education and social services are the result. More than two-thirds of the residents of an area like Whiteaker (where tenants comprise 77% of the population) will have moved on within two years. Out of 100 children who are enrolled in elementary school, only 26 will return in September.

A promising solution to these and other problems is the housing cooperative corporation. Such a corporation can own housing and make it available to a specific section of the population, or make it open to anyone, with significant benefits for residents.

The Housing component has developed a strategy that would allow low-income people to become home owners through a cooperative corporation. Although the corporation, with a democratically-elected board, would actually own and manage the properties involved, members of the cooperative would receive tax deductions for property tax and interest paid like any homeowner.

A major economic advantage to cooperative corporations over rental housing, condominiums, or single-family ownership is the elimination of transfer charges. A cooperative corporation often carries the same mortgage intact for 30 or 40 years, whereas the other types of housing ownership change hands on the average of every 7 years. When a new mortgage is issued, about 2½% of the purchase price must be paid for transfer costs, plus a 6% fee if a real estate agent is involved. Even on a minimal $35,000 house, these costs would total $2920, or (at a 9½% discount rate financing) an additional $47 on the mortgage payment every month. The cooperative avoids these transfer charges, and over a period of years will become increasingly less costly compared with comparable housing on the speculative market.

In fact, through this means alone, cooperative housing is so advantageous that it would cost only one-half the rental of comparable housing within 10 years!

This and other advantages make cooperative housing ideal for provision of low- to moderate-income units. The Housing component is moving ahead with two main objectives: construction of an energy-efficient 50-unit housing cooperative complex, and conversion of existing rental units to cooperative forms of ownership. The project has already attracted considerable resources, such as a two-year Community Development Block Grant commitment for full-time staff and $87,000 to leverage financing for land acquisition. Still needed are funds to continue outreach efforts and development costs, such as legal aid, accounting and architectural consulting, survey fees, soils tests, loan initiation costs, etc. Monies raised for these purposes could establish a revolving fund, to be financed within the mortgage and thus recaptured for use in subsequent projects.

In addition to the benefits of low-cost housing, the Housing component has the greatest potential for social dividends related to the other components. Land will be made available for urban farm programs; materials will be collected in an efficient form for the marketing of recyclables; and units will be prime candidates for cost-effective weatherization and retrofitting of renewable energy devices. Finally, the increased family stability which comes from property ownership rather than month-to-month rental will enhance the livability of the entire area.

Fig. 2: RESIDENTIAL COMMUNITY. Cooperative housing can be moderately dense and still be pleasant when residents have a say in the management of the facilities. This earth-sheltered cluster is in Minneapolis.

Health

The Health component is designed to:

1) lower health costs by keeping people well rather than treating them only in crises;

2) giving people the information they need to keep themselves healthy;

3) organize a Health Action Council of both consumers and producers of health care to advocate more effective health services.

Poor health is one of the major problems afflicting low-income families. Because health care is expensive, disadvantaged households often postpone treatment of illness until the matter becomes serious, thereby multiplying their costs. A 1976 survey of Whiteaker, where the majority of residents are low-income, found that 90.3% of all households had a significant "health need" in a two-year period. Money that must be spent on medical crises often keeps families from being able to afford other basic necessities.

The Health component seeks to establish an alternative to the traditional "subsidy approach" of merely paying crisis intervention costs for families that have chronic health problems due to poverty. Since emergency care is much more costly than prevention, traditional social services in health care put a heavy burden on available revenues. "Band-aid treatment" returns needy patients to a situation where they may get sick again from poor nutrition, inadequate sanitation, cold and damp housing, or the stress of despair.

Instead, the Health component is establishing the feasibility of preventive health care for low- and moderate-income communities. Under this plan, a family, their employer, or a social services agency would pay a set fee for primary health care needs. The program would emphasize regular check-ups, counseling on health problems, and "prescriptions" on self-help methods to avoid illness.

There is considerable evidence that this approach saves money and could be adopted community-wide or nationwide to dramatically cut health expenditures. Today, health care expenditures total $181 billion dollars per year--almost 10% of the Gross National Product! This is a 1,400% increase since 1950. Businesses paid more than 5% of their payroll to health insurance plans in 1977, according to the U.S. Chamber of Commerce, up from 3.8% in 1973.

One type of preventive pre-paid health plan, known as a Health Maintenance Organization (HMO), has been found to reduce outpatient visits by 15% and hospitalization by 30% to 50%. Enrollment in HMO's would represent a $3 billion savings for employees of the "Fortune 500" companies alone.

To determine the potential acceptance of a pre-paid preventive plan, the Community Health & Education Center (CHEC) conducted a survey for the Health component in cooperation with SelectCare, a Federally-qualified HMO. The survey was conducted in two parts: one for businesses, the other for household residents. The results were surprising.

The household survey, conducted with a random sample of Whiteaker residents, found that 71.4% of those interviewed stated "myself" as their primary health provider when they are ill. This contrasts with 12.0% who named a family member, and only 15.0% who named a physician. When a pre-paid health approach was described, 93.5% said they would like to receive more information, and 83.8% said they would use a neighborhood health organization as the primary clinic for such a plan. 11.4% of the households had had a member use the clinic at

CHEC, which has operated for the past two years a neighborhood program emphasizing preventive care. And of that number, a remarkable 100% said they had been satisfied with the service (total sample: 137).

In the business survey, employers stated that the benefits they would most like which are not covered by their present health insurance plans were: 1) prevention and 2) routine well-care treatment. This interest may be due in part to the fact that employers cited colds/flu and stomach ailments as their first and third most common reasons for medical absenteeism. (Both may be related to poor general fitness.) Close behind were high blood pressure, alcohol/drug misuse, and anxiety/depression/stress ("accidents" was rated second). All of these complaints may be susceptible to reduction by preventive care.

The Health component, while seeking funds to establish a model community-wide preventive health service, is moving ahead to educate low-income residents and set up a health advocacy group. CHEC has been operating a successful series of health workshops, has delivered brochures on simple self-care steps anyone can take (some brochures were delivered door-to-door by project outreach workers), and trained block residents to act as health resource people for their neighbors.

At midpoint in the planning phase, a Health Action Council had been formed to take steps leading to better health services. Consisting of 12 residents and 6 health providers, the council includes representatives from the county health department and medical society, the nurse for neighborhood schools, the area-wide health planning agency, and Sacred Heart Hospital, a regional health facility. Since the elderly and young mothers make up a majority of the low-income in the neighborhood (and have many of the most-reported health problems), they have representatives on the council as well.

The Health component is embarking upon a major feasibility study that may result in a decision to establish a pre-paid plan by March 31, 1980. The decision will take into account staffing needs, Federal regulations, financing, and corporate structure--with the object of creating a comprehensive and effective health program that meets the needs of residents regardless of income.

PROJECT INTEGRATION

The Health component affects:

The Food component by developing information on good nutrition to improve health;

The Housing and Energy components by stressing the need for warm and dry shelter for good health;

The Recycling component by educating residents on proper sanitation and treatment of solid wastes.

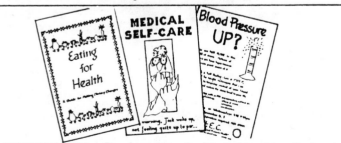

Fig. 3: PREVENTIVE CARE. The Community Health & Education Center has produced a variety of self-help brochures, part of the Health work.

Recycling

The Recycling component is designed to create jobs by re-processing materials that would otherwise be thrown away.

A major headache for large metropolitan areas is the management of the tons of solid waste produced each year by residents. The most widely-used system for coping with these piles of garbage is to bury them in a central sanitary landfill. But so much waste is being produced that it is starting to overwhelm many cities. More than half of the nation's metropolitan areas will have filled their sanitary landfills by the year 1982, and it is becoming practically impossible to get public support for new landfill sites in any existing neighborhoods.

A solution to this problem has been demonstrated repeatedly on a small scale: source separation of recyclables from garbage. About 58% of what people throw away consists of metal cans, glass containers, and fibers such as newspaper and cardboard--all of which can be very valuable as raw materials. The remainder of the waste stream includes a large amount of foodstuff, which can be made into precious soil conditioner as petrochemical-based fertilizers become exceedingly expensive. The remainder of garbage is just that--irreducible garbage. But the value of the saleable portion is forcing urban areas to examine whether they should be paying to bury materials they could be selling, such as metals at $600 to $900 per ton.

For these and other reasons, source separation is going big scale. Pilot projects are underway to reclaim every possible pound

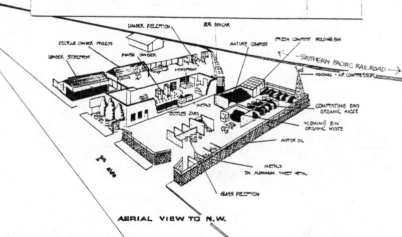

WHITEAKER RECYCLING CENTER PROPOSED SITE SYSTEMS LAYOUT

AERIAL VIEW TO N.W.

Fig. 4: RE-PROCESSING CENTER. Sections of the city comprising several neighborhoods would be served by a center such as this, handling a full-line of recyclable products, including metals, glass, fibers, organic materials, and lumber.

of metal, glass, fibers, and organic waste generated by entire cities. Instead of one large garbage can, neighborhood residents are provided a can and several smaller buckets. A typical municipal system would pick up the garbage can once a week, and pick up a bucket of metals, glass, or fibers once or twice a month.

A collection system such as this has been functioning in Whiteaker since 1977. Each household places its organics in a compost bucket that is picked up weekly along with regular garbage, puts glass and metal containers in separate plastic mesh bags which are picked up twice a month, and bundles fibers with twine for monthly pickups. A consumer-owned business named Garbagio's comes regularly to make the collections, and charges the same fee to pick up the garbage as regular garbage haulers. The only savings people might receive is that it is possible they would reduce their waste from two cans to one (garbage haulers usually charge a fee per can), and they must do some extra work to put materials in the right can or bucket. Yet people are so happy to see their waste materials put to use that Garbagio's has had a customer renewal rate of 96%!

With this success in collecting about 15% of the solid waste from Whiteaker, the Recycling component is expanding to bigger and better things. Preliminary feasibility studies show that a major recycling center (such as the one shown at left) would financially break even within four years of operation. Cash flow projections show a $20,000 cash drain in the center's first year, and cash inflows thereafter from $7,000 in Year Two to $36,000 in Year Five.

These projections are based on prices prevailing in the current marketplace, and were used to arrive at a quarterly sales volume of $25,000, which would be achieved in the fifth quarter of operation. This could be considered a conservative estimate, since both prices and costs of materials can reasonably be expected to rise.

The recycling center would serve several neighborhoods within the city itself, and would expand beyond Garbagio's present line of collectables to include resale of lumber salvaged from demolition projects. The compost Garbagio's now collects will be processed into a marketable soil conditioner by a forced air system that speeds the sanitary decomposition of the materials. This method was selected over methane digestion which currently has a poor market potential, and digestion by worm beds, which requires a longer time period. Other solid materials will be marketed directly from the proposed center, as well.

Initial capital requirements for the center will be approximately $140,000, excluding the acquisition of a site. About $20,000 is available in the form of a dump truck, front loader, shredder, air fans for composting, and lumber to renovate such a facility, all of which are being donated by residents and units of local government. Another $72,000 will be financed through a commercial loan. This leaves an estimated figure of $48,000 to be raised, in addition to a facility. Local businesses in the Whiteaker industrial area are considering donations of property or in-kind support for such a site.

PROJECT
INTEGRATION

The Recycling component affects:

The Energy component by saving energy;

The Housing component by picking up waste and converting it to marketable materials;

The Food component by picking up compostables and supplying soil conditioner;

The Health component in the area of solid waste sanitation.

Food

The Food
component aims
to increase the
availability
of inexpensive
food to inner-
city residents
by:

1) providing
for the maximum
use of
available
urban land for
food produc-
tion, and

2) developing
preservation
centers for
food, to take
advantage of
surplus times
when food is
cheap.

The price of food has been increasing rapidly alongside every other basic necessity these days. Food is such a basic source of sustenance that its gradual transition from an essential to something of a precious commodity is especially deplorable for low-income families.

The Food component intends to make food more available to urban residents by taking advantage of whatever land and resources can be assembled within neighborhoods. To make food affordable, urban residents must relearn the skills their rural counterparts have always used: gathering foods when they are plentiful and cheap, and storing them against the time when they are costly.

Building on the interest of Whiteaker residents in backyard gardening, the Food component is studying ways that neighborhood people can supply more of their own food needs from within the city itself.

Using existing maps and charts, aerial photos, and foot surveys, the entire neighborhood was inventoried for ownership patterns, soil quality, and location of existing food-bearing trees and gardens. As a result several areas were identified as desirable and feasible as sites for urban agriculture. Neighborhood residents met to hear these recommendations and determine priorities for land use. It was decided to focus efforts to acquire a large parcel for community garden use on vacant land adjacent to a riverfront park. The land is owned by a school district, which has indicated willingness to consider its use as an urban farm site.

In addition, two demonstration sites have been identified, with the help of Outreach Workers, for block gardens. One block, which consists mostly of home-owners, will develop the interior space of the block into a cooperative garden area. The other site involves mostly renters who have made a committment to the project, and will include garden, compost, chicken and rabbit raising on the fringe of an industrial zone. Representatives from each block will attend the Master Gardening Program, a ten-week training workshop offered by the Agricultural Extension Division. In return for the free classes, these master gardeners will become resources for other neighborhood gardeners, able to offer personalized advice and consultation.

Another aspect of food activities involved contacting schools and day care centers in Whiteaker to promote school gardens and to offer to develop lesson plans about food-raising and nutrition that could be incorporated into neighborhood children's learning experiences. Neighborhood schools agreed to participate one hundred per cent, and the food component has worked with teachers to develop curricula and locate garden sites. Site plans are being developed by University landscape architecture students for all sites involved.

A preliminary feasibility study exploring the possibility of developing a neighborhood business pruning, harvesting, and marketing the fruit and nuts from neglected trees and shrubs has been completed.

From the data, given condition, age, and the scattered location of the trees, it is not likely for such a business to be economically viable. However, the food value accrued to neighborhood residents through rehabilitation of these old trees still may be worth the investment. Accordingly the food component will be determining probable yields, and exploring other strategies for maintaining the trees and distributing their fruits.

Finally, the food component has studied the food delivery system as it presently exists in the neighborhood, including wholesale, retail, and service outlets, and how these relate to neighborhood needs and values. A proposal has been developed to establish two food preservation centers as community bases for such cooperative efforts as canning, gleaning, workshops and demonstrations on food preservation, coordinating transportation to U-pick farms, and making loans to homes that need the use of canning equipment. The initial two facilities would promote enough support for a minimum of 100 families to preserve 10,000 quarts of food. Included in the centers will be pressure canners, water bath canners, guage testers, dehydraters, and other necessary equipment. Through these efforts Whiteaker neighbors will become more self-reliant in meeting their food needs.

PROJECT INTEGRATION

The Food component affects:

The Housing component by coordinating ideal food-production sites with housing acquisition;

The Energy component through the use of solar greenhouses;

The Health component in the area of nutritional information;

The Recycling component by exchanging compostables for soil conditioner.

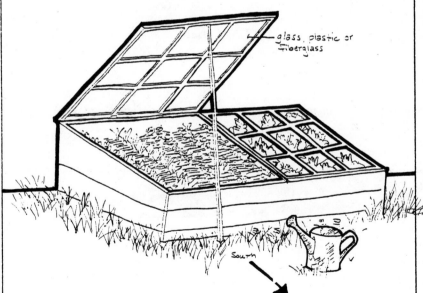

glass, plastic or fiberglass

South

Fig. 5: URBAN FOOD PRODUCTION. Simple cold frames such as this one fit into limited space, make food more readily accessible, and extend the growing season.

Chapter 7

East Skinner's Butte

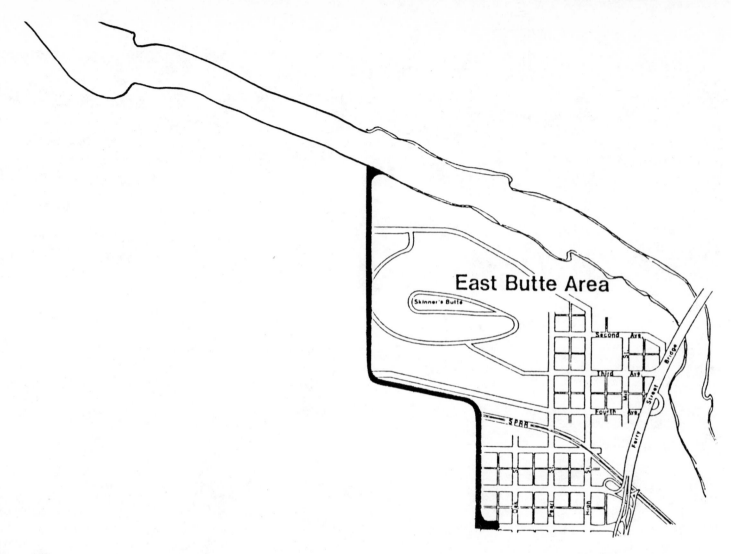

East Butte Area

Skinner's Butte

Second Ave.

Third Ave.

Fourth Ave.

SPAN

Ferry Street

Bridge

SECTOR 1
THE EAST SKINNER'S BUTTE AREA
HISTORIC DISTRICT · PRESERVING THE PAST...

EAST BUTTE (East Skinner's Butte sub-neighborhood Sector 1)

East Skinner's Butte neighborhood context is designated as a historic zone. The majority of residents are seniors, living in two major senior housing projects, Ya·Po·Ah Terrace (with a population of 285) and Parkview Terrace (with a population of 200±). These two building complexes are situated on two edges of East Skinner's Butte neighborhood and serve as the focus for gathering together the seniors' activities in and adjacent to the buildings. While Parkview has the indoor-related Campbell Senior Center adjacent to it and directly adjoining Skinner's Butte Park, until last spring and summer, Ya·Po·Ah Terrace seniors had no similar accommodations.

The blocks of single family housing (low density) in East Skinner's Butte are few and diminishing rapidly (with the exception of those designated as historic by the City of Eugene). Transitions in land use can be expected (with attendant housing density increase) in this area due to its close proximity to the downtown area. With trends toward decreased automobile reliance and the fixed and declining incomes of seniors, one of our assumptions for this work was the desirability for greater reliance on walking to any basic services needed by the East Butte residents. The essence of this report deals with food and food related services, so the remainder of this section describes desirable goals for these projects.

The goals and objectives for this neighborhood overlap somewhat with issues related to automobile routes through the neighborhood, since our research (see full explanation of data under West Butte) indicates a growing concern in the scientific community about the health effects of automobiles on human and plant life. Given these data, we concluded that the best course of action for East Butte neighbors was to proceed with all possible speed to close down major traffic throughways in this sector and simultaneously increase the quantity and density of oxygen-producing green plants.

With the goal of neighborhood self-reliance in food, we felt it would be a disservice to the neighborhood not to point out the health consequences of growing food within the city. Before locating and developing urban block farms or neighborhood farms, we decided to begin the process of simultaneously decreasing automobile impact in all areas of the sector. As can be seen from the accompanying air photo, major traffic routes both bound and bisect the East Skinner's Butte area. The major traffic boundary is on the east side of the neighborhood -- Ferry Street Bridge (on the right side of the air photo). Important buffering of this major auto route by coniferous trees is highly recommended. Any tall tree buffer will help decrease airborne pollutants, especially particulate matter such as asbestos brake lining particles and airborne noise levels. Decible readings indicate 90+ d.b. during peak rush hours at the Ferry Street Bridge, and as much tree planting as possible along this corridor, and the possible inclusion of a "sound" wall when the bridge is rebuilt, is highly recommended.

Trees (from seedlings to 2' high) are available from the U.S. Forest Service and B.L.M., and some planting has been accomplished this spring by students and faculty of the Department of Landscape Architecture. We have already purchased 1000 trees (Douglas Fir) for $60.00 from the Forest Service and they have promised to donate them in lots of 1000 after we planted the first 1000. This tree planting occurred in early May, 1980. Cooperating merchants in the East Butte area are Midgeley's Glass and Mill (Dan Marshall and the cooperative), and

Bill Emerey, owner of His Lordships Antiques at the corner of the railroad tracks and the Fifth Street Market. Bill has eliminated two parking places directly adjacent to his shop, and planted a large cedar, intended to be the first of another pollutant buffer along the railway corridor. Small starts, but important, since those who plant these trees have agreed to water them and nurture them to maturity. The continuation of planting belts immediately adjacent to major transportation corridors is highly recommended.

Urban forests suitable for small scale logging and/or other wood uses could evolve from these forested corridors, utilizing rights of way owned by the railroad, the City and State Highway Commissions, and others to serve future needs of the neighborhood's economy while simultaneously reducing the impact of automobiles on neighborhood air, water, and soil quality.

The major traffic bisecting East Skinner's Butte neighborhood is on High Street. This traffic moves off the Ferry Street Bridge onto Third Street and turns right onto High Street and circles Skinner's Butte by way of Cheshire and goes through West Skinner's Butte neighborhood by way of Lawrence, and ultimately merges with First Street which merges into River Road. From both the viewpoint of East Butte and West Butte, it is highly recommended to close the Third Street offramp from the Ferry Street Bridge changing the character of the traffic on High Street from thoroughfare to local "park circling" traffic. This effectively eliminates the "slicing" of East Butte neighborhood into two parts and provides less noise, danger, and airborne particulate matter near the park and the river.

Both with and without these traffic changes, two major areas and one minor area in East Butte are favorable for urban agricultural purposes. Both of these major areas serve population concentrations in large buildings whose residents have no immediately adjacent land to which they can relate. As density increases in this sector (such as the new condominium complex immediately adjacent to the river by Sedor, Unthank and Poticha, Architects) it will become more and more important to have open space within the neighborhood -- in our case, we recommend this land be used for neighborhood-scale food production. The first open land useful for both community building and self-reliant food production is at Ya·Po·Ah Terrace senior housing. Ya·Po·Ah Terrace is a twenty story building with 285 senior citizen residents over the age of 62 who are retired and/or permanently disabled. There is no grocery store with fresh produce within a mile radius. Although there is fresh produce sold at the apartments one day a week presently, the program was discontinued after the first of the year (1980). We surveyed the residents of the building with a questionnaire to determine their interest in developing a vegetable garden. Half the residents answered favorably to the idea of establishing garden plots at Ya·Po·Ah. We also noticed vegetable plants tucked in among the immaculately maintained flower beds on the property.

Previously at Ya·Po·Ah, people were not encouraged and places were not provided for outdoor activities. There were no places for people to play horseshoes, or lawn games, no individual gardens (flower or vegetable) and there was little opportunity for walking on the grounds. The residents were limited to activities within the building and in other areas of the city.

Nationwide tax revolts are resulting in social service cutbacks and pressure is presently being felt in the Eugene area. Particularly hard hit in Lane County are the Senior Services, one example being the loss of funding for the delivery of fresh produce to Ya·Po·Ah.

YA·PO·AH TERRACE

NORTH

PROPOSAL

CLOSE ACCESS ROAD TO MOTOR VEHICLES

WORKSHOP MAINTAIN SERVICE ACCESS

MAINTAIN SERVICE ACCESS

EAST SKINNER'S BUTTE HISTORICAL COMMUNITY CULTURAL CENTER

CRAFT WORKSHOP

REMOVE PARKING. RE-GRADE SLOPE TO RECREATE ORIGINAL SLOPES.

FUTURE ADDITIONAL PARKING CUT INTO SLOPE USE BASALT FOR TERRACING

PARKING TO BE ON GROUND LEVEL INSIDE ON THIS NORTH SIDE. MOVE DINING TO SOUTH SIDE.

STUDIO APARTMENTS BUILT INTO HILLSIDE

YA·PO·AH TERRACE RETIREMENT APARTMENTS (ON THE BUTTE)

PATH TO THE NEIGHBORHOOD AND THE PARK

FRUIT TREES ON SOUTH SLOPE

RELOCATED HOUSE

EXTENDED DECK UPPER FLOOR

PATIO

NEIGHBORHOOD ORCHARD

SOLAR GREENHOUSE

HANDICAPPED GARDENS

FIRST EXPERIMENTAL GREENHOUSE

GREENHOUSE

FOOTPATH

EAST SKINNER'S BUTTE NEIGHBORHOOD GARDENS

FLOWERS

WITH ANIMALS = FARM

GRAPES

MOVE 4th AVE. PARALLEL TO RAILWAY TRACKS

NEIGHBORHOOD LAWN A PLACE TO PLAY

BIKE PATH

RELOCATE HOUSES

HOUSES NEEDING HOMES CAN HELP TO REBUILD THESE UNITS AS A PART OF THE NEIGHBORHOOD.

In attempting to generate food production at Ya·Po·Ah Terrace, we had three major goals we wanted to meet. The first was to provide a source of fresh local produce. Community gardens at Ya·Po·Ah would make fresh produce available on the site and help compensate for the demise of the past program through Lane County Senior Services. Building a greenhouse on the site to be used for starting plants in early spring could be expanded into a program of winter food production in the future. This would produce a year-round supply of fresh food for the residents.

Secondly, we wanted to create outdoor physical activity spaces where seniors could meet other people in addition to gaining physical exercise. Day-to-day working in the garden, weeding, hoeing and seasonal tillings are activities most people can participate and do in varying degrees. Young and old persons can visit and mingle, discussing neutral topics such as healthy cabbages and wilted tomatoes. The benches provide a place for resting from garden work and another chance for people to meet and interact with other gardeners. The creation of a path from the building to the garden granted people permission to walk and participate in their landscape. The path also makes walking outside the building and the parking lot navigable by persons in wheelchairs and with walkers.

The third major goal was to establish a situation where people could take part in helping themselves. The garden could remove the middleman and take the burden from outsiders who presently are relied upon to bring produce into the building. Instead, seniors are taking on the responsibility themselves and are beginning to cultivate their own sense of self-worth and community self-reliance.

Preliminary soil testing indicated high clay content, but hard work by neighborhood residents and volunteers produced a series of basalt stone terraces and raised bed garden plots. In April of this year, eight fruit trees were planted along the terraces providing for fruit crops in two or three years, as well as the more immediately available garden crops. Working with the seniors was difficult and it required endurance and perseverance, but the interchange between senior adults and junior adults during the work phase of the project made exchange of learning possible. This project is a continuing one, and urban farm student labor has already added more raised beds and retaining walls as requested by the seniors. This project could be followed by a similar project at Parkview Terrace, utilizing the Campbell Senior Center as a community focus and developing raised beds on the land directly adjacent to the Campbell Senior Center. Permission will have to be granted to the seniors by Eugene Parks and Recreation to start small gardens within walking distance of Parkview and the new condominiums, but the land is there and is suitable for food production (class 1 malabon, silty clay loam). Due north of the Campbell Senior Center is approximately 2-1/2 acres of land unused by the Parks Department except as a continuing maintenance cost for mowing the grass. This site on the river terrace, and that at Ya·Po·Ah Terraces, jointly comprise about five acres of land suitable for an evolving neighborhood farm. Food and food-related activities (several food preservation workshops were given by the Edible City Resource Center staff at the Campbell Senior Center during this grant period, and negotiations were begun for use of facilities for neighborhood food preservation activities) are of importance to the East Butte seniors who now have assumed major responsibility for food growing in this area. Continued attention by neighbors to this activity is probable, and follow-up projects should be planned to facilitate this land parcel aggregation to a five-acre neighborhood farm. The health benefits of outdoor work for seniors are well documented in the urban agriculture literature, especially mental health by Dr. Karl Menninger of the Menninger Foundation in Topeka, Kansas.

It is important to aggregate small-scale successes (such as Ya·Po·Ah Terrace's gardens and orchard) into a larger organizational system, especially if food production becomes an even more pressing need among the senior people in Whiteaker neighborhood. Ya·Po·Ah Terrace's gardens and orchards require a sustaining coordinator, who, in the short run, can consolidate efforts of individuals into a mutually-supportive network of neighbors. This person, by starting small and using the gardens as a semi-public "commons" for group support activities, could coordinate food production activities with another coordinator at Parkview. This decentralization of responsibility and localization of authority should lead to more personal investment in the land use within East Skinner's Butte sub-neighborhood. Examples of gains generated from this involvement might be: bulk-buying of seeds, starts, plants or trees for future orchards; organized compost systems which re-use family garbage for soil amendments; food-buying in bulk for other neighbors with less time on their hands, etc., etc.

Two other projects in East Butte were accomplished during this grant period, but both are still in the "developing" stage. The first was the demolition and recycling of Herschel's tool shed on Third Street to provide wood material for Project Self-Reliance's use. The materials were stockpiled for use by neighbors to build recycling bins and/or compost bins throughout the neighborhood, with energy (labor) as payment for materials used. This project, accomplished in two days, opened up land suitable for a small garden directly across the street from the block we felt had the highest potential for a block farm in East Butte neighborhood. Herschel, on a fixed income and declining eyesight, must use his land to his maximum advantage since he, like many other seniors, had been approached by real estate speculators three times this past winter to sell his house. We recommended to Herschel that he build a small cottage and food producing gardens on the former site of his tool sheld, as an alley house and garden, and draw help from his neighbors. He needs help and could move into the smaller cottage and rent the larger house. His lot is a corner lot and can legally be developed as a duplex. If someone from his block or sector will help him, the Edible City Resource Center, Inc. has an architecture and planning parent, Edible Earth, Inc., who will donate architectural services to him.

The second project in the East Butte area which we undertook was the development of the block we felt had the highest potential for community interaction and the promise for block-scale food production. Resident concern was evident for maintaining and improving sub-neighborhood quality both socially and physically, and we felt that supporting these activities would aid the whole sub-area. One of this block's residents, Martha Filer, had previously helped build the gardens at Ya·Po·Ah Terraces in the block immediately adjacent to this one. It was our hope that the physical development of this block could serve as a bridge which connected the downtown area from Willamette Street and the Growers' Market to the Willamette River as primarily a pedestrian way serving the whole sub-neighborhood and the downtown. This pedestrian way would include basalt terraces, weaving their way from Ya·Po·Ah to the River, punctuated by flower and vegetable gardens, orchards, arbors, trellises, swings, benches, and people weaving their way through the

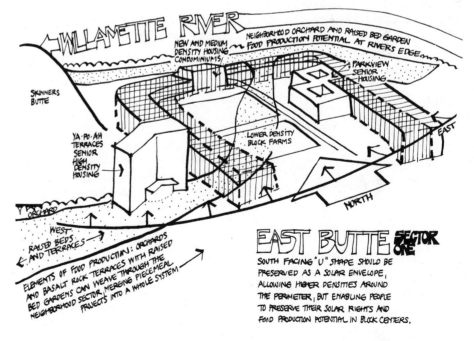

WILLAMETTE RIVER

NEIGHBORHOOD ORCHARD AND RAISED BED GARDEN
FOOD PRODUCTION POTENTIAL AT RIVERS EDGE

NEW AND MEDIUM
DENSITY HOUSING
CONDOMINIUMS

SKINNERS BUTTE

PARKVIEW SENIOR HOUSING

YA·PO·AH TERRACES SENIOR HIGH DENSITY HOUSING

LOWER DENSITY BLOCK FARMS

EAST

Ya·Po·Ah ORCHARD

NORTH

WEST RAISED BEDS AND TERRACES

ELEMENTS OF FOOD PRODUCTION: ORCHARDS
AND BASALT ROCK TERRACES WITH RAISED
BED GARDENS CAN WEAVE THROUGH THE
NEIGHBORHOOD SECTOR, MERGING PIECEMEAL
PROJECTS INTO A WHOLE SYSTEM

EAST BUTTE SECTOR ONE

SOUTH FACING "U" SHAPE SHOULD BE
PRESERVED AS A SOLAR ENVELOPE,
ALLOWING HIGHER DENSITIES AROUND
THE PERIMETER, BUT ENABLING PEOPLE
TO PRESERVE THEIR SOLAR RIGHTS AND
FOOD PRODUCTION POTENTIAL IN BLOCK CENTERS.

neighborhood on foot. This would involve narrowing of Third Street to through traffic and repaving with brick or stone. The block itself could preserve its character, maintain its open center, and develop extensions of the small family-size gardens into a slightly larger-scale organization which could significantly affect a block's expenditures for food. Also existing in this block is a historic garden (formerly owned by Mrs. Dailey who now resides at Ya·Po·Ah Terrace) which Paul Swenby and Alan Boner and family are renovating. This improvement, involving design work and the promotion necessary to have this open space classified as historic by the City of Eugene, is underway.

The remainder of effort in this block (known as Martha Filer's block) by Paul Swenby and Garth Ruffner with some help from Noel Prchal, came to no immediate response by the neighbors. Block parties were held on at least two occasions, and the minimal turnout was disconcerting to the two "catalyst" workers who ultimately discontinued their "block-scale" organizing efforts. Small gardens dotted through this block seemed to indicate an opportunity for larger-scale coordination, but this proved not to be the case. Perhaps at some future time?

Some notes by Paul Swenby regarding their efforts at block-scale organizing might be instructive to future planning efforts in this sub-neighborhood:

"Garth Ruffner's and my strategy was to do for the block only what the neighbors were interested in. The issues we first brought to them concerned transportation, community gardens, preservation of deteriorating old houses, building of new houses, and parking. The only ones met with enthusiasm were traffic problems on Third and High Streets and buying out absentee landlords. As is immediately apparent, the only solution to these problems is neighborhood power. Therefore, our planning and design work centered on bringing the block together. Our strategy in doing site analysis was based on mutual needs or concerns of the neighbors.

The structuring and organizing of meetings is an important tool for any civic involvement and neighbor participation. At our first block party, we discovered the difference between the landowners and renters. The block's prime movers had the sole concern of saving degenerating houses, either by political control or by buying the property."

"Traffic was the major concern of the renters, who couldn't easily visualize the concerns of neighborhood power and control. An easy out was 'What can I do in the short time I'm here that will actually last?'"

"Our first block meeting had a lack of cohesive direction, due to differing ideas and time scales. Although we presented various site analysis maps and some issues, there was a lack of enthusiasm, due in part to a lack of faith in our abilities and in the neighborhood's survival.

For the second meeting, we decided on a specific topic for discussion with a guest speaker. Noel Prchal lectured on composting with worms. This worked and if we would have carried on with meetings that dealt with specific concerns, we would have gradually met the whole block. This did result in our sole construction. Noel and I built the compost bin in the center of the block.

THIS WALK SHOULD TAKE ABOUT 20 MINUTES. YOU'LL NEED 10 SMALL PIECES OF PAPER, A PENCIL, AND A WRITING SURFACE (A HARDCOVER BOOK?).

BEGIN AT THE ALLEY INTERSECTION AND FOLLOW THE ROUTE SHOWN ON THE MAP. STOP AT EACH NUMBERED POINT AND EXAMINE YOUR BLOCK. RECORD YOUR IMPRESSIONS AND SUGGEST SOME THINGS THAT COULD BE DONE TO IMPROVE THE AREA. TRY NOT TO SPEND MORE THAN TWO MINUTES AT EACH STOPPING POINT.

AT POINT #9, IMAGINE YOURSELF TO BE A YA-PO-AH TERRACE RESIDENT WHO WOULD LIKE TO WALK TO CELESTE CAMPBELL SENIOR CENTER. DRAW THE ROUTE YOU WOULD TAKE ON THE MAP. WALK THE ROUTE FOR 2 MINUTES. STOP AND RECORD YOUR IMPRESSIONS OF THE WALK AND WHAT COULD MAKE IT BETTER.

PLEASE BRING YOUR NOTES TO THE MEETING, TUESDAY.

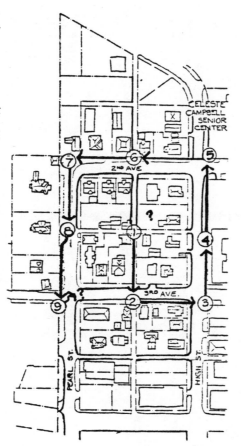

Before the meeting, we went door-to-door, passing out this exercise, which was designed to further our view of issues. Another method would be to tour the neighborhood as a group. This also reinforces the idea of how various people move through the area. The old follow different paths than the bicyclists."

BLOCK PLANNING WORKSHOP

MONDAY, NOV. 19 - 7:00 P.M.
CELESTE CAMPBELL CENTER

WE'LL BE TRYING TO REACH A CONSENSUS OF OPINION ON OUR GOALS AS WELL AS EVALUATING THE WAYS IN WHICH THEY MIGHT BE ACHIEVED. WHILE MANY OF THE SOLUTIONS WE DISCUSS CANNOT BE IMPLEMENTED IN THE NEAR FUTURE, OTHERS, SUCH AS A BLOCK LAUNDROMAT OR A BLOCK COMPOST BIN, CAN BE STARTED IMMEDIATELY IF ENOUGH PEOPLE ARE INTERESTED.

THERE WILL BE A DESSERT POTLUCK TOO, IF YOU'D LIKE TO BRING A DESSERT OR REFRESHMENTS, AND A FORK AND A PLATE.

ON THE FOLLOWING PAGE IS AN EXCERCISE TO HELP YOU GENERATE IDEAS FOR THE MEETING. IT IS NOT A REQUIRED ACTIVITY, BUT YOU MIGHT BE BETTER PREPARED FOR THE MEETING IF YOU DO IT.

Chapter 8

West Skinner's Butte

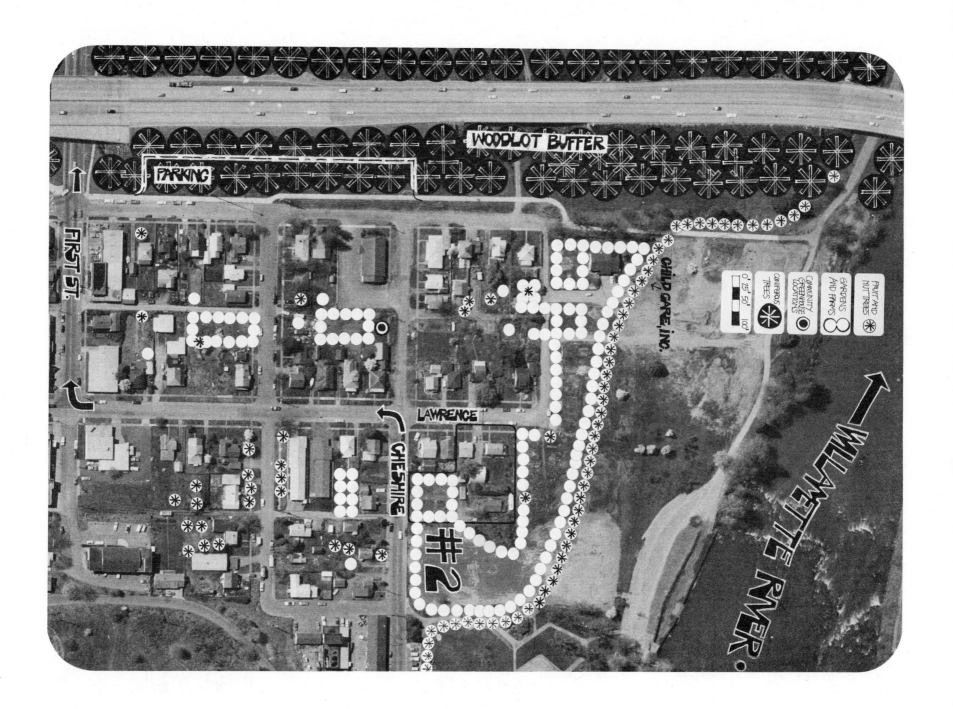

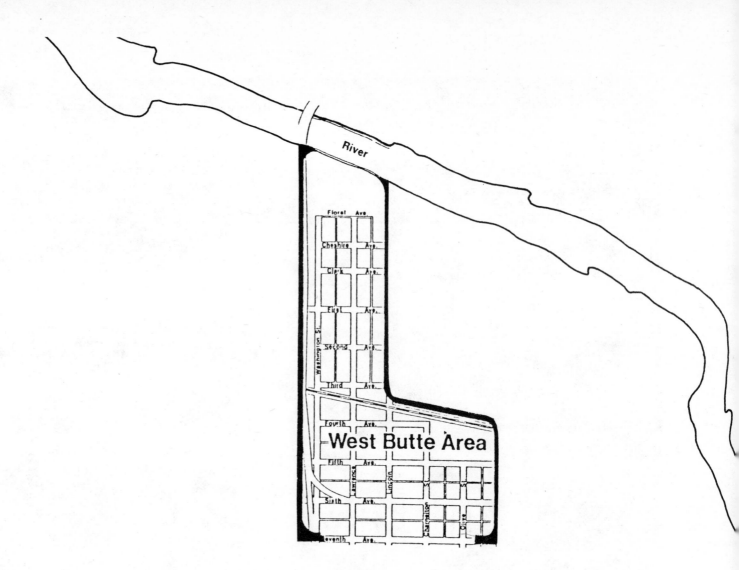

The map shows the West Butte Area with the following labeled streets:

River

Floral Ave.
Cheshire Ave.
Clark Ave.
First Ave.
Second Ave.
Third Ave.
Fourth Ave.
Fifth Ave.
Sixth Ave.
Seventh Ave.

West Butte Area

SECTOR 2
THE WEST SKINNER'S BUTTE AREA
EXPERIMENTAL DISTRICT · WELCOMING THE FUTURE...

WEST BUTTE (West Skinner's Butte sub-neighborhood Sector 2)

West Butte as a neighborhood is largely decimated. Various factors at play in this area have isolated this sector into two pieces of a neighborhood, one north of First Street and the other just south of the railroad tracks. Between the two pieces lies a zone of industrial and "industrializing" land, <u>dividing the neighborhood not only physically but economically as well</u>. Additionally, West Butte has also been the focus for intensive land speculation, especially during the time it was considered as an "opportunity area", and one of the potential sites for the Performing Arts Center. Following the typical path of inner-city "development" common in most urban areas in the U.S.A., this would have permanently displaced the folks who now live there. Nonetheless, in anticipation of this possible boon, land was acquired (Bob Woods' property for example) and <u>residents were displaced leaving this neighborhood sector in the worst physical condition of any in Whiteaker neighborhood</u>.

West Butte has other major problems as well. Convenient transportation provided to the city-at-large (or to the region as some say) has resulted in the Washington-Jefferson freeway on the west edge of this neighborhood sector and the "slicing" in half again by the Cheshire-Lawrence-First Street-River Road convenient cutoff from the Ferry Street Bridge (discussed earlier in East Butte section). These traffic volumes impinge dramatically upon the quality of life for West Skinner's Butte residents.

During the earliest phases of the grant, Marlene Nelson, the West Butte outreach worker, pointed out the severity of the circumstances in West Butte and a decision was made to consolidate the energies available between the food component (Richard Britz) and the housing component coordinator (Jim McCoy) and try to do something. <u>A site was identified for a demonstration project which would integrate at least food production and cooperative housing, but which also would ultimately demonstrate appropriate technologies of recycling and sound energy use</u>. This site was purchased from W. Mizell on August 16, 1979, with the cooperation and support of Jean Tate from Jean Tate Realty, representing Joe Richards and Joyce Benjamin, both attorneys for the City of Eugene. The following two pages graphically describe the site and acquisition principles.

LAND ACQUISITION & RETENTION

WILLAMETTE RIVER FLOW

1. INVESTIGATE PUBLIC OWNERSHIP OF SHARED HOLDINGS. THRU: A. NON-PROFIT CORPORATIONS B. LAND TRUSTS

2. INVESTIGATE THE "WATERFALL EFFECT" (I.E.) MAINTAINING THE STRUCTURE OF "THE GARDEN" AND BEAUTY OF ITS APPEARANCE WHILE ANTICIPATING HIGH PEOPLE MOBILITY THRU THE NEIGHBORHOOD. FOOD IS BASIC. IF FOOD IS BASIC TO THE NEIGHBORHOOD, AT LEAST MAINTAIN ENOUGH OPEN LAND TO PROVIDE DIRECT ACCESS TO FOOD PRODUCING LAND.

3. IDENTIFY LAND OWNERSHIP CURRENT AND PROJECTED VALUE ($) OF THE NEIGHBORHOODS' "WASTED" LAND. (UNDERUTILIZED)

4. ACQUIRE LAND FOR "COMMUNITY" AND NEIGHBORHOOD OWNERSHIP.

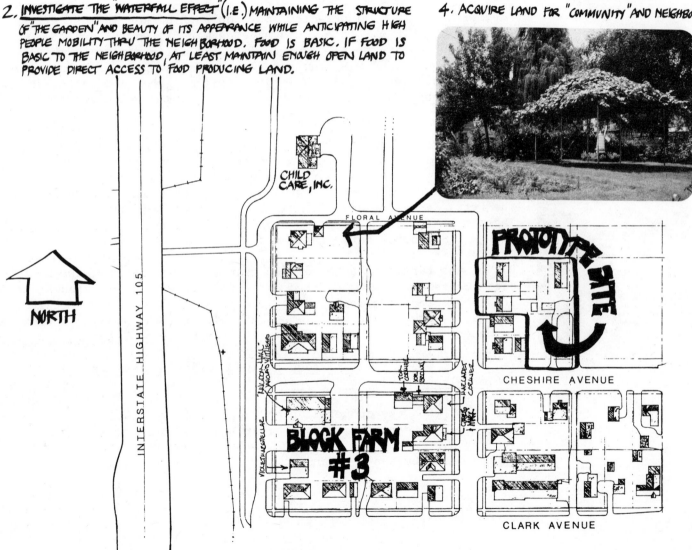

NORTH

INTERSTATE HIGHWAY 105

CHILD CARE, INC.

FLORAL AVENUE

PROTOTYPE SITE

CHESHIRE AVENUE

BLOCK FARM #3

CLARK AVENUE

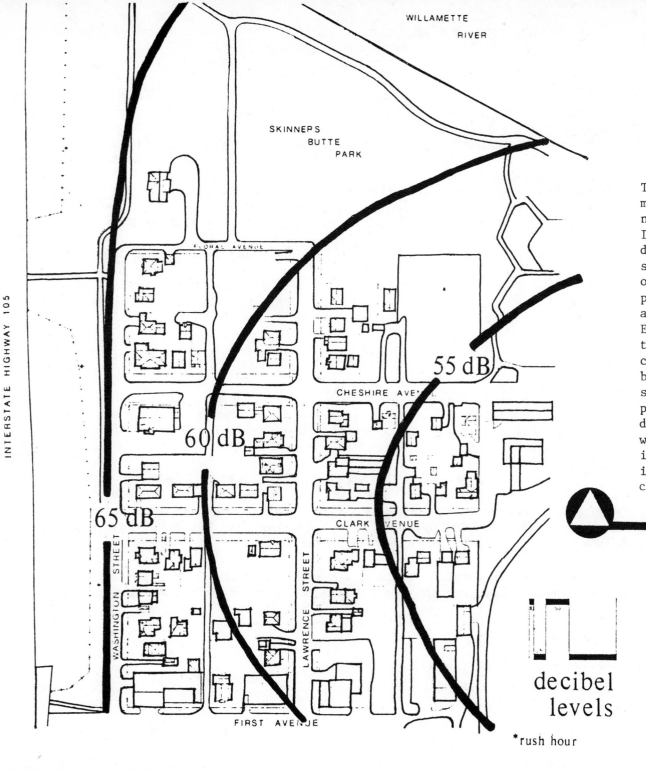

WILLAMETTE RIVER

SKINNERS BUTTE PARK

INTERSTATE HIGHWAY 105

FLORAL AVENUE

CHESHIRE AVENUE

CLARK AVENUE

55 dB

60 dB

65 dB

WASHINGTON STREET

LAWRENCE STREET

FIRST AVENUE

decibel levels

*rush hour

The greatest single environmental factor in West Butte neighborhood is the noise. It is a continuous roar during rush hours and a constant problem from the "halo" of lead and other airborne particulate emissions from automobiles and trucks. Every measure should be taken to reduce this problem including sound and particle buffering walls at the Interstate Highway 105 edge, more planting of coniferous and deciduous trees on the highway berm, and extremely well-insulated and noise suppressing (mass) housing when new construction takes place.

It was the original intention to develop this site as a demonstration project of opportunities within this decimated neighborhood and reverse some of the declining life support capabilities.

Transportation issues also loomed large in this decision to act during the planning phase of our work, since the site the three of us acquired was planned to be used by the City of Eugene Parks and Recreation Department as the pivotal piece of land for a 102-car parking lot. We purchased the site with the intention of producing an experiment in land use combining the goals and objectives of the sub-neighborhood, the Parks Department, and ourselves and have developed alternative proposals for parking and traffic as a result of many meetings with Parks personnel. Along with land as leverage, the strategy of neighborhood capital reinvestment has been an important tool for economic development. For myself, all of the money paid to me during the grant period has gone directly into the West Butte property, plus most of the capital I could raise from other sources. Cash into land is one overall strategy we also recommend to other neighbors to stabilize land ownership and build a community based on improving the land and the life support capability of that part of Whiteaker.

A developing proposal for this site is included in the appendix to this section and is labelled Experimental Block Farm, a Phased Development Proposal.

Two obvious circumstances were (and remain) apparent in West Butte, both of which require some action simultaneous with the development of food production within this sector.

First, traffic throughways need to be changed and some early attempts at buffering the impact of the freeway need to occur.

Second, individuals need to consult with one another and develop some cohesiveness of group strength and visions of how they want to live in this beleagured neighborhood sector. Growing food, whether it be small animals, vegetable gardens, or fruit and nut tree crops is directly dependent on what kind of pollution is in the air; and also, group strength and cooperation are mandatory for large scale output.

Important developments in both of these areas have occurred during, and as a result of, this grant activity. First, West Butte neighbors have requested a traffic diverter on Cheshire, are continuing to develop esprit de corps (they have named the group the Northwest Butte Neighborhood), and will put into place private gardens and a start on a neighborhood farm on the Lawrence and Cheshire land. Skeeter Duke has moved into the neighborhood and has begun to do some work in helping organize this effort.

Access to Sladden area will be enhanced by the removal of storage material under the freeway, and transient sleeping accommodations will be diminished. This proposal needs to be implemented by the Parks Department. (See photo on next page.)

PARKING ←   → PARKING

Parking serving the residents of this area and Skinner's Butte Park should also fit on either side of this underpass, rather than where Parks Department has proposed it. The land where Parks Department has designed this parking will better serve the neighborhood as a neighborhood farm. This neighborhood land can be a very definite economic asset in years to come, both from the standpoint of decreasing Parks Department maintenance costs, and from the standpoint of using part of the land for food production. Following this page is a page showing the location of the proposed parking lot (A) and a proposal (B) showing the alternative proposed. Of course, some cooperation from the State Highway Commission is required, but the benefits of the alternative proposal should be clear.

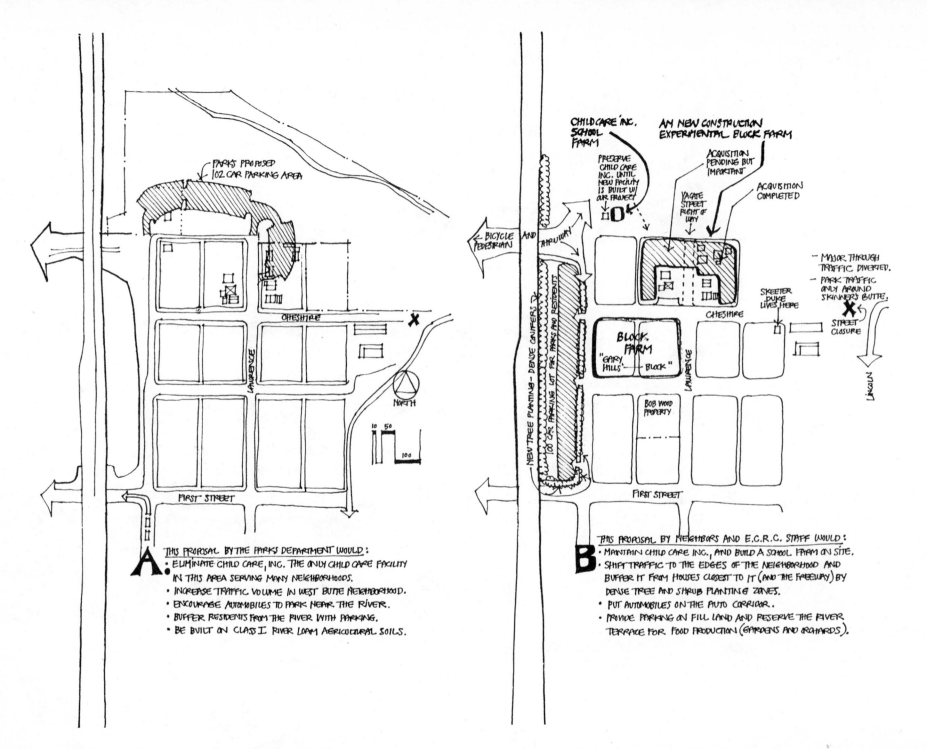

A. THIS PROPOSAL BY THE PARKS DEPARTMENT WOULD:
- ELIMINATE CHILD CARE, INC. THE ONLY CHILD CARE FACILITY IN THIS AREA SERVING MANY NEIGHBORHOODS.
- INCREASE TRAFFIC VOLUME IN WEST BUTTE NEIGHBORHOOD.
- ENCOURAGE AUTOMOBILES TO PARK NEAR THE RIVER.
- BUFFER RESIDENTS FROM THE RIVER WITH PARKING.
- BE BUILT ON CLASS I RIVER LOAM AGRICULTURAL SOILS.

B. THIS PROPOSAL BY NEIGHBORS AND E.C.R.C. STAFF WOULD:
- MAINTAIN CHILD CARE INC., AND BUILD A SCHOOL FARM ON SITE.
- SHIFT TRAFFIC TO THE EDGES OF THE NEIGHBORHOOD AND BUFFER IT FROM HOUSES CLOSEST TO IT (AND THE FREEWAY) BY DENSE TREE AND SHRUB PLANTING ZONES.
- PUT AUTOMOBILES ON THE AUTO CORRIDOR.
- PROVIDE PARKING ON FILL LAND AND RESERVE THE RIVER TERRACE FOR FOOD PRODUCTION (GARDENS AND ORCHARDS).

Implementation of these proposals, one for moving the proposed Parks Department parking lot from its indicated location to along the freeway edge; and the second one, a new-construction experimental block farm on the Lawrence and Cheshire site, will aid the West Butte neighborhood in the following ways:

1. Moving the parking lot frees class 1 agricultural soils for neighborhood farm #2. This land, currently managed by the Parks Department can form an edge to the housing and provide up to five acres of vegetable gardens (see (A) on *page 223* -- the area surrounded by *white* dots!)

2. Opportunities for green fingers of the park, gardens and fruit and nut trees can penetrate the blocks south of the park by converting the existing alleys to pedestrian through-ways.

3. Massive automobile traffic moving north on Lawrence to the Parks Department parking lot will not further bisect the neighborhood.

4. Existing patterns of fruit and nut trees (✳ dots on *page 223*) can be extended as an orchard belt parallel to the river.

5. The experimental block farm residents can coordinate full neighborhood food production including that on the Lawrence and Cheshire site, and provide technical and labor aid to surrounding neighbors, specifically those on "Gary Hill's block", the proposed site of a retrofit block farm.

The accompanying photograph outlines the "Gary Hill Block", site of one proposed block farm. This block, bounded by Cheshire and Clark, Washington and Lawrence, was identified as having the greatest potential in the Northwest Butte sector. It was identified because 1) it had the largest concentration of concerned and active neighbors, 2) the non-absentee majority ownership of the block made it stable, and 3) the food production area in the center of the block was available and somewhat utilized.

Early meetings between neighbors, planners, and designers arranged by Marlene Nelson, the West Butte outreach worker, developed proposals to re-organize the block interior's land use, including fence modification, provision of visual security for Violet Marstellar's back yard (a senior living alone), pooling of semi-public/ semi-private space and sharing of facilities were all considered. Gary Hoyt and Fred Patch built several scale working models of the block and used drawings, photographs, and discussion to jointly evolve the following development proposal for this block farm. One element of this proposal included block shade patterns in both summer and winter which indicated where food crops might maximize their exposure to the sun. A three-phase development proposal was developed to consolidate portions of the private parcels of back lots into an aggregated and shared economic unit which could lead toward greater block cohesiveness and economic stability. In phase one and two, it should be noted

that housing infill has also taken place (Margaret and Tom Cormier's corner lot), and increased the density of the block. As time passes, this block plan increases the intensity and quantity of (biomass) food production through the use of raised bed gardens, fruit and nut trees, grape arbors, block-scale organic waste composting, and a block-scale neighborhood solar greenhouse for winter growing.

Some progress on this block has been made. Since attitudes change first before physical changes, it is anticipated that more connectedness will begin to appear in the West Butte blockscape as demand for low-cost food becomes a more central issue in people's lives.

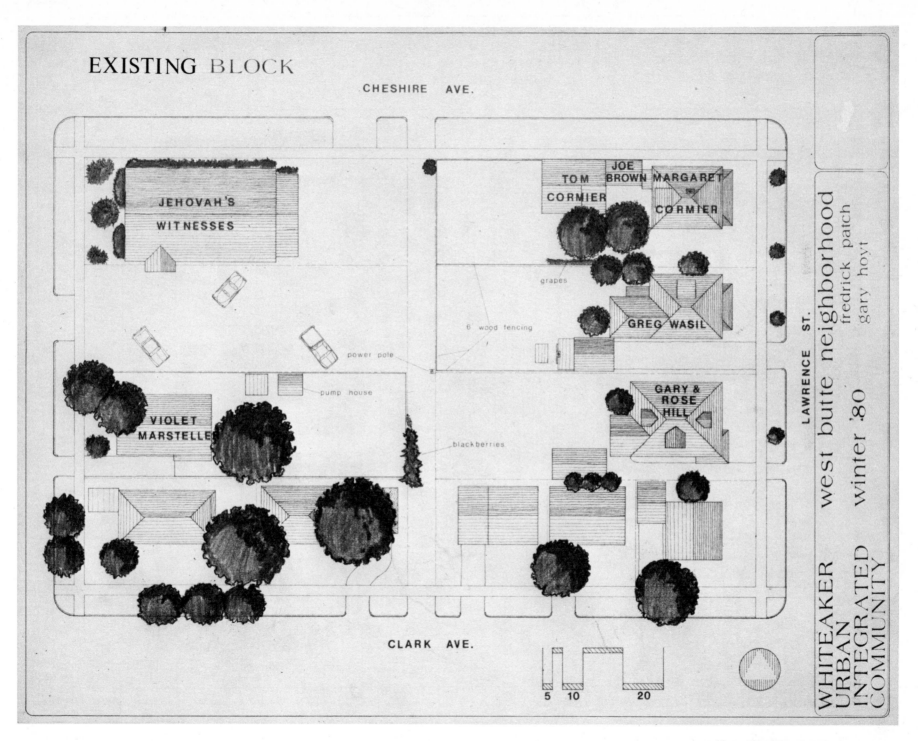

EXISTING BLOCK

CHESHIRE AVE.

JEHOVAH'S WITNESSES

TOM CORMIER

JOE BROWN

MARGARET CORMIER

grapes

GREG WASIL

6' wood fencing

power pole

pump house

VIOLET MARSTELLER

blackberries

GARY & ROSE HILL

CLARK AVE.

LAWRENCE ST.

WHITEAKER URBAN INTEGRATED COMMUNITY

west butte neighborhood
fredrick patch
gary hoyt

winter '80

5 10 20

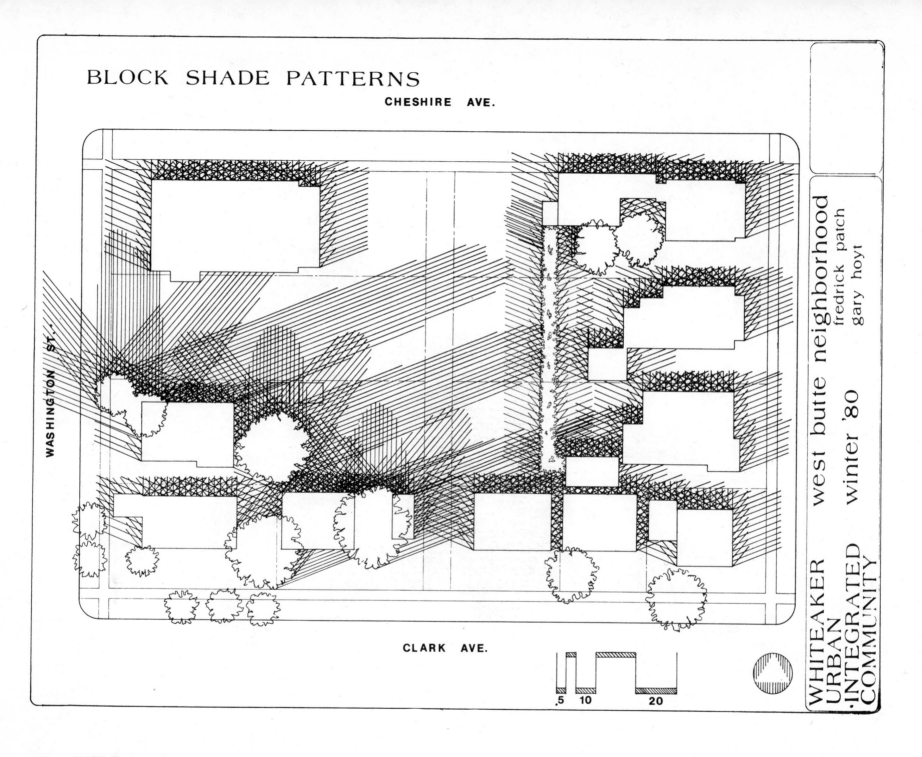

BLOCK SHADE PATTERNS

CHESHIRE AVE.

WASHINGTON ST.

CLARK AVE.

.5   10   20

WHITEAKER   west butte neighborhood
URBAN   fredrick patch
·INTEGRATED   gary hoyt
COMMUNITY   winter '80

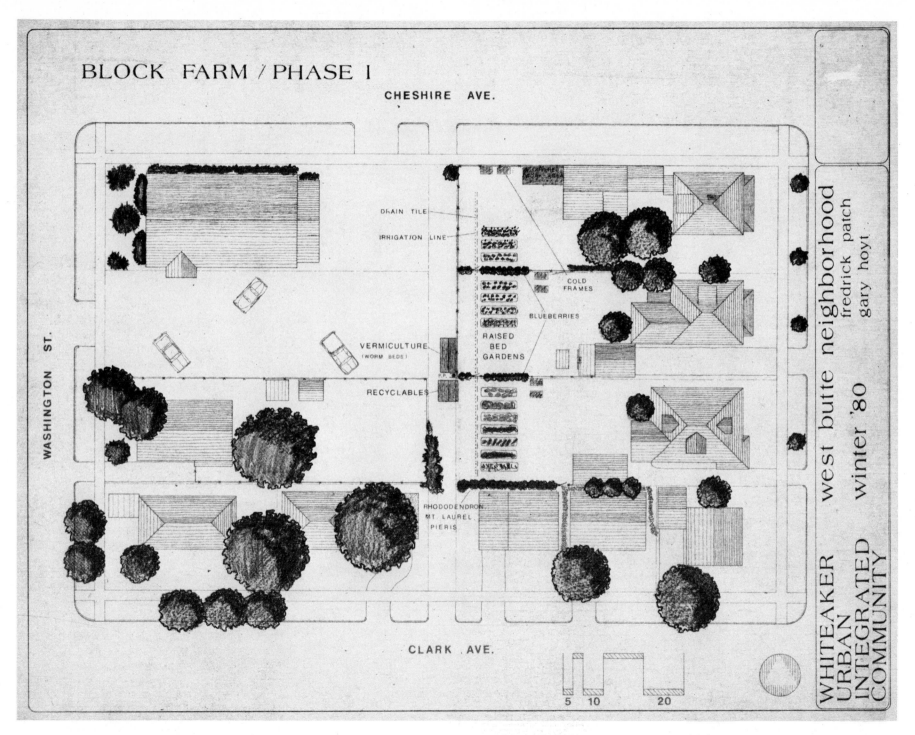

BLOCK FARM / PHASE I

CHESHIRE AVE.

WASHINGTON ST.

DRAIN TILE

IRRIGATION LINE

COLD FRAMES

BLUEBERRIES

VERMICULTURE
(WORM BEDS)

RAISED
BED
GARDENS

P.P.

RECYCLABLES

RHODODENDRON
MT. LAUREL
PIERIS

CLARK AVE.

5   10   20

WHITEAKER
URBAN
INTEGRATED
COMMUNITY

west butte neighborhood
fredrick patch
gary hoyt

winter '80

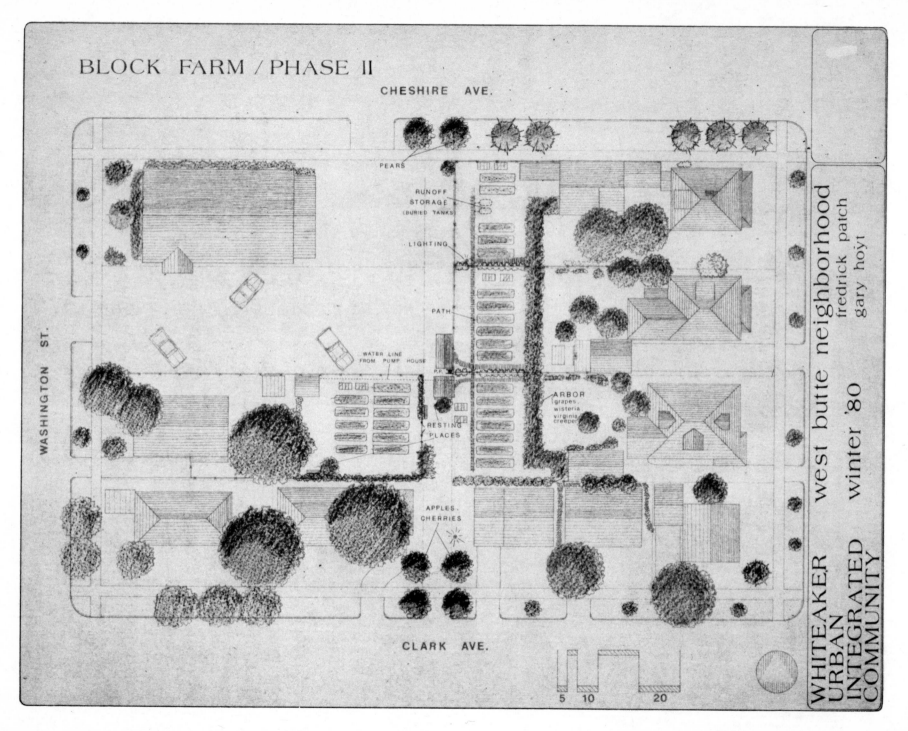

BLOCK FARM / PHASE II

CHESHIRE AVE.

PEARS

RUNOFF
STORAGE
(BURIED TANKS)

LIGHTING

PATH

WATER LINE
FROM PUMP HOUSE

P.P.

RESTING
PLACES

ARBOR
(grapes,
wisteria,
virginia
creeper)

APPLES,
CHERRIES

WASHINGTON ST.

CLARK AVE.

5 10 20

WHITEAKER   west butte neighborhood
URBAN   fredrick patch
INTEGRATED   gary hoyt
COMMUNITY   winter '80

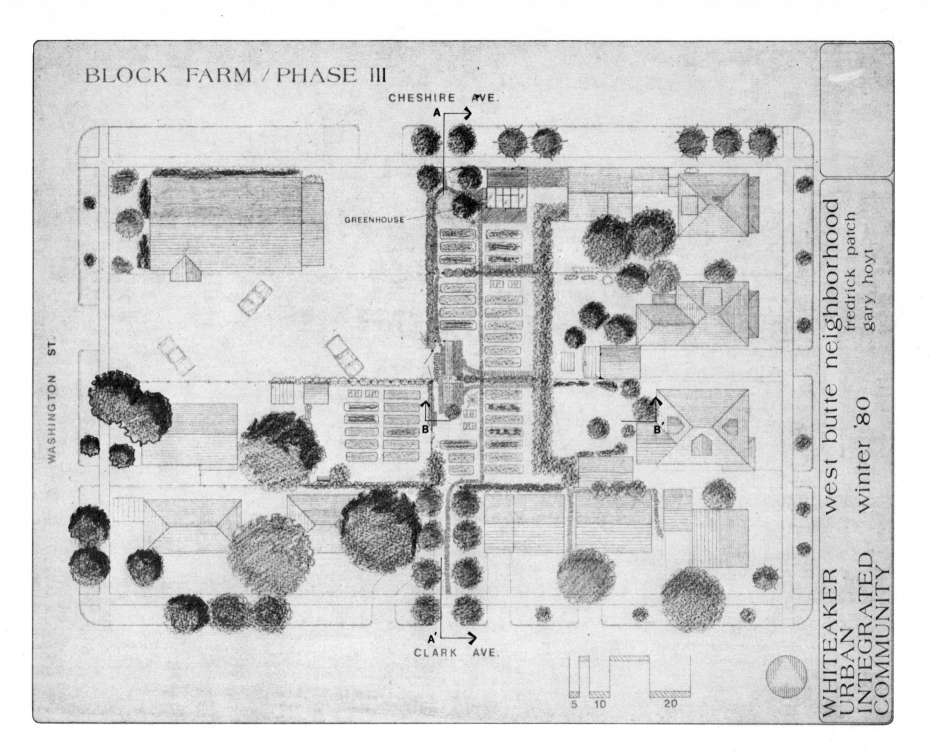

BLOCK FARM / PHASE III

CHESHIRE AVE.

A

GREENHOUSE

WASHINGTON ST.

B

B'

A'

CLARK AVE.

5 10 20

WHITEAKER west butte neighborhood
URBAN fredrick patch
INTEGRATED gary hoyt
COMMUNITY winter '80

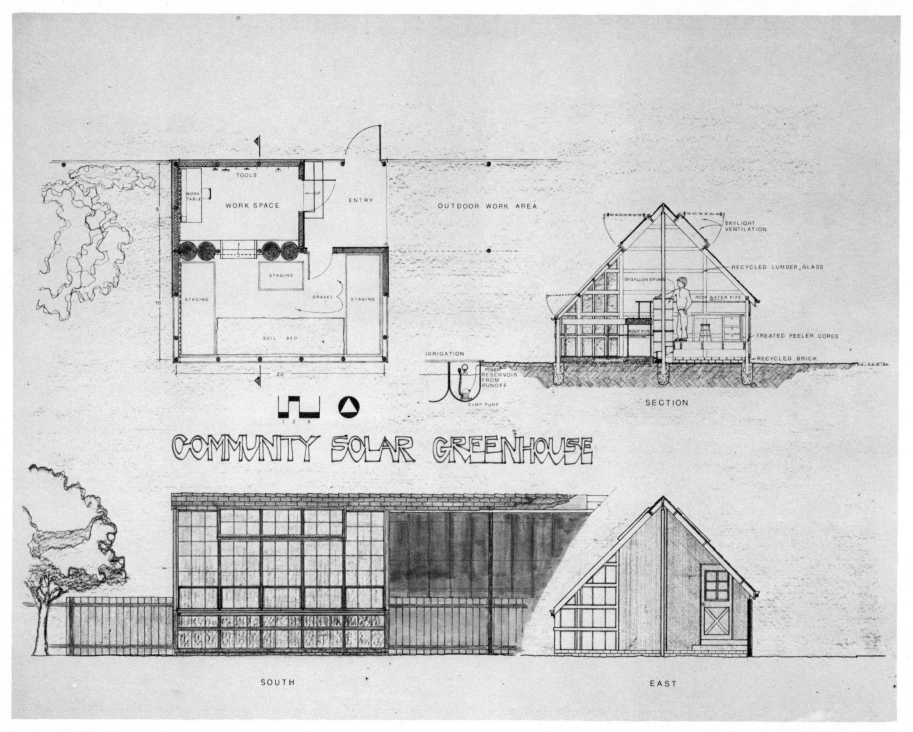

WORK SPACE

TOOLS

WORA TABLE

UP

ENTRY

OUTDOOR WORK AREA

8

STAGING

STAGING

GRAVEL

STAGING

SOIL BED

10

20

SKYLIGHT VENTILATION

RECYCLED LUMBER, GLASS

50 GALLON DRUMS

ROOF WATER PIPE

ROOF PIPE

TREATED PEELER CORES

RECYCLED BRICK

IRRIGATION

POWER RESERVOIR FROM RUNOFF

SUMP PUMP

SECTION

1 2 4

COMMUNITY SOLAR GREENHOUSE

SOUTH

EAST

A final project in West Butte is a Child Care, Inc. school farm which could stabilize Child Care, Inc. (since the Parks Department owns the land upon which they sit) and provide a seed opportunity for gardening and fruit trees on public land to serve the neighborhood. This site is pivotal in the maintenance of "a foot in the door" to lead toward some connection with other food-producing land in the edge between recreation and housing in West Butte. Step by step, land in small pieces can be opened up to apartment dwellers or those whose landlords prohibit food production activities. Along the edge of the park, or on reclaimed street right-of-ways, food production land can account for a minimum of five acres -- just enough to serve West Butte's needs, assuming adequate management. Work parties from the neighborhood and the U of O Urban Farm have begun the Child Care, Inc. school farm by providing loads of chicken manure and volunteers who raised the beds and organized the gardens with the teachers and children of Child Care, Inc.

As of June, 1980, garden construction had been completed, including a large worm composting raised platform bed and eight raised beds. Mary Truax and Skeeter Duke continued to work with the staff of Child Care, Inc. to insure transfer of the curriculum from Edible City Resource Center staff to the teachers and students. A recent fire at Child Care, Inc. has temporarily closed the building, and activities there are on hold waiting for response from the City of Eugene and insurance company. There is some fear that the building will come down and the site turned into the parking lot mentioned earlier in this section. Neighborhood support of Child Care, Inc. will never be more sorely needed than at this time.

Real Estate/Hugh Prichard

Why a Good House Isn't Worth the Land It's Built On

I sat in recently on a panel discussion of "The Housing Crunch," sponsored by the Neighborhoods Housing Resource Center. The research I did to prepare for that talk shocked me. The grim statistics in today's column graphically demonstrate Eugene's housing problem.

In order to buy a two bedroom house in West Eugene in 1976, your income needed to be at least $8400. To buy the exact same house in West Eugene today, your income needs to be $21,120. In other words, it needed to increase 151 per cent over the three-year period, or a steady 36 per cent per year. In dollars, you needed to make an additional $4240 each year.

In 1976, that two bedroom house cost $20,000. Today, the same house costs $45,000. That's an increase of 125 per cent over the three-year period, or 30 per cent per year. The payments on a loan for 80 per cent of the purchase price at the prevailing rate of interest rose from $125 per month in 1976 to $387 in 1979, an increase of 140 per cent, or 47 per cent per year.

What caused these gigantic increases? It is not primarily inflation as you might expect, nor is it the ever-increasing cost of borrowing. It is simply the cost of land in Eugene.

The lot on which the two bedroom house is sitting increased in value from $4,000 in 1976 to $20,000 in 1979. That's an increase of 400 per cent, or 75 per cent per year. This amazing increase in the value of land is the chief culprit in Eugene's housing problem. This rate of increase has levelled off recently, but the damage has been done. It will take years of inflation for our incomes to catch up. In the meantime, moderately-priced housing is an impossibility.

Assuming that you can build a new house for $40 per foot, a 1000-square-foot house on a $20,000 lot will cost $60,000. You need

$12,000 as a down payment, and your monthly payment will exceed $600 per month. To qualify for this loan payment at any institution in town, you need to gross $2400 per month, which is almost $29,000

Construction costs are up, but it's our residential land shortage that is chiefly responsible for Eugene's absurdly high housing prices.

per year. That's $29,000 per year for a modest, 1000-square-foot home.

If land had appreciated at only 15 per cent per year, which is approximately the rate at which the improvements have been appreciating, our situation would still be awful, but not intolerable. Our 1000-square-foot home would cost $46,000, requiring an income of $22,000 per year.

To put these income figures into perspec-

tive, the median individual income in Lane County is considerably less than $8000 per year. The median household income in Lane County is $17,700. This income of $17,700 will support the purchase of a

home with a maximum purchase price of $34,000. A decent $34,000 home in Eugene does not exist. As these figures indicate, the average Eugene household is priced out of the housing market, right now, today. On a national scale, the head of the Federal Home Loan Bank Board recently announced that 85 per cent of potential U.S. home buyers are priced out of the market.

The true picture may be even gloomier

than these figures show. Talking about incomes is only half the financing picture. The other half is the down payment. Where does it come from? I just conducted a completely unscientific survey of a specific Eugene population in order to answer the down payment question.

I contacted 37 couples and individuals with whom I have worked in the past three years. These 37 households had purchased their first home and had qualified for financing of one kind or another; 32 out of 37 had borrowed all or part of the down payment from parents of friends! Of the remaining five, only one had what would today be considered an adequate down payment.

I realize this sample isn't random, but it does serve to illustrate one more problem in the housing picture. Even if your income qualifies you for high monthly payments, a down payment is hard to come by unless your parents or friends are sitting on a comfortable pile of cash.

As we've seen, Eugene's housing problem stems largely from the high cost of land. Stated another way, there is a great demand for residential land, and a very small supply. The demand side is not likely to change by any acceptable political means. So the only workable way to reduce the cost of land is on the supply side. It looks like we have two choices: Either abandon the urban services boundary to allow development on currently unavailable land, or increase density.

Increasing density is an immediately workable alternative. Creative and aesthetic infilling would be a boon to the city in every area from public transportation to energy conservation. In addition, a close look at the urban services boundary reveals several areas where high density housing could be intelligently sited.

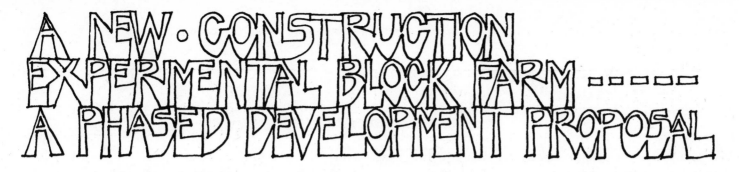

A NEW·CONSTRUCTION EXPERIMENTAL BLOCK FARM ▫▫▫▫▫ A PHASED DEVELOPMENT PROPOSAL

This proposal recommends the design and construction processes for affordable houses on the land at Lawrence and Cheshire without displacing the existing housing. In a phased approach, new clusters of small units can be built on this site while maintaining the food production capability of some of the land, and building around and between some of the existing buildings. It is intended that the structures will blend into the earth, reinforcing the existing rolling (bermed) landscape of the Skinner's Butte Park.

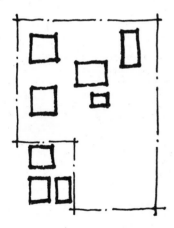

PHASE 1   Existing Site

1.   Acquire site

2.   Organize financing

3.   Organize management

4.   Develop preliminary design/construction proposal

5.   Negotiate with Parks Department for <u>overall</u> West Butte alternative development proposal.

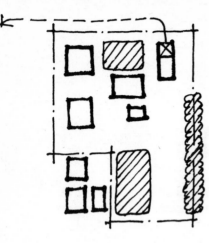

PHASE 2

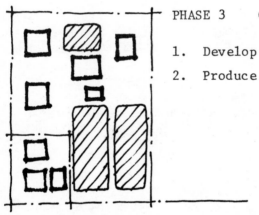

1. Residents put in gardens (Soil improves)

2. Structure recycled from lean-to on the back of green house. (2x4's become grape arbor in the Sladden area.) *BUS REMOVED FROM SITE.*

3. Grass mowing, clean up, and shrub transplant joins site to park. (Shrubs moved) Land maintenance schedule planned.

4. Produce and present alternative proposal (following pages) *A NEW CONSTRUCTION EXPERIMENTAL BLOCK FARM--- A PHASED DEVELOPMENT PROPOSAL*

5. Finance the other 50% of the land cost through <u>long-term</u> loan.

6. Grant proposal written to finance construction feasibility and architects' working drawings— *TO NATIONAL SCIENCE FOUNDATION*

PHASE 3   (Begun summer 1980)

1. Develop full scale mock-up and develop construction techniques.

2. Produce architects' preliminary design and some working drawings.

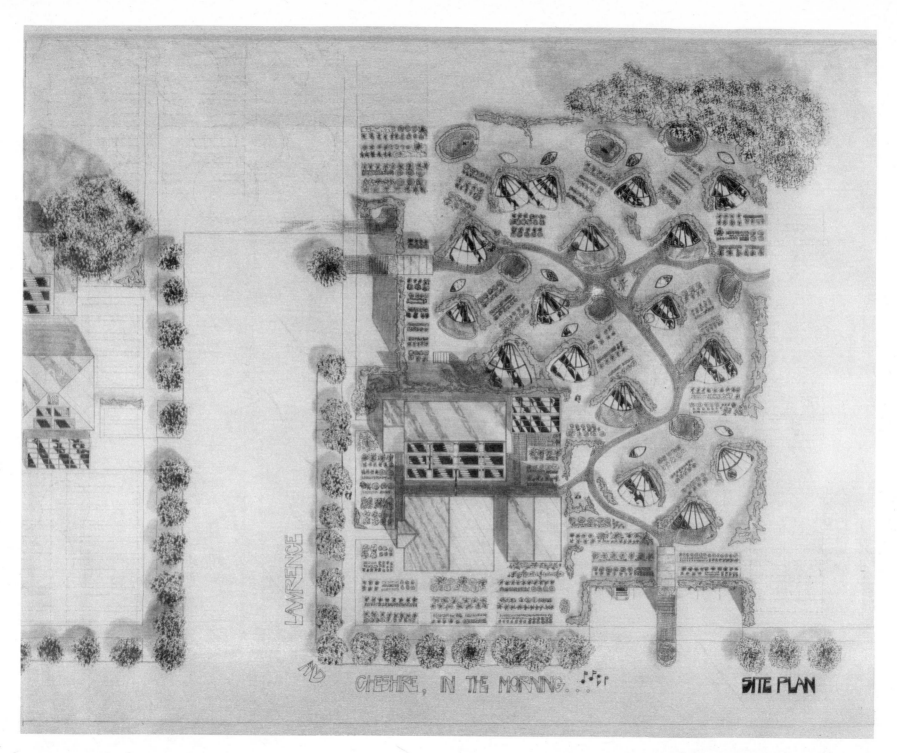

LAWRENCE

CHESHIRE, IN THE MORNING... ♪♪♪♪

SITE PLAN

"RECYCLING INDUSTRIAL "WASTE" INTO ENERGY EFFICIENT LOW COST SELF-HELP CO-OPERATIVE HOUSING - FOOD CLUSTERS

AN EXPERIMENTAL BLOCK FARM ...

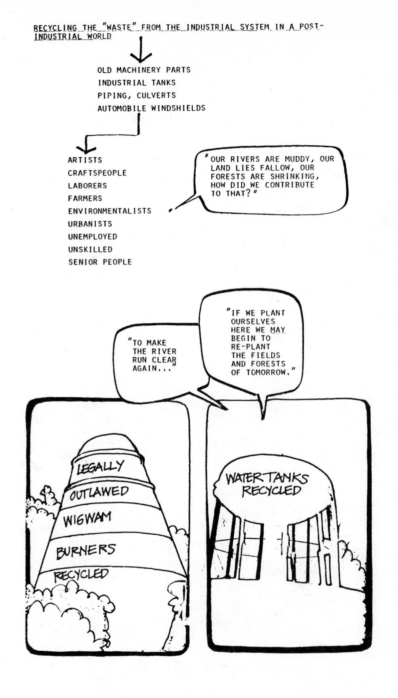

RECYCLING THE "WASTE" FROM THE INDUSTRIAL SYSTEM IN A POST-INDUSTRIAL WORLD

OLD MACHINERY PARTS
INDUSTRIAL TANKS
PIPING, CULVERTS
AUTOMOBILE WINDSHIELDS

ARTISTS
CRAFTSPEOPLE
LABORERS
FARMERS
ENVIRONMENTALISTS
URBANISTS
UNEMPLOYED
UNSKILLED
SENIOR PEOPLE

"OUR RIVERS ARE MUDDY, OUR LAND LIES FALLOW, OUR FORESTS ARE SHRINKING, HOW DID WE CONTRIBUTE TO THAT?"

"TO MAKE THE RIVER RUN CLEAR AGAIN..."

"IF WE PLANT OURSELVES HERE WE MAY BEGIN TO RE-PLANT THE FIELDS AND FORESTS OF TOMORROW."

LEGALLY OUTLAWED WIGWAM BURNERS RECYCLED

WATER TANKS RECYCLED

MATERIALS RESOURCES...

THE PRIMARY INDUSTRIES IN OREGON IN THE 20TH CENTURY HAVE BEEN THE NATURAL RESOURCE-BASED INDUSTRIES OF TIMBER, AGRICULTURE, AND FISHERIES. DURING THE TENURE OF THESE INDUSTRIES, DECLINING YIELD FROM FORESTS, FARMLAND, AND FISHERIES HAS GENERATED ENORMOUS WASTE MATERIAL IN THE FORM OF EQUIPMENT, UNUSED FACILITIES AND GENERAL "SALVAGE". SIMULTANEOUSLY, THE GROWTH OF THE AUTOMOBILE INDUSTRY HAS IMPORTED TO THE STATE AN UNTOLD WEALTH OF NATURAL RESOURCES IN THE FORM OF AUTOMOBILIES, MANY OF WHICH LAY "FALLOW" IN REST-ING PLACES OF ALL SORTS THROUGHOUT OUR STATE.

THIS PORTION OF OUR WORK HAS AS ITS BASIC INTENT THE GENERATION OF AN EXPERIMENTAL AND PROTOTYPICAL HOUSING-FOOD CLUSTER WHICH UTILIZES MANY OF THESE "WASTES".

·BOILER TUBES,
·WATER TANKS,
·AUTOMOBILE WINDSHIELDS,
·UNUSED CHAIN LINK FENCING,
·STEEL PLATE,
·VENTILATOR SHAFTS,
·ETC.,
·ETC.,
CAN BE RECYCLED AND WELDED INTO REUSABLE STEEL FORMWORK FOR CONCRETE SHELL CONSTRUCTION WHICH UTILIZES LOCAL SAND AND GRAVELS FROM ALONG THE WILLAMETTE RIVER.

FIREPROOF, SOUNDPROOF, AND WELL INSULATED (RECYCLED STYROFOAM) CON-STRUCTION WILL RESULT IN SHELTERS NESTLED COMFORTABLY IN THE EARTH, AND CLUSTERED TOGETHER COMFORTABLY TO PRESERVE GOOD AGRICULTURAL SOILS.

METALS OF THIS SORT CAN EASILY BE WELDED INTO SECTIONAL FORMWORK SIZED ACCORDING TO TRUCK TRANSPORT TOLERANCES AND/OR THE SIZE OF THE GROUP BUILDING THESE HOUSING-FOOD CLUSTERS.

CONCRETE SHELL CONSTRUCTION IS RECOMMENDED FOR ITS EFFICIENCY OF STRUCTURE AND MATERIALS WHICH TRANSLATES SIMPLY INTO AFFORDABLE HOUSING. SECOND TO WOOD, WHICH HAS BEEN SERIOUSLY DEPLETED IN OREGON, CONCRETE (THAT IS, SAND AND GRAVELS) IS AVAILABLE AND PLENTIFUL AS A BUILDING MATERIAL, AND ITS MINING BY-PRODUCT, THE SAND AND GRAVEL PIT, CAN ALSO BE UTILIZED FOR ECOLOGICALLY AND ECONOMICALLY USEFUL PURPOSES. EARLIER IN THIS WORK, IT WAS NOTED THAT THESE SAND AND GRAVEL PITS WOULD WELL SERVE THE EUGENE-SPRINGFIELD METROPOLITAN AREA AS MANAGED FISH PONDS, PROVIDING A RIVER CLEANSING FUNCTION AS WELL AS A LOCAL FOOD SUPPLY.

THE UTILIZATION OF "WASTE" MATERIALS AND "WASTED" LANDSCAPES FOR BENEFICIAL PURPOSES ALSO CALLS OUR ATTENTION TO REPAIR AND RENO-VATE OUR FORESTS AND OUR RIVERS. IT MIGHT BE USEFUL TO PONDER THE OVERENTHUSIASM OF CUTTING WHILE REORGANIZING FOR MASSIVE REPLANTING OF NATIVE TREE SPECIES, WHILE SIMULTANEOUSLY SHIFTING OUR CONSTRUCTION BURDEN FROM WOOD TO CONCRETE.

LOCAL WOOD "WASTES" MEANWHILE, CAN SUSTAIN USAGE FOR LINING AND PANELING THE INTERIORS OF THESE "SHELLS" ESPECIALLY SINCE ONLY ·NON-SELECT OR SCRAP MILL ENDS WILL CONFORM TO THEIR CURVATURES. INGENUITY AND CREATIVITY SHOULD FLOW AS INTERIOR SPACES ARE SHAPED BY CHAIN SAWS, HAND TOOLS, AND INDIGENOUS SCULPTORS, RE-CLAIMING THE RIGHT TO CONSTRUCT ONE'S OWN HOME WITH THE HELP OF FRIENDS.

LOCAL CLAYS, FIRED CAREFULLY AND INFREQUENTLY IN HOME-BUILT KILNS CAN PROVIDE MUCH OF THE POTTERY NOW IMPORTED FROM DISTANT SOURCES, WITH THEIR ENERGY INEFFICIENCIES AND ESCALATING COSTS. TABLEWARE, SINKS, TUBS AND OTHER CERAMIC CREATIONS CAN SIT BESIDE HOME-CARVED (A NEIGHBORHOOD INDUSTRY) FURNITURE OF LOCAL "TRASH" TREES SUCH AS ALDER AND ASH.

CARPETS AND TAPESTRIES SPUN FROM LOCAL WOOLS FROM LOCAL SHEEP (WHICH BY THEIR GRAZING KEEP DOWN PARKS' MAINTENANCE COSTS) WILL ADD SOFTNESS AND WARMTH TO OUR PERMANENT DWELLINGS.

WHILE THE ABOVE REDUCTION OF COSTS FOR BUILDING CAN BE SIGNIFICANT, IT IS ALSO IMPORTANT TO IDENTIFY NEW OPPORTUNITIES FOR LABOR-COST-SAVINGS IN HOME CONSTRUCTION. THE REVISIONS TO THE STATE BUILDING CODE NOW UNDER CONSIDERATION, SUGGEST THAT ECONOMIES CAN BE IM-PROVED WITH AN OWNER-BUILDER CODE.

LABOR RESOURCES...

LOCAL CRAFTSPERSONS AND ARCHITECTS WORKING COOPERATIVELY WITH OWNER-BUILDERS WILL SHAPE AND ORGANIZE SPACES ACCORDING TO FAMILY SIZE AND TYPE AND PREFERRED LIFE STYLE, BUT WITHIN FAIRLY RIGID GUIDE-LINES OF SOLAR AND THERMAL EFFICIENCY. HIGH UNEMPLOYMENT ENABLES TRANSFER OF LABOR INTO CAPITAL. IT IS ANTICIPATED THAT LABOR OF ALL TYPES, SENIORS, CHILDREN AND OTHER MEN AND WOMEN OF ALL AGES MIGHT PARTICIPATE IN CREATING THESE HOUSING-FOOD CLUSTERS.

DEAD LOAD + LIVE LOAD

"LIGHT?? LOT'S!"

DOUBLE SLEEPING POD WITH STOR-AGE UNDER

EXISTING GRADE

BALANCE OF COM-PRESSION OF EARTH FILL IN #/Φ (OR INCH) IS EQUALLY RE-SISTED BY UPLIFT GAINED FROM FLOATING SHELLS ON THE WATER TABLE.

FORCE DISTRIBUTED OVER A DOUBLE CURVATURE MORE EFFICIENTLY TRANSMITS STRESSES THROUGH THE STRUCTURE COMPRISING THE DOUBLE CURVATURE.

STRESS DISTRIBU-TION FROM NORM-ALLY OVERBURDENED FOOTING OF FOUN-DATION IS MORE EQUALLY DISTRIB-UTED OVER A MUCH LARGER BEARING SURFACE ENABLING THIN SHELL TO WORK.

NORTH

NESTLING IN THE EARTH

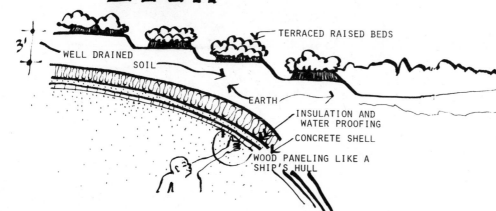

3'

WELL DRAINED SOIL

TERRACED RAISED BEDS

EARTH

INSULATION AND WATER PROOFING

CONCRETE SHELL

WOOD PANELING LIKE A SHIP'S HULL

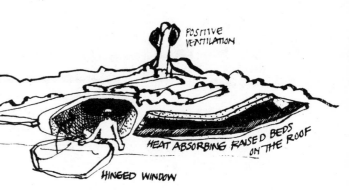

POSITIVE VENTILATION

HEAT ABSORBING RAISED BEDS ON THE ROOF

HINGED WINDOW

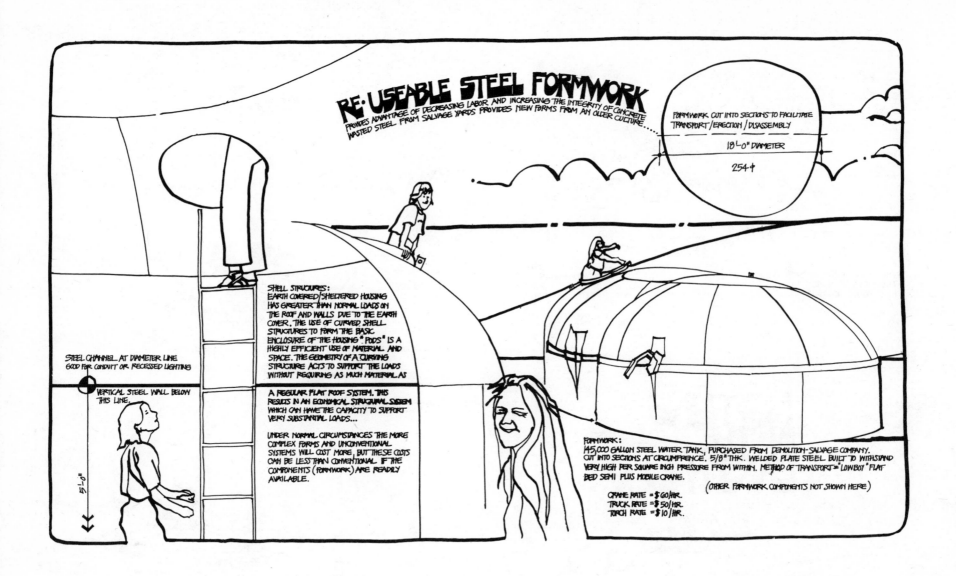

RE·USEABLE STEEL FORMWORK

PROVIDES ADVANTAGE OF DECREASING LABOR AND INCREASING THE INTEGRITY OF CONCRETE WASTED STEEL FROM SALVAGE YARDS PROVIDES NEW FORMS FROM AN OLDER CULTURE...

FORMWORK CUT INTO SECTIONS TO FACILITATE TRANSPORT/ERECTION/DISASSEMBLY

18'-0" DIAMETER

254¢

STEEL CHANNEL AT DIAMETER LINE
GOOD FOR CONDUIT OR RECESSED LIGHTING

VERTICAL STEEL WALL BELOW THIS LINE.

5'-0"

SHELL STRUCTURES:
EARTH COVERED/SHELTERED HOUSING HAS GREATER THAN NORMAL LOADS ON THE ROOF AND WALLS DUE TO THE EARTH COVER. THE USE OF CURVED SHELL STRUCTURES TO FORM THE BASIC ENCLOSURE OF THE HOUSING "PODS" IS A HIGHLY EFFICIENT USE OF MATERIAL AND SPACE. THE GEOMETRY OF A CURVING STRUCTURE ACTS TO SUPPORT THE LOADS WITHOUT REQUIRING AS MUCH MATERIAL AS

A REGULAR FLAT ROOF SYSTEM. THIS RESULTS IN AN ECONOMICAL STRUCTURAL SYSTEM WHICH CAN HAVE THE CAPACITY TO SUPPORT VERY SUBSTANTIAL LOADS...

UNDER NORMAL CIRCUMSTANCES THE MORE COMPLEX FORMS AND UNCONVENTIONAL SYSTEMS WILL COST MORE, BUT THESE COSTS CAN BE LESS THAN CONVENTIONAL IF THE COMPONENTS (FORMWORK) ARE READILY AVAILABLE.

FORMWORK:
145,000 GALLON STEEL WATER TANK, PURCHASED FROM DEMOLITION-SALVAGE COMPANY. CUT INTO SECTIONS AT CIRCUMFRENCE. 5/8" THK. WELDED PLATE STEEL BUILT TO WITHSTAND VERY HIGH PER SQUARE INCH PRESSURE FROM WITHIN. METHOD OF TRANSPORT = "LOWBOY" FLAT BED SEMI PLUS MOBILE CRANE.

(OTHER FORMWORK COMPONENTS NOT SHOWN HERE)

CRANE RATE = $ 60/HR.
TRUCK RATE = $ 50/HR.
TORCH RATE = $ 10/HR.

YOU WOULD PROBABLY WANT SOME OF THOSE GOOSE-DOWN-FILLED WINDOW CURTAINS TOO! THE ONES WITH THE INSULATING QUALITY THAT KEEPS WARMTH IN THE HOUSE AT NIGHT AS WELL AS THE LIGHT (AND ITS HEAT, TOO). MAKING THESE LOCALLY MAKES SENSE, (ANY GOOD SEAMSTRESS CAN MAKE THEM) AND DUCKS (FOR DOWN) CAN BE GROWN IN THE CANALS CONNECTING THE URBAN NEIGHBORHOODS (THE EMERALD CANAL IS THE NEXT FOLLOWING THE MILLRACE EXAMPLE). OTHER EXTERNAL INSULATING PANELS CLOSED UP AFTER 4:00 P.M. ACT AS "EYELIDS" FOR THESE SHELTERS CONSERVING AS MUCH HEAT AS POSSIBLE.

THESE PRINCIPLES ACCOMPLISH TWO THINGS. MOST IMPORTANTLY, THEY ENABLE US TO DECREASE OUR DEPENDENCE ON BOTH NUCLEAR-FUEL-GENERATED ELECTRICITY AND DECREASE OUR DEPENDENCE ON WOOD BURNING AS A HEATING SOURCE. ADDITIONALLY, THE "TUNING" OF OUR LIVING ENVIRONMENTS TO SEASONAL AND CLIMATIC SHIFTS ENCOURAGES US TO APPRECIATE OUR MILD OREGON CLIMATE TO ITS FULLEST, AND TO PARTICIPATE IN THE FUNCTIONING OF OUR SHELTERS.

THE INTEGRATION OF FOOD PRODUCTION WITH HOUSING (SO INTIMATELY AS TO HAVE IT ON YOUR ROOF!) ENABLES US TO ATTEND TO OUR RENEWABLE RESOURCE BASE OF FOOD AND AGRICULTURAL SOILS AS VEHEMENTLY AND EMPHATICALLY AS WE CAN, AND HOPE THAT OUR MODEL GENERALIZES TO THE LARGER CULTURE.

AUTOMOBILES, BICYCLES AND WALKING

AUTOMOBILES WILL TAKE SECOND PRIORITY AND RELINQUISH THE SURFACE OF THE SOIL FOR FOOD GROWING PURPOSES, AND OLDER OR LESS-EFFICIENT AUTOMOBILES WILL DONATE THEIR (SAFETY PLATE TEMPERED GLASS) WINDSHIELDS FOR SOUTH FACING SOLAR GREENHOUSE ENCLOSURES, INTERMIXED WITH STAINED GLASS BUILT BY LOCAL CRAFTSPERSONS.

ENERGY RESOURCES...
AN ENERGY-EFFICIENT PROTOTYPE

THE INCOMING SUN EQUALS DIRECT CONVERSION INTO STORAGE AND IMMEDIATE HEAT GAIN FOR WELL-INSULATED SOUTH OPENING HOUSING. THE MAJORITY OF HEAT REQUIRED DURING THE EVENING IS FROM RADIATED (STORED) HEAT OUT FROM THE STORAGE MEDIUM; BODY WARMTH (200 BTU/HR/PERSON AT REST); AND THE HEAT FROM INDANDESCENT LIGHTING. IF THE VOLUME OF THE MAJOR SPACES IS REASONABLY LOW (YOU WOULD PROBABLY WANT TO KEEP EACH DWELLING WAY UNDER 1000ϕ (METERS), SAY, NOT MORE THAN 750ϕ (METERS), THEN DIRECT GAIN PASSIVE SOLAR HEATING PLUS BODY WARMTH, PLUS LIGHTING (AND MAYBE A CANDLE OR TWO!) WOULD KEEP YOU ADEQUATELY WARM AT NIGHT WITHOUT AN AIR POLLUTING WOODSTOVE.

WEST → ← EAST

COURTYARD

. MAXIMUM USEFUL SUN ARC DURING WINTER IS ABOUT FROM 10:00AM TO 4:00PM DAILY...

"AND THINK OF HOW MUCH OF FUTURE FORESTS WE WOULD BE SAVING !!..."

"AND DON'T FORGET THE AIRSHED QUALITY IMPROVEMENT FROM FEWER WOOD STOVES!"

"AND REDUCED SINUS INFECTIONS TOO !!"

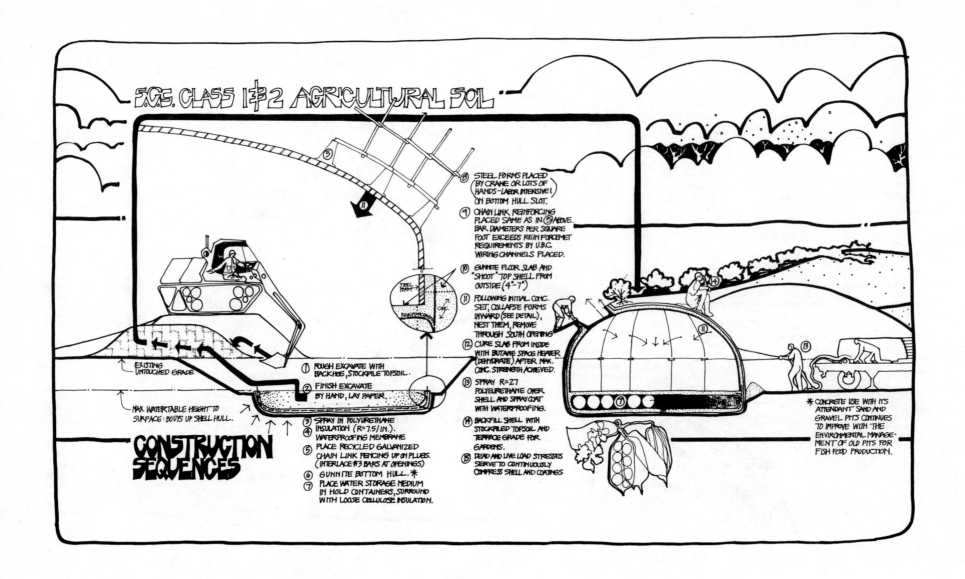

S.C.S. CLASS 1 & 2 AGRICULTURAL SOIL

CONSTRUCTION SEQUENCES

EXISTING UNTOUCHED GRADE

MAX. WATERTABLE HEIGHT TO SURFACE: BUOYS UP SHELL HULL.

① ROUGH EXCAVATE WITH BACKHOE, STOCKPILE TOPSOIL.

② FINISH EXCAVATE BY HAND, LAY PAPER.

③ SPRAY IN POLYURETHANE INSULATION (R=7.5/IN.).

④ WATERPROOFING MEMBRANE

⑤ PLACE RECYCLED GALVANIZED CHAIN LINK FENCING UP ON PLUGS. (INTERLACE #3 BARS AT OPENINGS)

⑥ GUNNITE BOTTOM HULL. ✳

⑦ PLACE WATER STORAGE MEDIUM IN HOLD CONTAINERS, SURROUND WITH LOOSE CELLULOSE INSULATION.

⑧ STEEL FORMS PLACED (BY CRANE OR LOTS OF HANDS - LABOR INTENSIVE) ON BOTTOM HULL SLOT.

⑨ CHAIN LINK REINFORCING PLACED SAME AS IN ⑤ ABOVE. BAR DIAMETERS PER SQUARE FOOT EXCEEDS REINFORCEMENT REQUIREMENTS BY U.B.C. WIRING CHANNELS PLACED.

⑩ GUNNITE FLOOR SLAB AND "SHOOT" TOP SHELL FROM OUTSIDE (4"-7").

⑪ FOLLOWING INITIAL CONC. SET, COLLAPSE FORMS INWARD (SEE DETAIL), NEST THEM, REMOVE THROUGH SOUTH OPENING.

⑫ CURE SLAB FROM INSIDE WITH BUTANE SPACE HEATER (DEHYDRATE) AFTER MAX. CONC. STRENGTH ACHIEVED.

⑬ SPRAY R=Z7 POLYURETHANE OVER SHELL AND SPRAY COAT WITH WATERPROOFING.

⑭ BACKFILL SHELL WITH STOCKPILED TOPSOIL AND TERRACE GRADE FOR GARDENS.

⑮ DEAD AND LIVE LOAD STRESSES SERVE TO CONTINUOUSLY COMPRESS SHELL AND COATINGS

STEEL FORM

REINFORCING   CONC.

✳ CONCRETE USE WITH ITS ATTENDANT SAND AND GRAVEL PITS CONTINUES TO IMPROVE WITH THE ENVIRONMENTAL MANAGEMENT OF OLD PITS FOR FISH FOOD PRODUCTION.

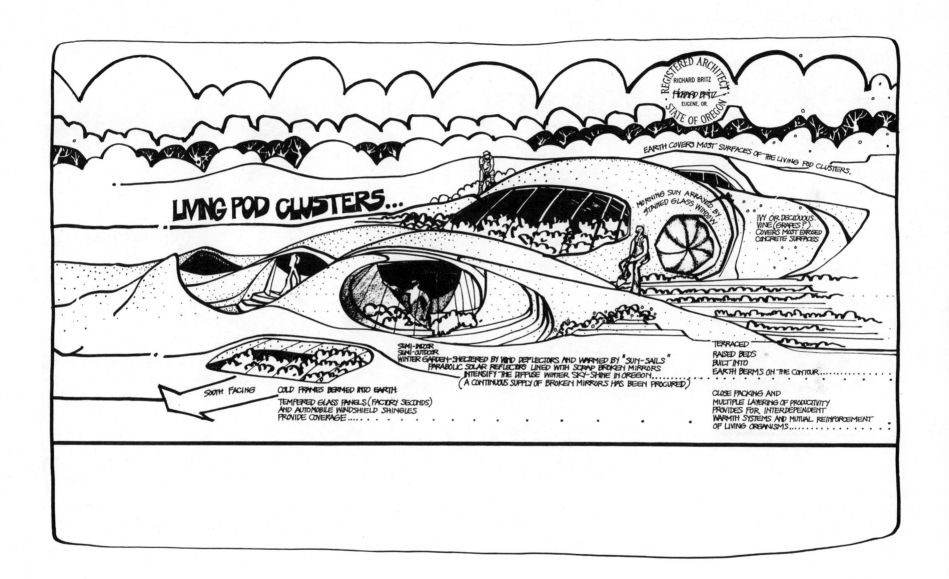

LIVING POD CLUSTERS...

REGISTERED ARCHITECT
RICHARD BRITZ
RICHARD BRITZ
EUGENE, OR.
STATE OF OREGON

EARTH COVERS MOST SURFACES OF THE LIVING POD CLUSTERS.

MORNING SUN ARRAYED BY STAINED GLASS WINDOW

IVY OR DECIDUOUS VINE (GRAPES?) COVERS MOST EXPOSED CONCRETE SURFACES

SEMI-INDOOR
SEMI-OUTDOOR
WINTER GARDEN-SHELTERED BY WIND DEFLECTORS AND WARMED BY "SUN-SAILS"
PARABOLIC SOLAR REFLECTORS LINED WITH SCRAP BROKEN MIRRORS
INTENSIFY THE DIFFUSE WINTER SKY-SHINE IN OREGON..........
(A CONTINUOUS SUPPLY OF BROKEN MIRRORS HAS BEEN PROCURED)

TERRACED
RAISED BEDS
BUILT INTO
EARTH BERMS ON THE CONTOUR..............

SOUTH FACING

COLD FRAMES BERMED INTO EARTH.
TEMPERED GLASS PANELS (FACTORY SECONDS)
AND AUTOMOBILE WINDSHIELD SHINGLES
PROVIDE COVERAGE..........

CLOSE PACKING AND
MULTIPLE LAYERING OF PRODUCTIVITY
PROVIDES FOR INTERDEPENDENT
WARMTH SYSTEMS AND MUTUAL REINFORCEMENT
OF LIVING ORGANISMS..............

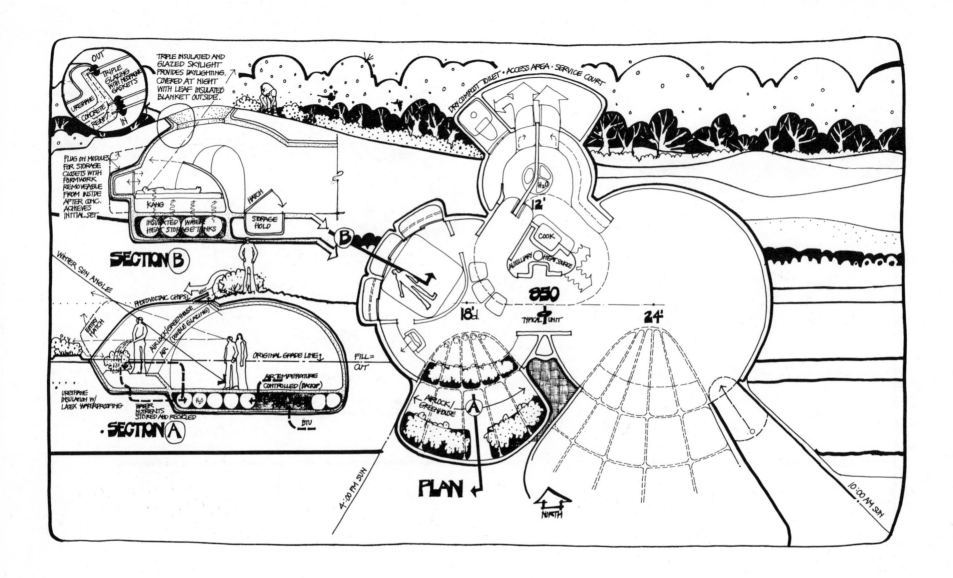

TRIPLE INSULATED AND GLAZED SKYLIGHT PROVIDES DAYLIGHTING. COVERED AT NIGHT WITH LEAF INSULATED BLANKET OUTSIDE.

OUT
TRIPLE GLAZING WITH NEOPRENE GASKETS
URETHANE
CONCRETE REINF.
IN

PLUG ON MODULES FOR STORAGE CLOSETS WITH FORMWORK REMOVEABLE FROM INSIDE AFTER CONC. ACHIEVES INITIAL SET.

KANG
HATCH
INSULATED WATER HEAT STORAGE TANKS
STORAGE HOLD

SECTION B

DRY COMPOST TOILET · ACCESS AREA · SERVICE COURT

H₂O
12'
COOK
AUXILIARY HEAT SOURCE

850

WINTER SUN ANGLE
PHOTOVOLTAIC CHIPS
AIRLOCK/GREENHOUSE (DOUBLE GLAZING)
ENTRY HATCH
AIR
ORIGINAL GRADE LINE
FILL = CUT
18'd
TYPICAL UNIT
24'

AIR TEMPERATURE CONTROLLED (BACKUP)

URETHANE INSULATION W/ LATEX WATERPROOFING
WATER NUTRIENTS STORED AND RECYCLED
H₂O
BTU

SECTION A

AIRLOCK/GREENHOUSE

PLAN

NORTH

4:00 PM SUN

10:00 AM SUN

IT IS DESIRABLE TO STORE MOST OF THE AUTOMOBILES, SINCE CLUSTERS OF HOUSING-FOOD WILL ALSO GROW IN THE URBAN AREA MOST LIKELY TO SUPPORT PEDESTRIAN AND BICYCLE-ORIENTED TRAVEL. A SECOND OPTION IS TO SELL THE MAJORITY OF AUTOMOBILES AND POOL THE CAPITAL TO PURCHASE A FEW SELECT UTILITY VEHICLES, AND UTILIZE THE CAPITAL SURPLUS TO ACQUIRE MATERIALS FOR CONSTRUCTION.

THE DECREASING USE OF AUTOMOBILES WILL CONTRIBUTE GREATLY TO THE AIR QUALITY OF THE SOUTHERN WILLAMETTE VALLEY, NOW FACING SATURATION FROM PARTICULATE MATTER ASSOCIATED WITH UNBURNT FUELS OF ALL SORTS, ESPECIALLY AUTOMOBILE EMISSIONS AND WOOD-BURNING STOVE INEFFICIENCIES. OIL CONSUMPTION WILL DROP, DECREASING THE NEED FOR RAPID DEPLETION OF WORLD OIL RESERVES, AND SUCH SMALL BENEFITS AS FEWER AUTOMOBILE INSURANCE PAYMENTS MIGHT EASE THE CASH-FLOW BURDEN OF NEIGHBORS THROUGHOUT EUGENE.

THIS PROPOSAL IS INTENDED TO MERGE WITH OTHER DEVELOPMENT INTERESTS IN THE WEST BUTTE AREA AND IS LOCATED IMMEDIATELY ADJACENT TO THE RIVER, AND SKINNER'S BUTTE PARK AND THE WILLAMETTE GREENWAY. IT LIES AT THE EDGE BETWEEN THE UNDEVELOPED PARK AND THE LOW-DENSITY HOUSING COMPRISING WEST SKINNER'S BUTTE NEIGHBORHOOD. ON 1/2 ACRE, CURRENTLY OWNED BY A COOPERATIVE HOUSING PARTNERSHIP, WE PROPOSE TO INTRODUCE PEOPLE TO A WORKING AND MAINTENANCE RELATIONSHIP TO THE LAND (COMPRISED OF CLASS I AGRICULTURAL SOILS), BUILD EXPERIMENTAL AND ENERGY-CONSERVING HOUSING, PROVIDE A NEW MIX OF HOUSING, FOOD PRODUCTION, PROFESSIONAL OFFICES, AND POSSIBLY A SMALL RESTAURANT SERVED BY THE GARDENS AND FARMS OF WHITEAKER NEIGHBORHOOD.

IN SHORT, WE PROPOSE TO DOUBLE POPULATION DENSITY ON THIS EXPERIMENTAL SECTION OF THE CITY (AND THUS PRESERVE THE URBAN SERVICES BOUNDARY), INTEGRATE PARKLAND WITHIN OUR SITE BY PROVIDING FOOD PRODUCTION SPACE, DEMONSTRATE AN ALTERNATIVE TRANSPORTATION PROGRAM BY MINIMIZING AUTOMOBILE DOMINANCE, DECREASE NEED FOR CITY SERVICES TO THIS AREA THROUGH MAJOR ENERGY CONSERVATION MEASURES, AND MAINTAIN DIVERSITY OF HOUSING ACCESS TO LOW AS WELL AS HIGHER INCOME GROUPS.

OVERALL, IT IS OUR INTENT TO EXPERIMENTALLY EXAMINE THE OPPORTUNITIES AND CONSTRAINTS OF PROTOTYPICAL MODELS FOR A RAPIDLY GROWING AND EXPANDING ECONOMY IN EUGENE, LANE COUNTY, AND THE WILLAMETTE VALLEY.

IN SUMMARY, IT IS OUR BELIEF THAT CONSTRUCTION OF THESE

EXPERIMENTAL BLOCK FARMS

WILL:

· CONSERVE CLASS I AGRICULTURAL SOILS AND ENABLE FOOD GROWING.

· ENABLE HOUSING DENSITY INCREASE IN THE CITY BY A FACTOR OF TWO TO TWO-POINT-FIVE: CLOSE PACKING.

· PROVIDE ENERGY-EFFICIENT PASSIVE SOLAR SEMI-UNDERGROUND HOUSING.

"SUN, WATER, SOIL, FOOD, & SHELTER."

· CLUSTER TIGHTLY FOR EFFICIENT LAND USE.

· EMPHASIZE FOOT AND BICYCLE TRAFFIC PATTERNS.

· PROVIDE AN AFFORDABLE ALTERNATIVE TO EXISTING HOUSING.

· STRESS SELF-HELP AND COOPERATIVE ACTIVITY TO BUILD AND MAINTAIN.

· UTILIZE INDUSTRIAL WASTES FOR MATERIALS FOR FORMWORK AND REINFORCING AND GLAZING.

· RECYCLE HOUSEHOLD WASTES AND COMPOST THEM FOR REUSE.

· UTILIZE LOCAL SAND AND GRAVEL FOR CONCRETE (FIRE AND SOUND-PROOF) CONSTRUCTION.

· DESIGN FOR TIGHT-LOOP CYCLING OF ORGANIC NUTRIENTS.

· PROVIDE SEMI-UNDERGROUND PARKING FOR A FEW VEHICLES.

· ENCOURAGE POOLING OF VEHICLES, SALES OF SURPLUS, RENOVATION OF FLEET.

· REORGANIZE STREET PATTERNS AND VACATE A FEW OF THEM.

· MINIMIZE OR ELIMINATE INSURANCE COSTS AND BUILDING MAINTENANCE (YEARLY UPKEEP) COSTS.

· REDUCE TO A MINIMUM ELECTRICITY AND WATER COSTS.

· UTILIZE CISTERNS FOR WATER STORAGE AND AQUIFER RECHARGE.

· SEEK MAXIMUM EFFICIENCY OF MATERIALS UTILIZED.

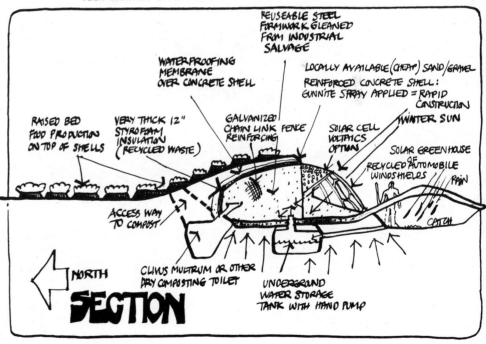

REUSEABLE STEEL FORMWORK GLEANED FROM INDUSTRIAL SALVAGE

WATERPROOFING MEMBRANE OVER CONCRETE SHELL

LOCALLY AVAILABLE (CHEAP) SAND/GRAVEL REINFORCED CONCRETE SHELL: GUNNITE SPRAY APPLIED = RAPID CONSTRUCTION

WINTER SUN

RAISED BED FOOD PRODUCTION ON TOP OF SHELLS

VERY THICK 12" STYROFOAM INSULATION (RECYCLED WASTE)

GALVANIZED CHAIN LINK FENCE REINFORCING

SOLAR CELL VOLTAICS OPTION

SOLAR GREENHOUSE OF RECYCLED AUTOMOBILE WINDSHIELDS

RAIN

ACCESS WAY TO COMPOST

CATCH

NORTH

SECTION

CLIVUS MULTRUM OR OTHER DRY COMPOSTING TOILET

UNDERGROUND WATER STORAGE TANK WITH HAND PUMP

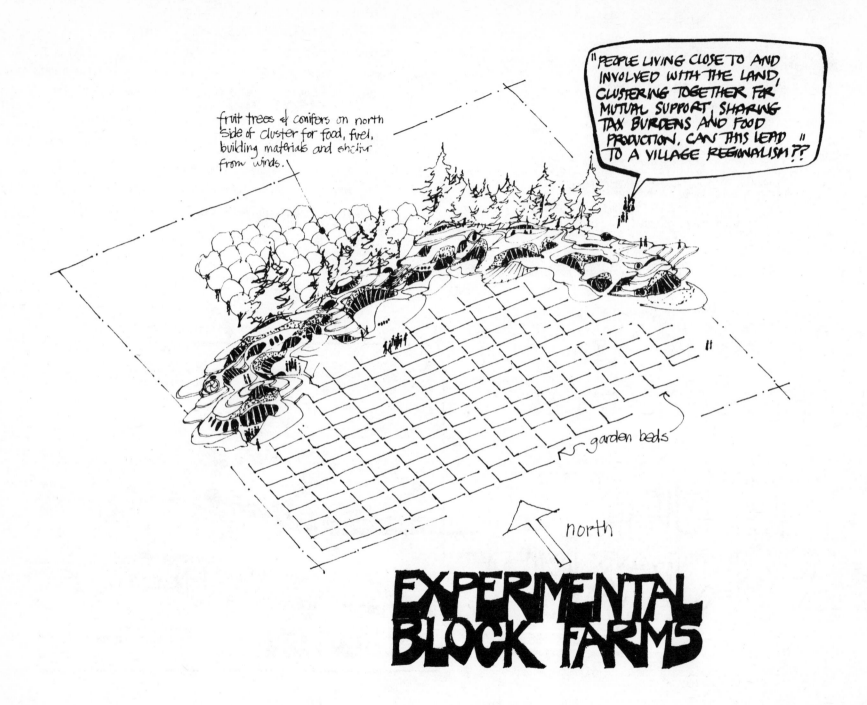

•ENCOURAGE LIFE-ENHANCING AND GROWTH-PRODUCTIVE NEIGHBORHOODS.

•BUILD IN TOOL SHEDS AND SHELTERS FOR BIRDS AND ANIMALS.

•MINIATURIZE SCALE (650¢ TO 750¢)TO CHILDREN'S SCALE.

•GENERATE INTERSECTIONS OF SURFACES THAT HAVE A BENDING SURFACE
 SO AS NOT TO BUMP YOUR SHOULDERS IN MOVING FROM ROOM TO ROOM.

•BE WARM IN WINTER, COOL IN SUMMER (MAXIMIZE THERMAL QUALITIES
 OF SOIL) AND MINIMIZE THE DRASTIC TEMPERATURE CHANGE DURING
 THE DAY TO NIGHT FLUCTUATION.

•INTERTWINE AND BLEND WITH THE SURROUNDING LANDSCAPE: TO BE IN
 MANY CASES UNNOTICEABLE TO THE GLANCING EYE.

•UTILIZE THE QUALITIES OF NATURAL LIGHT ENTERING THE ROOMS.

•MAINTAIN A PROTECTIVE, ENCLOSED, QUIET, PEACEFUL FEELING WITHIN.

•EMPHASIZE THE CHANGE OF SCALE AND SPACE BY MOVING FROM CLOSE
 IN TIGHT, COZY SPACE TO OUTDOORS.

•ORGANIZE SOCIAL ORIENTATION TO THE SUN AND SUN-PATIOS AND
 GARDENS RATHER THAN THE STREET.

•RECONSIDER THE SQUANDERING OF SQUARE FOOTAGE WITHIN OUR HOMES
 AND SUGGEST DUAL USAGE OF AREAS.

•ENCOURAGE DURABILITY   AND RESPONSIVENESS
        •LONG-LASTING SHELLS   •FLEXIBILITY OF PEOPLE
        •BUILT-INS   (MOBILITY)
                                     •MODERATE INTERIOR
                                      CHANGES POSSIBLE
                                     •LOW MAINTENANCE

•EXAMINE ALTERNATIVES AND POSSIBILITIES IN AFFORDABLE NEW
 HOUSING.

•ALLOW EXPERIMENTATION WITH THE CONSTRUCTION OF CLIMATE-
 RESPONSIVE SHELTERS, TEST THEM AGAINST CONVENTIONAL STRUCTURES,
 AND ANALYZE THEIR CAPABILITIES AND CONSTRAINTS.

•CONSIDER LONG AND SHORT TERM COSTS: EARTH-SHELTERED HOUSING IS
 LABOR INTENSIVE ORIGINALLY AND LOW COST TO MAINTAIN. -- LONG
 TERM ECONOMIES.

•RECOMMEND AN APPROPRIATENESS OF LIVING WITHIN THE EARTH AND
 ENCOURAGE A NESTED QUALITY WITH THE LAND AND HER FRUITS.

"UNDERGROUND" RESOURCES:

BLASER, WERNER. ROCK IS MY HOME. SWITZERLAND: WEMA-VERLAG, 1976

    CONTAINS EXAMPLES OF STONE HOUSES LOCATED THROUGHOUT EUROPE
WITH SUCH FEATURES AS STONE WALLS, STEPS, PORCHES AND LOFTS,
STRUCTURES LIKE THE TRULLO AND CORBEL DOME SHOW THE BUILDER'S
FEELING FOR THE NATIVE MATERIALS. NOTES THAT THESE HOUSES
ARE CONSIDERED TO BE BUILT WITHIN THE GROUND SINCE CONSOL-
IDATED STONE WAS USED. ILLUSTRATIONS AND PHOTOGRAPHS.

BOERICKE, ART AND SHAPIRO, BARRY. HANDMADE HOUSES - A GUIDE TO THE
WOODBUTCHER'S ART. SAN FRANCISCO, CALIFORNIA: SCRIMSHAW
PRESS, 1973-
A COLLECTION OF PHOTOGRAPHS ON HANDMADE STRUCTURES (A-
FRAMES, SAUNAS, CABINS, A MUSIC STUDIO AND EVEN A TEA HOUSE
OVER A RIVER) THAT REFLECT THE ARTIST/BUILDER'S SPIRIT,
INGENUITY AND PERSONALITY. ONE CAN SEE THE GENUINE LOVE
AND CRAFTSMANSHIP PUT INTO A VARIETY OF BUILDING MATERIALS
(SALVAGED LUMBER, DRIFTWOOD, HANDMADE BEAMS AND ADOBE
BRICKS). INTERESTING, REFRESHING PICTURES OF UNIQUE HOMES.

BROWN, JAMES T. AND GANNON, DAVID L. UNDERGROUND HOMES - PRIMER
TO EARTH SHELTERED LIVING POURTSMOUTH, OHIO: BY THE AUTHORS,
700 MASONIC BUILDING, 1979.

    BOOKLET WITH ACCOMPANING INFORMATION CONCERNING EARTH COVER-
ED LIVING. AN INTRODUCTION TO THE ADVANTAGES OF ENERGY
EFFICIENT HOUSING DESIGNS, THEIR CONSTRUCTION RESTRAINTS,
AND EXAMPLES OF A FEW BASIC STYLES. PUBLICATION OF AN
UNDERGROUND HOME DESIGN COMPANY.

CURRENT, WILLIAM. PUEBLO ARCHITECTURE OF THE SOUTHWEST. AUSTIN,
TEXAS: UNIVERSITY OF TEXAS PRESS, FOR THE AMON CARTER
MUSEUM OF WESTERN ART, 1971.

    COMPREHENSIVE PHOTOGRAPHIC ESSAY OF THE PUEBLO ARCHITECTURE
IN THE AMERICAN SOUTHWEST. THE BUILDINGS ARE PRESENTED TO
SHOW THE STRUCTURES AS THEY EXIST IN THEIR NATURAL ENVIRON-
MENT AND ANNOTATED WITH HISTORICAL INFORMATION. AMONG THOSE
FEATURED ARE THE NAVAJO NATIONAL MONUMENT IN ARIZONA, THE
AZTEC NATIONAL MONUMENT IN NEW MEXICO, AND MESA VERDE IN
COLORADO.

DEPT. OF HOUSING AND URBAN DEVELOPMENT, RESIDENTIAL ENERGY CONSUM-
PTION. SPRINGFIELD, VIRGINIA" NATIONAL TECHNICAL INFORMATION
SERVICE, 1972.

    DIAGRAMS, CHARTS AND PHOTOGRAPHS ASSIST IN REPORTING THE
FINDINGS ON THIS BALTIMORE, MARYLAND AND WASHINGTON, D.C.
AREA STUDY OF PRESENT AND PROJECTED RESIDENTIAL ENERGY
CONSUMPTION. CHARACTERISTIC HOUSE DESIGNS, CONSTRUCTION
TECHNIQUES, AIRFLOW PATTERNS, AS WELL AS INDIVIDUAL ENERGY
CONSUMPTION FOR HEATING, COOLING, MAJOR APPLIANCES AND LIGHTS
ARE DESCRIBED. A BIBLIOGRAPHY AND QUESTIONAIRE ARE ALSO
INCLUDED.

DITCHFIELD, P.H. PICTURESQUE ENGLISH COTTAGES AND THEIR DOORWAY
GARDENS. PHILADELPHIA, PENNSYLVANIA: THE JOHN C. WINSTON
CO, 1905.

    VARIETY OF INTIMATE ENGLISH COTTAGES, SHOPS AND INNS THAT
SEEM TO BLEND WITH THE LANDSCAPES. PHOTOGRAPHS ENHANCE THE

SUBTLE RELATIONSHIP OF THE COTTAGES TO THEIR SURROUNDINGS.
METHODS OF CONSTRUCTION (WEATHER PROTECTION, STONE WALLS,
THERMAL LAYERING) THE ROOFS, CHIMNEYS, WINDOWS, GARDENS AND
EVEN FOLKLORE ARE PRESENTED. ALSO DISCUSSED ARE THE EVOLU-
TION AND FOREIGN INFLUENCES ON THE ENGLISH COTTAGE ARCHI-
TECTURE.

EUGENE REGISTER GUARD. "WHILE OTHERS GO UNDERGROUND", FEB. 26, 1980.

    THIS RECENT ARTICLE LISTS BRIEFLY THE ADVANTAGES OF PEOPLE
USING THE EARTH AS AN INSULATOR IN HOUSING. NOTES THAT THE
TECHNIQUES OF UNDERGROUND SPACE USING ARE AMONG THE SIMP-
LEST, MOST COST EFFECTIVE AND READILY AVAILABLE MEANS OF
ENERGY CONSERVATION.

GRAY, VIRGINIA, MACRAE, ALLAN AND MCCALL, WAYNE. MUD, SPACE AND
SPIRIT: HANDMADE ADOBES. SANTA BARBARA, CALIFORNIA" CAPRA
PRESS, 1976.

    PHOTOGRAPHS AND INTERVIEWS WITH THE BUILDERS OF HANDMADE
ADOBE HOMES. EACH HOUSE IS UNIQUE BUT ALL ARE MADE FROM
COMMON MATERIALS FOUND IN THE SOUTHWEST'S HIGH DESERT WHERE
THE ENVIRONMENT (SUN/WIND) IS QUITE SEVERE. THESE INDI-
VIDUALS WORKING WITH MUD HAVE PRODUCED AN IMAGINATIVE
SEVEN LEVEL HOUSE, CLIFF DWELLINGS, A PIT HOUSE AND A VAR-
IETY OF FIREPLACE TYPES. AN AWARENESS OF THE SY AND THE USES
OF THE SUN ARE SHOWN THROUGH THE APPLICATION OF "GROW HOLES",
GREEN HOUSES AND SOLARIUMS.

FABOS, JULIUS G.Y., MILDE, GORDON T. AND WEINMAYR, V MICHAEL.
FREDERICK LAW OLMSTED, SR.: FOUNDER OF LANDSCAPE ARCHITECTURE
IN AMERICA. AMHERST, MASS: U. OF MASS PRESS, 1968.

    THE AMERICAN SOCIETY OF LANDSCAPE ARCHITECTS DEVELOPED A
TRIBUTE TO FREDERICK LAW OLMSTED BY CREATING AN EXHIBITION
OF HIS WORK THAT TOURED THE UNITED STATES IN 1964. THIS
EXCELLENT BOOK GOVES ALL THE COLLECTED GRAPHIC MATERIAL
(PLUS LISTS OF SIGNIFICANT WORKS AND IMPORTANT DATES IN
HIS LIFE) GATHERED A PERMANENT FORM. QUALITY VISUAL
ESSAY ON THE FOUNDER OF AMERICAN LANDSCAPE ARCHITECTURE.

FEIN, ALBERT. FREDERICK LAW OLMSTED AND THE AMERICAN ENVIRONMENTAL
TRADITION. NEW YORK, NEW YORK: GEORGE BRAZILLER, INC.
1972.

    DEFINES, EXPLAINS AND ILLUSTRATES A LARGE AMOUNT OF THE LATE
FREDERICK LAW OLMSTED'S CONTRIBUTIONS TO AMERICAN CITIES
DURING THE 19TH CENTURY. GIVES PICTORAL AND HISTORICAL
INFORMATION ON HIS ENVIRONMENTAL PLANNING AND DESIGN; IN-
CLUDING URBAN OPEN SPACES, PLANNED RESIDENTIAL COMMUNITIES,
CAMPUS PLANS AND 'NATIONAL PARKS'. THIS BIOGRAPHY PORTRAYS
OLMSTED'S LIFE-LONG GOAL OF IMPROVING THE ENVIRONMENTS OF
THOSE WHO LIVE IN THE CITIES OF AMERICA.

FERGUSON, FRANCIS. ARCHITECTURE, CITIES AND THE SYSTEMS APPROACH.
NEW YORK: NEW YORK: GEORGE BRAZILLER, INC. 1975.

    FERGUSON OFFERS SYSTEMS ANALYSIS AS A POSSIBLE APPROACH TO
SOLVING THE CURRENT PROBLEMS AND DEMANDS OF ARCHITECTURAL
AND URBAN PLANNING. HE EXPLAINS THE SYSTEMS IDEA THEORET-
ICALLY AND DESCRIBES ITS APPLICATION IN PROFESSIONAL PRAC-
TICE. TECHNICAL READING AND DRAWINGS FOR THE ARCHITECT AND
PLANNER.

FISCHER, ROBERT E., ED. ARCHITECTURAL ENGINEERING: NEW STRUCTURES.
SAN FRANCISCO, CALIFORNIA: MCGRAW HILL BOOK CO, 1963.

COMPREHENSIVE PRESENTATION OF WHAT THE NEW STRUCTURES IN
ARCHITECTURE ARE, HOW THEY ARE BEING INTEGRATED INTO CURRENT
BUILDING DESIGN, AND HOW THEY ARE ACTUALLY PUT TO USE AND
BUILT. THIN SHELLS, SPACE STRUCTURES, RECTILINEAR FRAMES
AND COMPONENTS SYSTEMS ARE THE MAJOR AREAS REPRESENTED.
INFORMATIVE DISCUSSION OF STRUCTURAL CONCEPTS, SYSTEMS AND
DESIGN PROCEDURES. LARGE AMOUNT OF PRACTICAL EXAMPLES AND
INTERESTING PHOTOGRAPHS.

FORDE-JOHNSON, J. PREHISTORIC BRITAIN AND IRELAND. LONDON, ENGLAND:
J.M. DENT AND SONS, LTD. 1976.

STUDY OF ANCIENT HOUSES AND SETTLEMENTS INCLUDING THEIR
MEGALITHIC TOMBS, BARROW, GRAVES AND CEREMONIAL SITES.
PRESENTS PREHISTORIC MAN AS A BUILDER (RATHER THAN A TOOL
MAKER) THROUGH THE STRUCTURES HE ERECTED FOR HIS ACCOMMO-
DATION, HIS WORSHIP AND HIS DEFENCE.

JACKSON, J.B. LANDSCAPES. SELECTED WRITINGS OF J.B. JACKSON. ED. BY
ERVIN H. ZUBE. AMHERST, MASS: U. OF MASS PRESS, 1970.

ANTHOLOGY OF PUBLISHED ESSAYS AND LECTURES GIVEN BY J.B.
JACKSON, AN INTERPRETER AND CRITIC OF THE AMERICAN LANDSCAPE.
SELECTIONS EXEMPLIFY HIS CONCERN THAT MAN BE A PART OF,
NOT APART FROM, THE 'HUMANIZED' LANDSCAPE.

MACKIE, EUAN. THE MEGALITH BUILDERS. NEW YORK, NEW YORK: E.P. DUTTON
1977.

THEORIES CONCERNING MEGALITHIC TOMBS; MASSIVE STONE CON-
STRUCTIONS REPRESENTING GREAT RESOURCES OF MANPOWER AND
OF TECHNICAL INGENUITY. VARIOUS PREHISTORIC, EUROPEAN
SITES ARE DESCRIBED, INCLUDING MALTA, IBERIA, AVEBURY,
SKARA BRAE AND SONEHENGE. IN COMPARISON, THE MAJORITY OF
THE COLLECTIVE OF COMMUNAL BURIAL SITES ON THE EUROPEAN
MAINLAND AND IN THE BRITISH ISLES WERE ARTIFICIAL STONE
CHAMBERS INSIDE BUILT MOUNDS, WHEREAS THE FURTHER EAST
IN THE MEDITERRANEAN AREA, THEY TENDED TO BE MOSTLY ROCK
CUT TOMBS. ACTUAL PHOTOGRAPHS.

MILLER, G. AND TYLER, JR. LIVING IN THE ENVIRONMENT: CONCEPTS, PROB-
LEMS AND ALTERNATIVES. BELMOST, CALIFORNIA: WADSWORTH PUB
LISHING COMPANY, INC., 1975.

GENERAL ECOLOGICAL PRINCIPLES AND THEIR APPLICATION TO
SUCH MAJOR ENVIRONMENTAL PROBLEMS AS POPULATION, RESOURCES,
POLLUTION, AND ENVIRONMENTAL ETHICS OR 'EARTHMANSHIP'.
SEPARATE 'ENRICHMENT STUDY' ON ENERGY SAVINGS IN BUILDINGS
PROPOSES TO REDUCE WASTES BY IMPROVING INSULATION AND BETTER
BUILDING DESIGN. PROPER PLACEMENT WITH RESPECT TO THE
SUN, USE OF NATURAL VENTILATION, SHIPT TO SOLAR HOT WATER
HEATERS, IMPROVING EFFICIENCY IN AIR CONDITIONERS, RE-
FRIGERATORS AND OTHER APPLIANCES ARE SUGGESTED SOLUTIONS.

MORELAND, FRANK L. ED. ALTERNATIVES IN ENERGY CONSERVATION; THE
USE OF EARTH COVERED BUILDINGS. PROCEEDINGS OF A CONFERENCE
HELD IN FORT WORTH, TEXAS: THE NATIONAL SCIENCE FOUNDATION,
JULY 9-12, 1975.

PROCEEDINGS OF A MAJOR CONFERENCE RECORDING ALL KINDS OF
UNDERGROUND BUILDING SUBJECT MATTER. AMONG THE TOPICS OF
THE SEMINARS AND PAPERS PRESENTED IN THIS REFERENCE BOOK
ARE THE USE OF EARTH COVERED BUILDINGS THROUGH HISTORY,
THE POLITICS, ECONOMICS, CONSERVATION, AND ENVIRONMENTAL
PSYCHOLOGY OF UNDERGROUND CONSTRUCTION. ALSO INCLUDED ARE
EXAMPLES, COMPARISONS, BIOGRAPHICAL SKETCHES OF THE AUTHORS
AND A 34 PAGE BIBLIOGRAPHY.

NATIONAL ACADEMY OF SCIENCES. CONFERENCE ON DESIGN FOR THE NUCLEAR
AGE. SPRINGFIELD, VIRGINIA: NATIONAL TECHNICAL INFORMATION
SERVICE, 1972.

ACCUMULATION OF WORKS BY AUTHORS PARTICIPATING IN A CON-
FERENCE HELD BY THE BUILDING RESEARCH INSTITUTE. ISSUES
ENCOMPASSING NEW FACTORS IN THE ENVIRONMENT, STRUCTURAL
DESIGN OF PROTECTED AREAS, AND THE DESIGN OF A NUCLEAR CITY
ARE PROPOSED.

PARKER, HARRY AND MACGUIRE, JOHN W. SIMPLIFIED SITE ENGINEERING
FOR ARCHITECTS AND BUILDERS. NEW YORK, NEW YORK: JOHN WILEY
AND SONS, INC. 1954.

ALTHOUGH MOST ARCHITECTS ARE SELDOM REQUIRED TO PERFORM
THE ACTUAL SURVEYING FOR A SET OF DRAWINGS FOR A BUILDING
PROJECT, THEY OFTEN NEED THE ABILITY TO PERFORM THE OFFICE
COMPUTATIONS THAT INVOLVE SURVEY DATA. TEXT FOR THE U. OF
O. SURVEY WORKSHOP COVERS THE ANALYSIS OF COUNTOUR LINES
AND THEIR USE IN SOLVING GRADING PROBLEMS. INCLUDING EX-
CAVATION, DRAINAGE, CUT AND FILL, MAXIMUM AND NIMIMUM
GRADES FOR DRIVEWAYS, SIDEWALKS AND PLAY AREAS.

"PASSIVE SOLAR: YESTERDAY IS TOMORROW". SUNSET MAGAZINE. FEBRUARY,
1979, PP. 76-80.

THE SAME PASSIVE SOLAR PRINCIPLE AS THE HEAT RETAINING
WALLS OF MESA VERDE IN COLORADO IS DISCUSSED IN THIS RECENT
ARTICLE. DESCRIBES THE THERMAL MASS THEORY, DIRECT AND IN-
DIRECT GAIN, AND SHOWS HOW THE SOLAR BUILDING ITSELF IS
THE SYSTEM.

RAMSEY, CHARLES G. AND SLEEPER, HAROLD R. EDITED BY JOSEPH N.
BOAZ. ARCHITECTURAL GRAPHIC STANDARDS. 6TH EDITION
NEW YORK:NEW YORK: JOHN WILEY AND SONS, INC. 1970.

STANDARD REFERENCE GUIDE USED BY ARCHITECTS, ENGINEERS,
DESIGNERS AND BUILDERS, DECORATORS, HOMEOWNERS, DRAFTS-
PERSONS, AND STUDENTS. PRACTICAL INFORMATION AND CURRENT
STANDARDS FOR CONSTRUCTION SPECIFICATIONS. RESEARCHED;
TECHNICAL.

RAPOPORT, AMOS. HOUSE FORM AND CULTURE. ENGLEWOOD CLIFFS, NEW
JERSEY:PRENTICE HALL, INC. 1969.

CONSIDERS HOW SOCIO-CULTURAL FACTORS (NEEDS, CHOICES,
RELATIONSHIP OF HOUSE TO SETTLEMENT, ETC.) CLIMATE, CON-
STRUCTION TECHNIQUES AND MATERIALS INFLUENCE HUMAN HABI-
TATS. ANALYSES HOW HOUSE DESIGN HISTORICALLY REFLECTS THE
THE PHYSICAL CONDITIONS OF THE ENVIRONMENT, AS WELL AS
CULTURAL PREFERENCES AND CAPABILITIES. (SOME ILLUSTRATIONS)

RAPOPORT, AMOS. "THE PUEBLO AND THE HOGAN" SHELTER AND SOCIETY.
ED. PAUL OLIVER. WASHINGTON, D.C.:FREDERIC A PRAEGER PUBL.
1969.

DETAILS THE HOGAN'S SUBTERRANEAN CONSTRUCTION AND DISCUSSES
THE NAVAJO'S SPIRITUAL FEELINGS IN RELATION TO THE SPACE.
THE PUEBLO RELIGION STRESSES A 'HARMONIOUS UNIVERSE WHERE
NATURE, GODS, PLANTS, ANIMALS AND MAN ARE ALL INTERDEPENDENT".
STUDIES THE PUEBLO ENVIRONMENT, AND THE EVOLUTION OF
TRIBAL AND LANGUAGE GROUPINGS, AND HOW THEIR SHELTERS
REFLECTED THE MANY FACETS OF THE PUEBLO SOCIETY.

REYNOLDS, JOHN S., LARSON, MILTON B, BAKER, M. STEVEN, MATHEW,
   HENRY, AND GRAY, ROBERT L.   THE ATYPICAL MATHEW SOLAR
   HOUSE AT COOS BAY OREGON.   EUGENE, OREGON:CENTER FOR
   ENVIRONMENTAL RESEARCH, SCHOOL OF ARCHITECTURE AND ALLIED
   ARTS, UNIVERSITY OF OREGON, 1976.

      LOCAL STUDY OF AN ATYPICAL SOLAR HOUSE WITH DATA PRESENTED
      THROUGH NUMEROUS CHARTS, GRAPHS AND DIAGRAMS.   AREAS OF
      INTEREST COVERED ARE STORAGE OF SOLAR ENERGY, COLLECTOR
      PERFORMANCE FACTORS, AND HEAT GAIN OR LOSS CALCULATIONS.
      THE PURPOSE OF THE RESEARCH IS TO DETERMINE THE EXTENT TO
      WHICH THE SOLAR ENERGY CONTRIBUTED TO THE SPACE HEATING
      NEEDS OF THE MATHEW'S HOUSE.

REYNOLDS, JOHN S.   SOLAR ENERGY FOR PACIFIC NORTHWEST BUILDINGS.
   EUGENE, OREGON: CENTER FOR ENVIRONMENTAL RESEARCH, SCHOOL
   OF ARCHITECTURE AND ALLIED ARTS, UNIVERSITY OF OREGON, 1976.

      RESEARCH PAPER DISCUSSING SOLAR ENERGY AND ITS APPROPRIATE-
      NESS IN REGIONAL ARCHITECTURE.   MAPS, GRAPHS, CHARTS,
      DRAWINGS AND PHOTOGRAPHS PRESENT INFORMATION ON SOLAR ENERGY
      AND SPACE HEATING PLUS VARIOUS WEATHER CONSIDERATIONS
      (TEMPERATURE, HUMIDITY, PRECIPITATION, SKY COVER, % OF
      POSSIBLE SUNSHINE, WIND FACTORS, ETC.).   COMPARES TWO OF
      THE EARLIEST, LIVED-IN DESIGNER-BUILDER-OWNER HOUSES IN
      THE PACIFIC NORTHWEST WITH OTHER SOLAR HOMES IN THE UNITED
      STATES.   ALSO INCLUDES BIBLIOGRAPHY ON SOLAR ENERGY IN
      ARCHITECTURE.

SCULLY, VINCENT.   PUEBLO/MOUNTAIN, VILLAGE, DANCE.   NEW YORK, NEW
   YORK: VIKING PRESS, 1972.

      THE AMERICAN INDIAN WORLD IS A PLACE WHERE THERE IS NO
      DIFFERENCE BETWEEN MAN AND NATURE--"ALL ARE LIVING THINGS
      ARE ONE AND ALL ARE LIVING; SNAKE, MOUNTAIN, CLOUDS, EAGLES
      AND MEN." THE PUEBLO ARCHITECTURE IS POSSIBLY THE MOST
      PERMANENT OF ALL NORTH AMERICAN INDIAN ARCHITECTURE.
      ILLUSTRATED HISTORIES OF THE TAOS, PICURIS, TEWA AND KERES
      VILLAGES, THE JEMEZ, PECOS AND SCANDIA SETTLEMENTS, THE
      HOPI TOWNS TO THE NAVAJO HOGAN.   THE INDIAN'S VERY CULTURE,
      LIFESTYLE AND HABITAT MIRRORED THE CONSTANT AWARENESS OF
      NATURE.

"SEE! PASSIVELY HEATED UNDERGROUND HOUSING CAN BE BEAUTIFUL, TOO!"
   MOTHER EARTH NEWS, MAY/JUNE 1978, PP. 101-103.

      SOLAR-EARTH ENERGY, INC., AND ARCHITECTURAL/CONSTRUCTION
      FIRM FORMED IN OHIO EXPRESSLY FOR THE DESIGNING AND BUILD-
      ING OF UNDERGROUND HOMES, EXHIBITS THEIR NEW EARTH SHELTERED
      DWELLING IN THIS ARTICLE.   EXAMPLE OF HOW FUNCTION AND
      BEAUTY CAN BE COMBINED TO CONSTRUCT A FIRST CALSS, ENERGY
      EFFICIENT HOME.

SHELTERS.   SERIATIM: JOURNAL OF ECOTOPIA, SUMMER, 1978, PP. 43-73.

      SEVERAL TOPICS OF INTEREST TO THE UNDERGROUND ENTHUSIAST
      ARE INCLUDED IN THIS ISSUE.   ARTICLES (ONE WRITTEN BY
      U OF O. PROFESSOR RICHARD BRITZ), NOTES AND INTERVIEWS
      ARE DIRECTLY CONCERNED WITH MOVING UNDERGROUND, EARTH
      COVERED SETTLEMENTS, PERTINENT UNIFORM BUILDING CODE IN-
      FORMATION, AND OTHER ALTERNATIVE TECHNOLOGIES.   INTEREST-
      ING, REGIONAL, SUBJECT MATTER.

STEADMAN, PHILIP.   ENERGY, ENVIRONMENT AND BUILDING.   NEW YORK,
   NEW YORK: CAMBRIDGE UNIVERSITY PRESS, 1975.

      OFFERS WAYS IN WHICH BUILDINGS MAY BE DESIGNED SO AS TO
      HAVE LESS DESTRUCTIVE IMPACT ON THE NATURAL ENVIRONMENT.
      OUTLINES METHODS FOR THE CONSERVATION OF ENERGY IN OTHER-
      WISE CONVENTIONALLY DESIGNED BUILDINGS.   A SURVEY OF NEW
      SOURCES OF ENERGY IS PRESENTED.   THESE ENERGY/ENVIRONMENTAL
      PRINCIPLES ARE APPLIED TO BUILDING DESIGNS WITH ILLUSTRATIONS
      DRAWN FROM ACTUAL PROJECTS.   SHORT HISTORIES, PAST DE-
      VELOPMENTS, DETAILED MAP OF BUILDING LOCATIONS, LISTS OF
      MANUFACTURERS, BIBLIOGRAPHY AND OTHER SOURCES ARE ALSO
      PROVIDED.

STILES, DAVID.   HUTS AND HIDEAWAYS.   CHICAGO, ILLINOIS: HENRY
   REGNERY CO., 1977.

      MR. STILES HAS ASSEMBLED DESIGNS OF BUILDINGS, SHELTERS,
      AND WEEKEND RETREATS THAT ARE BOTH RELATIVELY INEXPENSIVE
      AND CAN BE MADE BY ONE PERSON OVER A RELATIVELY SHORT PER-
      IOD OF TIME.   HUTS AND HIDEAWAYS INCOURAGES INNOVATION
      AND EXPERIMENTATION IN UNCONVENTIONAL CONSTRUCTION TECH-
      NIQUES.   EMPHASIS IS PLACED ON CLARITY AND SIMPLICITY IN
      BOTH ILLUSTRATIONS AND INSTRUCTIONS.   COSTS OF MATERIALS,
      TOOLS AND OTHER PRACTICAL CONSIDERATIONS ARE PRESENTED.
      HELPS THE PART-TIME BUILDER TO UNDERSTAND BASIC ENVIRON-
      MENTAL DESIGN AND TO BE ABLE TO ADAPT TO IT IN VARIOUS
      SITE CONDITIONS AND/OR LIMITATIONS OF BUILDING MATERIALS.

TUAN, YI-FU.   TOPOPHILIA: A SURVEY OF ENVIRONMENTAL PERCEPTION,
   ATTITUDES AND VALUES.   ENGLEWOOD CLIFFS, NEW JERSEY:
   PRENTICE -HALL, INC. 1974.

      TUAN DEFINES TOPOPHILIA AS 'SPECIFIC MANIFESTATIONS OF
      THE HUMAN LOVE OF A PLACE--ALL OF THE HUMAN BEING'S AFFEC-
      TIVE TIES WITH THE MATERIAL ENVIRONMENT.'   GEOGRAPHIC STUDY
      OF HOW ECONOMY, LIFESTYLE AND PHYSICAL SETTING AFFECT
      ENVIRONMENTAL PERCEPTION, ATTITUDES AND VALUES.   FURTHER
      DISCUSSES THE ESSENCE OF WHAT HAVE BEEN AND WHAT ARE OUR
      ENVIRONMENTAL IDEALS.

UNDERGROUND SPACE CENTER, UNIVERSITY OF MINNESOTA.   EARTH SHELTERED
   HOUSING DESIGN: GUIDELINES, EXAMPLES AND REFERENCES.   NEW
   YORK, NEW YORK: VAN NOSTRAND REINHOLD CO. 1979.

      THREE PART HANDBOOK ON EARTH SHELTERED HOUSING.   SPECIFIC
      DESIGN CONSIDERATIONS ARE GIVEN (SITE PLANNING, ARCHITECT-
      URAL DESIGN, ENERGY EFFICIENCY, PUBLIC POLICY ISSUES, AND
      SO ON).   EXISTING DESIGNS, LOCATED IN BOTH COLD AND WARM
      CLIMATES, ARE REPRESENTED BY FLOOR PLANS, DETAILS, SECTIONS
      AND PHOTOGRAPHS.   ENERGY CALCULATIONS AND COMPARISONS,
      PRODUCT DIRECTORY AND BIBLIOGRAPHY ARE AMONG THE ADDITIONAL
      INFORMATION COMPILED BY THE UNIVERSITY OF MINNESOTA.

WALSH, FRANK, AND HALLIDAY, WILLIAM R. DISCOVERY AND EXPLORATION
OF THE OREGON CAVES. GRANTS PASS, OREGON: TE-CUM-TOM
ENTERPRAISES, 1971.

RESEARCHES THE CHRONOLOGICAL HISTORY OF THE OREGON CAVES
NATIONAL MONUMENT IN THE WILLIAMS CREEK VALLEY, SOUTH OF
GRANTS PASS OREGON. WALSH RETELLS THE 19TH CENTURY STOR-
IES OF THE DISCOVERY OF THE CAVES BY ELIJAH J. DAVIDSON IN
THE FALL OF 1874 AND OF THE CAVES' EARLY EXPLORATION AND
EXPLOITATION. HALLIDAY WRITES OF THE 20TH CENTURY CONSER-
VATION EFFORTS TO SAME THE CAVES FOR 'ALL THE PEOPLE'. A
SPELEOGIST'S MAP (A PERSON WHO SYSTEMATICALLY STUDIES
CAVES) IS INCLUDED AS WELL AS A LIST OF HISTORICAL HIGH-
LIGHTS. EASY, INTERESTING READING AON NATURAL UNDERGROUND
FORMATIONS.

WELLS, MALCOLM. "UNDERGROUND ARCHITECTURE". THE COEVOLUTION QUARTERLY
FALL, 1976, PP. 84-93.

ARTICLE BY ARCHITECT, WRITER AND SOLAR CONSULTANT ON WHY
TO BUILD WITH THE EARTH COVERED STRUCTURE. DESCRIBES AND
ILLUSTRATES HIS OWN OFFICE AND HIS REASONS FOR GOING UNDER-
GROUND. ANSWERS TEN OF THE MOST FREQUENTLY ASKED QUESTIONS
ABOUT BUILDING WITHIN THE EARTH.

WELLS, MALCOLM. UNDERGROUND DESIGNS. BREWSTER, MASSACHUSETTS:
BY THE AUTHOR, BOX # 1149, 1977.

COLLECTION OF DESIGNS BY A WELL KNOWN AUTHOR AND ARCHITECT
OF UNDERGROUND STRUCTURES. SITE PLANS, ELEVATIONS AND
PERSPECTIVES ARE HUMOROUSLY AS WELL AS FACTUALLY ANNOTATED.
ADDITIONAL SECTIONS ARE INCLUDED COVERING BUILDING CODES,
AND COSTS, WATERPROOFING, INSULATION AND LANDSCAPING.

Chapter 9

The Sladden Area

Legend:
- FRUIT AND NUT TREES
- GARDENS AND FARMS
- COMMUNITY GREENHOUSE LOCATIONS
- CONIFEROUS TREES

0' 25' 50' 100'

WILLAMETTE RIVER.

RIVER HOUSE

ROSE GARDEN

3
2
1

#3

NORTH

4J

MORENA

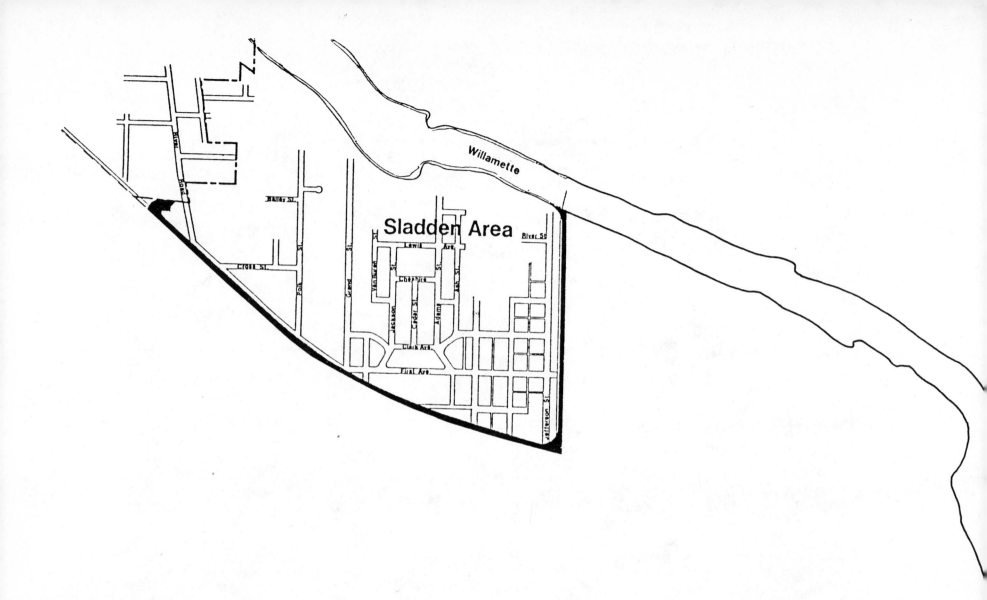

Willamette

Sladden Area

Bailey St.

Cigar St.

Polk St.

Grand St.

Van Buren St.

Jackson St.

Cedar St.

Lewis Ave.

Chaplies St.

Adams St.

Ash St.

River St.

Clark Ave.

First Ave.

Jefferson St.

SECTOR 3
THE SLADDEN AREA

SLADDEN (Sector 3)

Just west of the Washington-Jefferson freeway between the railroad tracks and the river is the Sladden area. Primarily comprised of reasonably-stabilized single family dwellings (R-1 zoning) this is not an area of tremendous need at the present time. However, several factors of major <u>potential</u> importance were present in the Sladden area which became apparent at the beginning of the grant period. First, during the inventory phase, we discovered the process used to acquire the land for and develop the river edge apartments -- developed by Mike Safely. This project was of major import, due to the nature of the acquisision of a huge parcel of land abutting the river for private development -- a major change in housing density and riverfront usage along the Willamette Greenway <u>and</u> Parks Department-owned land. In other words -- a precedent. This investigation gave precedent for other increased density housing along the river and materially influenced the acquisition of the Mizell site in West Butte sub-neighborhood (Sector 2).

The second major consequence noted during the early stages of research and inventory pointed us to the large, yet scattered quantity of land owned by the Oregon State Highway Commission throughout both the Sladden and the Polk areas of the neighborhood. We felt than and do now that the possibilities of neighborhood acquisition of some of this land would be a good idea, yet Highway Commission policies preclude the sale of any of this land initially to anyone other than County or City governmental bodies. Much time during the initial grant stage was spent on the Sladden area, probably at least three months in total, when the students first started working on the grant (mostly during August, September and October, 1979). A lot of this time was finally lost but the possible benefits to the neighborhood could have been high. Most of this time concerned the lands owned by the Oregon State Highway Commission which during previous land acquisition strategies, they had acquired in anticipation of running a freeway parallel to the Willamette River off of the Washington-Jefferson freeway. The general notion came up as a suggestion from several neighbors on Ash Street that the neighbors might want to acquire the rental houses they were living in for permanent ownership and thus more greatly stabilize the owner-occupied portion of Whiteaker. Direct negotiations were opened up with members of the Highway Commission and larger tracts of land were also discussed for possible neighborhood acquisition -- especially the Briarcliff site (further down river) for future food production as one of the six neighborhood farms.

The results of these negotiations eventually involved the City of Eugene Parks and Recreation Department, the Neighborhood Advisory Group, numerous individual neighbors and developed a strong understanding in some of us how hard these lands would be to acquire. But what it did do, however, was to uncover even more plans which the Parks Department had for further expansion of their parking lots near the river.

The major projects upon which we concentrated during and just after the grant period were the School District 4-J site; the Community Gardens' River House site; and the small-scale block farm beginning at the houses of Shawn Boles, Melva Edrington, et al; and the second "seniors" project at Ivorena Care Center adjacent to 4-J; and a small private demonstration garden and arbor at the individual family scale.

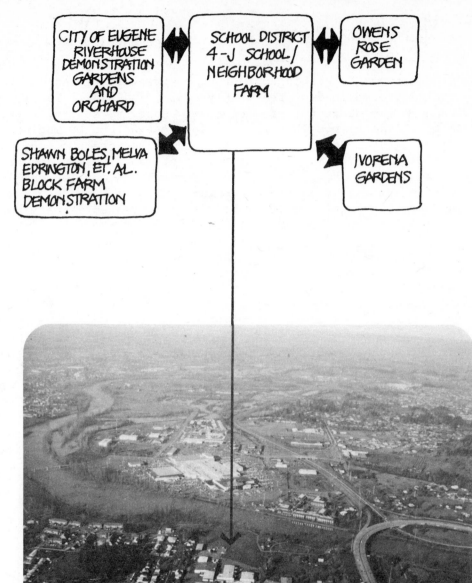

These projects are intended to intertwine and form a system of urban agriculture which uses the School District 4-J site as an ultimate focus for Sladden area food production.

The close adjacency of these projects and the clustering of effort at these sites enabled us to build and mutually reinforce the integrative strategies of government agencies working on similar projects, old and young working together in food related activities, and the possibilities inherent in families working together to generate a greater food production self-sufficiency.

In turn, then, the following was accomplished:

SCHOOL FARM / NEIGHBORHOOD FARM #3

The school farm/neighborhood farm #3 site was identified early as having great possibilities for river edge food production due to its good soils, excellent sun exposure and probable availability. At the same time, this land was under increasing threat of parking lot expansion by Rose Garden workers and other recreational users since automobiles had begun to park on the land.

The School District maintenance and storage facility also had just recently been expanded. Therefore, we decided to approach both the 4-J School District and Looking Glass Youth Services with a mutually-advantageous proposal which would maintain the land in open space, engage youth in agriculturally-related jobs, and get a toe-hold on the land for eventual neighborhood use.

In Cleveland, Ohio, thousands of students grades 3-12 annually grow as much as $800,000 worth of vegetables. In Weston, Mass., a 20-acre youth farm grows 100 tons of food every year. All across our nation schools are developing farm and garden programs, which not only produce food, but also provide employment and educational opportunities for the students involved. Here in Eugene, cooperation between interested citizens and the city government has produced a model community gardens program, but a comprehensive school gardens program has yet to materialize. With the incentive of skyrocketing food costs and a solid base of community support, the time is ripe for Eugene's school district to initiate a program which could employ and educate students in socially-useful and fun activity.

In contrast to existing community gardens, a school farm would utilize an organized system of food production. Students, youth workers and community volunteers would be responsible for planting, nurturing, harvesting, and marketing crops. This "hands-on" type of educational experience would not only provide student-workers with training in labor-intensive horticultural practices, but also employ them both part-time and full-time.

The initial 2-3 years of the program would focus on summertime production. Curriculum opportunities would include spring classes and/or workshops involving soil preparation, starting seedlings, planting times and techniques, and crop scheduling, and in the fall, crop harvest, processing, marketing, and planting winter cover crops. Ideally, students demonstrating the most interest and initiative could be worked into the summer job program. As the farm became established, curriculum could be expanded to include a winter gardening program. This would not only increase the educational opportunities of the program, but also improve the quality and quantity of produce from the farm.

Classes utilizing the school farm could include:

-- biology field trips and workshops

-- horticulture classes using the farm for outdoor classes, an experiment station

-- social studies concerning the world food crisis and the viability of labor intensive vs. energy intensive agriculture

-- energy conservation programs dealing with energy efficient food production and distribution

-- health and nutrition classes concerning diet, food production, and processing

Phase I -- A pilot project could do initial testing and development on the site. A crew from the Community Improvement Project (funded by CETA) could test soil viability, feasibility of different types of cultivation, irrigation, and crops, and markets for produce. Food produced in this phase would mostly be donated to needy organizations and individuals, with enough sold to test markets, and supplement the program with any proceeds. This trial period would allow for careful analysis of the program, including economic feasibility and costs, and gradual integration of curriculum and student workers.

Phase II -- In this phase, School District 4-J would take over administration, management, and funding of the program. The program would begin focusing on the goal of self-sufficiency by selling enough produce to pay for summer student worker positions. Some equipment, supplies, and transportation would have to be provided by the school district. Curriculum would be formalized and allowed to expand.

Phase III -- The school farm would expand and diversify within the limits of what it could support. The school district would control curriculum and management, but discontinue funding.

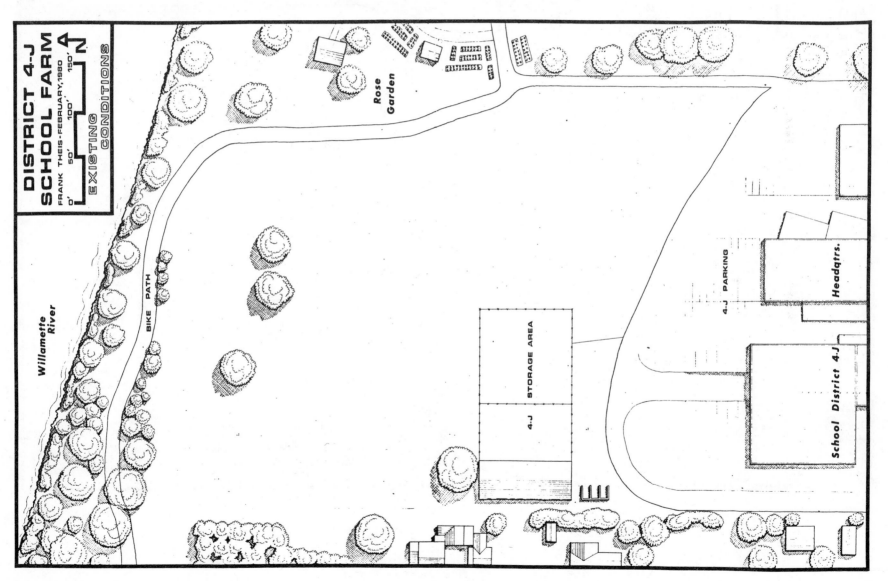

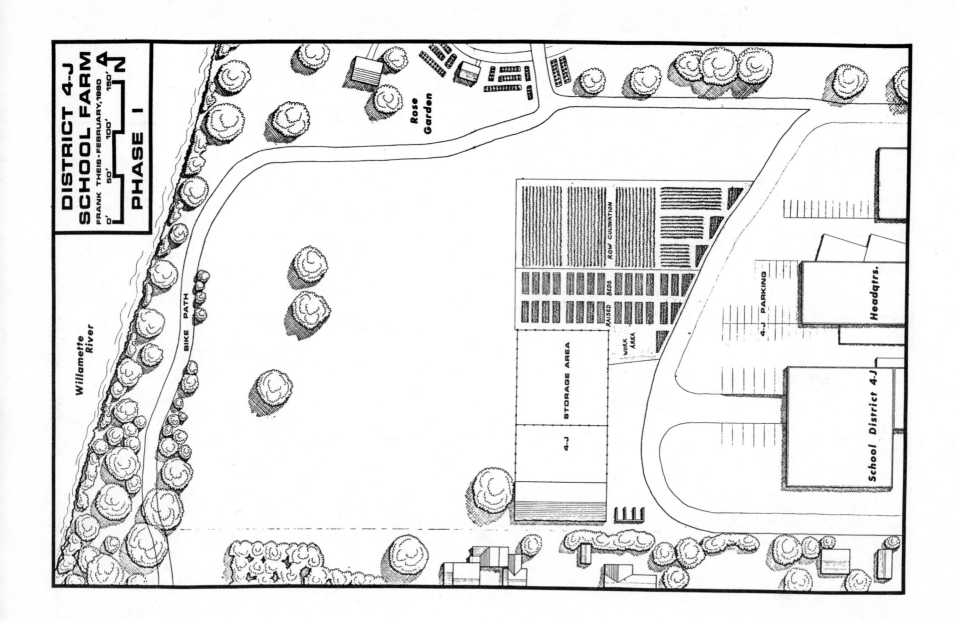

DISTRICT 4-J
SCHOOL FARM
FRANK THEIS – FEBRUARY, 1980

PHASE I

0' 50' 100' 150'

N

Willamette River

BIKE PATH

Rose Garden

ROW CULTIVATION

RAISED BEDS

WORK AREA

4-J STORAGE AREA

4-J PARKING

Headqtrs.

School District 4-J

PROGRAM ANALYSIS — PHASE I

SCHEDULE

CROPS:	JAN	FEB	MAR	APR	MAY	JUN	JUL	AUG	SEP	OCT	NOV	DEC
ONIONS				▓	▓	▓	▓	▓	▓			
LETTUCE					▓	▓	▓					
BROCCOLI					▓	▓	▓	▓				
CORN					▓	▓	▓	▓	▓			
TOMATOES					▓	▓	▓	▓	▓	▓		
EMPLOYEES:												
FULL-TIME: LEADWORKER				▓	▓	▓	▓	▓	▓	▓	▓	▓
2 YOUTH WORKERS				▓	▓	▓	▓	▓	▓	▓	▓	▓
PART-TIME: 4 YOUTH WORKERS				▓	▓	▓	▓	▓	▓	▓		

NOTES:
* CROPS USED FOR ANALYSIS ARE A HYPOTHETICAL CROSS-SECTION, THERE WOULD ACTUALLY BE A MUCH HIGHER LEVEL OF DIVERSITY.
* SALARIES WOULD BE PAID BY C.E.T.A. IN PHASES 1 AND 2 THROUGH THE COMMUNITY IMPROVEMENT PROJECT

EXPENSES

NO.	TYPE	UNIT COST	TOTAL
	SALARIES		
1	LEADWORKER	$4.00/HR.	$4500.00
2	YOUTHWORKERS	$3.50/HR.	$7300.00
4	PART-TIME	$3.50/HR.	$7300.00
	TOOLS		
	SUPPLIED BY THE COMMUNITY IMPROVEMENT PROJECT		
	SEED & FERT.		
	SEED + STARTS FOR 3/4 ACRE		$150.00
600 LBS. COMMERCIAL 16-16-16	$80.00		
500 LBS. MANURE / 500 LBS. ROCK PHOSPHATE	$20.00 $70.00		$170.00
	TRANSPORT		
	PROVIDED BY C.I.P.		
	MISC.		
	GARDEN STAKES + CRATES... INSURANCE (C.I.P.)		$180.00
	TOTAL		$19,600.00

INCOME

CROPS	ROW FT.	LBS.	$/LB.	LBS. SOLD	$	SURPLUS
ONIONS	1500	30 BUSHELS	$4.00/BUSHEL	20 BUSHELS	$80.00	10 BUSHELS
LETTUCE	1500	1200	$.15/LB.	400 LBS.	$60.00	800 LBS.
BROCCOLI	1500	1500	$.15/LB.	700 LBS.	$105.00	800 LBS.
CORN	3000	3000 EARS	$.05/EAR	2000 EARS	$100.00	1000 EARS
TOMATOES	500	50 BUSHELS	$9.00/BUSHEL	20 BUSHELS	$180.00	30 BUSHELS
TOTAL	8000				$525.00	

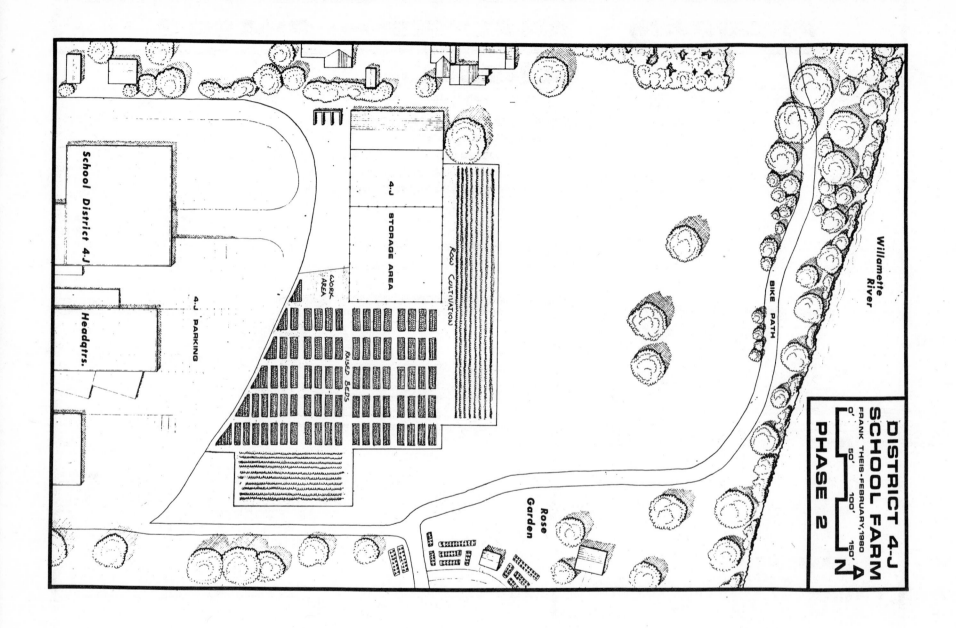

DISTRICT 4-J
SCHOOL FARM

FRANK THEIS · FEBRUARY, 1980

PHASE 2

0'   50'   100'   150'

N

School District 4-J

Headqrs.

4-J PARKING

WORK AREA

4-J STORAGE AREA

ROW CULTIVATION

RAISED BEDS

Rose Garden

BIKE PATH

Willamette River

PROGRAM ANALYSIS — PHASE 2

SCHEDULE

CROPS:		JAN	FEB	MAR	APR	MAY	JUN	JUL	AUG	SEP	OCT	NOV	DEC
ONIONS				▬	▬	▬	▬	▬	▬				
LETTUCE	EARLY			▬	▬	▬	▬	▬	▬				
BROCCOLI	EARLY				▬	▬	▬	▬					
CORN	EARLY					▬	▬	▬	▬	▬			
TOMATOES						▬	▬	▬	▬	▬			
EMPLOYEES:													
FULLTIME:													
1 LEADWORKER				▬	▬	▬	▬	▬	▬	▬			
3 YOUTHWORKERS				▬	▬	▬	▬	▬	▬	▬			
PART-TIME:													
4 YOUTHWORKERS					▬	▬	▬	▬	▬	▬			

NOTES:
*ALL PRODUCTION DATA WAS EXTRAPOLATED USING LANE Co. EXTENSION SERVICE DATA FOR A HOME VEGETABLE GARDEN.
*RESEARCH DONE SINCE 1972 BY ECOLOGY ACTION PALO ALTO, CAL., HAS PROVEN THAT PRODUCTION MAY BE 20 TIMES GREATER USING BIODYNAMIC/FRENCH INTENSIVE CULTIVATION

EXPENSES

NO.	TYPE	UNIT COST	TOTAL
	SALARIES		
1	LEADWORKER	$4.00/HR.	$5200.00
3	YOUTHWORKERS	$3.50/HR.	$12,900.00
4	PART-TIME	$3.50/HR.	$8000.00
	TOOLS		
	PROVIDED BY C.I.P.		
	SEED & FERT.		
	SEED + STARTS FOR 1¼ ACRES	$250.00	$250.00
600 LBS. COMMERCIAL 16-16-16	$80.00		
1000 LBS. MANURE / 1000 LBS. ROCK PHOSPHATE	$40.00 / $140.00	$260.00	
	TRANSPORT		
	PROVIDED BY C.I.P.		
	MISC.		
	GARDEN STAKES + CRATES. INSURANCE (C.I.P.)		$100.00
	TOTAL		$26,710

INCOME

CROPS	ROW FT.	LBS.	$/LB.	LBS. SOLD	$	SURPLUS
ONIONS (EARLY)	2000	40 BUSHELS	$4.00/BUSHEL	25 BUSHELS	$100	15 BUSHELS
ONIONS (LATE)	2000	40 BUSHELS	$4.00/BUSHEL	25 BUSHELS	$100	15 BUSHELS
LETTUCE (EARLY)	1500	1200	.15/LB.	400 LBS.	$60	800 LBS.
LETTUCE (LATE)	1500	1200	.15/LB.	400 LBS.	$60	800 LBS.
BROCCOLI (EARLY)	1500	1500	.15/LB.	700 LBS.	$105	800 LBS.
BROCCOLI (LATE)	1500	1500	.15/LB.	700 LBS.	$105	800 LBS.
CORN	4000	4000 EARS	.05/EAR	3000 EARS	$150	1000 EARS
TOMATOES	1000	100 BUSHELS	9.00/BUSHEL	40 BUSHELS	$360	60 BUSHELS
TOTAL	15000				$1040.00	

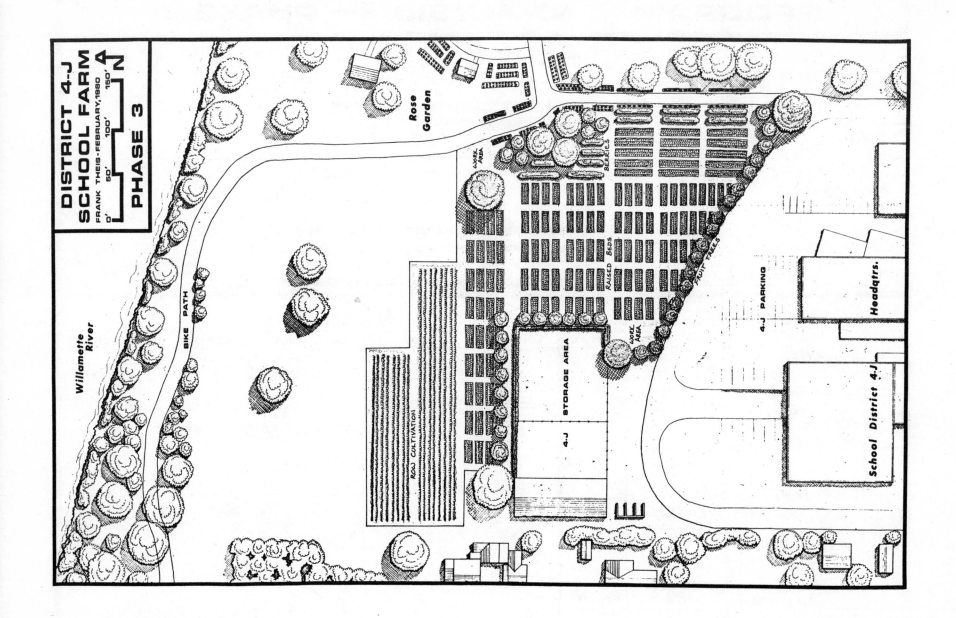

DISTRICT 4-J
SCHOOL FARM
FRANK THEIS · FEBRUARY, 1980

PHASE 3

0'   50'   100'   150'

N

Rose Garden

Willamette River

BIKE PATH

WORK AREA

BERRIES

RAISED BEDS

FRUIT TREES

WORK AREA

ROW CULTIVATION

4-J STORAGE AREA

WORK AREA

4-J PARKING

Headqtrs.

School District 4-J

PROGRAM ANALYSIS — PHASE 3

SCHEDULE

CROPS:		JAN	FEB	MAR	APR	MAY	JUN	JUL	AUG	SEP	OCT	NOV	DEC
ONIONS													
LETTUCE	EARLY												
BROCCOLI	EARLY												
CORN	EARLY												
TOMATOES													
EMPLOYEES:													
FULLTIME:													
2 LEADWORKERS													
6 STUDENT-WORKERS													
PART-TIME:													
12 STUDENT-WORKERS													

NOTES:
* SCHOOL DISTRICT TAKES OVER TOTAL ADMINISTRATION AND FUNDING IN PHASE 3. A GREAT DEAL OF POTENTIAL EQUIPMENT SHARING AND JOB-SHARING IS NOT TAKEN INTO ACCOUNT
* STUDENTS IN WORK-SHOPS AND CLASSES COULD DO A GOOD AMOUNT OF WORK
* FRUIT TREES AND BERRIES AREN'T AT PRODUCTION MATURITY.

EXPENSES

NO.	TYPE	UNIT COST	TOTAL
	SALARIES		
2	LEADWORKERS	$5.00/HR.	$12,800.00
6	STUDENT-WORKERS	$3.50/HR.	$10,200.00
12	PART-TIME	$3.50/HR.	$16,800.00
	TOOLS		
20	SHOVELS, RAKES, FORKS, HOES	$12.50	$250.00
4	WHEELBARROWS	$50.00	$200.00
	HOSES, HANDTOOLS, ETC.		$150.00
	SEED & FERT.		
	SEED + STARTS FOR 2½ ACRES		$500.00
1000 LBS. COMMERCIAL 16-16-16 $120.00 / 2000 LBS. MANURE $60.00 / 2000 LBS. ROCK PHOSPHATE $200.00			$380.00
	TRANSPORT GAS + MAINTENANCE 2 TRUCKS PART-TIME		$1000.00
	MISC. GARDEN STAKES, CRATES... INSURANCE (4-J)		$200.00
	TOTAL		$42,280

INCOME

CROPS	ROW FT.	LBS.	$/LB.	LBS. SOLD	$	SURPLUS
ONIONS (EARLY)	4000	80 BUSHELS	4.00/BUSHEL	55 BUSHELS	$220	25 BUSHELS
ONIONS (LATE)	4000	80 BUSHELS	4.00/BUSHEL	55 BUSHELS	$220	25 BUSHELS
LETTUCE (EARLY)	3000	2400	.15/LB.	1200 LBS.	$180	1200 LBS.
LETTUCE (LATE)	3000	2400	.15/LB.	1200 LBS.	$180	1200 LBS.
BROCCOLI (EARLY)	3000	3000	.15/LB.	2000 LBS.	$300	1000 LBS.
BROCCOLI (LATE)	3000	3000	.15/LB.	2000 LBS.	$300	1000 LBS.
CORN	8000	8000 EARS	.05/EAR	7000 EARS	$350	1000 EARS
TOMATOES	2000	200 BUSHELS	9.00/BUSHEL	100 BUSHELS	$900	100 BUSHELS
TOTAL	30,000				$2650.00	

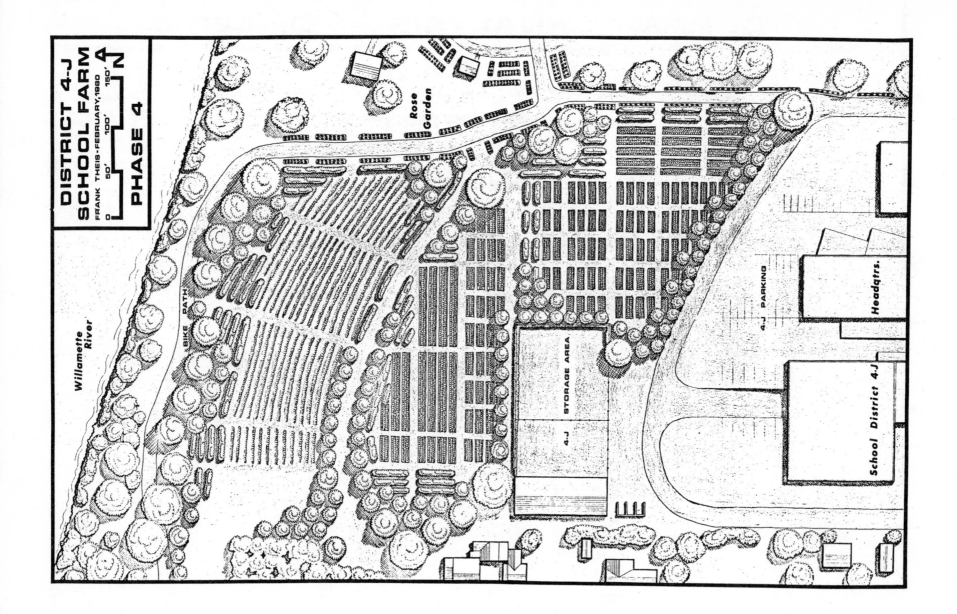

DISTRICT 4-J
SCHOOL FARM
FRANK THEIS · FEBRUARY, 1980

0 50' 100' 150'
N

PHASE 4

Willamette River

BIKE PATH

Rose Garden

STORAGE AREA

4-J

4-J PARKING

School District 4-J

Headqtrs.

PROGRAM ANALYSIS — PHASE 4

SCHEDULE

CROPS:		JAN	FEB	MAR	APR	MAY	JUN	JUL	AUG	SEP	OCT	NOV	DEC	NOTES:
ONIONS														* WINTER GARDENING PROGRAM BEGINS, POSSIBLY WITH A GREENHOUSE, DEFINITELY WITH COLD FRAMES. IT SHOULD BE HANDLED BY ONE FULL-TIME FARM MANAGER AND STUDENTS IN CLASSES AND WORKSHOPS.
LETTUCE	EARLY													
BROCCOLI	EARLY/LATE													
CORN	EARLY/LATE													* FRUIT, NUT, AND BERRY PRODUCTION IS FROM PHASE 3 PLANTINGS.
TOMATOES														
EMPLOYEES:														
FULL-TIME:														
2 LEADWORKERS														
6 STUDENT-WORKERS														
PART-TIME:														
12 STUDENT-WORKERS														

EXPENSES

NO.	TYPE	UNIT COST	TOTAL
	SALARIES		
2	LEADWORKERS	$5.00/HR.	$16,000.00
6	STUDENT-WORKERS	$3.50/HR.	$10,200.00
12	PART-TIME	$3.50/HR.	$16,800.00
	TOOLS		
20	SHOVELS, RAKES, FORKS, HOES	$12.50	$250.00
4	WHEELBARROWS	$50.00	$200.00
	MISCELLANEOUS		$300.00
	SEED & FERT.		
	SEED + STARTS FOR 4 ACRES		$700.00
1000 LBS.	COMMERCIAL 16-16-16	$120.00	
2 TONS	MANURE	$120.00	$640.00
2 TONS	ROCK PHOSPHATE	$400.00	
	TRANSPORT		
	GAS + MAINTENANCE 2 TRUCKS PART-TIME		$1500.00
	MISC.		
	GARDEN STAKES, CRATES... INSURANCE (4-J)		$300.00
	TOTAL		$46,690.00

INCOME

CROPS	ROW FT.	LBS.	$/LB.	LBS. SOLD	$	SURPLUS
ONIONS (EARLY)	6400	130 BUSHELS	4.00/BUSHEL	100 BUSHELS	$400	30 BUSHELS
ONIONS (LATE)	6400	130 BUSHELS	4.00/BUSHEL	100 BUSHELS	$400	30 BUSHELS
LETTUCE (EARLY + LATE)	8000	6400	.15/LB.	4400 LBS.	$660	2000 LBS.
LETTUCE (MIDSEASON)	5000	4000	.15/LB.	4400 LBS.	$660	2000 LBS.
BROCCOLI (EARLY + LATE)	8000	8000	.15/LB.	6000 LBS.	$900	2000 LBS.
BROCCOLI (MIDSEASON)	5000	5000	.15/LB.	3500 LBS.	$525	1500 LBS.
CORN	15,000	15,000 EARS	.05/EAR	13,000 EARS	$650	2000 EARS
TOMATOES	3200	320 BUSHELS	9.00/BUSHEL	200 BUSHELS	$1800	120 BUSHELS
FRUIT, NUTS AND BERRIES	40 FRUIT TREES 10 NUT TREES 100 BERRY BUSHES	15,000 LBS.	.15/LB.	10,000 LBS.	$1500	5000 LBS.
TOTAL	57,000				$7495.00	

THE RIVERHOUSE DEMONSTRATION GARDENS

Next door to the School District 4-J site is an adjacent and connected parcel of land owned by the Eugene Parks Department and managed by the Department of Parks and Recreation Community Gardens program. This parcel of land had only a fledgeling and disorganized private garden on it as well as an overlooked and non-maintained small Spanish filbert orchard on it at the early stages of this grant. Today, at the conclusion of the formal grant peiod, the gardens are renovated into beds, plans have been drawn up for further development, a small solar greenhouse is built and available to the neighborood, and the orchard has been pruned and is moving back toward full production.

The goals and directions for this project -- a joint effort (and a head-banger) were developed by Marc Russell and Richard Wilen. Major goals and focus for the project included:
  · development of this land to meet community needs

  · development of a demonstration urban garden

  · year-round gardening in the greenhouse plus workshops

  · south facing (warm) area for public use

  · development of arbor system for blackberries, raspberries

  · maintenance of existing walnut trees on site

  · new plantings of dwarf and semi-dwarf trees

  · street closure of Ash at northern end

  · development of local workshops during winter 1980

Two of the sheets of the full project are included here.

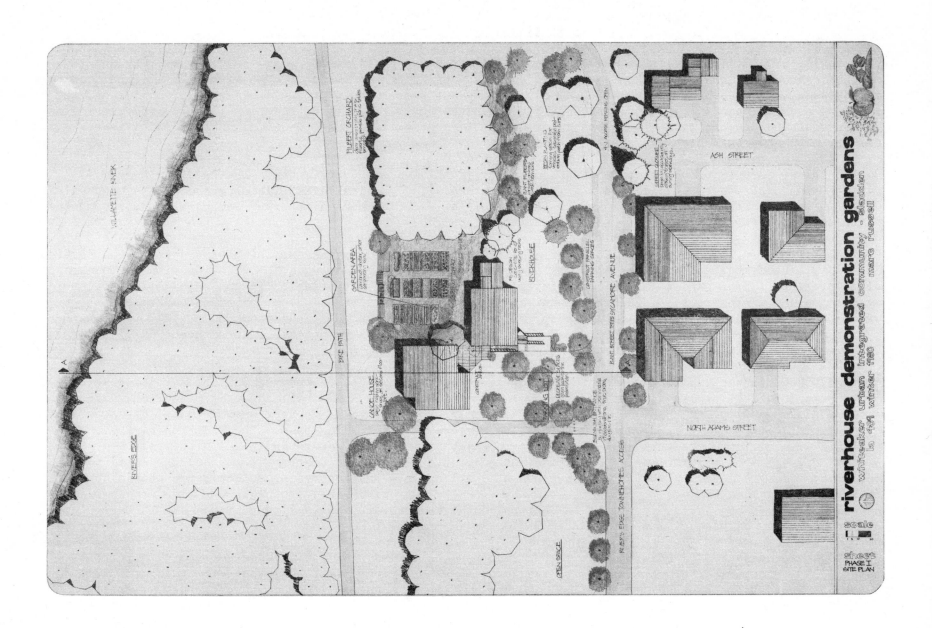

To the east side of the School District 4-J, and immediately adjacent to it is

IVORENA CARE CENTER GARDENS

This facility, privately run, has as its clients 24-hour care seniors from all walks of life. The majority of the remainder of the seniors' lives is presently spent indoors in various rooms with little connection to the outside landscape or heritage of the land. It was our goal to help generate access to, and use of, the land and climate by our most handicapped neighbors. Growing your own food, planting flowers, and helping make places beautiful surely rests in the hands of our seniors as well as our juniors, and we hoped to generate the optimism of people preparing healthy landscapes for those yet to come.

During the winter, 1980, we developed the proposals for the Ivorena Care Center. The facility houses 70-100 residents who are generally aged and infirm. Severe mental or physical disabilities are common among residents. The purpose of the proposal is the enhancement of existing medical therapy and care by improvement of the therapeutic value of the residents' whole environment.

Our initial contact with residents and staff was through a series of workshops which were intended to develop interest in gardening and allow us to learn more about the residents' needs. The first session included a slide presentation of garden images and a brief description of our intentions. In the second workshop, held in the Ivorena dining room, residents planted seeds in earth-filled trays which were then placed on a hotbed we had built outside. At the third workshop, residents examined the seedlings from their first planting effort and planted more seeds of later-season flowers and vegetables. When the plants became large enough, they were transplanted into raised garden beds.

Our experience in working with residents of Ivorena reinforced our belief that within the nursing home environment there should be places supportive of a wide range of activities and abilities. They include raised garden beds accessible to those confined to a wheelchair or unable to bend over, and, in some instances, seating with the garden bed, allowing residents to converse while tending beds. The existing patio can be enclosed with glass to create a protected place to enjoy views of the garden and passing bicyclists or pedestrians. Indoor plants can be tended in this enclosure during cold weather. A second protected place can be provided by a garden shelter which is open to the sun's warmth on two sides while the remaining two walls break unwanted winds. To allow interaction with the surrounding neighborhood, the old five-foot fence can be replaced with a three-foot wood fence combined with a masonry wall, allowing residents to have views out and to visit with passing neighbors.

The immediate adjacency of this project to the School District 4-J site and the Rose Garden also suggests connections to other age groups and to the landscape along the river. The workshops, the raised beds, and the handicapped planters that already have been built at Ivorena suggest that continuing catalytic action by concerned "outsiders" can, in fact, have a dramatic impact on the quality of life that most of our disabled neighborhood seniors now presently enjoy.

The following four sheets show the existing place and the phased project including its spatial relationship to the other adjacent food production projects.

IVORENA CARE CENTER · LOOKING NORTHEAST

GARDEN SITE · LOOKING NORTH

GARDEN SITE · LOOKING EAST

GARDEN SITE · LOOKING SOUTH

IVORENA GARDENS PHASE I

neighborhood health.

LA 469–BRITZ
WINTER 1980
DEPT. OF LANDSCAPE ARCHITECTURE
UNIVERSITY OF OREGON *don high* &
garth rufsner

IVORENA GARDENS PHASE II

neighborhood health:

LA 489—BRITZ WINTER 1980 DEPT. OF LANDSCAPE ARCHITECTURE UNIVERSITY OF OREGON don hugh & garth ruffner

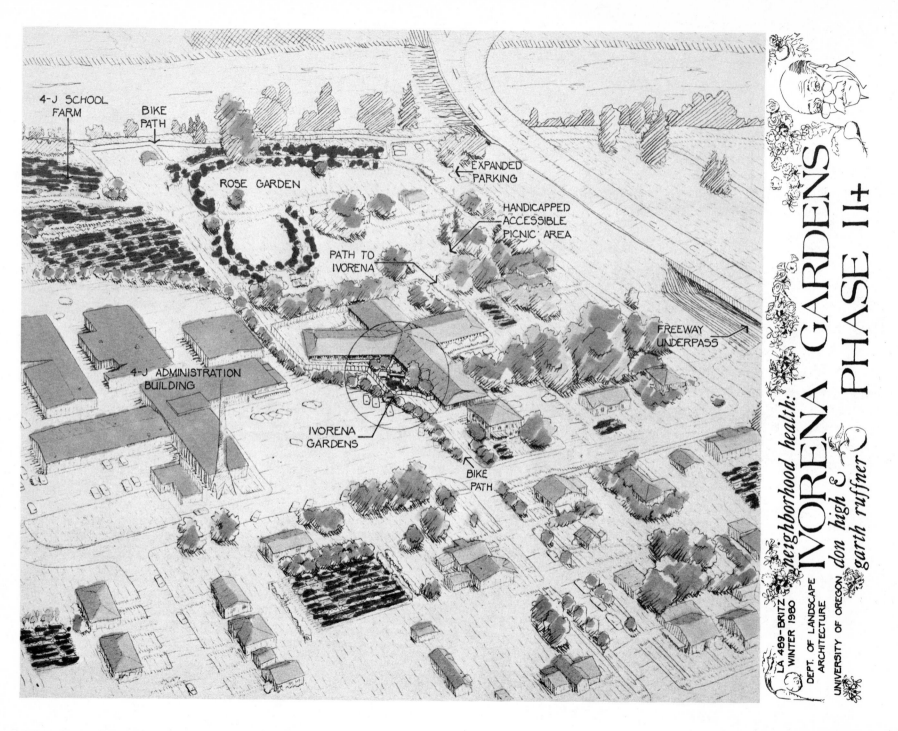

4-J SCHOOL FARM

BIKE PATH

ROSE GARDEN

EXPANDED PARKING

HANDICAPPED ACCESSIBLE PICNIC AREA

PATH TO IVORENA

4-J ADMINISTRATION BUILDING

FREEWAY UNDERPASS

IVORENA GARDENS

BIKE PATH

neighborhood health:

IVORENA GARDENS
PHASE II+

LA 489 - BRITZ
WINTER 1980

DEPT. OF LANDSCAPE ARCHITECTURE

UNIVERSITY OF OREGON don high &
garth ruffner

ASH STREET BLOCK FARM DEMONSTRATION PROJECT

The last project we concentrated on in the Sladden area was the block farm development on Ash Street. A group of neighbors requested help in establishing cooperative strategies for sharing incomes, resources and land, initially involving three houses on the end of their block. This site also became important in the Sladden area overall since one of the houses and its lot abutted the School District 4-J site previously mentioned. The following <u>very crude</u> design sketch was done <u>on</u> the site <u>with</u> the neighbors to enable them to participate in the planning and design process, while also enabling the designers to participate in the social and economic considerations being decided.

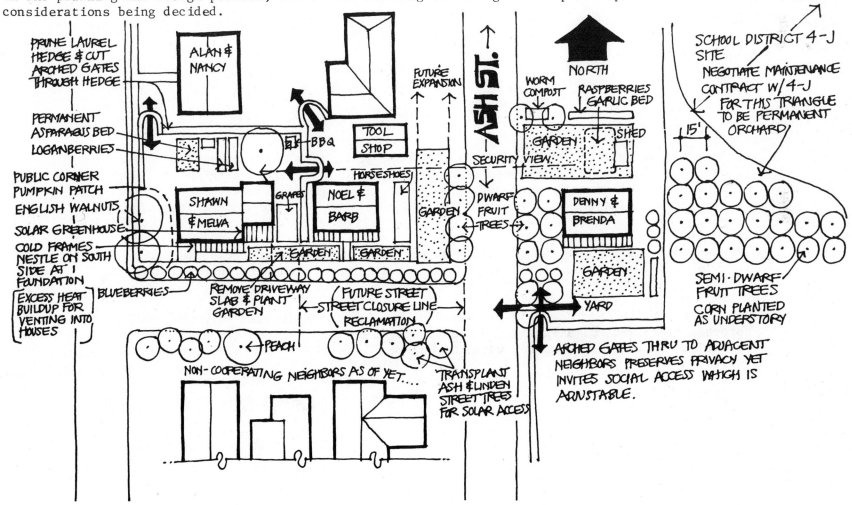

This project is one of the most significant results of the planning effort so far, since it is already built, planted, and connected to the larger social fabric of the neighborhood. Infill housing (garage and workshop converted to studio apartment); recognition of solar rights; planned extension of the seasons using an additional (design in process) solar greenhouse and cold frames nestled right up to the house; cooperative land use; gardens that span "private" property lines; cooperative planting schedule and location of vegetables, herbs, berries and orchard; provisions for joint storage of tools and shop facilities; and security interdependencies of visual access from one "yard" to another by opening gates and reduced hedge heights, are all components of this project.

Maximum advantage is taken of the biological diversity principle taking into account the optimum production from the land as well as the overall biomass increase. Former "lawns" are converted to raised beds yielding maximum economic return from labor invested in the garden, rather than labor and capital loss due to normal (and increasing) maintenance costs (such as mowing excessive amounts of lawn). Localized water percolation (rather than runoff) is also maximized by open and aerated soil.

Additional support is being sought from the U.S. Department of Energy Mini-Grants Appropriate Technology Program for this project as testified by the support letter below.

DOE Grant Reviewers:

I have had an opportunity to help plan the food production and landscaping for the Urban Farm Cluster being implemented by the applicants and their neighbors. Part of this plan includes a greenhouse which would allow them to increase food self-sufficiency during the winter months and adequately prepare for summer gardening needs.

Funding of this proposal would allow for further development of their cluster and provide an excellent model for other families in the area who are interested in similar endeavors.

In addition, I am particularly impressed by the funding mechanism suggested in the proposal entitled Multi-Family Greenhouse, Solar Heating, and Studio Apartment Combination, as this is consonant with the universal concept of recycling at a local level.

Respectfully,

Richard Britz

Richard Britz, Architect
University of Oregon
Landscape Architecture Dept.

Chapter 10

The Briarcliff-Polk Area

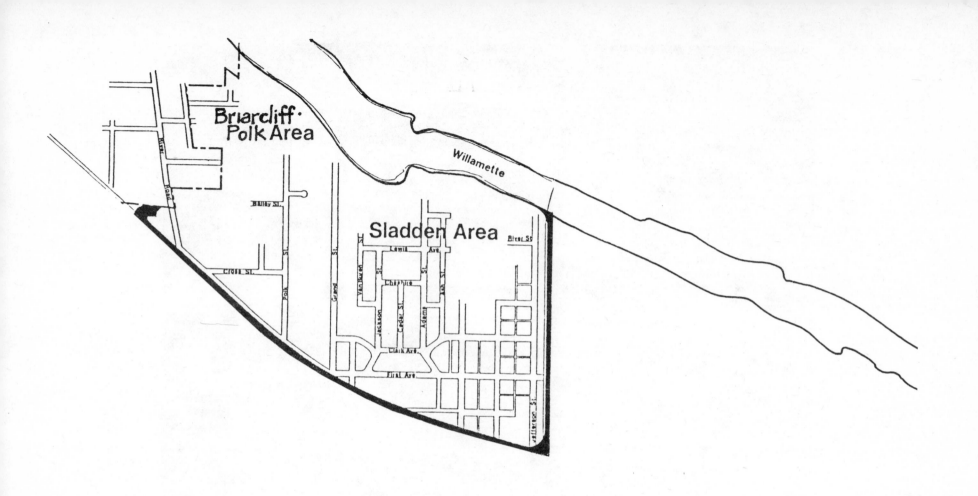

Briarcliff·
Polk Area

Willamette

Sladden Area

SECTOR 4
THE BRIARCLIFF·POLK AREA

BRIARCLIFF·POLK AREA (Sector 4)

Just west of the Sladden area and downriver past the new Rivers Edge apartments (townhouses/condominiums) is the last sector of Whiteaker neighborhood that abuts the river. Considerable amounts of open land were present before the grant period began and now most of the open land is either subdivision housing or scheduled by the Eugene Parks and Recreation Department for use as another large riverfront parking lot. Much of the early grant period effort was put into Neighborhood Advisory Group meeting in an attempt to help residents become more aware of the impending uses for which these lands were headed. Neighborhood participation in this area is low and being just at the edge of the city limits, land and private property become synonymous terms. Suburbanization of the land and displacement of the former agricultural uses including gardens and fruit and nut orchards is common. Density increases through lot subdivision are not taking into account the soil productivity inherent in the rich fertile river loam, nor are there trends toward increasing local food production. Two main sites in this area are still available for large scale food production and both abut the river. One site, the Polk Street Community Gardens, is available to neighborhood residents but also serves the whole City of Eugene with the attendant automobile commuter traffic through the neighborhood. During the grant period, Jeff Wilson, Laurel Lyon and Charlie Sundberg were primarily responsible for this area. Jeff Wilson and myself attempted to deal with City Parks and Recreation to save the Gum barn adjacent to the Polk Street Community Gardens, both for its historic and character value, and also as a remnant agricultural resource for the neighborhood. Proposals were made to the Parks Department to utilize the barn for community tool storage near the community gardens and to incorporate toilet facilities for urban gardeners. When proposed to the Neighborhood Advisory Group, this proposal was deemed another added burden rather than a resource, and despite backup strategies to recycle the wood from the barn for the neighborhood recycling component, the barn was demolished and removed to the Glenwood Dump. Non-responsiveness and seeming irrelevancy of the issue removed resource material from the neighborhood, decreased the possible neighborhood use of the area and simplified the landscape structure of the riverfront park. Continuing the concept of biological diversity and increased richness in this area, we concentrated on the Polk Street Community Garden. Proposals were drawn up in consultation with Richard Wilen, coordinator of the City of Eugene Community Gardens Program to add an orchard to the south side of the Polk Street gardens. This project, described elsewhere, recalls and replaces orchard stock once abundant in this sector of the neighborhood.

After this incident, we felt that it was necessary to work more directly with the neighbors to aid in the generation of more urban agriculture as well as prevent the large open Briarcliff site from being paved as regional parking. It is sadly ironic that the City of Eugene Parks and Recreation Department is the chief proponent of parking lots whose traffic will even further slice apart the Whiteaker neighborhood.

The main project we chose to aid during the planning stage of this grant was the Helios Development Company

BLOCK INTERIOR SOLAR INFILL

This project, a solar housing infill strategy, built by Mary Beck (the Chicken Lady), Gary Spivack, Mark Johnson and others, serves as one form of a potential urban block farm. Laurel Lyon and Charlie Sundberg developed the following proposal with Helios, which resolves many of the site planning problems at Helios and generates alternatives to the expensive future re-paving costs to senior residents on Cross Place, as well as the developers of Helios. Maximum consideration was given in this proposal to food production, cost staging, block resident participation and recreation. The houses on Cross Place were also identified as potential district heating case studies by placing large quantities of remote solar collectors on the roof of the immediately adjacent Consumer Warehouse.

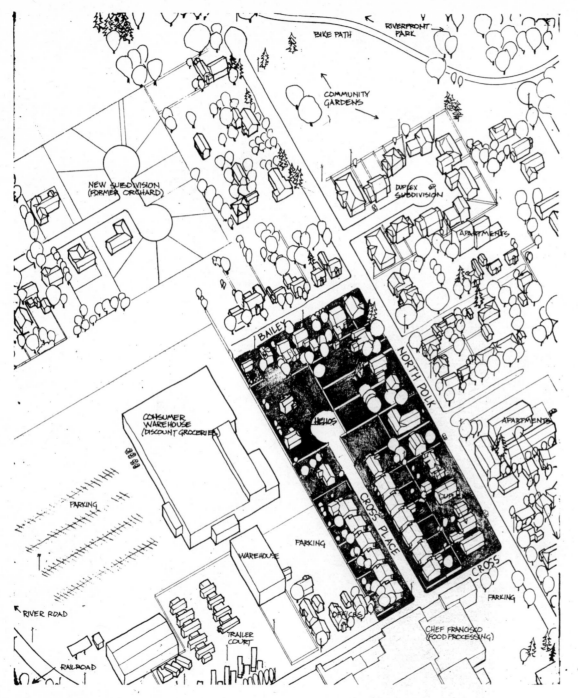

cross place/ helios

CROSS PLACE/HELIOS IS AN ISLAND CAUGHT BETWEEN INDUSTRY, AUTO-ORIENTED COMMERCIAL ACTIVITY AND INCREASING REPLACEMENT OF HOUSES BY APARTMENTS. A BACK EDDY IN THE CITY'S CIRCULATION SYSTEM, IT IS CLOSE TO A MAJOR ARTERIAL AND RAILROAD AND WITHIN WALKING DISTANCE OF A RIVERFRONT PARK, A MAJOR BIKEWAY AND A REGIONAL SHOPPING CENTER. EXCEPTING OCCASIONAL REMNANT ORCHARD TREES AND AN ACTIVE COMMUNITY GARDEN, THE PRIME AGRICULTURAL SOILS OF THE AREA ARE NEGLECTED.

GENERAL CHANGES IN THE ECONOMY, ZONING CHANGES ON CROSS PLACE AND DEVELOPMENT OF HELIOS' SOLAR HOUSING HAVE RAISED SEVERAL ISSUES —

• HOW CAN STREETS BE USED FOR A VARIETY OF PURPOSES, BEYOND AUTO MOVEMENT & STORAGE?

• HOW CAN AGRICULTURAL SOILS WITHIN THE CITY BE BETTER UTILIZED?

• HOW CAN FOOT & BIKE CIRCULATION BE IMPROVED WITHIN A RIGID GRID OF STREETS & LOT LINES?

• HOW CAN MORE VARIED AND USEFUL PUBLIC, SEMI-PUBLIC AND PRIVATE AREAS BE CREATED WITHIN THE STANDARD FRAMEWORK OF LAND DIVISION AND DEVELOPMENT?

CROSS PLACE OFFERS AN OPPORTUNITY FOR ENRICHING LINEAR SPACE LIKE THAT CREATED BY HIGHER DENSITY ROW HOUSING, WHILE THE HELIOS DEVELOPMENT PRESENTS A COMPLEMENTARY END OF THE ROAD OPENING UP THE CENTER OF A BLOCK.

WHITEAKER
URBAN
INTEGRATED
COMMUNITY

LA 489
WINTER/80

L. LYON
C. SUNDBERG

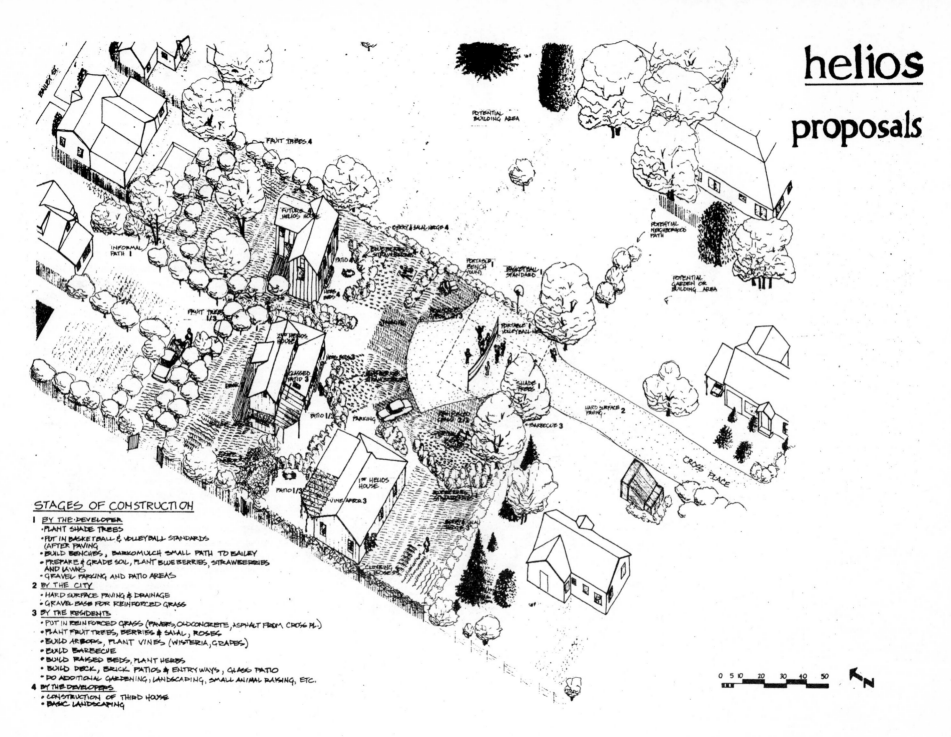

helios
proposals

STAGES OF CONSTRUCTION

1 BY THE DEVELOPER
- PLANT SHADE TREES
- PUT IN BASKETBALL & VOLLEYBALL STANDARDS (AFTER PAVING)
- BUILD BENCHES, BARKMULCH SMALL PATH TO BAILEY
- PREPARE & GRADE SOIL, PLANT BLUEBERRIES, STRAWBERRIES AND LAWNS
- GRAVEL PARKING AND PATIO AREAS

2 BY THE CITY
- HARD SURFACE PAVING & DRAINAGE
- GRAVEL BASE FOR REINFORCED GRASS

3 BY THE RESIDENTS
- PUT IN REINFORCED GRASS (PAVERS, OLD CONCRETE, ASPHALT FROM CROSS PL.)
- PLANT FRUIT TREES, BERRIES & SALAL, ROSES
- BUILD ARBORS, PLANT VINES (WISTERIA, GRAPES)
- BUILD BARBECUE
- BUILD RAISED BEDS, PLANT HERBS
- BUILD DECK, BRICK PATIOS & ENTRY WAYS, GLASS PATIO
- DO ADDITIONAL GARDENING, LANDSCAPING, SMALL ANIMAL RAISING, ETC.

4 BY THE DEVELOPERS
- CONSTRUCTION OF THIRD HOUSE
- BASIC LANDSCAPING

helios

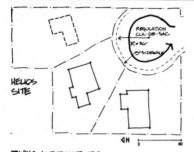

HELIOS SITE

PAVING ALTERNATIVES:

THE HELIOS DEVELOPMENT IS AN ISLAND WITHIN AN ISLAND — NEW AND VERY DIFFERENT HOUSING TURNED IN UPON ITSELF. STANDARD LOT DIVISION AND STREET DEDICATION HAVE PRODUCED PROBLEMATIC ACCESS, POOR RELATIONSHIPS BETWEEN HOUSES AND A SOON-TO-BE PAVED NO MAN'S LAND ALMOST LARGE ENOUGH FOR ANOTHER HOUSE. FENCES SURROUND THE DEVELOPMENT AND PARTITION NEIGHBORS' YARDS, BLOCKING WHAT USED TO BE A NEIGHBORHOOD PATH TO THE ONLY FOOD STORE WITHIN WALKING DISTANCE.

ISLAND

- LIMITED PAVING, PROVIDES ADEQUATE TURNING ROOM FOR FIRETRUCK
- UNPAVED AREA ISOLATED, TRAFFIC FORCED TO CIRCUMFERENCE

HAMMERHEAD

- MINIMUM PAVING FOR TRUCK TURNAROUND — AWKWARD
- EXTRA SPACE CUT INTO ODD AREAS

PLAZA / SQUARE

- WELL DEFINED OPEN SPACE WHICH IMPLIES PUBLIC USE
- INCREASED PAVING, EXPENSE AND WATER RUNOFF
- EASILY USED FOR PARKING IF NOT CLEARLY MARKED

GAME SPACES

.... VOLLEYBALL
------ BASKETBALL (1/2 COURT)

- SEVERAL POPULAR GROUP SPORTS CAN BE PLAYED IN THE SPACE OF THE CUL DE SAC
- NON-COURT GAMES ALSO POSSIBLE — CATCH, FRISBEE, ROLLERSKATING, DANCING, ETC.

COMBINATION

REINFORCED GRASS
MARKED VOLLEYBALL COURT WITH REMOVEABLE NET STANDARDS
MARKED BASKETBALL AREA WITH NET

- HARD SURFACE ON DRIVING AREAS, SOFT BUT DRIVEABLE SURFACE FOR TRUCK TURNAROUND
- NO SIDEWALK
- CORNER AREAS FOR SITTING, EATING, SUN, SHADE, RESTING
- ROOM FOR SHADE TREE PLANTING IN SOUTHEAST CORNER

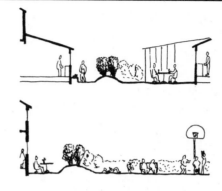

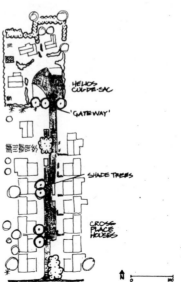

HELIOS CUL-DE-SAC

'GATEWAY'

SHADE TREES

CROSS PLACE HOUSES

SMALL SCALE HOUSING ON TIGHT LOTS ALMOST DEMANDS THE CREATION OF 'OUTDOOR ROOMS.' SEASONAL EXTENSIONS OF LIVING SPACE SUPPORTING MANY ACTIVITIES. LINES OF SIGHT AND BOUNDARIES OF SPACES MUST BE CAREFULLY CONSIDERED TO PROVIDE PRIVACY WITHOUT SEPARATING HOUSES FROM THE COMMON AREAS THEY SHARE AND WATCH OUT OVER.

BETTER FENCES MAKE BETTER NEIGHBORS — BOUNDARIES CAN BE PLACES OF CONNECTION AND SHARING RATHER THAN WALLS BUILT OVER UNNATURAL LINES.

PATHWAYS NEEDN'T COME BETWEEN NEIGHBORS EITHER — LOCATION, VISIBILITY, SIZE AND SURFACE MATERIAL ALL AFFECT HOW AND BY WHOM A PATH IS USED. STRANGERS AVOID INFORMAL WALKWAYS ON THE FAR SIDE OF A DRIVEWAY. BUSY TRAFFIC STICKS TO WIDER, WELL WORN WAYS.

CONNECTIONS BETWEEN PLACES CAN RECOGNIZE THEIR DIFFERENCES AS WELL AS CREATE CONTINUITY. THE HELIOS CUL-DE-SAC CAN PROVIDE AN OPEN ACTIVITY AREA WHICH COMPLEMENTS THE STRAIGHT LINES OF CROSS PLACE — ROOM FOR ACTIVITIES WHICH WON'T FIT ELSEWHERE. PHYSICAL AND VISUAL CONTINUITY CAN BE DEVELOPED BY USING THE SAME PLANTING MATERIALS IN SIMILAR WAYS, AND BY ESTABLISHING A PATTERN OF CONNECTED SPACES (WITH SHADE TREES, PAVING, PLANTING BEDS, ETC.). A CLOSER KNITTING OF THE NEIGHBORHOOD'S FABRIC IS TO EVERYONE'S BENEFIT.

plant list

Cross Place/Helios is in an area of former orchards. The soil is acidic silty clay loam. The flat terrain and compacted soils slow surface drainage. The following list of plants will adapt well to these conditions, and offer ornamental as well as food-producing value.* By mixing organic matter** with the soil, a wide variety of herbs & vegetables can be successfully cultivated.

	maximum height	fruit-bearing age	annual yield	cultural requirements soil · water · sun	Remarks · suggested varieties, cross pollination planting specs.
shade trees					
Japanese Persimmon (Diospyros kaki)	30'x30'	4-5 yrs.	75-100 lbs.	takes heavy soil, pest free	buy 1-2 yr. old plant in early spring; self pollinator.
Japanese Pagoda Tree (Sophora japonica)	30-40'	—	—	not soil particular, pest free	creates dappled shade
fruit & nut trees					
Apple (Malus)	semi-dwarf 12-15'	3-4 yrs.	5-10 bush.	takes heavy soil	plant varieties ripening at same time. A good espalier Aug-Gravenstein, Lodi. Sept: Wealthy, MacIntosh. Oct-Chehalis, Melrose
Fig (Ficus carica)	6-20'	3-5 yrs.	40-50 lbs.	plant in sun for ripe fruit	protect fruit from birds reco. varieties; Latterula, BrownTurkey, Desert King
Filbert (Corylus avellana)	5-20'	3 yrs.	10 lbs.	need enriched soil	plant 2 varieties in fall or winter
Pear (Pyrus communis)	standard 12-15'	4-6 yrs.	1½ bushels	takes heavy soil	plant 2 varieties for cross-pollination Bartlett & Anjou. can be espaliered to wall or wire fence (see pruning chart.)
shrubs					
Blueberry (Vaccinium corymbosum)	5-6'x3-4'	3 yrs.	6-8 pts.	add organic matter to soil need good drainage	buy 2 yr old plant, cut back ½ way + plant 4-6' apart
Currant (Ribes)	3-5'x3-5'	2-3 yrs.	4-6 qts.	take full sun to light shade	thin to keep open & avoid fungus infection
Gooseberry (Ribes)	2-4'	2 yrs.	5-10 qts.	same as currant	plant deep in soil, reco. variety: Pixwell
Salal (Gaultheria shallon)	2-10'	2-3 yrs.	varies	water & sun determine growth habit. acid soil.	ripens in fall. good for jam, cooking cut back old growth in April every few years.
ground covers & vines					
Grapes (Vitus labrusca)	vines	2-3 yrs	varies	deep watering needs heavy pruning (see chart)	Concord & hybrids take heavy soils best for wines: consult extension agent for varieties suitable
Rhubarb (Rheum rhaponticum)	3-4'	2-3 yrs.	6-8 lbs in 6 wks	need rich soil . shade loving	plant crowns 3-4" deep harvest after second year reco. varieties: MacDonald, Valentine.
Roses (Rosa spp.)	Climbing vine	—	—	water deeply, mulch mound for drainage . need sun	buy bare root. prune only after 2-3 yrs. Aphids, fungus mildew Spring + sporadic bloomers, many colors, fragrances
Strawberry (Fragaria spp.)	6-8"	1st year	1 pt./plant	need rich well drained soil	plant 2' apart in rows 4' apart. Pinch off flowers first year everbearers & single crop both replace after 3rd harvest year.
Wisteria (Wisteria japonica)	vines	—	—	water well, mound for drainage prune for form & to control growth	many varieties - flowers white - pink - purple needs firm support; tie up with plastic tape.

* ornamentals only: Japanese Pagoda tree, Roses & Wisteria   ** organic matter — loam, sawdust, leaf mold, sand, compost, manure.

References: Time-Life Encyclopedia of Gardening·Vegetables & Fruits·James Crockett Western Garden Book - Sunset Magazine, The Self-Sufficient Gardener·John Seymair

cross place

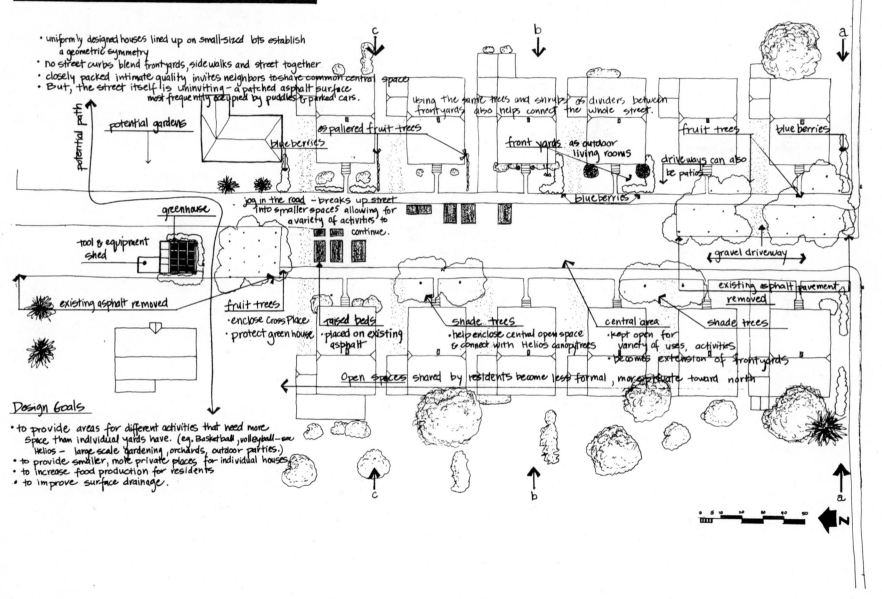

- uniformly designed houses lined up on small-sized lots establish a geometric symmetry
- no street curbs blend frontyards, sidewalks and street together
- closely packed intimate quality invites neighbors to share common central space
- But, the street itself is uninviting – a patched asphalt surface most frequently occupied by puddles & parked cars.

potential path

potential gardens

greenhouse

tool & equipment shed

existing asphalt removed

blueberries

espaliered fruit trees

Using the same trees and shrubs as dividers between frontyards also helps connect the whole street.

front yards as outdoor living rooms

driveways can also be patios

fruit trees

blue berries

blueberries

jog in the road – breaks up street into smaller spaces allowing for a variety of activities to continue.

gravel driveway

existing asphalt pavement removed

fruit trees
- enclose cross place
- protect green house

raised beds
- placed on existing asphalt

shade trees
- help enclose central open space & connect with Helios canopy trees

central area
- kept open for variety of uses, activities
- becomes extension of frontyards

shade trees

Open spaces shared by residents become less formal, more private toward north

0 5 10 20 30 40 50

N

Design Goals

- to provide areas for different activities that need more space than individual yards have. (eg. Basketball, volleyball—see Helios — large scale gardening, orchards, outdoor parties.)
- to provide smaller, more private places for individual houses
- to increase food production for residents
- to improve surface drainage.

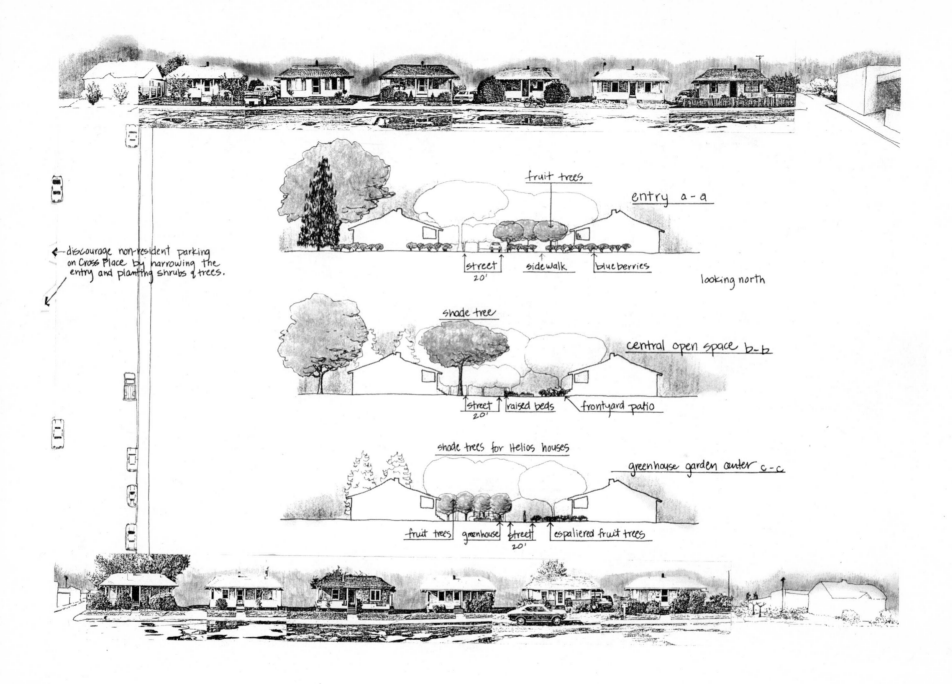

discourage non-resident parking on Cross Place by narrowing the entry and planting shrubs & trees.

entry a-a

fruit trees

street 20' — sidewalk — blueberries

looking north

central open space b-b

shade tree

street 20' — raised beds — frontyard patio

greenhouse garden center c-c

shade trees for Helios houses

fruit trees — greenhouse — street 20' — espaliered fruit trees

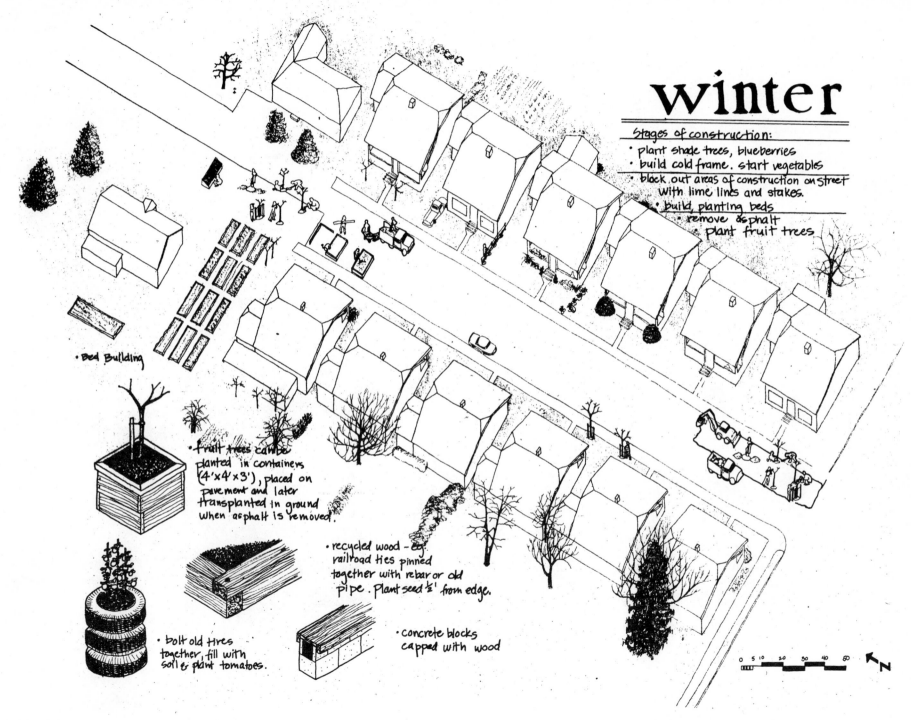

winter

Stages of construction:
- plant shade trees, blueberries
- build cold frame. start vegetables
- block out areas of construction on street with lime lines and stakes.
- build planting beds
- remove asphalt
- plant fruit trees

• Bed Building

• Fruit trees can be planted in containers (4'×4'×3'), placed on pavement and later transplanted in ground when asphalt is removed.

• recycled wood - e.g. railroad ties pinned together with rebar or old pipe. Plant seed ½' from edge.

• bolt old tires together, fill with soil & plant tomatoes.

• concrete blocks capped with wood

0 5 10 20 30 40 50

N

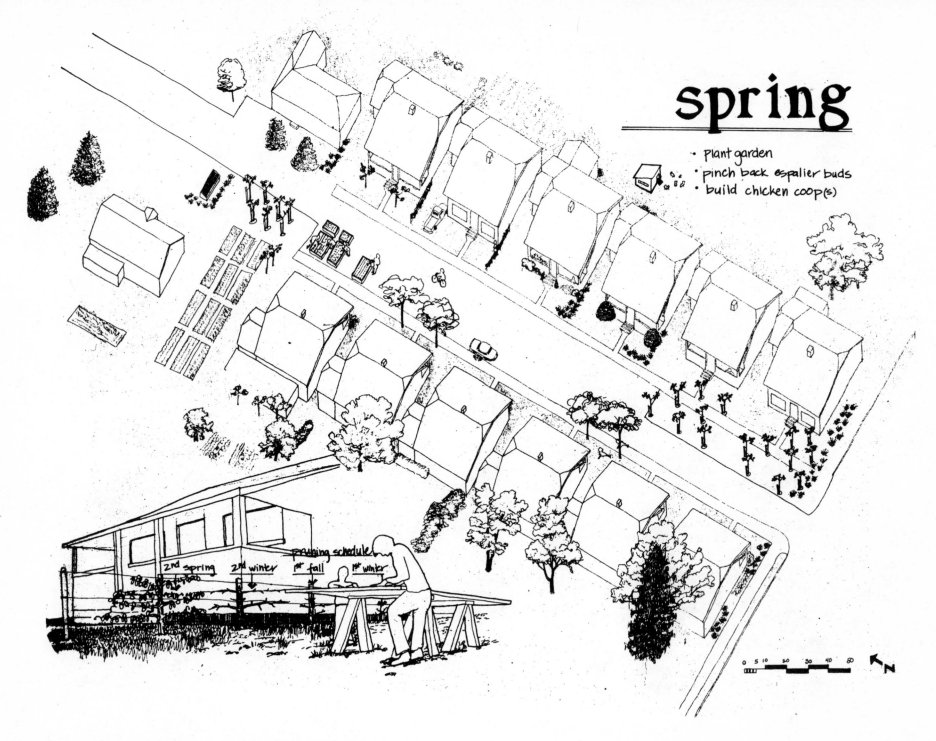

spring

- plant garden
- pinch back espalier buds
- build chicken coop(s)

pruning schedule

2nd spring 2nd winter 1st fall 1st winter

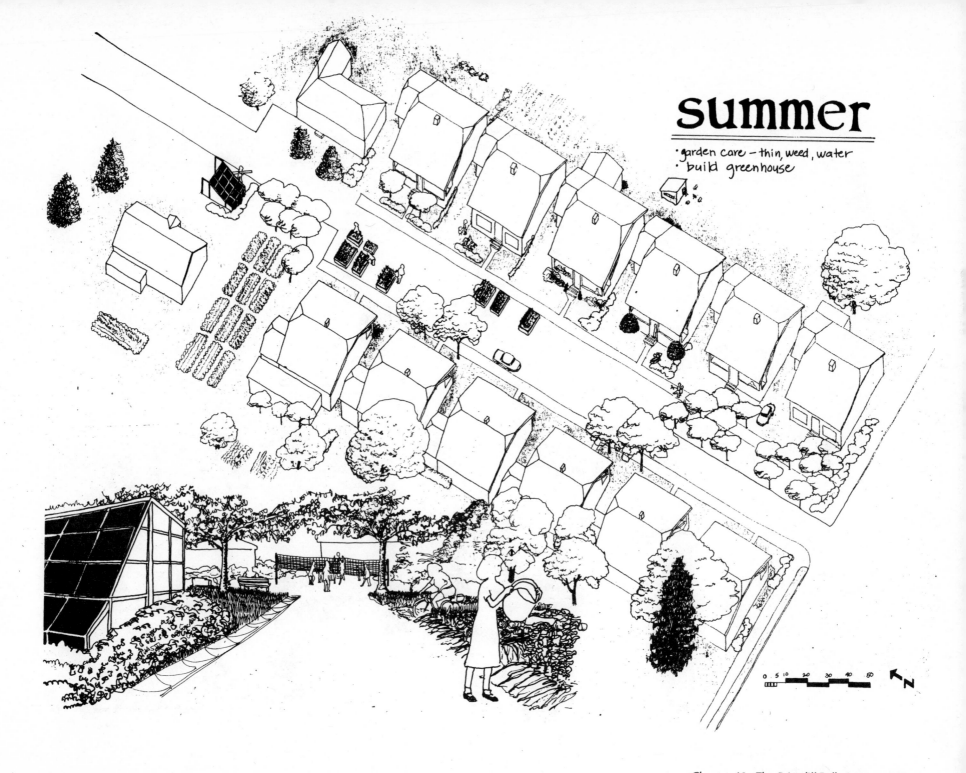

summer

- garden care — thin, weed, water
- build greenhouse

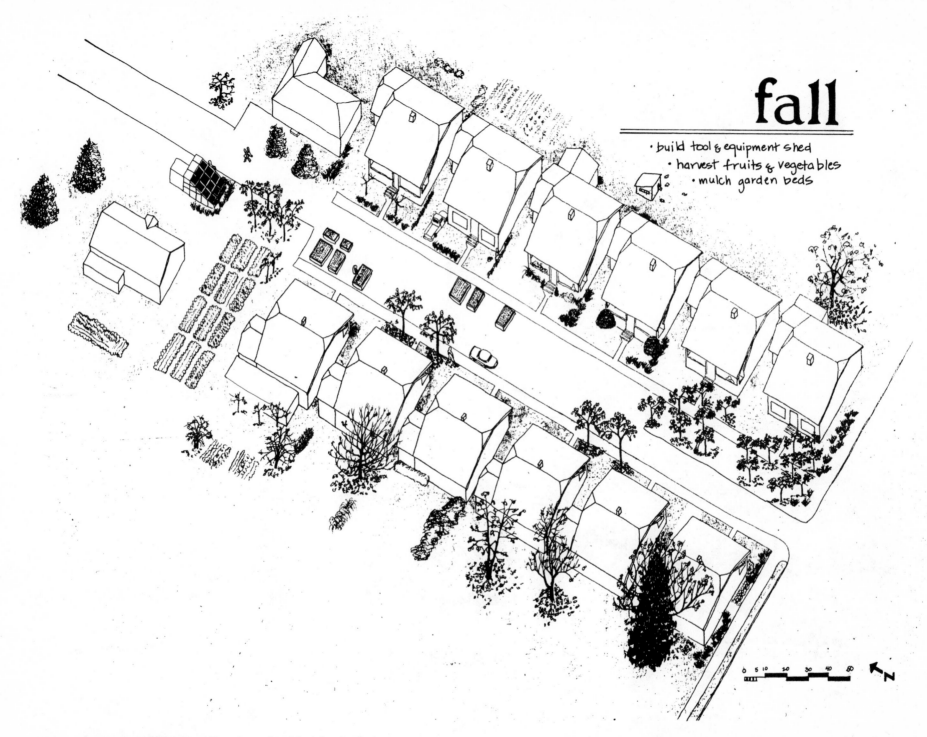

fall

- build tool & equipment shed
- harvest fruits & vegetables
- mulch garden beds

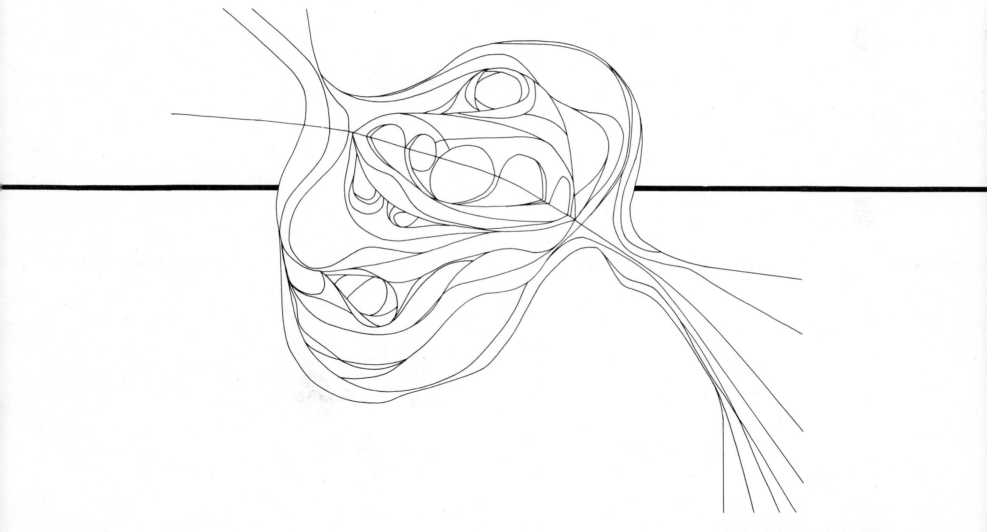

Chapter 11

The Blair Area

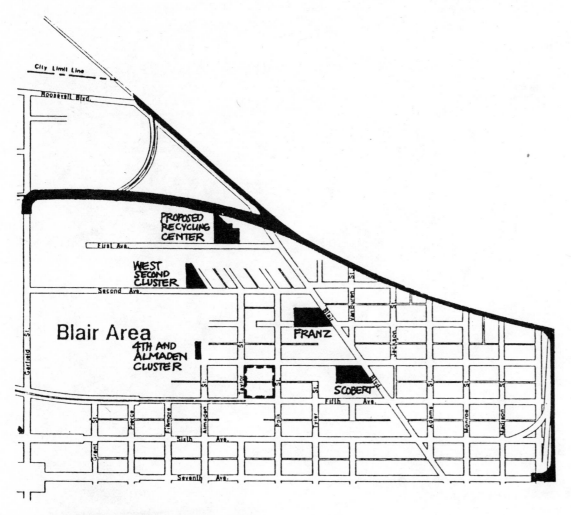

City Limit Line

Roosevelt Blvd.

PROPOSED RECYCLING CENTER

First Ave.

WEST SECOND CLUSTER

Second Ave.

Blair Area

4TH AND ALMADEN CLUSTER

FRANZ

SCOBERT

Fifth Ave.

Sixth Ave.

Seventh Ave.

SECTOR 5
THE BLAIR AREA

BLAIR

The Blair area is best characterized as mixed use: commercial, industrial and residential. It is divided approximately in half by Blair Boulevard, a diagonal street in a landscape of street grids. Clustering around Blair Boulevard are small shops and businesses which have marginal economic stability. On either side of Blair Boulevard are East and West Blair sub-neighborhoods, rather cleanly divided by the traffic flow on Blair on its way to River Road, and equally, back toward 6th and 7th Streets. Both residential sectors have extremely high resident transciency rates, and it is clear that land and property speculation rates are high also. Blair Boulevard in certain ways exemplifies the whole neighborhood in that a major traffic artery cleanly divides one sub-sector from another, further reducing personal contact between residents and spawning an automobile landscape dominant over other modes of transport. While certain new businesses, themselves of marginal stability, have sought the foot and bicycle oriented people, the dominant factor is the through-the-neighborhood traffic on Blair Street. Other planning efforts have concentrated on the redevelopment of this sector including housing cooperative property purchases and a development plan for the entire Blair area.

Given this distribution of energy to this sector, the work undertaken in this area consisted of an inventory of probably and potentially available sites for food production.

After the initial inventory and evaluation of the land in this area, two sites were identified as having utility for larger scale food production. The first site, here shown as being filled with plum, apple, and pear trees, with much of the food going to waste on the land, is the Whitney Scobert property at the corner of Blair and Fifth. This property and its block, retain food producing trees and is connected by a grassy alley to numerous gardens within the block. It is "a natural" location for an urban block farm which integrates a neighborhood nursery (one owner raises rhododendrons), increases housing density on the north side of the block, and preserves the south facing sunny gardens already existing. Early in the initial stages of this work it was strongly

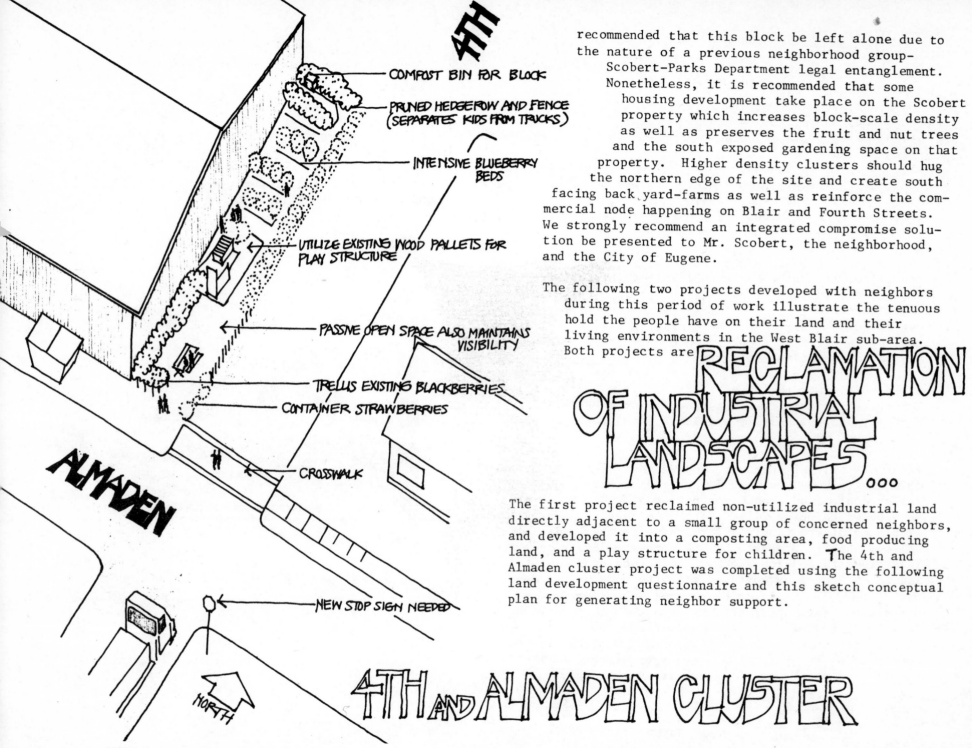

4TH

- COMPOST BIN FOR BLOCK
- PRUNED HEDGEROW AND FENCE (SEPARATES KIDS FROM TRUCKS)
- INTENSIVE BLUEBERRY BEDS
- UTILIZE EXISTING WOOD PALLETS FOR PLAY STRUCTURE
- PASSIVE OPEN SPACE ALSO MAINTAINS VISIBILITY
- TRELLIS EXISTING BLACKBERRIES
- CONTAINER STRAWBERRIES
- CROSSWALK
- NEW STOP SIGN NEEDED

ALMADEN

NORTH

recommended that this block be left alone due to the nature of a previous neighborhood group-Scobert-Parks Department legal entanglement. Nonetheless, it is recommended that some housing development take place on the Scobert property which increases block-scale density as well as preserves the fruit and nut trees and the south exposed gardening space on that property. Higher density clusters should hug the northern edge of the site and create south facing back yard-farms as well as reinforce the commercial node happening on Blair and Fourth Streets. We strongly recommend an integrated compromise solution be presented to Mr. Scobert, the neighborhood, and the City of Eugene.

The following two projects developed with neighbors during this period of work illustrate the tenuous hold the people have on their land and their living environments in the West Blair sub-area. Both projects are

RECLAMATION OF INDUSTRIAL LANDSCAPES ooo

The first project reclaimed non-utilized industrial land directly adjacent to a small group of concerned neighbors, and developed it into a composting area, food producing land, and a play structure for children. The 4th and Almaden cluster project was completed using the following land development questionnaire and this sketch conceptual plan for generating neighbor support.

4TH AND ALMADEN CLUSTER

ALMADEN-4TH STREET LAND DEVELOPMENT QUESTIONNAIRE

A group of neighbors have been meeting to discuss the development of the unused land located at the northwest corner of the Almaden-4th Street intersection.

Its present condition represents a hazard: 1) unsafe play, 2) pest shelter, 3) esthetically displeasing.

From the discussions, three uses were proposed for the site: play, compost, and intensive berry beds. Please overview our basic concept and respond to the following questions: (check one)

A. Priority Ratings:

1) The basic plan is fine.............

2) Emphasize play more................

3) Emphasize berry growth.............

4) Emphasize compost more.............

5) If your ideas are not included, please tell us.

B. Involvement

1) I would like to be involved.........

a) Planning.........................

b) Donation (a Federal grant may possibly be obtained at $200.00. However, your contribution would be appreciated.)

c) Labor for development............

d) Labor for maintenance............

e) Other contributions..............

2) I would not like to be involved.....

C. Comments:

1) Any additional comments, concerns, suggestions, etc. would be welcome.

D. Further information:

1) If you need specifics on any matter, please contact your block representative:

1300 W. 4th------D. Peircy
1400 W. 4th------Glenn Baker, 345-7548
442 Almaden------Kim Davidson, 342-6409

(Please feel free to mark up the print with your proposals.)

A second potential food producing site is the Franz Bakery site, the last piece of open land remaining in the Blair area. It was and continues to be recommended as a sector-neighborhood open space. Since this land could serve as an additional neighborhood farm serving this sector (both east and west Blair), high priority should be given to acquisition of (or access to use of) this land prior to its loss to other development.

Other lands in this sector are also under rapid transformation. Block center open space is being sub-divided and developed with infill housing or cheaply built speculative apartments leading to the paving over of <u>all</u> the soil from lot line to lot line. One proposal which could prevent this continuing quality of life deterioration is the following:

A TAX REDUCTION PROPOSAL

Between 5th and 4th, Polk and Taylor is a typical example of the need for <u>some kind of guarantee that urban land already in food production</u> (generally long-term senior people) <u>remain available</u> (solar rights guaranteed, limited land speculator pressure, etc.) <u>for food production</u> <u>in perpetuity</u>.

This proposal recommends that the back 1/2 of existing urban lots (typically over 100' deep) be sub-divided and pooled into ownership by a block-scale community corporation (non-profit) and held as <u>a public commons</u> for food production and recreational purposes. The re-design of private property boundaries into semi-public, semi-private lands (the block interiors) is a restructuring which alleviates economic woes, provides <u>direct</u> availability of fresh food (as macro-systems collapse) and can be sensitively designed.

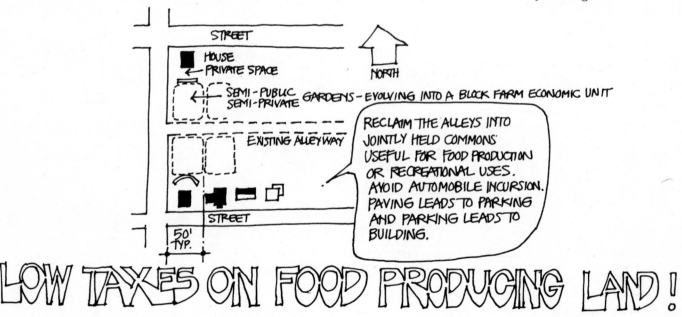

STREET

HOUSE
PRIVATE SPACE

SEMI-PUBLIC
SEMI-PRIVATE GARDENS — EVOLVING INTO A BLOCK FARM ECONOMIC UNIT

NORTH

EXISTING ALLEYWAY

RECLAIM THE ALLEYS INTO JOINTLY HELD COMMONS USEFUL FOR FOOD PRODUCTION OR RECREATIONAL USES. AVOID AUTOMOBILE INCURSION. PAVING LEADS TO PARKING AND PARKING LEADS TO BUILDING.

STREET

50' TYP.

LOW TAXES ON FOOD PRODUCING LAND!

The last project, on <u>West Second Avenue</u>, involved five houses clustered in a group completely surrounded on all four sides (for blocks) by industrial land including chemical and petroleum storage, log mills and storage, and various heavy equipment yards. Just north of this site is the Eugene Mission, accommodations for Eugene's transient population. This was the most difficult project of the period of this work, but exemplifies the essence of low-income environments. The land on which the families live has been decimated by extensive automobile and truck-related oils and chemicals and is heavily compacted gravel. High transience of renters with only one homeowner contribute to the overall environmental deterioration. During this work our goals, developed with Lyle, Wally and Diane, the homeowner and two stable renters, were to survey and define property lines, plan the reclamation of the soil, establish what could grow and where it could be grown, accommodate outside play areas, re-route vehicular traffic, and generate cooperative strategies for tool sharing, cluster composting, security, solar grains by using south facing air locks, and maximum food production in minimum space. While this project has yet to develop through full neighbor initiative, some cooperation has evolved to establish a worm composting system for the use of all the cluster. The following phased development plan suggests directions to go and the air photo following that shows the proximity this cluster has to the last project accomplished during this work -- the Whiteaker Neighborhood Recycling Center. While the Whiteaker Neighborhood Recycling Center proposal is developed under separate cover, its function is fundamentally important to this portion of the work since its operation could provide a continuing source of compost for continual replenishment of the soil with organic nutrients. Its composting windrows for organic waste conversion can process and stabilize organic wastes in 30 days (if shredded) thus providing a continuing source of income for the soils which support the larger scale urban agricultural processes in the neighborhood. From waste come resources.

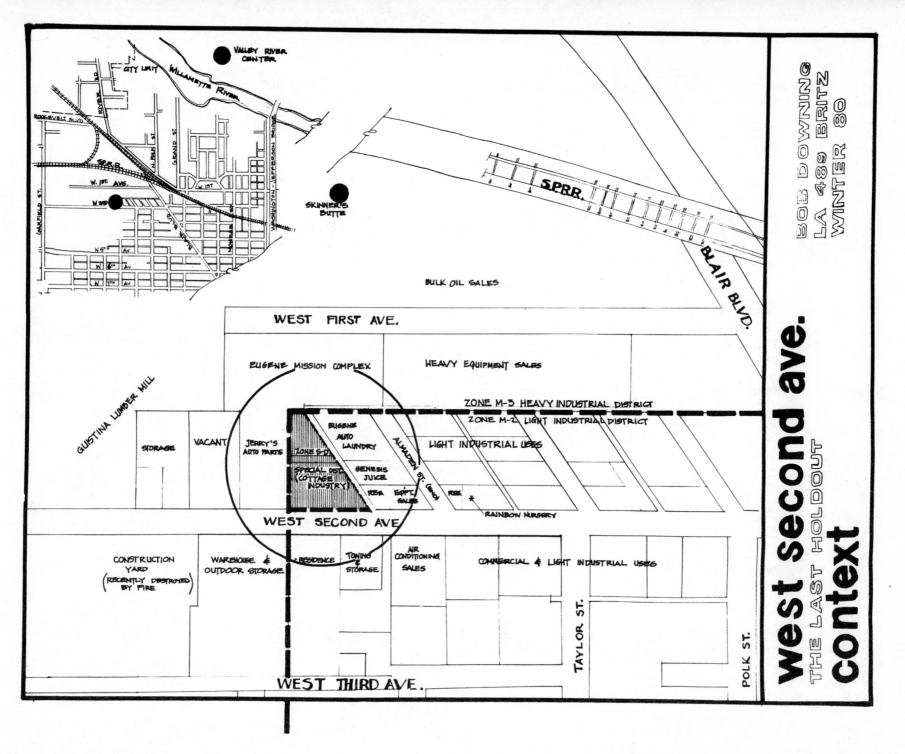

VALLEY RIVER CENTER

CITY LIMIT WILLAMETTE RIVER

ROOSEVELT BLVD.

SPRR

W. 1ST. AVE.

W 2ND

SKINNER'S BUTTE

SPRR.

BLAIR BLVD.

BULK OIL SALES

WEST FIRST AVE.

EUGENE MISSION COMPLEX

HEAVY EQUIPMENT SALES

GUISTINA LUMBER MILL

ZONE M-3 HEAVY INDUSTRIAL DISTRICT

ZONE M-2 LIGHT INDUSTRIAL DISTRICT

STORAGE

VACANT

JERRY'S AUTO PARTS

EUGENE AUTO LAUNDRY

LIGHT INDUSTRIAL USES

ZONE S-D SPECIAL DIST. (COTTAGE INDUSTRY)

GENESIS JUICE

ALMADEN ST. (ZONE)

RES. EQPT. SALES

RES.

RAINBOW NURSERY

WEST SECOND AVE

CONSTRUCTION YARD (RECENTLY DESTROYED BY FIRE)

WAREHOUSE & OUTDOOR STORAGE

RESIDENCE

TOWING & STORAGE

AIR CONDITIONING SALES

COMMERCIAL & LIGHT INDUSTRIAL USES

TAYLOR ST.

POLK ST.

WEST THIRD AVE.

west second ave.
THE LAST HOLDOUT
context

BOB DOWNING
LA 489 BRITZ
WINTER 80

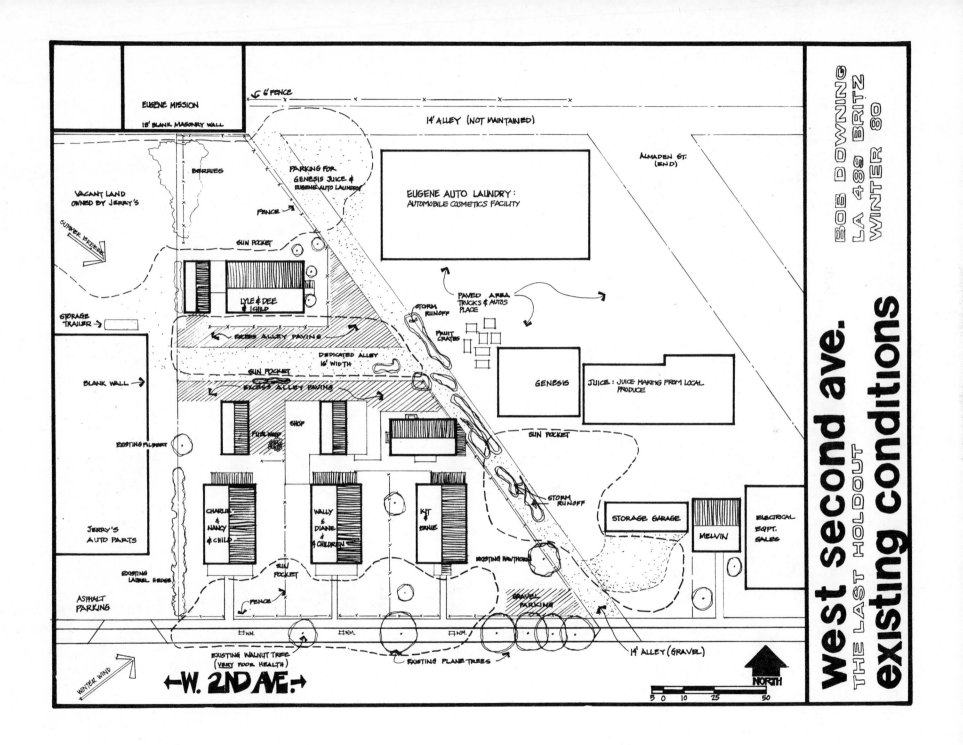

EUGENE MISSION
18' BLANK MASONRY WALL

6' FENCE

14' ALLEY (NOT MAINTAINED)

ALMADEN ST. (END)

BERRIES

VACANT LAND
OWNED BY JERRY'S

SUMMER BREEZE

PARKING FOR
GENESIS JUICE &
EUGENE AUTO LAUNDRY

FENCE

EUGENE AUTO LAUNDRY:
AUTOMOBILE COSMETICS FACILITY

SUN POCKET

STORAGE
TRAILER →

LYLE & DEE
& 1 CHILD

PAVED AREA
TRUCKS & AUTOS
PLACE

STORM
RUNOFF

FRUIT
CRATES

EXCESS ALLEY PAVING

DEDICATED ALLEY
16' WIDTH

BLANK WALL →

SUN POCKET

EXCESS ALLEY PAVING

GENESIS

JUICE: JUICE MAKING FROM LOCAL
PRODUCE

EXISTING FILBERT

SHOP

FULL HOOP

SUN POCKET

STORM
RUNOFF

JERRY'S
AUTO PARTS

CHARLIE
&
NANCY
& CHILD

WALLY
&
DIANE
& 4 CHILDREN

KIT
&
ERNIE

EXISTING HAWTHORN

STORAGE GARAGE

MELVIN

ELECTRICAL
EQPT.
SALES

EXISTING
LAUREL HEDGE

SUN
POCKET

FENCE

GRAVEL
PARKING

ASPHALT
PARKING

W.M.

N.M.

N.M.

EXISTING WALNUT TREE
(VERY POOR HEALTH)

EXISTING PLANE TREES

14' ALLEY (GRAVEL)

WINTER WIND

← W. 2ND AVE. →

NORTH

5 0   10   25   50

west second ave.
THE LAST HOLDOUT
existing conditions

BOB DOWNING
LA 489 BRITZ
WINTER 80

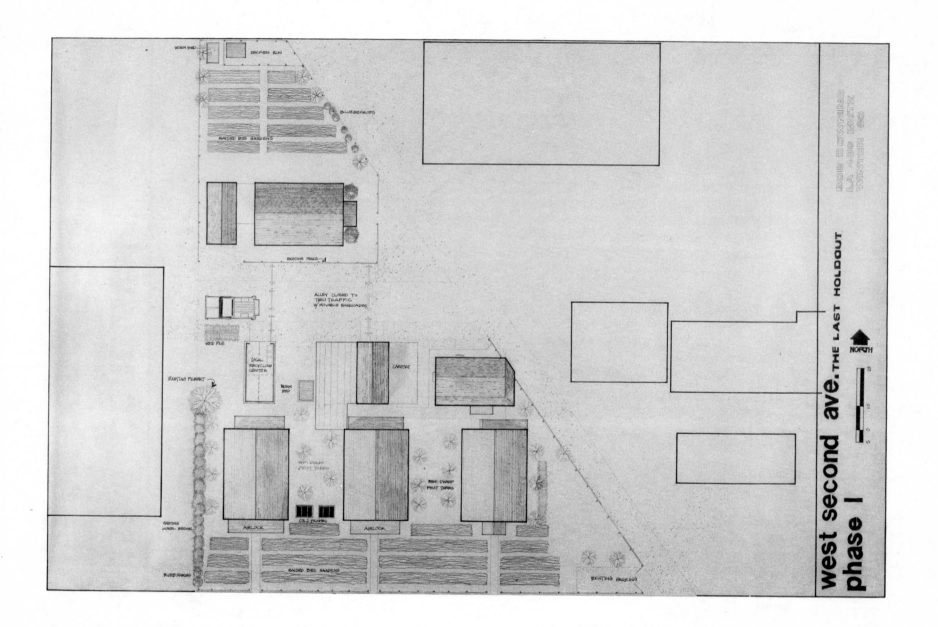

WORM BED

CHICKEN RUN

BLUEBERRIES

RAISED BED GARDENS

EXISTING FENCE

ALLEY CLOSED TO
THRU TRAFFIC
W/ MOVABLE BARRICADES

WOOD PILE

LOCAL
RECYCLING
CENTER

CARPORT

WORM
BED

EXISTING FILBERT

SEMI-DWARF
FRUIT TREES

SEMI-DWARF
FRUIT TREES

EXISTING LAUREL HEDGE

AIRLOCK

COLD FRAMES

AIRLOCK

RAISED BED GARDENS

BIKES PARKING

EXISTING PARKING

NORTH

5 0   10   15

west second ave. THE LAST HOLDOUT

phase I

BOB BENNING
LA 439 BLITZ
WINTER 80

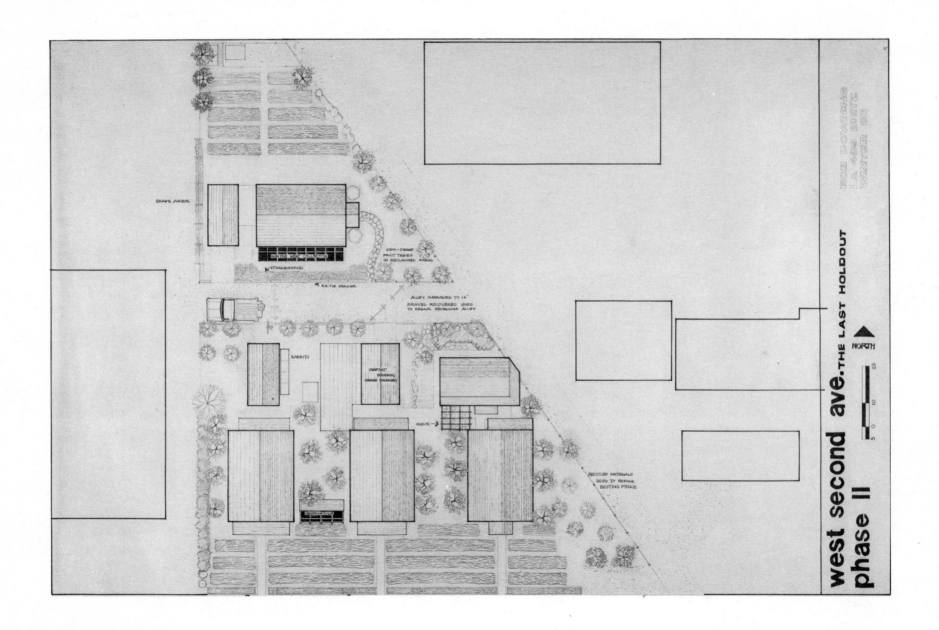

west second ave. THE LAST HOLDOUT
phase II

GRAPE ARBOR.

WINTER HEAT COLLECTOR

STRAWBERRIES

SEMI-DWARF
FRUIT TREES
IN RECLAIMED AREAS

R.R.TIE HEADER.

ALLEY NARROWED TO 14'
GRAVEL RECOVERED USED
TO REPAIR REMAINING ALLEY

RABBITS

CARPORT
REMOVED,
GARAGE ENLARGED

ARBOR →

GREENHOUSE

RECYCLED MATERIALS
USED TO REPAIR
EXISTING FENCE

NORTH

5   0   10   25

BOB DOWLING
LA 482 BLITZ
WINTER 80

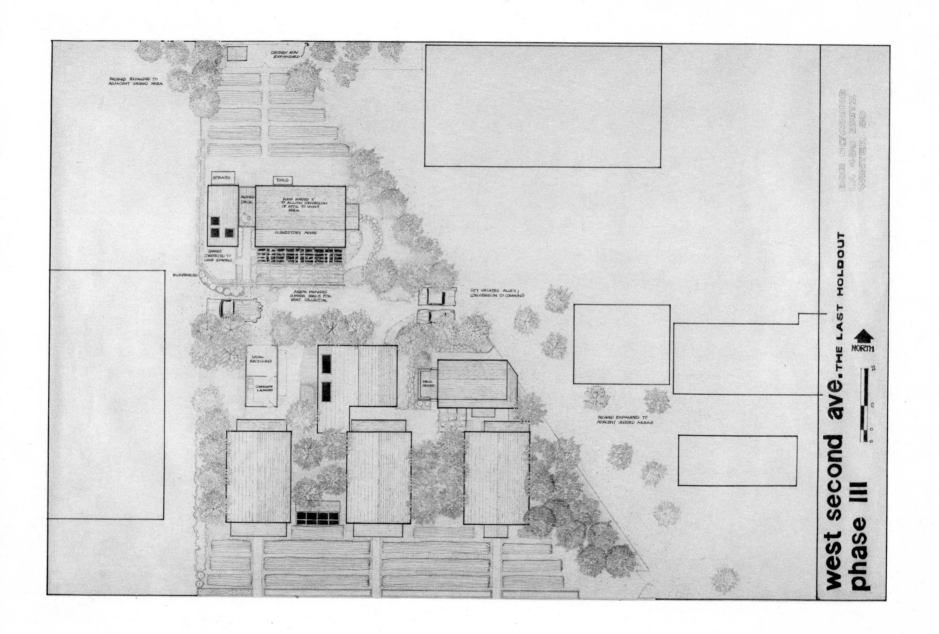

west second ave. THE LAST HOLDOUT
phase III

BOB DOWNING
L.A. 450 FRITZ
WINTER 80

NORTH

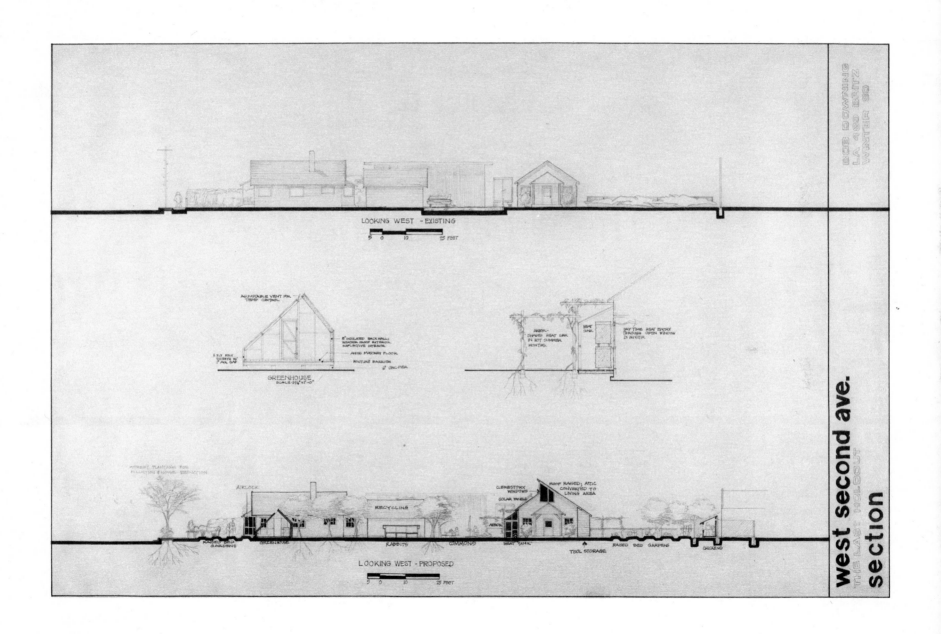

LOOKING WEST - EXISTING

5 0 10 25 FEET

GREENHOUSE
SCALE

LOOKING WEST - PROPOSED

5 0 10 25 FEET

west second ave.
THE LAST HOLDOUT
section

Chapter 12

The Urban Orchard

THE URBAN ORCHARD:
TRANSFORMING OUR PUBLICLY OWNED LANDS FROM LIABILITIES TO ASSETS...

This, the last section in the book, concludes our work to date with an optimistic note. By moving from gardens to farms (with animals) and from farms to tree crops, we are seeing greater investment in the future. Direct engagement in the healthy transformation of the places where we live has grown from a three-month commitment of time, energy, and heart to a summer garden, to a sixty- or a hundred-year cycle of fruit and nut trees near our places of dwelling. With this much belief in the future, can't we also expect our neighborhoods and our dwellings to also flourish and bloom?

DEVELOPING AN URBAN NEIGHBORHOOD & "OOPS!" ORCHADS PROGRAM

-- LLOYD LINDLEY III

THE OVERALL DEVELOPMENT CONCEPT FOR THE WHITEAKER NEIGHBORHOOD/ URBAN ORCHARD PROGRAM IS BASED ON THE ASSESSMENT OF NEIGHBORHOOD CONSTRAINTS AND OPPORTUNITIES, EXPRESSED CONCERNS BY NEIGHBORHOOD SPOKESPERSONS, NEEDS FOR DEVELOPING NEIGHBORHOOD SELF-RELIANCE, AND THE NEED FOR REVENUE GENERATION WITHIN THE NEIGHBORHOOD ADVISORY GROUP.

THE PLAN PROPOSES AN "IDENTIFIABLE NEIGHBORHOOD" BY DEMONSTRATING INTENSE URBAN FOOD PRODUCTION AND NEIGHBORHOOD SELF-RELIANCE. THE DEVELOPMENT INCLUDES THE UTILIZATION OF ALLEYS, VACANT LOTS, NEIGHBORHOOD/PARK BOUNDARIES, AND PRIVATE LANDS WILLING TO PARTICIPATE. DWARF AND SEMI-DWARF FRUIT TREE VARIETIES SHOULD BE USED TO PROMOTE QUICK AND EFFICIENT HARVESTING PRACTICES. DWARF FRUIT TREE VARIETIES ALSO DEMONSTRATE THE CONCEPT OF CLOSE PACKING, WITH TREES BEING PLANTED AT TEN FEET ON CENTER AND AS MANY AS 441 PER ACRE. CLOSE-PACKING ALLOWS GREATER PRODUCTION PER ACRE WHICH IN TURN GENERATES GREATER REVENUE NEEDED FOR OTHER NEIGHBORHOOD DEVELOPMENT PROJECTS.

AN URBAN ORCHARD PROGRAM CAN AID IN DEVELOPING ECONOMIC SUPPORT FOR THE WHITEAKER NEIGHBORHOOD. THE FOLLOWING IS AN EXAMPLE OF HOW ORCHARDING HAS WORKED IN AN URBAN AREA IN MEDFORD AND HOW ORCHARDING CAN WORK ECONOMICALLY WITHIN THE WHITEAKER NEIGHBORHOOD.

THE CITY OF MEDFORD LEASED A 36-ACRE PEAR ORCHARD TO A LOCAL PEAR FARMER WHO MANAGED THE ORCHARD AND MADE IT PRODUCTIVE FOR HIMSELF AND THE CITY OF MEDFORD. DUE TO URBAN PRESSURES THE ORCHARD IS NOW A CITY PARK AND IS NO LONGER IN FULL PRODUCTION.

DURING THE AGRICULTURAL SEASON OF 1976, BARTLETT AND BOSC VARIETIES OF PEAR WERE GROWN AND HARVESTED. 8952 LUGS OF BARTLETT PEARS WERE DELIVERED TO MARKET WITH EXTRA FANCY GRADE RECEIVING $18,512.29 AND $760.91 FOR CULLS. THIS BROUGHT A NET YIELD OF $19,273.20 FOR THE 1976 SEASON OF BARTLETT PEARS. FROM THE NET PROFIT THE CITY OF MEDFORD RECEIVED 20% OR $3854.64, LEAVING 80%, OR $15,418.56 FOR MANAGEMENT. BOSC PEARS WERE SCARCE DURING THIS SEASON; THEREFORE THEY WERE WORTH A BIT MORE ON THE MARKET THAN THE BARTLETT VARIETY. OUT OF 6168 LUGS, $32,795.13 WAS PAID FOR EXTRA FANCY, AND THERE WERE NO CULLS. THE GROSS PROFITS WERE $32,795.13, WITH THE CITY OF MEDFORD RECEIVING 26% OR $8,526.73 OF THAT GROSS PROFIT. THE REMAINING $24,286.40 WENT TO THE ORCHARD MANAGEMENT.

FOR THE 1976 PEAR SEASON, THE CITY OF MEDFORD RECEIVED $12,381.37 IN LEASE FEES, AND THE PEAR FARMER RECEIVED $39,686.96 FROM THE PEAR SALES.

WITHIN THE EUGENE AREA, FRUIT TREE FARMING IS A COMMON FORM OF AGRICULTURE. SOME OF THE MORE COMMON VARIETIES OF FRUIT PRODUCTION ARE SWEET CHERRIES, SOUR CHERRIES, AND FILBERTS. SWEET CHERRIES HAVE A VERY GOOD MARKET VALUE AT PRESENT, ALTHOUGH FOR NEW ORCHARDS PAY-BACK PERIODS ARE SLOW. SWEET CHERRIES CAN BE EXPECTED TO BEGIN

PAYING FOR THEMSELVES WITHIN ABOUT TEN TO FIFTEEN YEARS. SWEET CHERRIES ON THE AVERAGE YIELD THREE-FIVE TONS PER ACRE, AND IN THE 1979 GROWING SEASON RECEIVED .25¢ PER POUND ON THE EUGENE MARKET. SOUR CHERRIES HAVE A MUCH FASTER RETURN DUE TO THEIR FAST GROWTH. SOUR CHERRIES REQUIRE ABOUT FOUR YEARS TO BE IN FULL PRODUCTION, AND WILL YIELD ON THE AVERAGE 2.5-3 TONS PER ACRE. ON THE EUGENE MARKET DURING THE 1979 GROWING SEASON, SOUR CHERRIES RECEIVED .40¢ PER POUND. SOUR CHERRY TREES HAVE A VERY COMPACT HABIT, SIMILAR TO A DWARF OR SEMI-DWARF VARIETY OF OTHER FRUIT TYPES. WITH THE DWARF FORM, MORE TREES PER ACRE CAN BE PLANTED, ALLOWING LESS IMPACT ON THE OVERALL ORCHARD DUE TO SINGLE TREE LOSSES. FILBERTS ARE SLOW GROWERS AVERAGING FIFTEEN YEARS TO FULL PRODUCTION CAPABILITY. ALTHOUGH ONCE ESTABLISHED, FILBERTS WILL RECEIVE TOP PER-POUND PRICES ON THE OPEN MARKET. THIS YEAR IN THE EUGENE MARKET, FILBERTS RECEIVED .49¢ PER POUND, AND ORCHARDS WERE YIELDING ON THE AVERAGE 1-1.5 TONS PER ACRE. FILBERTS HAVE A GREATER LONGEVITY THAN EITHER OF THE CHERRY VARIETIES, THEREFORE ALLOWING A LONGER PERIOD BEFORE REINVESTMENT ON SUCCESSIVE ORCHARDS.

USING TEN ACRES AS A MODEL FOR FRUIT TREE PRODUCTION, THE FOLLOWING RESULTS SHOW THE VIABILITY OF THE URBAN ORCHARD PROGRAM:

VARIETY	PRICE/LB.	PRODUCTION/ PER ACRE	TOTAL/ACRE
SWEET CHERRIES	.25¢	5	$ 2,500
SOUR CHERRIES	.40¢	3	2,400
FILBERTS	.49¢	1.5	1,470

THE DOLLAR RANGE FOR FRUIT PRODUCTION, BASED ON A 10-ACRE PLOT, IS $25,000 TO $14,700. EXPENSES BREAK DOWN AS FOLLOWS:

LEASE @ 20%	$ 5,000
HARVEST @ .06¢/LB	3,600 CUSTOM
MAINTENANCE @ $500/ACRE	5,000
WAGES	6,000
TOTAL	$19,600
GROSS INCOME	$25,000
EXPENSES	- 19,600
NET PROFIT	$ 5,400

THERE WILL BE INITIAL COSTS IN DEVELOPING THE URBAN ORCHARD PROGRAM SUCH AS TREES AND PLANTING COSTS, AND FOR THE YEARS PRIOR TO FULL PRODUCTION. USING THE ABOVE DATE, THE LEASE FOR EACH YEAR WOULD BE BASED ON THE AMOUNT OF PRODUCTION. WITH VERY LOW PRODUCTION, INITIAL INVESTMENT MONEY WOULD BE REQUIRED TO EMPLOY A MANAGER, PAY FOR MAINTENANCE, MATERIALS AND TOOLS. TREE PLANTING WOULD COST AS FOLLOWS:

TREES @ $3.50 x 441/AC x 10	$15,435
PLANTING @ $2/HOLE x 441 x 10	8,820
TOTAL MAINTENANCE	$24,255

GENERAL DESIGN STRATEGIES

1. URBAN ORCHARD DEVELOPMENT WILL AID IN PROMOTING NEIGHBORHOOD SELF-RELIANCE.

2. PRODUCE WILL BE SOLD ON THE OPEN MARKET (AGRIPAC, BRUNNER DRYER, CHEF FRANCISCO).

3. FOOD WILL BE PROVIDED FOR NEIGHBORHOOD FAMILIES WITH THE LEAST MEANS OF SUPPORT.

4. REVENUES GENERATED BY THE PROGRAM WILL BE DISTRIBUTED BY THE NEIGHBORHOOD GROUP IN CHARGE OF NEIGHBORHOOD DEVELOPMENT PROGRAMS.

5. THE URBAN ORCHARD PROGRAM IS DEPENDENT UPON AND PROMOTES NEIGHBORHOOD ACTIVITY AND PARTICIPATION.

6. URBAN ORCHARD DEVELOPMENT PROVIDES BOTH FULL AND PART-TIME EMPLOYMENT.

7. THE ORCHARD MANAGER WILL BE A FULL-TIME PAID EMPLOYEE FROM THE COMMUNITY AND WILL BE RESPONSIBLE FOR ORCHARD OPERATIONS.

8. FRUIT TREES NOT ONLY PROVIDE FOOD BUT ALSO LANDSCAPE CONTINUITY, WHICH PROMOTES THE "IDENTIFIABLE NEIGHBORHOOD".

9. DWARF VARIETIES WILL BE USED TO DEMONSTRATE CLOSE PACKING, ACHIEVING A HIGHER PRODUCTION PER ACRE THAN CONVENTIONAL FRUIT ORCHARDS.

10. ORCHARDS WILL BE SITED SUCH THAT SHADOW CASTS WILL NOT INFRINGE ON THE RESIDENTS' SOLAR RIGHTS.

11. SIDEWALK PLANTING STRIPS, BIKEWAYS, ALLEYS AND STREETS WILL BE LINED WITH FRUIT-PRODUCING TREES.

THE PURPOSE FOR DEVELOPING THIS PORTION OF OUR WORK IS TO PROVIDE GUIDANCE AND DIRECTION FOR THE PROPOSED WHITEAKER NEIGHBORHOOD/URBAN ORCHARD PROGRAM. THE GUIDELINES ARE PRESENTED IN GRAPHIC AND WRITTEN FORM AND DEMONSTRATE HOW ORCHARD DEVELOPMENT WILL ADD TO ACHIEVING NEIGHBORHOOD SELF-RELIANCE. IT IS THE INTENT OF THE GUIDELINES TO: 1) IDENTIFY AREAS WITHIN AND ON THE BOUNDARIES OF WHITEAKER NEIGHBORHOOD THAT ARE SUITABLE FOR ORCHARD DEVELOPMENT, 2) DEMONSTRATE THE ECONOMIC FEASIBILITY OF THE PROGRAM, 3) PROTECT THE PRIVACY OF RESIDENTS WHILE ATTAINING MAXIMUM LAND USE POTENTIAL, 4) INTEGRATE NEIGHBORHOOD FOOD PRODUCTION WITH FOOD NEEDS WITHIN THE NEIGHBORHOOD, AND 5) COORDINATE THE PROGRAM WITH THE CITY AND COUNTY PARK AND RECREATION GOALS AND GUIDELINES IN ORDER TO MAINTAIN A GOOD WORKING RELATIONSHIP. THE MAINTENANCE OF THIS SET OF VALUES IN THE PLANNING AND DESIGN OF THE URBAN ORCHARD PROGRAM HAS BEEN OF PRIMARY CONSIDERATION.

THE AGENCIES AND POLICIES SECTION IS DEDICATED TO OUTLINING GOVERNMENTAL POLICIES ON STATE, REGIONAL, AND LOCAL LEVELS. THE FOLLOWING DATA HAS BEEN GATHERED FROM PLANNING DOCUMENTS PROVIDED BY THE CONCERNED AGENCIES. THIS IS NOT AN ENTIRE ACCOUNT OF STATE, REGIONAL, AND LOCAL POLICIES, BUT A COMPILATION OF DATA WHICH MAY AFFECT OR BE AFFECTED BY THE URBAN ORCHARD PROGRAM.

OREGON LAND CONSERVATION AND DEVELOPMENT COMMISSION, GOAL 5, OPEN SPACES, SCENIC AND HISTORIC AREAS, AND NATURAL RESOURCES

GOAL: TO CONSERVE OPEN SPACE AND PROTECT NATURAL AND SCENIC RESOURCES.

PROGRAM SHALL BE PROVIDED THAT WILL: 1) INSURE OPEN SPACE, 2) PROTECT SCENIC AND HISTORIC AREAS AND NATURAL RESOURCES FOR FUTURE GENERATIONS, AND 3) PROMOTE HEALTHY AND VISUALLY ATTRACTIVE ENVIRONMENTS IN HARMONY WITH THE NATURAL LANDSCAPE CHARACTER. THE LOCATION, QUALITY AND QUANTITY OF THE FOLLOWING RESOURCES SHALL BE INVENTORIED: (THIS INVENTORY ADDRESSES THOSE RESOURCES THAT WILL AFFECT OR BE AFFECTED BY THE PROPOSED URBAN ORCHARD PROGRAM).

1. LAND NEEDED OR DESIRABLE FOR OPEN SPACE;

2. CULTURAL AREAS.

WHERE NO CONFLICTING USES FOR SUCH RESOURCES HAVE BEEN IDENTIFIED, SUCH RESOURCES SHALL BE MANAGED SO AS TO PRESERVE THEIR NATURAL CHARACTER. WHERE CONFLICTING USES HAVE BEEN IDENTIFIED THE ECONOMIC, SOCIAL, ENVIRONMENTAL, AND ENERGY CONSEQUENCES OF THE CONFLICTING USES SHALL BE DETERMINED AND PROGRAMS DEVELOPED TO ACHIEVE THE GOAL.

CULTURAL AREA REFERS TO AN AREA CHARACTERIZED BY EVIDENCE OF AN ETHNIC, RELIGIOUS OR SOCIAL GROUP WITH DISTINCTIVE TRAITS, BELIEF AND SOCIAL FORMS.

OPEN SPACE CONSISTS OF LANDS USED FOR AGRICULTURAL AND FOREST USES, AND ANY LAND AREA THAT WOULD, IF PRESERVED AND CONTINUED IN ITS PRESENT USE: (THE FOLLOWING USES PERTAIN TO THE URBAN ORCHARD PROGRAM).

1. CONSERVE AND ENHANCE NATURAL OR SCENIC RESOURCES;

2. PROTECT AIR OR STREAM OR WATER SUPPLY;

3. PROMOTE CONSERVATION OF SOILS;

4. CONSERVE LANDSCAPED AREAS, SUCH AS PUBLIC OR PRIVATE GOLF COURSES, THAT REDUCE AIR POLLUTION AND ENHANCE THE VALUE OF ABUTTING OR NEIGHBORING PROPERTY;

5. ENHANCE THE VALUE TO THE PUBLIC OF ABUTTING OR NEIGHBORING PARKS, FORESTS, WILDLIFE PRESERVES, NATURE RESERVATIONS OR SANCTUARIES OR OTHER OPEN SPACES.

GUIDELINES

THE FOLLOWING GUIDELINES ARE THOSE WHICH AFFECT OR WILL BE AFFECTED BY THE URBAN ORCHARD PROGRAM.

1. THE NEED FOR OPEN SPACE IN THE PLANNING AREA SHOULD BE DETERMINED, AND STANDARDS DEVELOPED FOR THE AMOUNT, DISTRIBUTION AND TYPE OF OPEN SPACE.

2. CRITERIA SHOULD BE DEVELOPED AND UTILIZED TO DETERMINE WHAT USES ARE CONSISTENT WITH OPEN SPACE VALUES AND TO EVALUATE THE EFFECT OF CONVERTING OPEN SPACE LANDS TO INCONSISTENT USES. THE MAINTENANCE AND DEVELOPMENT OF OPEN SPACE IN URBAN AREAS SHOULD BE ENCOURAGED.

IMPLEMENTATION

1. DEVELOPMENT SHOULD BE PLANNED AND DIRECTED SO AS TO CONSERVE THE NEEDED AMOUNT OF OPEN SPACE.

2. THE CONSERVATION OF BOTH RENEWABLE AND NON-RENEWABLE NATURAL RESOURCES AND PHYSICAL LIMITATIONS OF THE LAND SHOULD BE USED AS A BASIS FOR DETERMINING THE QUANTITY, QUALITY AND LOCATION RATE AND TYPE OF GROWTH IN THE PLANNING AREA.

WILLAMETTE RIVER GREENWAY

USES AND ACTIVITIES ARE SUBJECT TO GREENWAY CONDITIONAL USE AND RURAL AREA EXTRAORDINARY EXCEPTION PERMITS...DEVELOPMENT PERMITS. GREENWAY CONDITIONAL USE AND RURAL AREA EXTRAORDINARY EXCEPTION DEVELOPMENT PERMITS SHALL BE REQUIRED FOR LAND USES AND ACTIVITIES AS PROVIDED IN THE PRELIMINARY WILLAMETTE RIVER GREENWAY PLAN. NEW INTENSIFICATIONS OR DEVELOPMENTS ALLOWED IN APPLICABLE ZONING DISTRICTS, INCLUDING PUBLIC IMPROVEMENTS AND INCLUDING PARTITIONS AND SUBDIVISIONS AS DEFINED IN Lc13.020, WHICH ARE PROPOSED FOR LANDS WITHIN THE BOUNDARIES OF THE WILLAMETTE GREENWAY ADOPTED AND AS REVISED FROM TIME TO TIME BY THE OREGON LAND CONSERVATION AND DEVELOPMENT COMMISSION, EXCEPT AS PROVIDED BELOW.

5. ACTIVITIES TO PROTECT, CONSERVE, ENHANCE AND MAINTAIN PUBLIC LANDS EXCEPT THAT A SUBSTANTIAL INCREASE IN THE LEVEL OF DEVELOPMENT OF EXISTING PUBLIC RECREATIONAL, SCENIC, HISTORICAL, OR NATURAL USES ON PUBLIC LANDS SHALL REQUIRE REVIEW AS PROVIDED BY THIS SUB-CHAPTER.

7. AGRICULTURE AS DEFINED IN ORS 215.203(2).

GREENWAY CONDITIONAL USE CRITERIA FOR DEVELOPMENT PERMIT APPROVAL: A DECISION TO APPROVE A GREENWAY CONDITIONAL USE PERMIT MAY BE GRANTED ONLY IF THE PROPOSAL CONFORMS TO ALL OF THE REQUIRED FINDINGS AS PROVIDED IN SUBSECTION 2, SECTION F OF THE PRELIMINARY WILLAMETTE RIVER GREENWAY PLAN, AND WILL REQUIRE FINDINGS THAT THE PROPOSED INTENSIFICATION OR DEVELOPMENT CONFORMS TO THE FOLLOWING CRITERIA, GUIDELINES AND SETBACK REQUIREMENTS.

1. CRITERIA: AFFIRMATIVE FINDINGS ARE MADE THAT, TO THE GREATEST POSSIBLE DEGREE:

    A. THE INTENSIFICATION OR DEVELOPMENT WILL PROVIDE THE MAXIMUM POSSIBLE LANDSCAPE AND OPEN SPACE OR VEGETATION BETWEEN THE ACTIVITY AND WILLAMETTE RIVER GREENWAY, AND

    B. NECESSARY PUBLIC ACCESS WILL BE PROVIDED TO AND ALONG THE RIVER BY APPROPRIATE LEGAL MEANS WHERE PRACTICAL.

2. GUIDELINES: CONSIDERATION OF THE FOLLOWING GUIDELINES:

    A. PRESERVATION AND MAINTENANCE OF LAND DESIGNATED "AGRICULTURE" IN THE ADOPTED WILLAMETTE RIVER GREENWAY PLAN FOR FARM USE, AND MINIMIZE INTERFERENCE WITH THE LONG-TERM CAPACITY OF LANDS FOR FARM USE.

B. PROTECTION, CONSERVATION OR PRESERVATION OF SIGNIFICANT SCENIC AREAS, VIEWPOINTS AND VISTAS.

J. MINIMIZING TO THE MAXIMUM EXTENT PRACTICAL, VANDALISM AND TRESPASS.

M COMPATIBILITY WITH THE SITE AND SURROUNDING AREA.

EUGENE, SPRINGFIELD, AND LANE COUNTY METRO-AREA GENERAL PLAN

ENVIRONMENTAL RESOURCES ELEMENT

THE ENVIRONMENTAL RESOURCES ELEMENT DEALS WITH THE NATURAL ASSETS AND HAZARDS OF THE METROPOLITAN AREA. THE ASSETS INCLUDE AGRICULTURAL LAND, CLEAN AIR AND WATER, FOREST LAND, SAND AND GRAVEL DEPOSITS, SCENIC AREAS, VEGETATION, WILDLIFE, AND WILDLIFE HABITAT. THE HAZARDS INCLUDE PROBLEMS ASSOCIATED WITH FLOODS, SOILS, AND GEOLOGY. THE EMPHASIS OF THE RECOMMENDATIONS IS DIRECTED TOWARD REDUCING URBAN IMPACTS ON WETLANDS THROUGHOUT THE METROPOLITAN AREA AND TOWARD PLANNING FOR THE NATURAL ASSETS AND CONSTRAINTS ON UNDEVELOPED LANDS ON THE URBAN FRINGE.

FINDINGS: THE FOLLOWING FINDINGS ARE THOSE WHICH RELATE TO THE URBAN ORCHARD PROGRAM AND ARE NOT INCLUSIVE OF THE TOTAL METRO PLAN FINDINGS.

1. OFFSETTING MEASURES CAN REDUCE THE NEGATIVE EFFECTS OF URBAN DEVELOPMENT ON WATER QUALITY AND QUANTITY PROBLEMS. EXAMPLES INCLUDE ON-SITE RETENTION OF STORM WATER, INCLUSION OF LANDSCAPED "BUFFER STRIPS" ADJACENT TO NEW DEVELOPMENTS AND CONSERVATION AND IMPROVEMENT OF STREAM-SIDE VEGETATION ALONG WATER COURSES.

2. THE STATE-WIDE AGRICULTURAL GOAL DEFINITION IS BASED UPON THE U.S. SOIL CONSERVATION SERVICE'S SOILS CAPABILITY CLASSIFICATION SYSTEM. THE MAJORITY OF LAND IN THE METROPOLITAN AREA IS LOCATED ON AGRICULTURAL SOILS RATED CLASS I THROUGH IV, AND MUCH OF THIS AREA HAS ALREADY EXPERIENCED URBAN DEVELOPMENT. THE HILLSIDE SOILS ARE GENERALLY INCLUDED IN NON-AGRICULTURAL RATED CLASSES V THROUGH VIII.

3. COMPACT URBAN GROWTH RESULTS IN PRESSURE ON OPEN SPACE WITHIN THE CURRENT URBAN GROWTH BOUNDARY. PROGRAMS FOR PRESERVING QUALITY OPEN SPACE WITHIN THE PROJECTED URBAN SERVICE AREA BECOME MORE IMPORTANT AS THE AREA GROWS.

4. OPEN SPACE PROVIDES MANY BENEFITS IN AN URBAN AREA INCLUDING: RETENTION OF HABITAT FOR WILDLIFE, FILTRATION OF POLLUTED WATER, ABSORPTION OF STORM RUNOFF FLOWS, PROTECTION OF SCENIC QUALITY, PROVISION OF RECREATION OPPORTUNITIES, REDUCTION OF ATMOSPHERIC TEMPERATURES, AND PERSONAL WELL-BEING.

5. URBAN AGRICULTURE, IN OTHER WORDS, BACKYARD AND COMMUNITY GARDENS, AND INTERIM USE OF VACANT AND UNDERDEVELOPED PARCELS, PROVIDES ECONOMIC, SOCIAL, AND ENVIRONMENTAL BENEFITS TO THE COMMUNITY.

GOALS:

1. PROTECT VALUABLE NATURAL RESOURCES AND ENCOURAGE THEIR WISE MANAGEMENT, PROPER USE AND REUSE, REFLECTING THEIR SPECIAL NATURAL ASSETS.

OBJECTIVES

1. MAINTAIN THE BENEFITS ASSOCIATED WITH ENVIRONMENTAL RESOURCES IN AN URBAN SETTING. THOSE RESOURCES INCLUDE AGRICULTURAL LANDS, CLEAN AIR AND WATER, FOREST LANDS, SAND AND GRAVEL DEPOSITS, SCENIC AREAS, WILDLIFE AND WILDLIFE HABITAT, AND VEGETATION. RECOMMENDATIONS DIRECTED TOWARD THESE RESOURCES MAY DIFFER DEPENDING UPON WHETHER THEY ARE LOCATED ON URBAN, RURAL, OR URBANIZABLE LANDS.

2. WHERE AGRICULTURAL LAND IS BEING CONSIDERED FOR URBANIZATION, THE LEAST PRODUCTIVE AGRICULTURAL LAND SHALL BE CONSIDERED FIRST. FACTORS OTHER THAN AGRICULTURAL SOIL RATINGS SHALL BE CONSIDERED WHEN DETERMINING THE PRODUCTIVITY OF AGRICULTURAL LAND. RELEVANT FACTORS INCLUDE OWNERSHIP PATTERNS, ADJACENT LAND USES, AGRICULTURAL HISTORY, AVAILABILITY OF IRRIGATION, AND MARKET AVAILABILITY.

3. AGRICULTURAL PRODUCTION SHALL BE CONSIDERED AN APPROPRIATE INTERIM AND TEMPORARY USE ON URBANIZABLE LAND AND ON VACANT AND UNDERDEVELOPED URBAN LAND.

4. CONTINUED LOCAL PROGRAMS SUPPORTING COMMUNITY GARDENS ON PUBLIC LAND AND PROGRAMS PROMOTING URBAN AGRICULTURE ON PRIVATE LAND SHALL BE ENCOURAGED. URBAN AGRICULTURE INCLUDES GARDENS IN BACKYARDS AND INTERIM USE OF VACANT AND UNDER-DEVELOPED PARCELS.

ENVIRONMENTAL DESIGN ELEMENT

THE FOLLOWING IS NOT INCLUSIVE OF THE ENVIRONMENTAL DESIGN ELEMENT YET ADDRESSES THOSE PORTIONS PERTAINING TO THE URBAN ORCHARD PROGRAM. THE ENVIRONMENTAL DESIGN ELEMENT IS CONCERNED WITH THAT BROAD PROCESS WHICH MOLDS THE VARIOUS COMPONENTS OF THE URBAN AREA INTO A DISTINCTIVE LIVABLE FORM THAT PROMOTES A HIGH QUALITY OF LIFE.

THE URBAN AREA SHOULD INCLUDE THE TYPE OF ENVIRONMENT IN WHICH PEOPLE CAN SEE VISUALLY ATTRACTIVE AND STIMULATING DETAILS, HEAR DISTINCTIVE SOUNDS, AND EXPERIENCE PLEASANT ODORS. BOTH THE FEELING OF EASE AND THE POSSIBILITY OF EXPERIENCING CHANGING SENSATIONS ARE FUNDAMENTAL TO ALLOW PEOPLE THE OPPORTUNITY TO RECOGNIZE OR IDENTIFY WITH THE VARIOUS PARTS OF AN URBAN AREA. WHEN THESE POSITIVE COMPONENTS ARE PRESENT, A LOCATION MAY DEVELOP DESIRABLE IDENTITY AND CHARACTER, AND THEREBY CONTRIBUTE TO A HIGHER QUALITY OF LIFE FOR ITS INHABITANTS.

BASED ON CONCERNS RELATED TO ENERGY CONSERVATION, ENVIRONMENTAL PRESERVATION, TRANSPORTATION, AND OTHER ISSUES, INCREASED DENSITY IS DESIRABLE. THIS INCREASES THE NEED FOR EFFECTIVE, DETAILED ENVIRONMENTAL DESIGN IN ORDER TO INSURE A HIGH QUALITY OF LIFE AND A HIGH DEGREE OF LIVABILITY IN AN INCREASINGLY DENSE URBAN ENVIRONMENT.

FINDINGS:

1. THE LOCATION AND DESIGN OF PUBLIC AND PRIVATE FACILITIES PLAY AN IMPORTANT ROLE IN GIVING DISTINCTIVE IDENTITY AND CHARACTER TO AN AREA. FOR EXAMPLE, AN AREA'S CHARACTER MAY BE DEVELOPED THROUGH ASSOCIATION WITH A PARTICULAR PARK, A LAND FORM, A PUBLIC BUILDING, AN AREA OF OLDER HOMES, VEGETATION, OR A DISTINCTIVE TYPE OF SUBDIVISION DESIGN.

2. THE USE OF BUFFER STRIPS AND OTHER DESIGN FEATURES CAN MINIMIZE THE NEGATIVE ENVIRONMENTAL IMPACT OF CERTAIN USES, SUCH AS ROADWAYS AND PARKING AREAS, WHILE PROTECTING ADJACENT LAND USES.

GOALS:

1. SECURE A SAFE, CLEAN, AND COMFORTABLE ENVIRONMENT WHICH IS SATISFYING TO THE MIND AND SENSES.

2. ENCOURAGE THE DEVELOPMENT OF THE NATURAL, SOCIAL, AND ECONOMIC ENVIRONMENT IN A MANNER THAT IS HARMONIOUS WITH OUR NATURAL SETTING AND MAINTAINS AND ENHANCES OUR QUALITY OF LIFE.

3. CREATE AND PRESERVE DESIRABLE AND DISTINCTIVE QUALITIES IN LOCAL AND NEIGHBORHOOD AREAS.

OBJECTIVES:

1. PROVIDE THE FACILITIES AND SERVICES NEEDED TO MAINTAIN OUR QUALITY OF LIFE. EXAMPLES INCLUDE EDUCATION, HOUSING, MEDICAL, PUBLIC TRANSPORTATION, AND RECREATIONAL FACILITIES.

2. ENCOURAGE A GREATER DIVERSITY OF LIVING EXPERIENCES AND ENVIRONMENTS.

3. ESTABLISH OR MAINTAIN A SENSE OF IDENTITY AND CHARACTER FOR LOCAL AND NEIGHBORHOOD AREAS.

4. SHAPE DEVELOPMENT TO SUIT NATURAL CONDITIONS AS MUCH AS POSSIBLE.

5. PRESERVE AND ENHANCE VIEWS AND PUBLIC USE OF RIVER CORRIDORS, DRAINAGEWAYS AND PROMINENT TOPOGRAPHIC FEATURES SUCH AS RIDGELINES, BUTTES, WHEN CONSISTENT WITH OTHER PLANNING POLICIES.

RECOMMENDED POLICIES

1. IN ORDER TO PROMOTE THE GREATEST POSSIBLE DEGREE OF DIVERSITY, A BROAD VARIETY OF COMMERCIAL, RESIDENTIAL, AND RECREATIONAL LAND USES SHALL BE ENCOURAGED WHEN CONSISTENT WITH OTHER PLANNING POLICIES.

2. THE PLANTING OF STREET TREES SHALL BE ENCOURAGED.

3. CAREFULLY DEVELOP SITES THAT PROVIDE VISUAL DIVERSITY TO THE URBAN AREA AND OPTIMIZE THEIR VISUAL AND PERSONAL ACCESSIBILITY TO RESIDENTS.

ANALYSIS AND ASSESSMENT

SOILS: SOILS ARE ALLUVIAL, SILTS, CLAY AND LOAM WITH MOST HAVING PRIME AGRICULTURAL QUALITIES. U.S.D.A. SOILS CONSERVATION SERVICE DATA IS AVAILABLE ALONG THE RIVER. MOST OF THE NEIGHBORHOOD HAS SMALL BACKYARD GARDEN PLOTS, AND FRUIT TREES ARE ABUNDANT. FILL SITES SHOULD BE INSPECTED TO ASSURE SUITABILITY IN RESPECT TO PLANT TYPES.

AIR QUALITY: AIR QUALITY IS QUESTIONABLE ALONG RAILROAD BOULEVARD. THIS IS DUE TO HIGH CO_2 LEVELS FROM DIESEL LOCOMOTIVES ALONG THAT CORRIDOR. OTHER NON-FOOD PRODUCING TREE TYPES MIGHT BE INTRODUCED TO DEVELOP A SOUND BARRIER AS WELL AS OXYGEN PRODUCERS.

SOLAR RIGHTS: PLACEMENT AND TREE TYPES SHOULD BE CONSIDERED IN RESPECT TO PASSIVE SOLAR HOMES AND TREE SHADOW ANGLE. TREE HEIGHTS AND LEAF-SHED TIME ARE PRIME CONCERNS IN RELATIONSHIP TO SOLAR ANGLE. SHADE FOR COOLING IN THE SUMMER, AND THE POTENTIAL FOR SOLAR GAIN IN THE WINTER, WOULD BE PROVIDED WITH DECIDUOUS TREEES.

WITH THE DEVELOPMENT OF NEIGHBORHOOD LOCAL SELF-RELIANT FOOD PRODUCTION, THE FIRST ENEMY IS THE POLLUTION DIRECTLY ADJACENT TO HIGH-VOLUME PRODUCERS SUCH AS AUTOMOBILE FREEWAYS AND RAILROAD CORRIDORS. TWO CHIEF OFFENDERS CAN BE DEALT WITH AND MODIFIED IN THEIR IMPACT: SOUND POLLUTION AND PARTICULATE EMMISSIONS.

BOTH MAY BE DEALT WITH BY PROVIDING AUDITORY, VISUAL, AND VISCERAL BUFFERING FROM NEIGHBORS BY PLANTING TREES, SHRUBS, AND GROUNDCOVERS AND ERECTING APPROPRIATELY LOCATED WOOD "SOUND" WALLS. WHERE POSSIBLE THESE REPARING AND RESTORING COSTS SHOULD BE BORNE BY THE OFFENDING AGENCIES, SUCH AS THE STATE HIGHWAY COMMISSION AND/OR THE FEDERAL DEPARTMENT OF TRANSPORTATION THROUGH APPEAL OR CLASS ACTION LEGAL SUITS, REGARDING HEALTH IMPACTS ON AUDITORY STRESS, RELATED NERVOUS DISORDERS, AND IMPACTS ON LEAD AND ASBESTOS PARTICLE FOOD POISONING. IMMEDIATE ACTION IS RECOMMENDED THROUGH THE FOLLOWING SELF-HELP PROCEDURES OF :

1. GENERATING DONATIONS FROM MAJOR TREE PRODUCERS

FOR THREE TO FIVE YEAR OLD TREES—TO BE PLANTED, WATERED, AND MAINTAINED BY THE NEIGHBORS ON CONTRACT WITH THE APPROPRIATE AGENCIES. MASSIVE REFORESTATION BY HOEDADS IN THE CITY STARTS PRECIDENT FOR THE URBAN FOREST

2. ESTABLISHING A NEIGHBORHOOD NATIVE PLANTS NURSERY

INITIAL TASKS ARE TO DIG, TRANSPORT, AND PLANT NATIVE TREE SPECIES FOR THE ABOVE PURPOSES. DUAL FUNCTIONS OF SOUND AND PARTICULATE ABATEMENT AND WOODLOT TREE PRODUCTION ARE ENCOURAGED.

EDIBLE TREES

INTRODUCTION

THE WHITEAKER NEIGHBORHOOD EDIBLE TREE PROGRAM IS NOT SIMPLY AN
OUTGROWTH OF A DESIRE FOR A FEW MORE BACKYARD FRUIT TREES.
RATHER, IT IS PREMISED UPON THE RECOGNITION OF LOCAL AND WORLD-
WIDE DEMOGRAPHIC AND RESOURCE TRENDS. THE HUMAN POPULATION IS
STILL GROWING AT EXPONENTIAL RATES DESPITE THE PROFUSION OF LITER-
ATURE AIMED AT FIRST WORLD COUNTRIES AND MASSIVE CAMPAIGNS FOR
BIRTH CONTROL ON THE INTERNATIONAL SCALE. THIS PHENOMENA IS
DIRECTLY TIED TO RESOURCE EXPLOITATION AND SUBSEQUENT DEPLETION.
ONE OF THOSE RESOURCES, AND UNDOUBTEDLY THE MOST IMPORTANT FOR
HUMAN SURVIVAL, IS FARMLAND. THESE SPECIAL LANDS WITH FERTILE SOILS
AND HARDY WATER SUPPLIES ARE VANISHING INTO DESERTS AND URBAN CEMENT
WASTELANDS AT ALARMING RATES.

HERE IN THE UNITED STATES, THE SO-CALLED BREAD BASKET OF THE ~~WORD~~ WORLD,
THOUSANDS OF ACRES OF PRIME FARM LAND IS LOST EACH WEEK TO EROSION,
NEGLECT AND URBAN EXPANSION. THE TOP 12 INCHES OF SOIL, AND OFTEN
MUCH LESS, SUPPORTS HUMAN LIFE ON THIS PLANET, AND THIS THIN SHEET
OF LIFE IS QUICKLY BEING LOST. THE UNITED STATES LOST ABOUT 3
MILLION TONS OF FERTILE TOPSOIL IN 1977 ALONE, FAR EXCEEDING THE
RATE OF NATURAL SOIL REGENERATION. MUCH OF THIS SOIL LOSS IS DUE
TO POOR LONG-RANGE PLANNING. THE CURRENT INTENSIVE MANAGEMENT
SCHEMES DESIGNED TO PRODUCE INCREASED YIELDS WITH HIGH DOSES OF
PETROCHEMICALS AND FERTILIZERS ARE DIMINISHING THE SOIL'S FERTILITY
AND TILTH, LEADING TO EROSIONAL PROBLEMS. THIS, HOWEVER, IS ONLY
A PORTION OF THE PROBLEM. ANOTHER 3 MILLION ACRES WAS LOST IN 1977
TO URBAN ENCROACHMENT ONTO FARMLANDS IN THE FORM OF URBAN SPRAWL.
THE SURROUNDING FARMS THAT ONCE SUPPORTED SMALL TOWNS ARE OVER-RUN,
AS THE TOWNS BECOME CITIES WITH NATIONAL AND INTERNATIONAL TRADE
CONNECTIONS. LAWNS, SWIMMING POOLS, STREETS, AND GOLF COURSES
COVER ONCE-PRODUCTIVE FARMLANDS, CRUSHING AND COMPACTING A VITAL
RESOURCE. THE RESOURCE BASE IS NO LONGER A FUNCTION OF LOCAL PRO-
DUCTION; RATHER, IT HAS EVOLVED INTO A SERIES OF INTERCONNECTED
TRADE NETWORKS, I.E. LETTUCE FROM CALIFORNIA, TOMATOES FROM MEXICO,
POTATOES FROM IDAHO, AND SOYBEANS AND WHEAT FROM THE MIDWEST.

THIS SYSTEM IS DIRECTLY RELATED TO ENERGY PRODUCTION AND CONSUMPTION.
AS OIL PRICES RISE AND SUPPLY DECREASES, THESE NETWORKS BECOME LESS
AND LESS VIABLE. THE TIME IS RAPIDLY APPROACHING WHEN LOCAL PRO-
DUCTION AND CONSUMPTION AND REGIONAL SELF RELIANCE WILL BE NECES-
SARY FOR FUTURE ECONOMIC STABILITY AND SURVIVAL. UNFORTUNATELY, OUR
PRIME FARMLANDS ARE COVERED BY THE CITIES WE LIVE IN!

THE WHITEAKER NEIGHBORHOOD IS A PERFECT CASE IN POINT. THE SOILS
UNDERLYING THE SPRAWLING SUBURBAN GROWTH ARE IN GENERAL, ALLUVIAL
OR RIVER-GENERATED SOILS OF HIGH QUALITY. ORCHARDS ONCE PROSPERED
HERE AND MANY OF THESE REMNANTS CAN BE SEEN TODAY IN BACK YARDS,
ALLEYS, AND PARKS.

THE WHITEAKER NEIGHBORHOOD EDIBLE TREE PROGRAM IS AN OUTGROWTH OF
THESE REALIZATIONS. THE NEED TO BEGIN THE WORK OF CREATING A SELF-
RELIANT COMMUNITY. A DESIRE TO RETURN A PORTION OF THESE URBAN
LANDSCAPES TO FRUITFUL AGRICULTURAL PRODUCTION AS A MEANS TO IN-
CREASE LOCAL FOOD SUPPLIES AND NEIGHBORHOOD COOPERATION.

THIS DESIRE FOR SELF RELIANCE, IN TERMS OF URBAN FOOD PRODUCTION, IS
NOT A NEW IDEA, BUT A REWORKING OF PAST IDEALS, FOR URBAN AGRICUL-
TURAL ACTIVITIES HAVE BEEN AROUND FOR A LONG TIME. THE FIRST
ORGANIZED ATTEMPT BEGAN EARLY IN THE 17TH CENTURY WHEN THE BRITISH
GOVERNMENT SET ASIDE URBAN ALLOTMENT TRACTS DEDICATED TO KITCHEN
GARDENS. THESE GARDENS WERE INTENDED TO HELP SUPPORT THE MASSES
OF RURAL PEASANTS MIGRATING TO THE CITIES IN SEARCH OF WORK AFTER
BEING DISPLACED FROM THEIR FARM LANDS. TODAY, THESE ALLOTMENTS ARE

A NATIONAL INSTITUTION AND HAVE SERVED THE PEOPLE WELL IN RECENT
CRISIS SITUATIONS. DURING W.W.II THE ALLOTMENT SYSTEM AND ITS
PARTICIPANTS HELP STAYED POSSIBLE STARVATION WHEN SUPPLY LINES WERE
CUT OFF. NOW THEY SERVE AS INFLATION HEDGES, AND IN SOME CASES
SURVIVAL GARDENS. THE COMMUNITY GARDEN SYSTEM SLOWLY DEVELOPING
IN THIS COUNTRY IS AN OUTGROWTH OF THIS HERITAGE. THE WHITEAKER
EDIBLE TREE PROGRAM ADDS ONE MORE DIMENSION -- THAT OF LONG-TERM
PERMANENCE, AS FRUIT TREES REQUIRE MORE TIME AND LONG-RANGE PLAN-
NING THAN A VEGETABLE GARDEN, AND THEY ARE QUITE LONG-LIVED.

WITH THESE PREMISES AND PRECEDENTS IN MIND, THE WHITEAKER EDIBLE
TREE PROGRAM PLAN IS ORGANIZED TO EDUCATE NEIGHBORHOOD RESIDENTS,
RENOVATE EXISTING TREES, AND PROMOTE ADDITIONAL PLANTINGS AND CARE
OF EDIBLE TREES.

ADOPT-A-TREE PROGRAM

THIS PROGRAM IS THE FIRST STAGE OF IMPLEMENTATION FOR THE STATED
GOALS OF THE EDIBLE TREE PROGRAM. IT IS PREMISED UPON SEVERAL EX-
ISTING CONDITIONS WITHIN THE NEIGHBORHOOD: 1) THE NEIGHBORHOOD HAS
A NUMBER OF EXISTING EDIBLE TREES, AND MANY COULD BE RESTORED TO
A REASONABLE CONDITION WITH ORGANIZED CONSISTENT CARE, 2) THE
MAJORITY OF THESE TREES ARE ON PRIVATE LAND AND THE OWNERS DO
NOT CARE OR DO NOT HAVE TIME TO TEND TO THEIR TREES, 3) MANY RESI-
DENTS DO NOT HAVE ACCESS TO FRUIT OR NUT TREES DUE TO THEIR HOUSING
SITUATIONS AND LOCATION, BUT WOULD LIKE TO BEGIN LEARNING ABOUT,
CARING FOR, AND HARVESTING THIS TYPE OF CROP.

BASICALLY, THE GOALS OF THIS PROGRAM ARE TWO-FOLD: 1) TO BRING ABOUT
AN AWARENESS OF THE BENEFITS OFFERED BY NEIGHBORHOOD EDIBLE TREES,
AND 2) TO ENHANCE NEIGHBOR INTERACTION AND COOPERATION THROUGH THE
RESTORATION AND CREATION OF NEW URBAN ORCHARDS.

THE PROGRAM REVOLVES AROUND AN EDUCATIONAL PACKAGE WHICH CONTAINS
THE BASIC METHODS, TOOLS AND RESOURCES NEEDED AND AVAILABLE FOR
EDIBLE TREE CARE. THIS PACKAGE IS INFORMATIONAL AND MOTIVATING IN
CHARACTER AND IS INTENDED FOR INDIVIDUAL NEIGHBOR USE. THE PACKAGE
IS DESIGNED TO FUNCTION AS A UNIT, AND ALSO AS SEPARATE HAND-OUT
SHEETS. IT CAN BE TAKEN BY OUTREACH WORKERS TO BLOCK PARTIES,
NEIGHBOR GATHERINGS, AND/OR STATIONED AT ONE LOCATION UNDER AN
ADOPT-A-TREE BULLETIN BOARD. ANOTHER PORTION OF THIS PROGRAM IS
DISCUSSED IN THE DEMONSTRATION URBAN ORCHARD SECTION. THIS SPECIFIC
PROJECT INTEGRATES WITH THE ADOPT-A-TREE PROGRAM AND SERVES TO EN-
HANCE AWARENESS OF THE POSSIBILITIES OFFERED BY FRUIT AND NUT
TREES UTILIZING AN ON-THE-GROUND DEMONSTRATION SITE. A MIXTURE OF
FRUIT, NUT, AND VINE VARIETIES APPROPRIATE FOR THIS AREA ARE
ARRANGED ILLUSTRATING THEIR DIFFERENT POTENTIALS AND USES WITHIN AN
URBAN AREA.

THE ACTUAL ADOPT-A-TREE PROGRAM FUNCTIONS AS A GO-BETWEEN AMONG
HAVES AND HAVE-NOTS. THOSE WHO MAY HAVE ONE OR SEVERAL FRUIT, NUT
OR VINE VARIETIES BUT ARE UNABLE TO CARE FOR THEM DUE TO TIME
LIMITATIONS, HEALTH OR AGE, SIMPLY REGISTER THEIR TREE/VINE ON THE
ADOPT-A-TREE BULLETIN BOARD. THOSE WHO WOULD LIKE TO ADOPT A TREE
ALSO REGISTER AT THE SAME LOCATION THEIR WANTS, RESOURCES AND
CAPABILITIES IN TERMS OF TIME AND ENERGY. SAMPLE CONTRACTS, LOCATED
WITH THE BOARD, CAN BE USED BY THE PARTICIPANTS TO ARRANGE THE
ADOPTION TO THE MUTUAL BENEFIT OF BOTH PARTIES. OF COURSE, THIS
BECOMES A FAIRLY HIGH LEVEL OF NEIGHBOR INTEREST AND ACTIVITY.
PERIODIC PROMOTIONAL ARTICLES IN THE WHITEAKER NEWSLETTER WOULD
HELP INFORM THE RESIDENTS OF THE PROGRAM. BASICALLY, IT IS UP TO
THE PROSPECTIVE ADOPTEE TO SEEK OUT AND FIND TREES THAT MIGHT BE
SUITABLE. THE PROGRAM IS MERELY AN OFFICIAL VEHICLE THAT LENDS A
SPECTOR OF RESPECTABILITY TO A ONCE-COMMON PRACTICE OF NEIGHBOR
INTERACTION AND MUTUAL COOPERATION.

A WHITEAKER FOOD OUTREACH WORKER, A FUNDED POSITION, WOULD FURTHER
ENHANCE THIS PROGRAM BY PROMOTING AND FACILITATING TRANSACTIONS,
CONDUCTING WORKSHOPS AND PARTICIPATING IN NEIGHBORHOOD OUTREACH.
SEE THE FOLLOWING FUNDING PROPOSAL FOR SPECIFIC INFORMATION.

A PROPOSAL FOR AN EDIBLE TREE COORDINATOR

GOAL: IMPLEMENT THE EDIBLE TREE REHABILITATION PROGRAM DESIGNED
DURING THE PLANNING PHASE OF THE WHITEAKER NEIGHBORHOOD'S
INTEGRATED URBAN COMMUNITY PROJECT.

PRINCIPAL TASKS:

1. RECRUIT NEIGHBOROOD BLOCK RESIDENTS AND TRAIN THEM IN EDIBLE
   TREE CARE.

   THE COORDINATOR WOULD ACTIVELY SEEK OUT BLOCK RESIDENT VOLUN-
   TEERS WHO WOULD SERVE AS EDUCATION/LABOR COORDINATORS FOR
   THEIR RESPECTIVE BLOCKS. THESE INDIVIDUALS WOULD BE TRAINED
   IN COORDINATION WITH COUNTY EXTENSION SERVICE AND PROJECT
   SELF-RELIANCE IN THE MAINTENANCE OF FRUIT AND NUT TREES.

2. PROMOTE A NEIGHBORHOOD-WIDE ADOPT-A-TREE PROGRAM DESIGNED
   DURING THE PLANNING PHASE, AND CONDUCT A FRUIT AND NUT TREE
   PLANTING PROGRAM.

   THE COORDINATOR WOULD ACTIVELY SOLICIT NEIGHBORS WITH NEGLECTED
   FRUIT AND NUT TREES TO LOAN THEIR TREE TO INDIVIDUALS AND/OR
   GROUPS INTERESTED IN CARING AND HARVESTING THE SELECTED TREES.

   THE COORDINATOR WOULD ALSO SOLICIT DONATIONS FROM APPROPRIATE
   SOURCES FOR TREE PLANTING CAMPAIGN DESIGNED DURING THE
   PLANNING PHASE.

IT IS INTENDED THAT THIS JOB DESCRIPTION WILL PROVIDE OCCUPATIONAL
AND DEVELOPMENTAL FUNDING FOR AN EXISTING WHITEAKER NEIGHBOR. THE
EDIBLE TREE PROGRAM COORDINATOR WOULD OPERATE OUT OF THE OFFICE OF
PROJECT SELF-RELIANCE WHICH ALSO HOUSES A TOOL LIBRARY STOCKED
WITH TREE MAINTENANCE MATERIALS AND TOOLS. THE JOB WOULD BE A
3/4-TIME POSITION, AND RUN FROM MARCH THROUGH DECEMBER. THIS
TIME PERIOD WOULD ALLOW FOR A COMPLETE CYCLE OF TREE CARE AND EDU-
CATIONAL ACTIVITIES. A SMALL BUDGET FOR SUPPLIES, PRINTING, AND
EQUIPMENT WOULD BE NEEDED TO COVER THE COSTS OF MISCELLANEOUS
OFFICE SUPPLIES, PAPER, STAMPS, ETC., AND BROCHURE/HANDOUT LITER-
ATURE, AND TOOL ACQUISITION AND REPAIR.

PROPOSAL SUMMARY FORM

PROPOSED BY: EDIBLE CITY RESOURCE CENTER

PROJECT TITLE: EDIBLE TREE PROGRAM COORDINATOR FOR THE WHITEAKER
                 NEIGHBORHOOD

TOTAL AMOUNT REQUIRED: $7,500

1. BRIEF DESCRIPTION OF THE PROJECT: THIS POSITION CALLS FOR
   ONE 3/4-TIME COORDINATOR WHO WOULD BASICALLY IMPLEMENT THE
   PROGRAM DESIGNED DURING THE PLANNING STAGE OF THE WHITEAKER
   NEIGHBORHOOD EDIBLE TREE PROGRAM. AN ATTACHED OUTLINE DE-
   SCRIBES THE PRINCIPLE GOALS AND RESPONSIBILITIES.

2. RELATIONSHIP TO NEIGHBORHOOD GOALS: THIS PROJECT IS IN LINE
   WITH THE NEIGHBORHOOD GOALS CALLING FOR INCREASED SELF-RELIANCE
   IN FOOD PRODUCTION. IT WILL ALLOW FOR INCREASED NEIGHBOR
   INTERACTION WHILE EDUCATING INDIVIDUALS AND GROUPS IN THE
   TECHNIQUES OF SMALL-SCALE FOOD PRODUCTION. THOSE BENEFITED
   INCLUDE LOW-INCOME RESIDENTS WHO REQUIRE SUPPLEMENTAL INCOME
   SOURCES AND THOSE APARTMENT DWELLERS WHO WOULD LIKE TO BE-
   COME MORE INVOLVED IN FOOD SELF-RELIANCE BUT DO NOT, AT
   PRESENT, HAVE ACCESS TO FRUIT OR NUT TREES.

BUDGET INFORMATION

ITEM	AMOUNT	
PERSONNEL (3/4 TIME @ $5/HR, 10 MOS.)	$ 6,000	
FRINGE (15% MAX.)	900	
CONSULTANTS	0	(EDIBLE CITY)
TRAVEL	0	RESOURCE CNTR)
TRAINING	0	
SPACE COSTS	0	(PROJECT SELF-
SUPPLIES	200	RELIANCE)
PRINTING	300	
EQUIPMENT	100	
OTHER DIRECT COSTS (EXPLAIN)	0	
INDIRECT COSTS (EXPLAIN)	0	
	$ 7,500	

SAMPLE CONTRACT

WHITEAKER NEIGHBORHOOD ADOPT-A-TREE PROGRAM

I, _____*Urban Farm*_____ HEREBY CONTRACT WITH
OWNER

_____*Eric Sloan*_____ AT _____*127½ N. 7th*_____
ADOPTOR   ADDRESS

TO ADOPT MY EDIBLE TREE(S) LOCATED AT _____SAME AS ABOVE_____
LOCATION

THE ADOPTEE WILL CARE FOR THE FOLLOWING EDIBLE TREE AND/OR VINE:

# AND KIND	# AND KIND
[1] APPLE(S)	[ ] FIG(S)
[ ] PEAR(S)	[ ] GRAPE(S)
[ ] PEACH(ES)	[ ] FILBERT(S)
[ ] PERSIMMON(S)	[ ] WALNUT(S)
[3] BLACKBERRY	[ ] OTHERS:
[3] RASPBERRY	
[3] BLUEBERRY	

THE	THE ADOPTER WILL PROVIDE
[ ] WATER	[X] WATER
[ ] FERTILIZER	[X] FERTILIZER
[ ] SPRAYS	[X] SPRAYS
[ ] PRUNING	[X] PRUNING
[ ] FROST PROTECTION	[X] FROST PROTECTION
[ ] TOOLS	[X] TOOLS

COMMENTS: _____
_____
_____

THE OWNER WILL RECEIVE	THE ADOPTER WILL RECEIVE
% OF HARVEST	% OF HARVEST
[ ] NONE	[ ] 10%
[X] 10%	[ ] 25%
[ ] 25%	[ ] 50%
[ ] 50%	[ ] 75%
[ ] 75%	[ ] 100%
[ ] OTHER REIMBURSEMENT:	[90] OTHER REIMBURSEMENT:
To be given	_____
to charity.	_____

THE MANNER IN WHICH CARE IS GIVEN SHALL BE:

[X] ORGANIC
[ ] SEMI-ORGANIC
[ ] OTHER ARRANGEMENT: _____
_____
_____

THE TERM OF THIS AGREEMENT SHALL BE FOR *As long as possible* BEGINNING
MONTHS/YEARS

ON _____*3-8-80*_____, AND SHALL BE OPEN FOR RENEWAL _____, OR
DATE

OTHER ARRANGEMENT: _____
_____.

THIS SPACE FOR PROPERTY MAP TO SHOW LOCATION OF SPECIFIC EDIBLE PLANTS
TO BE ADOPTED.

COMMENTS: *This is my aunt's place, I rent the apt. while I'm going to school. She's getting a bit old 83 but she'll take care of them after I graduate.*

BY _____*Bil Marshall Jr.*_____   BY _____*Eric T Sloan*_____
OWNER   ADOPTER

DATE_____*3-8-80*_____

EDIBLE TREE PROGRAM
GOALS
OBJECTIVES
ACTIVITIES

GOAL: RE-ESTABLISHMENT AND DEVELOPMENT OF LOCALIZED TREE AND SHRUB BASED FOOD PRODUCTION SYSTEMS INTEGRATED INTO URBAN RESIDENTIAL PATTERNS.

OBJECTIVE A: IDENTIFICATION OF EXISTING EDIBLE TREES AND SHRUBS AND POTENTIAL EDIBLE TREES AND SHRUBS LOCATIONS.

ACTIVITIES:

1. ACCUMULATION OF EXISTING DATA INCLUDING SURVEYS DONE FOR MUNICIPAL STREET TREE PROGRAMS.

2. COMPILATION OF AERIAL PHOTOGRAPHS AT A SCALE OF 1"= 100'-0".

3. "GROUND REALITY" SURVEYS IDENTIFYING EXISTING EDIBLE TREES, VERIFYING POTENTIAL PLACEMENT SITES, AND IDENTIFYING RESIDENT INTEREST AND ACTIVITY.

OBJECTIVE B: DEVELOPMENT OF POTENTIAL TREE AND SHRUB DEVELOPMENT INTENSITIES.

ACTIVITIES:

1. ANALYSIS OF DATA COLLECTED UNDER OBJECTIVE A.

2. DEVELOPMENT OF MAPS IDENTIFYING EXISTING AND POTENTIAL INTENSITIES ON THREE LEVELS: INDIVIDUAL TREES AND SHRUBS; BLOCK FARM MINI-ORCHARDS; SMALL SCALE INTRA-NEIGHBORHOOD COMMERCIAL ORCHARDS AND NURSERIES.

3. GENERATION OF AN ECONOMIC FEASIBILITY STUDY FOR A STRONG POTENTIAL IDEA FROM 2 ABOVE.

OBJECTIVE C: DEVELOPMENT OF TREE AND SHRUB CULTIVATION AND MAINTENANCE METHODS RESOURCE MATERIALS.

ACTIVITIES:

1. RESEARCH AND COMPILE APPLICABLE MATERIALS AND RESOURCE PEOPLE IN THE NEIGHBORHOOD APPROPRIATE TO THE REHABILITATION, MAINTENANCE, AND HARVESTING, AND SITE INFLUENCE POTENTIALS FOR EDIBLE TREES AND SHRUBS.

2. DEVELOP RESOURCE MATERIALS FOR EACH DEVELOPMENT INTENSITY (SEE OBJECTIVE B, ACTIVITY 2 ABOVE).

3. CAUSE TO OCCUR MODEST OR PROTOTYPICAL PILOT PROJECTS INVOLVING EDIBLE TREES AND SHRUBS.

OBJECTIVE D: GENERATION OF STRATEGIES FOR EACH DEVELOPMENT INTENSITY FOR REHABILITATION AND DEVELOPMENT.

ACTIVITIES:

1. MAPPING OF LAND OWNERSHIP PATTERNS.

2. DEVELOPMENT OF SITE PLANS BY OWNERSHIP.

3. IDENTIFY AND CULTIVATE POTENTIAL TREE AND SHRUB ACQUISITION RESOURCES.

4. GENERATE DEVELOPMENT PROPOSAL PACKETS AND "HOW TO" STRATEGIES.

OBJECTIVE E: DEVELOPMENT OF COMPREHENSIVE EDIBLE TREE AND SHRUB PROGRAM AND INTEGRATION INTO THE WHITEAKER OUTREACH OF THE EDIBLE CITY RESOURCE CENTER.

ACTIVITIES:

1. DEVELOP ACTION PLANS FOR IMPLEMENTATION OF DEVELOPMENT STRATEGIES AT EACH DEVELOPMENT INTENSITY.

2. DEVELOP REHABILITATION PROJECT INCLUDING (POSSIBLE) C.E.T.A.-BASED WORK TEAMS.

3. DEVELOP EDUCATION PROGRAMS FOR TREE CULTURE AND SHRUB PRODUCE USE (POST HARVEST).

4. DEVELOP RESOURCE MATERIALS INTO PUBLIC RESOURCE FILE INCLUDING WHITEAKER TOOL LIBRARY ADDITIONS.

5. DEVELOP JOB DESCRIPTION FOR EDIBLE CITY RESOURCE CENTER OUTREACH WORKER.

INVENTORY AND ANALYSIS

INTRODUCTION

EACH OF THE EDIBLE TREES HAS BEEN INVENTORIED BY VARIETY AND LOCA-
TED ON 100-SCALE MAPS. THE PRINCIPLE SPECIES IN THE NEIGHBORHOOD
(APPLES, FILBERTS, CHERRIES, AND BLACKBERRIES) ARE DISCUSSED WITH
RESPECT TO THEIR EXISTING CONDITION, RENOVATION PROCEDURES AND
POSSIBILITIES, AND MARKET INFORMATION. ALSO, A PROPOSAL FOR A
NEIGHBORHOOD ORCHARDIST WHO WOULD BE RESPONSIBLE FOR IMPLEMENTING
RENOVATION AND NEIGHBORHOOD EDUCATION IS INCLUDED.

SUMMARY OF ANALYSIS

THE WHITEAKER NEIGHBORHOOD CONTAINS APPROXIMATELY SEVEN ACRES OF
FRUIT AND NUT TREES (IF THEY WERE CONTIGUOUS PLANTINGS IN A
STANDARD ORCHARD OPERATION). IN ADDITION, THERE ARE APPROXIMATELY
THREE ACRES OF BLACKBERRIES SCATTERED THROUGHOUT THE NEIGHBORHOOD.
ALL TOTAL, THERE ARE TEN ACRES OF EDIBLE TREES AND VINES. IF ALL
OF THE ITEMS WERE CONTIGUOUS, IT MIGHT BE A VIABLE FARM OPERATION
FOR ONE FARM FAMILY AND ONE EMPLOYEE. HOWEVER, THE PRESENT STATE
OF THE INDIVIDUAL SPECIES, THE DISPERSED NATURE OF THE PLANTINGS
ON A VARIETY OF PUBLIC AND PRIVATE PROPERTIES WOULD MAKE IT IM-
POSSIBLE TO RENOVATE AND MAINTAIN THE ENTIRE RESOURCE AS A PRO-
DUCTIVE UNIT.

GIVEN THE AVAILABILITY OF FUNDING AND THE CONDITION OF THE EXISTING
RESOURCE, AN EDIBLE TREE CARE BUSINESS OR A SMALL SCALE FARMING
ENTERPRISE WOULD NOT BE VIABLE. THE BEST USE OF THE EXISTING RE-
SOURCE WOULD BE BY AND FOR INDIVIDUALS AND GROUPS IN A HOME/BLOCK
OPERATION. TO ACCOMPLISH THIS, AN EDIBLE TREE COORDINATOR WOULD
BE NEEDED TO CONDUCT EDUCATIONAL WORKSHOPS AND FACILITATE NEIGHBOR
INTERACTION PERTAINING TO FRUIT AND NUT PRODUCTION. A DETAILED
PROPOSAL FOLLOWS.

APPLES

EXISTING

THE MAJORITY OF THE EDIBLE TREES IN WHITEAKER ARE APPLES. THERE
ARE APPROXIMATELY 275 APPLIES OF VARIOUS VARIETIES SCATTERED IN
MINI-ORCHARDS OR ISOLATED. IN GENERAL THE MAJORITY OF THESE TREES
ARE UNCARED FOR AND PRODUCING FAR LESS THAN WELL-MAINTAINED ORCHARD
TREES WOULD PRODUCE. THE AVERAGE OREGON APPLE TREE CAN PRODUCE
UP TO 260 POUNDS OF FRUIT PER YEAR. THIS OF COURSE INVOLVES A
GOOD DEAL OF MAINTENANCE AND CAREFUL ATTENTION TO DISEASE PROBLEMS
AND CLIMATIC CONDITIONS. THE WITHEAKER TREES DOE NOT RECEIVE
THIS CARE, AND AS A RESULT ARE IN A DILAPIDATED STATE.

RENOVATION

MANY OF THE TREES ARE ISOLATED, WHICH INHIBITS FRUIT SET DUE TO
LOW POLLINATION ACTIVITY. THIS MAY BE DUE TO AGE OF THE TREE,
MIXURE OF VARIETIES, OR THE LACK OF BEES IN URBAN AREAS. THE
DAMAGED AND DISEASED TREES SHOULD BE REMOVED, WHILE OTHERS MIGHT
BE RESTORED. RESTORATION OF AN OLD APPLE TREE TAKES THREE OR
FOUR YEARS. IT MIGHT BE BEST TO BEGIN RESTORATION WHILE PLANTING
SEMI-DWARF APPLES NEARBY. AS THE SEMI-DWARF COMES INTO PRODUCTION
THE OLDER TREE MIGHT BE REMOVED OR ALLOWED TO REMAIN DEPENDING ON
THE SUCCESS OF THE REJUVENATION PROCESS. (SEE SUCCESSIONAL PLANT-
ING PLAN IN THE URBAN FARM SECTION.)

MARKET

THE BEST MARKET POTENTIAL FOR WHITEAKER APPLES IS THE JUICE MARKET.
GENESIS JUICE COOPERATIVE WILL PAY 6¢/LB FOR ORGANICALLY CERTIFIED
APPLES AND 5¢/LB FOR OTHERS. THEY WOULD PREFER THAT THE APPLES
BE FREE OF LEAVES OR BRANCHES, AND FRESHLY HAND PICKED. WINDFALLS
WILL ONLY BE ACCEPTED ON A DAY-TO-DAY BASIS. SCAB AND MISSHAPEN
APPLES ARE ACCEPTABLE. BRUISED, SOFT AND WORM-RIDDEN FRUIT IS NOT.
THE PRICE WILL FLUCTUATE THROUGHOUT THE SEASON.

FILBERTS

EXISTING

FILBERT NUT TREES ARE THE SECOND MOST DOMINANT EDIBLE TREE IN THE
NEIGHBORHOOD. THERE ARE APPROXIMATELY 150 FILBERT TREES OR BUSHES
SCATTERED INDIVIDUALLY OR IN MINI-ORCHARDS. THE MAJORITY OF THE
TREES ARE IN A STATE OF DISREPAIR AND IN NEED OF ATTENTION. A
HEALTHY TREE CAN PRODUCE UP TO 13 POUNDS OF NUTS PER SEASON. EVEN
IN A LESS-THEN-PERFECT MAINTENANCE SITUATION, THEY WILL CONTINUE TO
PRODUCE NUTS.

RENOVATION

FILBERTS ARE RELATIVELY EASY TO RENOVATE. THEY CAN BE COMPLETELY
CUT BACK AND WILL SPROUT NEW GROWTH FROM THE ROOT STOCK. THEY CAN
BE MAINTAINED AS A TREE OR AS A BUSH; THE FORMER, THOUGH, IS A
BETTER FORM FOR HARVEST AND CARE. IT IS EXTREMELY IMPORTANT THAT
ALL CUTS IN A TREE WHICH ARE LARGER THAN TWO INCHES BE SEALED WITH
A LATEX SEALER TO PREVENT DISEASE PROBLEMS.

MARKET PROSPECTS

FILBERTS CURRENTLY RETAIL FOR $1.00/LB. IF ALL THE WHITEAKER TREES WERE PRODUCING MAXIMUM CROPS, A $2,000.00 GROSS PROFIT MIGHT BE REALIZED. HOWEVER, MAINTENANCE, HARVESTING AND THE DISPERSED NATURE OF THE PLANTING LIMITS THE POSSIBILITY OF A TRULY PROFITABLE OR VIABLE ENTERPRISE.

CHERRIES

EXISTING

CHERRIES ARE THE THIRD MOST PLENTIFUL EDIBLE TREE IN THE NEIGHBORHOOD. MANY OF THESE TREES ARE 30-40 YEARS OLD, AND IN VARIOUS STATES OF DISREPAIR. IDEALLY, A SWEET CHERRY TREE IN A WELL-MAINTAINED SYSTEM SHOULD PRODUCE APPROXIMATELY 120 POUNDS PER TREE. VERY FEW OF THE 108 CHERRY TREES IN THE NEIGHBORHOOD COULD MATCH THIS OUTPUT. CHERRIES REQUIRE A GOOD DEAL MORE CARE THAN APPLES AND ARE MORE SUSCEPTIBLE TO DISEASE AND WEATHER FLUCTUATIONS. THEY ALSO SUFFER FROM ISOLATION AND LOW FRUIT SET. AS WITH APPLES, INCREASED BEE ACTIVITY WOULD BE HELPFUL.

RENOVATION

THE EXISTING CHERRY TREES MIGHT BE RENOVATED TO A REASONABLY PRODUCTIVE STATE BY CAREFUL PRUNING AND MAINTENANCE. HOWEVER, MANY ARE BEYOND REPAIR AND REPLACEMENT IS NECESSARY (POSSIBLY WITH ANOTHER VARIETY OR SPECIES OF FRUIT, AS THE MARKET FLUCTUATES TOO GREATLY TO EXPECT ANY COMMERCIAL FEASIBILITY FOR CHERRIES AS A CASH CROP.

MARKET

THE SWEET AND SOUR CHERRY ORCHARDS IN OREGON BEGAN WHEN THE ORCHARDS OF THE EAST WERE DAMAGED BY DISEASE AND WEATHER CONDITIONS. HOWEVER, THE EASTERN PRODUCTION IS ONCE AGAIN GAINING MOMENTUM, AND WILL EVENTUALLY REPLACE OREGON'S ORCHARDS IN NATIONAL DISTRIBUTION. THE LOCAL MARKET REMAINS, WITH CURRENT ROADSIDE PRICES FOR SWEET CHERRIES ABOUT $1.00/POUND. AN ENTERPRISING RESIDENT MIGHT UTILIZE THIS EXISTING RESOURCE AT A REASONABLE PROFIT.

WALNUTS

EXISTING

THERE ARE APPROXIMATELY 87 ENGLISH WALNUT TREES AND 40 BLACK WALNUT TREES IN THE WHITEAKER NEIGHBORHOOD. MANY OF THESE TREES ARE IN A REASONABLE CONDITION; HOWEVER, FEW ARE PRODUCING VAST QUANTITIES OF NUTS. THE ENGLISH WALNUT WILL OUT-PRODUCE THE BLACK WALNUT IN TERMS OF NUTS. THE BLACK WALNUT IS USED COMMERCIALLY AS A HARDWOOD SOURCE. ALTHOUGH THE WILLAMETTE VALLEY WAS PLANTED IN WALNUTS AT ONE TIME, IT IS NO LONGER CONSIDERED A VIABLE CROP. THE CLIMATE AND DISEASE PROBLEMS INHIBIT NUT PRODUCTION. MANY OF THE NEIGHBORHOOD TREES EXAMPLIFY THIS FACT BY LOW FRUIT SET, WORM DAMAGE, AND ROOT ROT.

RENOVATION

THOSE ENGLISH WALNUTS GRAFTED ONTO HARDY CARPATHIAN OR MANREGIAN WALNUT STOCK CAN BE MAINTAINED AS IS, WITH MINOR PRUNING AND APPROPRIATE INSECT/DISEASE MAINTENANCE. SOME OF THE AILING BLACK WALNUTS MIGHT BE HARVESTED AND SOLD AS A HARDWOOD LUMBER RESOURCE. HOWEVER, THIS LAST OPTION IS SOMEWHAT LESS THAN DESIRABLE TO MANY RESIDENTS AT THIS TIME, BUT SHOULD BE CONSIDERED IN LONG-TERM URBAN PRODUCTION STRATEGIES.

MARKET

WALNUTS ARE CURRENTLY SELLING FOR APPROXIMATELY $1.00/POUND, CLEANED, DRIED AND ROT FREE. THE BEST USE OF THE WHITEAKER NUT PRODUCTION MIGHT BE AN INTERNAL TRADE/BARTER SYSTEM RATHER THAN THE COMMERCIAL MARKET SYSTEM.

BLACKBERRIES

EXISTING

THE TOTAL VOLUME OF BLACKBERRY PRODUCTION IS SOMEWHAT DIFFICULT TO QUANTIFY. THERE ARE APPROXIMATELY THREE ACRES OF VINES SCATTERED IN VARIOUS LOCATIONS: VACANT LOTS, ALLEYS, ALONG THE RAILROAD, ALONG THE RIVER, AND ON SKINNER'S BUTTE. IDEALLY, 7,000 POUNDS OF BERRIES MIGHT BE PRODUCED FROM A WELL-MAINTAINED ACRE. THE VINES IN WHITEAKER ARE GENERALLY WILD AND UNMAINTAINED, AND AS SUCH, DIFFICULT TO HARVEST, MAINTAIN, OR ESTIMATE YIELD.

RENOVATION

RENOVATION OF THE BLACKBERRIES WOULD BE TIME CONSUMING AND EXPENSIVE. THE VINES COULD BE TRAINED ON TRELLISSES TO ALLOW ACCESS TO THE BERRIES. PRUNING AND PEST CONTROL WOULD BE REQUIRED REGULARLY. IF CULTIVATION OF THE RAILSIDE BERRIES IS DESIRED, SOME SORT OF PROTECTION FROM THE ENGINE SOOT WOULD DEFINITELY BE REQUIRED.

MARKETS

DURING THE LAST FIVE YEARS THE BERRY MARKET HAS SKYROCKETED. BERRIES THIS YEAR SOLD FOR AS MUCH AS 55¢/POUND ON THE RETAIL MARKET. ROADSIDE STANDS, LOCAL MARKETS AND SPRINGFIELD CREAMERY PROVIDE READY OUTLETS FOR THIS CROP. GIVEN THE POTENTIAL YIELDS AND PROLIFIC NATURE OF THE VINE, THIS CROP HOLDS THE BEST POTENTIAL FOR A NEIGHBORHOOD CASH GENERATOR.

VINES

EXISTING

THE REMAINING VINES AND FRUIT TREES ARE SCATTERED IN BACK YARDS AND VACANT LOTS. THEIR OVERALL NUMBERS ARE SO LOW THAT A COMMERCIAL MARKET FOR THE FRUIT IS IMPOSSIBLE. MOST OF THE TREES ARE MAINTAINED BY OWNERS OR OTHER INDIVIDUALS. THOSE IN A DERELICT CONDITION MIGHT BE RENOVATED IF AN INDIVIDUAL OR GROUP WAS INTERESTED. THESE CROPS MAY INCREASE IN NUMBER AND DIVERSITY AS THE EDIBLE TREE PROGRAM CONTINUES.

EDIBLE TREES FOR THE WHITEAKER NEIGHBORHOOD

THE FOLLOWING IS A LIST OF EDIBLE TREE SPECIES ADAPTED TO THE SOUTHERN WILLAMETTE VALLEY'S GROWING CONDITIONS. SOME ARE QUITE COMMON, WHILE OTHERS ARE EXOTIC BUT STILL TASTY. THERE ARE A NUMBER OF POSSIBLE VARIETIES OF EACH SPECIES AVAILABLE (SEE CHART), AND NEW ONES ARE CONTINUALLY APPEARING IN NURSERIES. SO CHECK WITH THE LANE COUNTY COOPERATIVE EXTENSION OFFICE NURSERY CATALOGS AND LOCAL NURSERIES FOR DETAILS.

EDIBLE TREES FOR THE WHITEAKER NEIGHBORHOOD

CROP	SPACE/ TREE FOOT	POLLEN- IZER	YEARS TO BEARING	SPRAY FOR
APPLE	5-40	SOMETIMES	3-10	CODLING MOTH, SCAB
BUTTERNUT	30-40	YES	6-7	NONE
CHERRY (SOUR)	14-20	NO	5-6	FRUIT FLY
CHERRY (SWEET)	20-35	YES	6-7	FRUIT FLY
CHESTNUT	20-40	YES	5-7	NONE
FIG	12-20	NO	5-6	NONE
FILBERT	15-20	YES	5-6	BACTERIAL BLIGHT FILBERT MOTH
HICKORY	20-40	YES	10-14	NONE
PAPAW	15-20	NO	12-14	NONE
PEACHES	12-15	NO	4-5	LEAF CURL, BORERS
PEARS	10-20	YES	5-7	
PERSIMMONS	15-20	YES	8-10	NONE
PLUMS & PRUNES	10-20	SOMETIMES	5-6	CROWN BORERS, BROWN ROT
WALNUTS (BLACK)	30-40	NO	10-12	HUSK FLY BLIGHT
WALNUTS (ENGLISH)	40-50	NO	10-12	HUSK FLY BLIGHT

VARIETY	APPROX. DATE OF MATURITY	COMMENTS
APPLES		
•SUMMER VARIETIES		
LODI	JULY 15-30	YELLOW
RED MELBA	AUG. 1-20	PARTLY SCAB RESISTANT
TYDEMAN'S RED	AUG. 20-30	LARGE, RED, ATTRACTIVE
•MID SEASON VAR.		
McINTOSH	SEPT.15-30	RED BLUSH, STORE 35 F
WEALTHY	SEPT.15-30	SMALL TREE
CORTLAND	SEPT.15-30	GOOD IN SALADS
DELICIOUS	SEPT.-OCT.	POLLINIZED BY GOLD. DEL.
GOLDEN DELICIOUS	SEPT.-OCT.	POLLINIZED BY DELICIOUS
MUTSU	OCT.	LARGE, CRISP
•LATE VARIETIES		
KING	OCT.	LARGE, RED STRIPES
ROME BEAUTY	OCT.20-NOV.15	RED, GOOD FOR PIE-BAKING
MELROSE	OCT.	STORES WELL
NEWTOWN	OCT.-NOV.	GREEN, STORES WELL
•SCAB-RESISTANT VARIETIES		
TYDEMAN'S EARLY RED	AUG. 20-30	RED, SLIGHTLY SOFT
PRIMA	SEPT. 1-15	NEW
CHEHALIS	SEPT. 1-15	YELLOW, SLIGHTLY SOFT
PRISCILLA	SEPT.10-20	NEW, STORES WELL
MACOUN	OCT. 1-15	SMALL, DARK RED
SPARTAN	OCT. 1-15	RED, SMALL, STORES WELL
CHERRIES		
•SOUR VARIETY		
MONTMORENCY	JULY	MICHIGAN STRAIN BEST
•SWEET VARIETIES		
ROYAL ANN	MID	WHITE, POLLINIZED BY CORUM
CORUM	EARLY	WHITE, POLLINIZED BY R. ANN
BING	MID	BLACK, POLL. BY VAN, CORUM
LAMBERT	LATE	BLACK, POLL. BY VAN, CORUM
VAN	EARLY	BLACK, POLL. BY BING, LAMBERT
SAM	MID	BLACK, POLL. BY LAMBERT

MORE EDIBLE TREES

VARIETY	APPROX. DATE OF MATURITY	COMMENTS
FIGS		
DESERT KING	AUGUST	GREEN, LARGE
LATTARULA	AUGUST	GREEN, GOLDEN INSIDE
FILBERTS		
BARCELONA	OCTOBER	STANDARD VARIETY, POLLINIZED BY DAVIANNA
DAVIANNA	OCTOBER	LIGHT PRODUCER, POLLINIZED BY BARCELONA
DUCHILLY	OCTOBER	LONG, FLATTENED, POLLINIZED BY BARCELONA
PEARS		
·EUROPEAN VARITIES		
BARTLETT	AUG. 15-30	POLLINIZED BY ANJOU, FALL BUTTER
ANJOU	SEPT. 5-20	POLLINIZED BY BARTLETT
SECKEL	AUG.20-SEPT.10	POLLINIZED BY ANJOU, BOSC, COMICE
BOSC	SEPT.10-30	POLLINIZED BY COMICE-- BEST IN SOUTHERN OREGON
COMICE	SEPT.20-30	POLLINIZED BY BOSC-- BEST IN SOUTHERN OREGON
·RED VARIETIES		
RED BARTLETT	AUG. 15-30	POLLINIZED BY ANJOU, FALL BUTTER
REIMER RED	SEPT.	POLLINIZED BY BARTLETT
RED ANJOU	SEPT.	POLLINIZED BY BARTLETT
STARKRIMSON	AUG. 1-15	POLLINIZED BY BARTLETT
PERSIMMONS		
FUYU	NOVEMBER	SEEDLESS JAPANESE
GARRETTSON	NOVEMBER	AMERICAN, SMALL
EARLY GOLDEN	NOVEMBER	AMERICAN, SMALL

PLUMS		
·EUROPEAN (PRUNES WHEN DEHYDRATED		
ITALIAN	SEPT. 10-30	TART, PURPLE PLUM
BROOKS	SEPT. 20-30	BEARS REGULARLY, LARGE
PARSONS	SEPT. 1-15	POLLINIZED BY ITALIAN, SWEET
MOYER PERFECTO	OCT. 1	BEST DRIED, SWEET
STANLEY	SEPT. 1-15	BEARS, BUT BROWN ROTS
WALNUTS		
·BLACK VARIETIES		
THOMAS	OCTOBER	SEEDLINGS INFERIOR BUY NURSERY STOCK.
OHIO	OCTOBER	
MYERS	OCTOBER	
·ENGLISH VARIETIES		
FRANQUETTE	LATE OCTOBER	STANDARD VARIETY, LIMITED HARDINESS
SPURGEON	LATE OCTOBER	LATE BLOOMER, HARDY
CHAMBERS #9	LATE OCTOBER	HEAVY PRODUCER, MODERATELY HARDY

ADAPTED FROM OREGON EXTENSION BULLETIN #819, MARCH 1979 BY ROBERT L. STEBBINS, AND "APPLE VARIETIES FOR THE WILLAMETTE VALLEY", BY ROBERT L. STEBBINS.

DISEASE RESISTANT APPLE VARIETIES

UNFORTUNATELY, INSECTS AND DISEASES ARE NOT EASY PROBLEMS TO DEAL WITH. IT TAKES A CONSISTANT EFFORT TO KEEP ANY PROBLEMS TO A MINIMUM. THIS IS COMPOUNDED BY THE FACT THAT THERE ARE NUMEROUS TYPES OF PESTS AND DISEASES THAT CAN ATTACK, MAR OR DESTROY FRUIT AND NUT TREES. HOWEVER, A FEW PRECAUTIONS WILL ENABLE YOU TO AT LEAST BEGIN ON A SUCCESSFUL GROWING SEASON.

1. MULCH THE SOIL WITH STRAW, SAWDUST OR COMPOST.

2. COMPOST OLD LEAVES; BURN PRUNINGS.

3. REMOVE DEAD OR DISEASED TREES.

4. FOLLOW THE GENERAL ALMANAC INSTRUCTIONS.

THE APPLE VARIETIES LISTED BELOW ARE RESISTANT TO APPLE SCAB, A PERSISTANT PROBLEM IN THE WILLAMETTE VALLEY. ALL ARE AVAILABLE ON DWARF, SEMI-DWARF AND STANDARD ROOTSTOCK. THEY ARE LISTED IN ORDER OF RIPENING. CHECK WITH LOCAL NURSERIES AND CATALOGS FOR AVAILABILITY AND PRICES.

TYDEMAN'S EARLY RED -- RIPENS AUGUST 20-30. AN ENGLISH VARIETY FROM A CROSS OF McINTOSH AND WORCESTER PEARMAIN. THE FRUIT IS LARGE, RED, AND SIMILAR TO McINTOSH IN SHAPE. IT HAS BEEN GROWN COMMERCIALLY IN THE NORTHWEST AND HAS BEEN WELL RECEIVED ON THE MARKET. THE TREE TENDS TO BEAR ON THE ENDS OF DROOPY BRANCHES, WHICH PRESENTS PROBLEMS IN PRUNING AND PROPPING. THE FRUIT RIPENS UNEVENLY THROUGH THE TREE. IT IS SLIGHTLY SOFT AND DOESN'T STORE WELL.

PRIMA -- RIPENS IN EARLY SEPTEMBER, ABOUT THREE WEEKS AHEAD OF JOHATHAN. AFTER 25 YEARS OF APPLE BREEDING, THIS VARIETY WAS INTRODUCED IN 1970 BY PURDUE UNIVERSITY, RUTGERS AND THE UNIVERSITY OF ILLINOIS, AS THE FIRST COMMERCIAL SCAB-RESISTANT APPLE. THE FRUIT IS ROUNDISH OBLATE, MOSTLY MEDIUM TO DARK RED, GLOSSY AND ATTRACTIVE. THE FLESH IS YELLOWISH, FINE GRAINED, FIRM, CRISP, JUICY, MILDLY SUBACID. THE TREE IS SPREADING AND VIGOROUS. IT BLOOMS A FEW DAYS BEFORE GOLDEN DELICIOUS.

CHEHALIS -- RIPENS IN EARLY SEPTEMBER, THREE TO FOUR WEEKS BEFORE GOLDEN DELICIOUS. IT ORIGINATED IN OAKVILLE, WASHINGTON, BY LLOYD LONBORG AND WAS INTRODUCED IN 1965. IT IS A MILD-FLAVORED GOLDEN DELICIOUS TYPE APPLE. THE SKIN IS THICK, SMOOTH, GREENISH-YELLOW TO FULL YELLOW WITH OCCASIONAL PINKISH BLUSH. THE FLESH IS WHITE TO CREAM COLORED, MEDIUM FINE TEXTURED, CRISP AND MODERATELY JUICY AT PRIME MATURITY, BECOMING MEDIUM SOFT. THE FLAVOR IS PLEASANTLY MILD, SUBACID. IT KEEPS WELL IN COLD STORAGE. THE TREE IS MODERATELY VIGOROUS, UPRIGHT, SPREADING, MODERATELY PRODUCTIVE. AVAILABLE IN LOCAL NURSERIES.

PRISCILLA -- RIPENS IN MID-SEPTEMBER, TWO WEEKS BEFORE DELICIOUS. IT WAS RELEASED IN 1972 BY PURDUE, RUTGERS, AND UNIVERSITY OF ILLINOIS. IT IS AN EXCELLENT DESSERT APPLE WITH A CRISP MEDIUM-GRAINED TEXTURE. FLESH IS WHITE TO SLIGHTLY GREENISH AND MILDLY SUBACID. ITS COLOR IS A BRIGHT RED OVER YELLOW WITH A SMOOTH, WAXY SKIN. THE TREE IS MODERATELY SPREADING AND VIGOROUS, AND THE FRUIT IS REPORTED TO STORE WELL. THE FRUIT IS LARGE, OFTEN OVER THREE INCHES IN DIAMETER.

MACOUN -- RIPENS IN EARLY OCTOBER. IT ORIGINATED IN GENEVA, NEW YORK, AS A CROSS BETWEEN McINTOSH AND JERSEY BLACK, AND WAS INTRODUCED IN 1923. THE FRUIT IS SIMILAR TO THE McINTOSH, BUT SMALLER, MORE RIBBED, AND DEVELOPS A BLACK-RED COLOR AT MATURITY. THE FLESH IS WHITE, RICHLY FLAVORED, AROMATIC, WITH HIGH DESSERT QUALITY. THE TREE IS UPRIGHT AND MAY REQUIRE LIMB SPREADING IN THE EARLY YEARS. IT HAS A STRONG TENDENCY TO BIENNIAL BEARING AND REQUIRES THINNING TO MAINTAIN GOOD FRUIT SIZE.

SPARTAN -- RIPENS IN EARLY OCTOBER WITH DELICIOUS. IT ORIGINATED AS A CROSS OF McINTOSH AND NEWTON IN SUMMERLAND, B.C., AND WAS INTRODUCED IN 1936 BY THE CANADIAN DEPARTMENT OF AGRICULTURE. IT IS GROWN COMMERCIALLY IN B.C. THE SKIN IS HIGHLY COLORED, SOLID DARK RED; FLESH IS FIRM, CRISP, WHITE, JUICY, QUALITY AND TEXTURE GOOD. THE TREE TENDS TO SET VERY HEAVY CROPS. THE FRUIT IS MEDIUM TO SMALL AND SHAPED SOMEWHAT LIKE DELICIOUS. IT STORES WELL IF PICKED AT PROPER MATURITY; OFTEN UNTIL MARCH OR APRIL. THE TREE IS VIGOROUS AND OF GOOD STRUCTURE, AND AN ANNUAL BEARER.

FROM "NEW SCAB-MILDEW RESISTANT APPLE VARIETIES AVAILABLE", BY ROBERT L. STEBBINS, EXTENSION HORTICULTURE SPECIALIST, OREGON STATE UNIVERSITY.

CHOOSING A SITE AND DESIGN FOR EDIBLE TREES

FRUIT TREES NEED A LOCATION THAT AFFORDS FULL SUN, GOOD WATER DRAINAGE AND IS SOMEWHAT HIGHER THAN THE SURROUNDING LAND TO AVOID FROST POCKETS. EXAMINE THE GENERAL SITE YOU HAVE AVAILABLE FOR SHADY SPOTS IN THE EARLY MORNING AND AFTERNOON. THEY ARE NOT GOOD FOR FRUIT OR NUT TREES. LOW SPOTS THAT COLLECT WATER AND ARE SHELTERED FROM ALL AIR PASSAGE ARE ALSO INAPPROPRIATE FOR EDIBLE TREES. WITHOUT A GOOD SUN EXPOSURE, THERE IS LITTLE HOPE FOR A GOOD CROP OF FRUIT OR NUTS; HOWEVER, IF YOU ARE FORTUNATE ENOUGH, SOME ALTERATIONS TO THE SITE WILL GREATLY IMPROVE AN URBAN ORCHARD'S PROSPECTS.

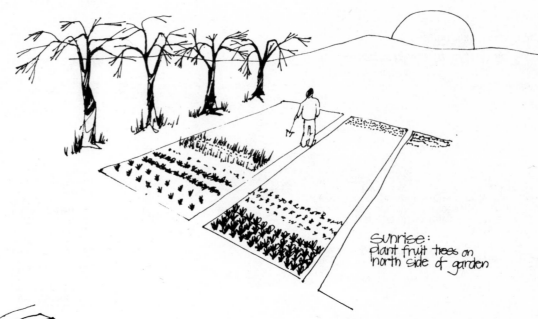

sunrise:
plant fruit trees on north side of garden

plant trees on slopes

cold air movement

bottom of slope frost pocket

DESIGN (CONTINUED)

FRUIT TREES CAN VARY IN SIZE FROM DWARF, SEMI-DWARF, TO FULL-SIZED STANDARD TREES. THEY ALL BEAR THE SAME SIZE FRUIT, ONLY THE OVER-ALL TREE IS DIFFERENT. DWARF AND SEMI-DWARF TREES ARE VERY POPULAR BECAUSE 1) THEY ARE EASY TO CARE FOR, 2) THEY ARE EASY TO HARVEST, 3) YOU CAN PLANT MORE TREES PER UNIT OF LAND, AND 4) THEY TEND TO BEAR FRUIT A FEW YEARS SOONER THAN STANDARD TREES. THEY ARE ALSO VERY APPROPRIATE FOR SITUATIONS WHERE SUNLIGHT AND SOLAR RIGHTS ARE IMPORTANT, IN THAT THEY DO NOT SHADE A LARGE AREA. THE PRINCIPLE DRAWBACK IS THAT THEY HAVE DROUGHT-RELATED PROBLEMS, AND GRAFTING DISORDERS. HOWEVER, IF YOU BUY FROM A REPUTABLE NURSERY AND FOLLOW THE INSTRUCTIONS ON MULCHING, THERE SHOULD BE NO PROBLEMS.

sun angles.
solar access and fruit trees

June 21

equinox

December 21

15 feet
dwarf & hedge zone

10 feet
semi-dwarf zone

dwarf: 8'-10'   semi-dwarf: 12'-15'   standard: 25'-30'

Fruit Tree Sizes:

A FRUIT TREE IS DWARFED BY SELECTING, OVER A PERIOD OF TIME, FOR DWARF CHARACTERISTICS AND GRAFTING THIS VARIETY ONTO A SPECIFIC DWARF ROOT STOCK. THE ROOT STOCK SIMPLY INHIBITS THE SIZE TO WHICH THE TREE CAN GO. THE UPPER PORTION WILL CARRY THE SAME SIZE FRUIT BUT IN FEWER NUMBERS. MORE TREES CAN BE PLANTED PER ACRE BECAUSE LESS SPACE IS NEEDED FOR EACH TREE. FOR THE URBAN ORCHARD-IST, THIS MEANS A NUMBER OF FRUIT TREES OF DIFFERENT VARIETIES CAN BE PLANTED IN THE SAME SPACE OCCUPIED BY A STANDARD TREE.

THE FULL DWARF IS APPROPRIATE FOR EXTREMELY TIGHT LOCATIONS WHERE A SMALL TREE IS DESIRED. IT WILL GROW ONLY TO 6 OR 8 FEET AND SHOULD BE PLACED ABOUT THAT DISTANCE APART. THE SEMI-DWARF IS THE BEST SIZE FOR ALL-AROUND URBAN ORCHARD USE. IT GROWS TO ABOUT 15 FEET, AND SHOULD BE PLANTED ABOUT 10 FEET APART. THIS PARTICULAR SIZE PRODUCES THE MOST FRUIT FOR THE AMOUNT OF LAND USED. IT STILL IS EASY TO MAINTAIN AND HARVEST, AND USUALLY ONLY A SMALL LADDER IS NEEDED. THE FULL SIZE OR STANDARD TREE IS USEFUL AS AN EDIBLE SHADE TREE. THESE CAN BE QUITE LARGE, FROM 20 TO 80 FEET, DEPENDING ON THE SPECIES. THEY REQUIRE A 20-FOOT SPACING AND ARE APPROPRIATE ONLY ON LARGER PARCELS OF LAND.

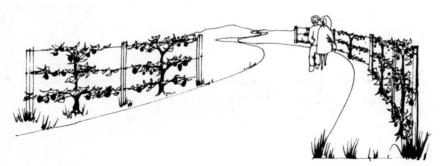

espaliered fruit trees along a path

GRAPE ARBOR AND TRELLIS

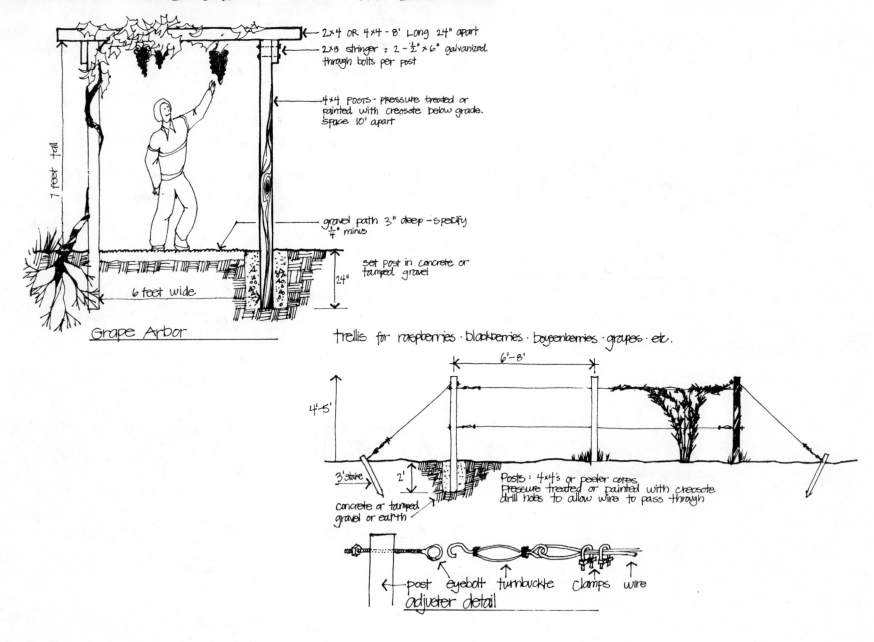

2×4 OR 4×4 - 8' Long 24" apart

2×8 stringer : 2 - ½" × 6" galvanized through bolts per post

4×4 POSTS - pressure treated or painted with creosote below grade. Space 10' apart

7 feet tall

gravel path 3" deep - specify ¼" minus

set post in concrete or tamped gravel

24"

6 feet wide

Grape Arbor

trellis for raspberries · blackberries · boysenberries · grapes · etc.

6'-8'

4'-5'

3' stake

2'

concrete or tamped gravel or earth

Posts : 4×4's or peeler cores.
Pressure treated or painted with creosote
drill holes to allow wire to pass through

post   eyebolt   turnbuckle   clamps   wire

adjuster detail

RESOURCES:

LOCAL NURSERIES

PADDOCK-BAUER NURSERY
1566 MAIN STREET
SPRINGFIELD, OREGON

REED & CROSS NURSERY
160 OAKWAY ROAD
EUGENE, OREGON

WOODRUFF NURSERY
1270 RIVER ROAD
EUGENE, OREGON

GRAY'S SEED AND GARDEN SUPPLY
737 W. 6TH
EUGENE, OREGON

ISLAND GARDENS NURSERY
701 GOODPASTURE ISLAND ROAD
EUGENE, OREGON

A AND D NURSERY
539 N. 18TH
SPRINGFIELD, OREGON

DECKER'S NURSERY
90808 B
ALVADORE, OREGON

BOOKS

BETTER SOIL, GENE LOGSDON, RODALE PRESS: EMMAUS, PA., 1975

BIODYNAMIC TREATMENT OF FRUIT TREES, BERRIES AND SHRUBS, E.E.
PFEIFFER, BIODYNAMIC FARMING AND GARDENING ASSOCIATION,
1957.

DWARFED FRUIT TREES, HAROLD BRADFORD TUKEY, CORNELL UNIVERSITY
PRESS, 1964

FERTILITY PASTURES AND COVER CROPS, NEWMAN TURNER, RATEAVER'S
PUBLICATIONS: PAUMA VALLEY, CAL. 92061

FOREST FARMING, J. SHOLTO DOUGLAS AND ROBERT A. DE J. HART,
RODALE, 1976

FRUIT TREES AND THE SOIL, DAVID M. RIGHT, FABER AND FABER, 1960

FRUITS FOR THE HOME GARDEN, KEN AND PAT KRAFT, WILLIAM MORROW, 1968

GROW YOUR OWN DWARF FRUIT TREES, KEN AND PAT KRAFT, CORNERSTONE
LIBRARY, 1974

GROWING UNUSUAL FRUIT, ALAN E. SIMMONS, WALKER, 1972

NUTS FOR THE FOOD GARDEN, LOUISE RIOTTE, GARDEN WAY, 1975

PEST CONTROL WITHOUT POISONS, L.D. HILLS, HENRY DOUBLEDAY RESEARCH
ASSOCIATION, 1964

PLANTS, PEOPLE AND ENVIRONMENTAL QUALITY, U.S. DEPT. OF INTERIOR,
1972

PRUNING, CHRISTOPHER BRICKELL, BITCHELL BEARLEY PUBLISHERS, LTD.
1979

SUCCESSFUL BERRY GROWING, GENE LOGSDON, RODALE PRESS, 1974

THE COMPLETE HANDBOOK OF PLANT PROPAGATION, R.C.M. WRIGHT, MAC-
MILLAN, 1973

THE EARTH MANUAL, MALCOLM MARGOLIN, HOUGHTON-MIFFLIN, 1975

THE LORETTE SYSTEM OF PRUNING, LOUIS LORETTE, RODALE PRESS, 1946

THE ONE-STRAW REVOLUTION, MASANOBU FUKUOKA, RODALE, 1978

THE ORGANIC WAY TO PLANT PROTECTION, ROGER B. YEPSEN (ED.),
RODALE PRESS, 1966

THE OWNER-BUILT HOMESTEAD, BARBARA AND KEN KERN, SCRIBNERS, 1977

TRAINED AND SCULPTURED PLANTS, BROOKLYN BOTANIC GARDEN, SUMMER,
1961

TREE CROPS, J. RUSSELL SMITH, HARCOURT BRACE, 1929

TREE MAINTENANCE, PASCAL P. PIRONE, OXFORD UNIVERSITY PRESS, 1978

TREES FOR THE YARD, ORCHARD, AND WOODLOT, ROGER B. YEPSEN, JR.
RODALE PRESS, 1976

SELECTED OREGON STATE UNIVERSITY EXTENSION SERVICE PUBLICATIONS:

1979 SPRAY PROGRAM FOR APPLES AND PEARS IN THE WILLAMETTE VALLEY,
#667, DECEMBER 1978

GROWING FILBERTS IN OREGON, LLOYD C. BARON AND ROBERT L. STEBBINS,
#628, OCTOBER 1976

GROWING WALNUTS IN THE PACIFIC NORTHWEST, ROBERT L. STEBBINS,
#974, JUNE 1979

TRAINING AND PURNING APPLE AND PEAR TREES, ROBERT L. STEBBINS,
PACIFIC NORTHWEST EXTENSION PUBLICATION, PNW 156, OCTOBER
1976

APPLE SCAB CONTROL IN HOME GARDENS, IAN C. MACSWAN, #FS, 85,
JUNE 1965

U.S. DEPT. OF AGRICULTURE PUBLICATIONS, AVAILABLE FROM THE LANE
COUNTY EXTENSION SERVICE OR FROM YOUR CONGRESSMAN.

COMMERCIAL BLUEBERRY GROWING, FARMERS BULLETIN #2254

CONTROL OF INSECTS ON DECIDUOUS FRUIT AND NUT TREES IN THE HOME
ORCHARD--WITHOUT INSECTICIDES, HOME AND GARDEN BULLETIN #211

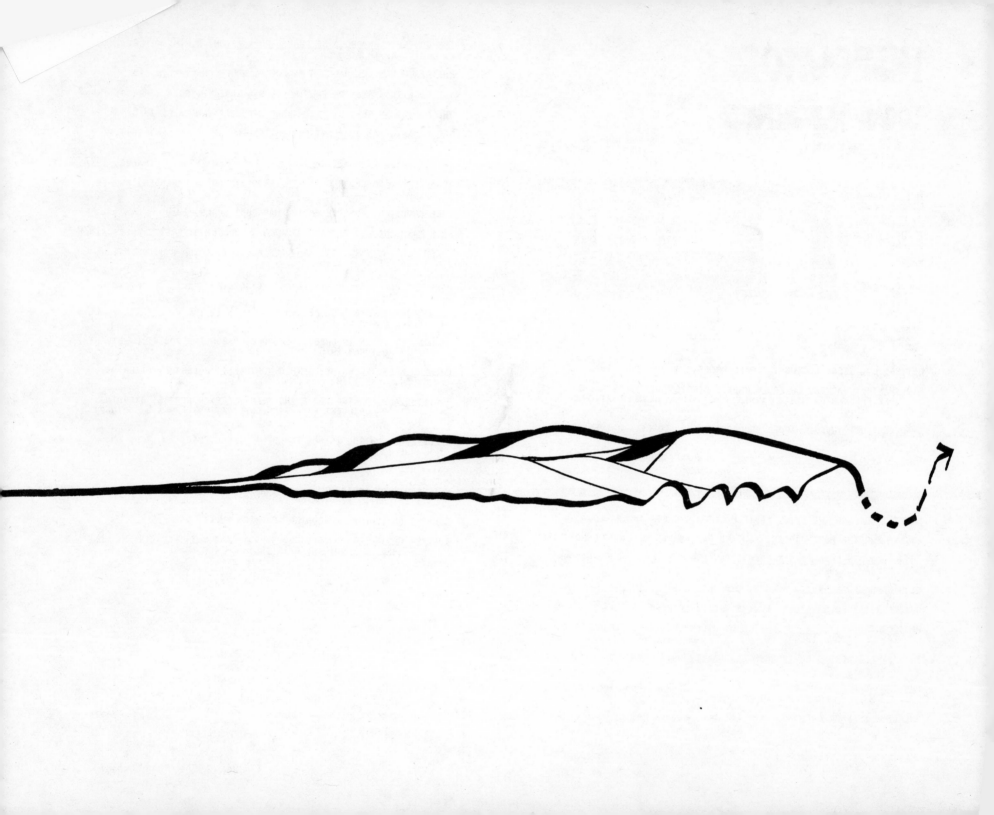

FEEDBACK ?

EDIBLE EARTH, INC.
P.O. Box 3258, University Sta.
Eugene, Oregon 97403

ENVIRONMENTAL
Research, Consultation
Architecture & Construction

Frontmatter composed in phototype
Optima by Publishing Services Center,
Los Altos, California. Printed
by George Banta Company
Menasha, Wisconsin, via offset
lithography on fifty pound
Glatfelter Ecolocote, in an
edition of ten thousand
copies.